Volume III: Chapters 13–16

Complete Solutions Guide to Accompany

Calculus

Fifth Edition

Larson/Hostetler/Edwards

Dianna L. Zook

The Pennsylvania State University

D. C. Heath and Company

Lexington, Massachusetts   Toronto

Address editorial correspondence to:
D. C. Heath and Company
125 Spring Street
Lexington, MA 02173

Preface

This solutions guide is a supplement to *Calculus, Fifth Edition,* by Roland E. Larson, Robert P. Hostetler, and Bruce H. Edwards. Solutions to every exercise in the text are given with all essential algebraic steps included. There are three volumes in the complete set of solutions guides. Volume I contains Chapters P–6, Volume II contains Chapters 7–12, and Volume III contains Chapters 13–16. Also available is a one-volume *Study and Solutions Guide to Accompany Calculus, Fifth Edition,* written by David E. Heyd, which contains worked-out solutions to *selected* representative exercises from the text.

I have made every effort to see that the solutions are correct. However, I would appreciate hearing about any errors or other suggestions for improvement.

I would like to thank the staff at Larson Texts, Inc. for their help in the preparation of this guide. I would also like to thank the students in my mathematics classes. Finally, I would like to thank my husband, Ed Schlindwein, for his support during the many months I have worked on this project.

Dianna L. Zook
Indiana University
Purdue University at Fort Wayne, Indiana 46805

CONTENTS

CHAPTER 13
Functions of Several Variables

Section 13.1   Introduction to Functions of Several Variables

1. $f(x, y) = \dfrac{x}{y}$

   a. $f(3, 2) = \dfrac{3}{2}$

   b. $f(-1, 4) = -\dfrac{1}{4}$

   c. $f(30, 5) = \dfrac{30}{5} = 6$

   d. $f(5, y) = \dfrac{5}{y}$

   e. $f(x, 2) = \dfrac{x}{2}$

   f. $f(5, t) = \dfrac{5}{t}$

2. $f(x, y) = 4 - x^2 - 4y^2$

   a. $f(0, 0) = 4$

   b. $f(0, 1) = 4 - 0 - 4 = 0$

   c. $f(2, 3) = 4 - 4 - 36 = -36$

   d. $f(1, y) = 4 - 1 - 4y^2 = 3 - 4y^2$

   e. $f(x, 0) = 4 - x^2 - 0 = 4 - x^2$

   f. $f(t, 1) = 4 - t^2 - 4 = -t^2$

3. $f(x, y) = xe^y$

   a. $f(5, 0) = 5e^0 = 5$

   b. $f(3, 2) = 3e^2$

   c. $f(2, -1) = 2e^{-1} = \dfrac{2}{e}$

   d. $f(5, y) = 5e^y$

   e. $f(x, 2) = xe^2$

   f. $f(t, t) = te^t$

4. $g(x, y) = \ln|x + y|$

   a. $g(2, 3) = \ln|2 + 3| = \ln 5$

   b. $g(5, 6) = \ln|5 + 6| = \ln 11$

   c. $g(e, 0) = \ln|e + 0| = 1$

   d. $g(0, 1) = \ln|0 + 1| = 0$

   e. $g(2, -3) = \ln|2 - 3| = \ln 1 = 0$

   f. $g(e, e) = \ln|e + e| = \ln 2e$

$$= \ln 2 + \ln e = (\ln 2) + 1$$

5. $h(x, y, z) = \dfrac{xy}{z}$

   a. $h(2, 3, 9) = \dfrac{(2)(3)}{9} = \dfrac{2}{3}$

   b. $h(1, 0, 1) = \dfrac{(1)(0)}{1} = 0$

6. $f(x, y, z) = \sqrt{x + y + z}$

   a. $f(0, 5, 4) = \sqrt{0 + 5 + 4} = 3$

   b. $f(6, 8, -3) = \sqrt{6 + 8 - 3} = \sqrt{11}$

7. $f(x, y) = x \sin y$

   a. $f\left(2, \dfrac{\pi}{4}\right) = 2 \sin \dfrac{\pi}{4} = \sqrt{2}$

   b. $f(3, 1) = 3 \sin 1$

8. $V(r, h) = \pi r^2 h$

   a. $V(3, 10) = \pi(3)^2(10) = 90\pi$

   b. $V(5, 2) = \pi(5)^2(2) = 50\pi$

9. $f(x, y) = \int_x^y (2t - 3)\, dt$

   a. $f(0, 4) = \int_0^4 (2t - 3)\, dt = \left[t^2 - 3t\right]_0^4 = 4$

   b. $f(1, 4) = \int_1^4 (2t - 3)\, dt = \left[t^2 - 3t\right]_1^4 = 6$

10. $g(x, y) = \int_x^y \frac{1}{t}\, dt$

   a. $g(4, 1) = \int_4^1 \frac{1}{t}\, dt = \left[\ln|t|\right]_4^1 = -\ln 4$

   b. $g(6, 3) = \int_6^3 \frac{1}{t}\, dt = \left[\ln|t|\right]_6^3 = \ln 3 - \ln 6 = \ln\left(\frac{1}{2}\right)$

11. $f(x, y) = x^2 - 2y$

   a. $\dfrac{f(x + \Delta x,\ y) - f(x,\ y)}{\Delta x} = \dfrac{[(x + \Delta x)^2 - 2y] - (x^2 - 2y)}{\Delta x}$

   $= \dfrac{x^2 + 2x(\Delta x) + (\Delta x)^2 - 2y - x^2 + 2y}{\Delta x} = \dfrac{\Delta x(2x + \Delta x)}{\Delta x} = 2x + \Delta x,\ \Delta x \neq 0$

   b. $\dfrac{f(x,\ y + \Delta y) - f(x,\ y)}{\Delta y} = \dfrac{[x^2 - 2(y + \Delta y)] - (x^2 - 2y)}{\Delta y} = \dfrac{x^2 - 2y - 2\Delta y - x^2 + 2y}{\Delta y} = \dfrac{-2\Delta y}{\Delta y} = -2,\ \Delta y \neq 0$

12. $f(x, y) = 3xy + y^2$

   a. $\dfrac{f(x + \Delta x,\ y) - f(x,\ y)}{\Delta x} = \dfrac{[3(x + \Delta x)y + y^2] - (3xy + y^2)}{\Delta x}$

   $= \dfrac{3xy + 3(\Delta x)y + y^2 - 3xy - y^2}{\Delta x} = \dfrac{3(\Delta x)y}{\Delta x} = 3y,\ \Delta x \neq 0$

   b. $\dfrac{f(x,\ y + \Delta y) - f(x,\ y)}{\Delta y} = \dfrac{[3x(y + \Delta y) + (y + \Delta y)^2] - (3xy + y^2)}{\Delta y}$

   $= \dfrac{3xy + 3x(\Delta y) + y^2 + 2y(\Delta y) + (\Delta y)^2 - 3xy - y^2}{\Delta y}$

   $= \dfrac{\Delta y(3x + 2y + \Delta y)}{\Delta y} = 3x + 2y + \Delta y,\ \Delta y \neq 0$

13. $f(x, y) = \sqrt{4 - x^2 - y^2}$

   Domain: $4 - x^2 - y^2 \geq 0$

   $x^2 + y^2 \leq 4$

   $\{(x, y): x^2 + y^2 \leq 4\}$

   Range: $0 \leq z \leq 2$

14. $f(x, y) = \sqrt{4 - x^2 - 4y^2}$

   Domain: $4 - x^2 - 4y^2 \geq 0$

   $x^2 + 4y^2 \leq 4$

   $\dfrac{x^2}{4} + \dfrac{y^2}{1} \leq 1$

   $\left\{(x, y): \dfrac{x^2}{4} + \dfrac{y^2}{1} \leq 1\right\}$

   Range: $0 \leq z \leq 2$

15. $f(x, y) = \arcsin(x + y)$

   Domain: $\{(x, y): -1 \leq x + y \leq 1\}$

   Range: $-\dfrac{\pi}{2} \leq z \leq \dfrac{\pi}{2}$

16. $f(x, y) = \arccos \dfrac{y}{x}$

   Domain: $\left\{(x, y): -1 \leq \dfrac{y}{x} \leq 1\right\}$

   Range: $0 \leq z \leq \pi$

17. $f(x, y) = \ln(4 - x - y)$

   Domain: $4 - x - y > 0$

   $x + y < 4$

   $\{(x, y): y < -x + 4\}$

   Range: all real numbers

18. $f(x, y) = \ln(4 - xy)$

   Domain: $4 - xy > 0$

   $xy < 4$

   $\{(x, y): xy < 4\}$

   Range: all real numbers

19. $z = \dfrac{x+y}{xy}$

Domain: $\{(x, y): x \neq 0 \text{ and } y \neq 0\}$
Range: all real numbers

20. $z = \dfrac{xy}{x-y}$

Domain: $\{(x, y): x \neq y\}$
Range: all real numbers

21. $f(x, y) = e^{x/y}$

Domain: $\{(x, y): y \neq 0\}$
Range: $z > 0$

22. $f(x, y) = x^2 + y^2$

Domain: $\{(x, y): x$ is any real number,
  y is any real number$\}$

Range: $z \geq 0$

23. $g(x, y) = \dfrac{1}{xy}$

Domain: $\{(x, y): x \neq 0 \text{ and } y \neq 0\}$
Range: all real numbers except zero

24. $g(x, y) = x\sqrt{y}$

Domain: $\{(x, y): y \geq 0\}$
Range: all real numbers

25. $f(x, y) = \dfrac{-4x}{x^2 + y^2 + 1}$

  a. View from the positive x-axis: $(20, 0, 0)$
  b. View where x is negative, y and z are positive:
    $(-15, 10, 20)$
  c. View from the first octant: $(20, 15, 25)$
  d. View from the line $y = x$ in the xy-plane:
    $(20, 20, 0)$

26. a. Domain: $\{(x, y): x$ is any real number,
  y is any real number$\}$

  b. Range: $-2 \leq z \leq 2$
  c. $z = 0$ when $x = 0$ which represents points on the
    y-axis.
  d. No. When x is positive, z is negative. When x
    is negative, z is positive. The surface does not
    pass through the first octant, the octant where y
    is negative and x and z are positive, the octant
    where y is positive and x and z are negative, and
    the octant where x, y, and z are all negative.

27. $z = e^{1-x^2-y^2}$

Level curves:
$$c = e^{1-x^2-y^2}$$
$$\ln c = 1 - x^2 - y^2$$
$$x^2 + y^2 = 1 - \ln c$$

Circles centered at $(0, 0)$
Matches (c)

28. $z = e^{1-x^2+y^2}$

Level curves:
$$c = e^{1-x^2+y^2}$$
$$\ln c = 1 - x^2 + y^2$$
$$x^2 - y^2 = 1 - \ln c$$

Hyperbolas centered at $(0, 0)$
Matches (d)

29. $z = \ln|y - x^2|$

Level curves:
$$c = \ln|y - x^2|$$
$$\pm e^c = y - x^2$$
$$y = x^2 \pm e^c$$

Parabolas
Matches (b)

30. $z = \cos\left(\dfrac{x^2 + 2y^2}{4}\right)$

Level curves:
$$c = \cos\left(\dfrac{x^2 + 2y^2}{4}\right)$$
$$\cos^{-1} c = \dfrac{x^2 + 2y^2}{4}$$
$$x^2 + 2y^2 = 4\cos^{-1} c$$

Ellipses
Matches (a)

31. $f(x,\ y) = 5$
Plane: $z = 5$

32. $f(x,\ y) = 6 - 2x - 3y$
Plane
Domain: entire xy-plane
Range: $-\infty < z < \infty$

33. $f(x,\ y) = y^2$
Since the variable x is missing,
the surface is a cylinder with
rulings parallel to the x-axis.
The generating curve is $z = y^2$.
The domain is the entire
xy-plane and the range is $z \geq 0$.

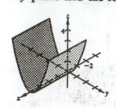

34. $g(x,\ y) = \frac{1}{2}x$
Plane: $z = \frac{1}{2}x$

35. $z = 4 - x^2 - y^2$
Paraboloid
Domain: entire xy-plane
Range: $z \leq 4$

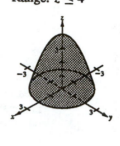

36. $z = \sqrt{x^2 + y^2}$
Cone
Domain of f: entire xy-plane
Range: $z \geq 0$

37. $f(x,\ y) = e^{-x}$
Since the variable y is missing,
the surface is a cylinder with
rulings parallel to the y-axis.
The generating curve is $z = e^{-x}$.
The domain is the entire
xy-plane and the range is $z > 0$.

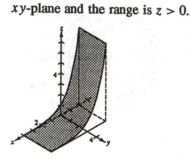

38. $f(x,\ y)$
$$= \begin{cases} xy, & x \geq 0, \quad y \geq 0 \\ 0, & \text{elsewhere} \end{cases}$$
Domain of f: entire xy-plane
Range: $z \geq 0$

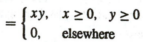

39. $z = y^2 - x^2 + 1$
Hyperbolic paraboloid
Domain: entire xy-plane
Range: $-\infty < z < \infty$

40. $f(x, y)$
$= \frac{1}{12}\sqrt{144 - 16x^2 - 9y^2}$
Semi-ellipsoid
Domain: set of all points
lying on or inside the ellipse
$(x^2/9) + (y^2/16) = 1$
Range: $0 \le z \le 1$

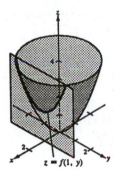

41. $f(x, y) = x^2 e^{(-xy/2)}$

42. $f(x, y) = x \sin y$

43. $f(x, y) = x^2 + y^2$

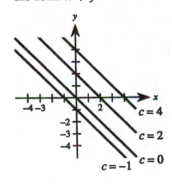

44. $z = f(x, y) = xy$

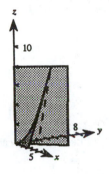

45. $z = x + y$
Level curves are parallel lines of
the form $x + y = c$.

46. $f(x, y) = 6 - 2x - 3y$
The level curves are of the form
$6 - 2x - 3y = c$ or
$2x + 3y = 6 - c$. Thus, the level
curves are straight lines with a
slope of $-\frac{2}{3}$.

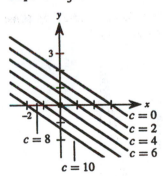

47. $f(x, y) = \sqrt{25 - x^2 - y^2}$
The level curves are of the
form $c = \sqrt{25 - x^2 - y^2}$,
$x^2 + y^2 = 25 - c^2$. Thus, the
level curves are circles of radius
5 or less, centered at the origin.

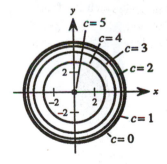

48. $f(x, y) = x^2 + y^2$

The level curves are of the form $x^2 + y^2 = c$. Thus, the level curves are circles centered at the origin.

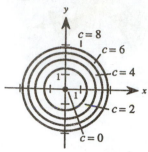

49. $f(x, y) = xy$

The level curves are hyperbolas of the form $xy = c$.

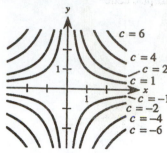

50. $f(x, y) = e^{xy}$

The level curves are of the form $c = e^{xy}$, $\ln(c) = xy$. Thus, the level curves are hyperbolas centered at the origin with the x- and y-axes as asymptotes.

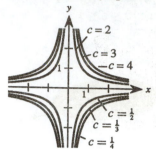

51. $f(x, y) = \dfrac{x}{x^2 + y^2}$

The level curves are of the form

$$c = \frac{x}{x^2 + y^2}$$

$$x^2 - \frac{x}{c} + y^2 = 0$$

$$\left(x - \frac{1}{2c}\right)^2 + y^2 = \left(\frac{1}{2c}\right)^2.$$

Thus, the level curves are circles passing through the origin and centered at $(1/2c, 0)$.

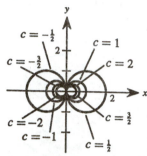

52. $f(x, y) = \ln(x - y)$

The level curves are of the form

$$c = \ln(x - y)$$

$$e^c = x - y$$

$$y = x - e^c.$$

Thus, the level curves are parallel lines of slope 1 passing through the fourth quadrant.

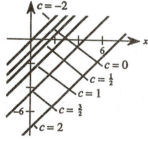

53. No

$$z = e^{-(x^2 + y^2)}$$

$$z = x^2 + y^2$$

54. $f(x, y) = \dfrac{x}{y}$

The level curves are the lines $c = x/y$ or $y = (1/c)x$. These lines all pass through the origin.

55. $f(x, y, z) = x - 2y + 3z$
$c = 6$
$6 = x - 2y + 3z$
Plane

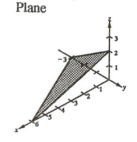

56. $f(x, y, z) = 4x + y + 2z$
$c = 4$
$4 = 4x + y + 2z$
Plane

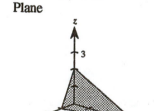

57. $f(x, y, z) = x^2 + y^2 + z^2$
$c = 9$
$9 = x^2 + y^2 + z^2$
Sphere

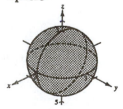

58. $f(x, y, z) = x^2 + y^2 - z$
$c = 1$
$1 = x^2 + y^2 - z$
Elliptic paraboloid
Vertex: $(0, 0, -1)$

59. $f(x, y, z) = 4x^2 + 4y^2 - z^2$
$c = 0$
$0 = 4x^2 + 4y^2 - z^2$
Elliptic cone

60. $f(x, y, z) = \sin x - z$
$c = 0$
$0 = \sin x - z$ or $z = \sin x$

61. $V(I, R) = 1000 \left[\dfrac{1 + 0.10(1 - R)}{1 + I} \right]^{10}$

	Inflation Rate, I		
Tax Rate, R	0	0.03	0.05
0	$2593.74	$1929.99	$1592.33
0.28	$2004.23	$1491.34	$1230.42
0.35	$1877.14	$1396.77	$1152.40

62. $A(r, t) = 1000e^{rt}$

	Number of years			
Rate	5	10	15	20
0.08	$1491.82	$2225.54	$3320.12	$4953.03
0.10	$1648.72	$2718.28	$4481.69	$7389.06
0.12	$1822.12	$3320.12	$6049.65	$11,023.18
0.14	$2013.75	$4055.20	$8166.17	$16,444.65

63. $N(d, L) = \left(\dfrac{d - 4}{4} \right)^2 L$

   a. $N(22, 12) = \left(\dfrac{22 - 4}{4} \right)^2 (12) = 243$ board-feet

   b. $N(30, 12) = \left(\dfrac{30 - 4}{4} \right)^2 (12) = 507$ board-feet

64. $W(x, y) = \dfrac{1}{x - y}, \quad y < x$

   a. $W(15, 10) = \dfrac{1}{15 - 10} = \dfrac{1}{5}$ hr $= 12$ min

   b. $W(12, 9) = \dfrac{1}{12 - 9} = \dfrac{1}{3}$ hr $= 20$ min

   c. $W(12, 6) = \dfrac{1}{12 - 6} = \dfrac{1}{6}$ hr $= 10$ min

   d. $W(4, 2) = \dfrac{1}{4 - 2} = \dfrac{1}{2}$ hr $= 30$ min

65. $T = 600 - 0.75x^2 - 0.75y^2$
The level curves are of the form
$$c = 600 - 0.75x^2 - 0.75y^2$$
$$x^2 + y^2 = \frac{600 - c}{0.75}.$$
The level curves are circles centered at the origin.

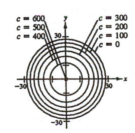

66. $f(x, y) = 100x^{0.6}y^{0.4}$

$f(2x, 2y) = 100(2x)^{0.6}(2y)^{0.4}$

$= 100(2)^{0.6}x^{0.6}(2)^{0.4}y^{0.4} = 100(2)^{0.6}(2)^{0.4}x^{0.6}y^{0.4} = 2[100x^{0.6}y^{0.4}] = 2f(x, y)$

67. $C = 0.75xy + \quad 2(0.40)xz \quad + 2(0.40)yz$

    base   +   front & back   +   two ends

    $= 0.75xy + 0.80(xz + yz)$

68. $V = \pi r^2 l + \dfrac{4}{3}\pi r^3 = \dfrac{\pi r^2}{3}(3l + 4r)$

69. $PV = kT, \quad 20(2600) = k(40)$

  a. $k = \dfrac{20(2600)}{40} = 1300$

  b. $P = \dfrac{kT}{V} = 1300\left(\dfrac{T}{V}\right)$

   The level curves are of the form:

    $c = 1300\left(\dfrac{T}{V}\right)$

     $V = \dfrac{1300}{c}T$

   Thus, the level curves are lines through the origin with slope $1300/c$.

70. $f(x, \ y) = 1.085x + 0.779y - 10.778$

  a. $f(15, \ 10) = 13.287$

  b. Because both x and y are first degree, x (revenue) has the greater influence since it has the larger coefficient.

71. a. Highest pressure at C
   b. Lowest pressure at A
   c. Highest wind velocity at B

72. Southwest

73. Latitude and land versus ocean location

Section 13.2   Limits and Continuity

1. $\displaystyle \lim_{(x,y)\to(a,b)} [f(x, \ y) - g(x, \ y)] = \lim_{(x,y)\to(a,b)} f(x, \ y) - \lim_{(x,y)\to(a,b)} g(x, \ y) = 5 - 3 = 2$

2. $\displaystyle \lim_{(x,y)\to(a,b)} \left[\frac{4f(x, \ y)}{g(x, \ y)}\right] = \frac{4\left[\displaystyle\lim_{(x,y)\to(a,b)} f(x, \ y)\right]}{\displaystyle\lim_{(x,y)\to(a,b)} g(x, \ y)} = \frac{4(5)}{3} = \frac{20}{3}$

3. $\displaystyle \lim_{(x,y)\to(a,b)} [f(x, \ y)g(x, \ y)] = \left[\lim_{(x,y)\to(a,b)} f(x, \ y)\right]\left[\lim_{(x,y)\to(a,b)} g(x, \ y)\right] = 5(3) = 15$

4. $\displaystyle \lim_{(x,y)\to(a,b)} \left[\frac{f(x, \ y) - g(x, \ y)}{f(x, \ y)}\right] = \frac{\displaystyle\lim_{(x,y)\to(a,b)} f(x, \ y) - \lim_{(x,y)\to(a,b)} g(x, \ y)}{\displaystyle\lim_{(x,y)\to(a,b)} f(x, \ y)} = \frac{5 - 3}{5} = \frac{2}{5}$

5. $\displaystyle \lim_{(x,y)\to(2,1)} (x + 3y^2) = 2 + 3(1)^2 = 5$

Continuous everywhere

6. $\displaystyle \lim_{(x,y)\to(0,0)} (5x + 3xy + y + 1) = 0 + 0 + 0 + 1 = 1$

Continuous everywhere

7. $\displaystyle\lim_{(x,y)\to(2,4)} \frac{x+y}{x-y} = \frac{2+4}{2-4} = -3$

Continuous for $x \neq y$

8. $\displaystyle\lim_{(x,y)\to(1,1)} \frac{x}{\sqrt{x+y}} = \frac{1}{\sqrt{1+1}} = \frac{\sqrt{2}}{2}$

Continuous for $x + y > 0$

9. $\displaystyle\lim_{(x,y)\to(0,1)} \frac{\arcsin(x/y)}{1+xy}$

$= \arcsin 0 = 0$

Continuous for $xy \neq -1$, $y \neq 0$, $|x/y| \leq 1$

10. $\displaystyle\lim_{(x,y)\to(\pi/4,2)} y\sin(xy)$

$= 2\sin\dfrac{\pi}{2} = 2$

Continuous everywhere

11. $\displaystyle\lim_{(x,y)\to(0,0)} e^{xy} = e^0 = 1$

Continuous everywhere

12. $\displaystyle\lim_{(x,y)\to(1,1)} \frac{xy}{x^2+y^2} = \frac{1}{2}$

Continuous except at $(0, 0)$

13. $\displaystyle\lim_{(x,y,z)\to(1,2,5)} \sqrt{x+y+z}$

$= \sqrt{8} = 2\sqrt{2}$

Continuous for $x + y + z \geq 0$

14. $\displaystyle\lim_{(x,y,z)\to(2,0,1)} xe^{yz} = 2e^0 = 2$

Continuous everywhere

15. $\displaystyle\lim_{(x,y)\to(0,0)} e^{xy} = 1$

Continuous everywhere

16. $f(x,\ y) = \dfrac{x^2}{(x^2+1)(y^2+1)}$

$\displaystyle\lim_{(x,y)\to(0,0)} \left(\frac{x^2}{(x^2+1)(y^2+1)}\right) = \frac{0}{(0+1)(0+1)} = 0$

Continuous everywhere

17. $\displaystyle\lim_{(x,y)\to(0,0)} \ln(x^2+y^2) = \ln(0) = -\infty$

The limit does not exist.
Continuous except at $(0, 0)$

18. $\displaystyle\lim_{(x,y)\to(0,0)} \left[1 - \frac{\cos(x^2+y^2)}{x^2+y^2}\right] = -\infty$

The limit does not exist.
Continuous except at $(0, 0)$

19. $\displaystyle\lim_{(x,y)\to(0,0)} \frac{xy}{x^2+y^2}$

Along $y = 0$:

$\displaystyle\lim_{(x,y)\to(0,0)} \frac{xy}{x^2+y^2} = \lim_{x\to 0} \frac{0}{x^2} = 0$

Along $y = x$:

$\displaystyle\lim_{(x,y)\to(0,0)} \frac{xy}{x^2+y^2} = \lim_{x\to 0} \frac{x^2}{2x^2} = \lim_{x\to 0} \frac{1}{2} = \frac{1}{2}$

Therefore, the limit does not exist.
Continuous except at $(0,\ 0)$

20. $\displaystyle\lim_{(x,y)\to(0,0)} \frac{y}{x^2+y^2}$

Along $y = 0$:

$\displaystyle\lim_{(x,y)\to(0,0)} \frac{y}{x^2+y^2} = \frac{0}{x^2} = 0$

Along $y = x$:

$\displaystyle\lim_{(x,y)\to(0,0)} \frac{y}{x^2+y^2} = \lim_{x\to 0} \frac{x}{2x^2}$

$= \displaystyle\lim_{x\to 0} \frac{1}{2x}$ does not exist.

Therefore, the limit does not exist.
Continuous except at $(0, 0)$

21. $f(x, y) = \dfrac{-xy^2}{x^2 + y^4}$

Along the line $x = y^2$:

$$\lim_{(x,y)\to(0,0)} \left(\frac{-xy^2}{x^2 + y^4}\right) = \lim_{y\to 0} \left(\frac{-y^4}{2y^4}\right) = -\frac{1}{2}$$

Along the line $x = -y^2$:

$$\lim_{(x,y)\to(0,0)} \left(\frac{-xy^2}{x^2 + y^4}\right) = \lim_{y\to 0} \left(\frac{y^4}{2y^4}\right) = \frac{1}{2}$$

Therefore, the limit does not exist.
Continuous except at $(0, 0)$

22. $\displaystyle\lim_{(x,y)\to(0,0)} \frac{2x - y^2}{2x^2 + y}$

Along $y = 0$:

$$\lim_{(x,y)\to(0,0)} \frac{2x - y^2}{2x^2 + y} = \lim_{x\to 0} \frac{2x}{2x^2}$$

$$= \lim_{x\to 0} \frac{1}{x} \text{ does not exist.}$$

Along $y = x$:

$$\lim_{(x,y)\to(0,0)} \frac{2x - y^2}{2x^2 + y} = \lim_{x\to 0} \frac{2x - x^2}{2x^2 + x}$$

$$= \lim_{x\to 0} \frac{2 - x}{2x + 1} = 2$$

Therefore, the limit does not exist.
Continuous except when $y = -2x^2$

23. $\displaystyle\lim_{(x,y)\to(0,0)} (\sin x +$
$\cos y) = 1$

24. $\displaystyle\lim_{(x,y)\to(0,0)} \left(\sin \frac{1}{x} + \cos \frac{1}{x}\right)$
Does not exist

25. $\displaystyle\lim_{(x,y)\to(0,0)} \frac{x^2 y}{x^4 + 4y^2}$
Does not exist

26. $\displaystyle\lim_{(x,y)\to(0,0)} \frac{x^2 + y^2}{x^2 y}$
Does not exist

27. $\displaystyle\lim_{(x,y)\to(0,0)} \frac{\sin(x^2 + y^2)}{x^2 + y^2} = \lim_{r\to 0} \frac{\sin r^2}{r^2} = \lim_{r\to 0} \frac{2r \cos r^2}{2r} = \lim_{r\to 0} \cos r^2 = 1$

28. $\displaystyle\lim_{(x,y)\to(0,0)} \frac{xy^2}{x^2 + y^2} = \lim_{r\to 0} \frac{(r\cos\theta)(r^2\sin^2\theta)}{r^2} = \lim_{r\to 0} (r\cos\theta\sin^2\theta) = 0$

29. $\displaystyle\lim_{(x,y)\to(0,0)} \frac{x^3 + y^3}{x^2 + y^2} = \lim_{r\to 0} \frac{r^3(\cos^3\theta + \sin^3\theta)}{r^2}$
$= 0$

30. $\displaystyle\lim_{(x,y)\to(0,0)} \frac{x^2 y^2}{x^2 + y^2} = \lim_{r\to 0} \frac{r^4 \cos^2\theta \sin^2\theta}{r^2}$
$= \lim_{r\to 0} r^2 \cos^2\theta \sin^2\theta = 0$

31. $f(x, y, z) = \dfrac{1}{\sqrt{x^2 + y^2 + z^2}}$
Continuous except at $(0, 0, 0)$

32. $f(x, y, z) = \dfrac{z}{x^2 + y^2 - 4}$
Continuous for $x^2 + y^2 \neq 4$

33. $f(x, y, z) = \dfrac{\sin z}{e^x + e^y}$
Continuous everywhere

34. $f(x, y, z) = xy \sin z$
Continuous everywhere

35. $f(t) = t^2$

$g(x, y) = 3x - 2y$

$f(g(x, y)) = f(3x - 2y)$

$\qquad = (3x - 2y)^2$

$\qquad = 9x^2 - 12xy + 4y^2$

Continuous everywhere

36. $f(t) = \dfrac{1}{t}$

$g(x, y) = x^2 + y^2$

$f(g(x, y)) = f(x^2 + y^2)$

$\qquad = \dfrac{1}{x^2 + y^2}$

Continuous except at (0, 0)

37. $f(t) = \dfrac{1}{t}$

$g(x, y) = 3x - 2y$

$f(g(x, y)) = f(3x - 2y) = \dfrac{1}{3x - 2y}$

Continuous for $y \neq \dfrac{3x}{2}$

38. $f(t) = \dfrac{1}{4 - t}$

$g(x, y) = x^2 + y^2$

$f(g(x, y)) = f(x^2 + y^2) = \dfrac{1}{4 - x^2 - y^2}$

Continuous for $x^2 + y^2 \neq 4$

39. $f(x, y) = x^2 - 4y$

a. $\displaystyle\lim_{\Delta x \to 0} \frac{f(x + \Delta x, y) - f(x, y)}{\Delta x} = \lim_{\Delta x \to 0} \frac{[(x + \Delta x)^2 - 4y] - (x^2 - 4y)}{\Delta x}$

$\qquad = \displaystyle\lim_{\Delta x \to 0} \frac{2x\Delta x - (\Delta x)^2}{\Delta x} = \lim_{\Delta x \to 0} (2x - \Delta x) = 2x$

b. $\displaystyle\lim_{\Delta y \to 0} \frac{f(x, y + \Delta y) - f(x, y)}{\Delta y} = \lim_{\Delta y \to 0} \frac{[x^2 - 4(y + \Delta y)] - (x^2 - 4y)}{\Delta y}$

$\qquad = \displaystyle\lim_{\Delta y \to 0} \frac{-4\Delta y}{\Delta y} = \lim_{\Delta y \to 0} (-4) = -4$

40. $f(x, y) = x^2 + y^2$

a. $\displaystyle\lim_{\Delta x \to 0} \frac{f(x + \Delta x, y) - f(x, y)}{\Delta x} = \lim_{\Delta x \to 0} \frac{[(x + \Delta x)^2 + y^2] - (x^2 + y^2)}{\Delta x}$

$\qquad = \displaystyle\lim_{\Delta x \to 0} \frac{2x\Delta x + (\Delta x)^2}{\Delta x} = \lim_{\Delta x \to 0} (2x + \Delta x) = 2x$

b. $\displaystyle\lim_{\Delta y \to 0} \frac{f(x, y + \Delta y) - f(x, y)}{\Delta y} = \lim_{\Delta y \to 0} \frac{[x^2 + (y + \Delta y)^2] - (x^2 + y^2)}{\Delta y}$

$\qquad = \displaystyle\lim_{\Delta y \to 0} \frac{2y\Delta y + (\Delta y)^2}{\Delta y} = \lim_{\Delta y \to 0} (2y + \Delta y) = 2y$

41. $f(x, y) = 2x + xy - 3y$

a. $\displaystyle\lim_{\Delta x \to 0} \frac{f(x + \Delta x, y) - f(x, y)}{\Delta x} = \lim_{\Delta x \to 0} \frac{[2(x + \Delta x) + (x + \Delta x)y - 3y] - (2x + xy - 3y)}{\Delta x}$

$\qquad = \displaystyle\lim_{\Delta x \to 0} \frac{2\Delta x + \Delta xy}{\Delta x} = \lim_{\Delta x \to 0} (2 + y) = 2 + y$

b. $\displaystyle\lim_{\Delta y \to 0} \frac{f(x, y + \Delta y) - f(x, y)}{\Delta y} = \lim_{\Delta y \to 0} \frac{[2x + x(y + \Delta y) - 3(y + \Delta y)] - (2x + xy - 3y)}{\Delta y}$

$\qquad = \displaystyle\lim_{\Delta y \to 0} \frac{x\Delta y - 3\Delta y}{\Delta y} = \lim_{\Delta y \to 0} (x - 3) = x - 3$

42. $f(x, y) = \sqrt{y}(y + 1)$

a. $\displaystyle\lim_{\Delta x \to 0} \frac{f(x + \Delta x, y) - f(x, y)}{\Delta x} = \lim_{\Delta x \to 0} \frac{\sqrt{y}(y + 1) - \sqrt{y}(y + 1)}{\Delta x} = 0$

b. $\displaystyle\lim_{\Delta y \to 0} \frac{f(x, y + \Delta y) - f(x, y)}{\Delta y} = \lim_{\Delta y \to 0} \frac{(y + \Delta y)^{3/2} + (y + \Delta y)^{1/2} - (y^{3/2} + y^{1/2})}{\Delta y}$

$\displaystyle\qquad = \lim_{\Delta y \to 0} \frac{(y + \Delta y)^{3/2} - y^{3/2}}{\Delta y} + \lim_{\Delta y \to 0} \frac{(y + \Delta y)^{1/2} - y^{1/2}}{\Delta y}$

$\displaystyle\qquad = \frac{3}{2} y^{1/2} + \frac{1}{2} y^{-1/2} \quad \text{(L'Hôpital's Rule)}$

$\displaystyle\qquad = \frac{3y + 1}{2\sqrt{y}}$

43. Since $\displaystyle\lim_{(x,y) \to (a,b)} f(x, y) = L_1$, then for $\epsilon/2 > 0$, there corresponds $\delta_1 > 0$ such that $|f(x, y) - L_1| < \epsilon/2$ whenever $0 < \sqrt{(x - x_0)^2 + (y - y_0)^2} < \delta_1$. Since $\displaystyle\lim_{(x,y) \to (a,b)} g(x, y) = L_2$, then for $\epsilon/2 > 0$, there corresponds $\delta_2 > 0$ such that $|g(x, y) - L_2| < \epsilon/2$ whenever $0 < \sqrt{(x - x_0)^2 + (y - y_0)^2} < \delta_2$. Let δ be the smaller of δ_1 and δ_2. By the triangle inequality, whenever $\sqrt{(x - x_0)^2 + (y - y_0)^2} < \delta$, we have

$$|f(x, y) + g(x, y) - (L_1 + L_2)| = |(f(x, y) - L_1) + (g(x, y) - L_2)|$$

$$\le |f(x, y) - L_1| + |g(x, y) - L_2| < \frac{\epsilon}{2} + \frac{\epsilon}{2} = \epsilon.$$

Therefore, $\displaystyle\lim_{(x,y) \to (a,b)} [f(x, y) + g(x, y)] = L_1 + L_2.$

44. Given that $f(x, y)$ is continuous, then $\displaystyle\lim_{(x,y) \to (a,b)} f(x, y) = f(a, b) < 0$, which means that for each $\epsilon > 0$, there corresponds a $\delta > 0$ such that $|f(x, y) - f(a, b)| < \epsilon$ whenever $0 < \sqrt{(x - a)^2 + (y - b)^2} < \delta$. Let $\epsilon = |f(a, b)|/2$, then $f(x, y) < 0$ for every point in the corresponding δ neighborhood since

$$|f(x, y) - f(a, b)| < \frac{|f(a, b)|}{2} \Rightarrow -\frac{|f(a, b)|}{2} < f(x, y) - f(a, b) < \frac{|f(a, b)|}{2}$$

$$\Rightarrow \frac{3}{2} f(a, b) < f(x, y) < \frac{1}{2} f(a, b) < 0.$$

45. True

46. False. Let $f(x, y) = \dfrac{xy}{x^2 + y^2}.$

See Exercise 19.

47. False. Let

$$f(x, y) = \begin{cases} \ln(x^2 + y^2), & x \ne 0, \ y \ne 0 \\ 0, & x = 0, \ y = 0 \end{cases}$$

See Exercise 17.

48. True

49. $f(x, y) = \dfrac{xy^3}{x^2 + 2y^6}$

Along $y = x$: $\displaystyle\lim_{(x,y) \to (0,0)} \frac{xy^3}{x^2 + 2y^6} = \lim_{x \to 0} \frac{x^4}{x^2 + 2x^6} = \lim_{x \to 0} \frac{x^2}{1 + 2x^4} = 0$

Along $y = \sqrt[3]{x}$: $\displaystyle\lim_{(x,y) \to (0,0)} \frac{xy^3}{x^2 + 2y^6} = \lim_{x \to 0} \frac{x^2}{x^2 + 2x^2} = \lim_{x \to 0} \frac{1}{3} = \frac{1}{3}$

Therefore, the limit does not exist.

Section 13.3   Partial Derivatives

1. $f(x,\ y) = 2x - 3y + 5$

$\quad f_x(x,\ y) = 2$

$\quad f_y(x,\ y) = -3$

2. $f(x,\ y) = x^2 - 3y^2 + 7$

$\quad f_x(x,\ y) = 2x$

$\quad f_y(x,\ y) = -6y$

3. $z = x\sqrt{y}$

$\quad \dfrac{\partial z}{\partial x} = \sqrt{y}$

$\quad \dfrac{\partial z}{\partial y} = \dfrac{x}{2\sqrt{y}}$

4. $z = x^2 - 3xy + y^2$

$\quad \dfrac{\partial z}{\partial x} = 2x - 3y$

$\quad \dfrac{\partial z}{\partial y} = -3x + 2y$

5. $z = x^2 e^{2y}$

$\quad \dfrac{\partial z}{\partial x} = 2x e^{2y}$

$\quad \dfrac{\partial z}{\partial y} = 2x^2 e^{2y}$

6. $z = x e^{x/y}$

$\quad \dfrac{\partial z}{\partial x} = \dfrac{x}{y} e^{x/y} + e^{x/y} = e^{x/y}\left(\dfrac{x}{y} + 1\right)$

$\quad \dfrac{\partial z}{\partial y} = x e^{x/y}\left(-\dfrac{x}{y^2}\right) = -\dfrac{x^2}{y^2} e^{x/y}$

7. $z = \ln(x^2 + y^2)$

$\quad \dfrac{\partial z}{\partial x} = \dfrac{2x}{x^2 + y^2}$

$\quad \dfrac{\partial z}{\partial y} = \dfrac{2y}{x^2 + y^2}$

8. $z = \ln\sqrt{xy} = \dfrac{1}{2}\ln(xy)$

$\quad \dfrac{\partial z}{\partial x} = \dfrac{1}{2}\dfrac{y}{xy} = \dfrac{1}{2x}$

$\quad \dfrac{\partial z}{\partial y} = \dfrac{1}{2}\dfrac{x}{xy} = \dfrac{1}{2y}$

9. $z = \ln\left(\dfrac{x+y}{x-y}\right) = \ln(x+y) - \ln(x-y)$

$\quad \dfrac{\partial z}{\partial x} = \dfrac{1}{x+y} - \dfrac{1}{x-y} = -\dfrac{2y}{x^2 - y^2}$

$\quad \dfrac{\partial z}{\partial y} = \dfrac{1}{x+y} + \dfrac{1}{x-y} = \dfrac{2x}{x^2 - y^2}$

10. $z = \dfrac{x^2}{2y} + \dfrac{4y^2}{x}$

$\quad \dfrac{\partial z}{\partial x} = \dfrac{2x}{2y} - \dfrac{4y^2}{x^2} = \dfrac{x^3 - 4y^3}{x^2 y}$

$\quad \dfrac{\partial z}{\partial y} = -\dfrac{x^2}{2y^2} + \dfrac{8y}{x} = \dfrac{-x^3 + 16y^3}{2xy^2}$

11. $h(x,\ y) = e^{-(x^2 + y^2)}$

$\quad h_x(x,\ y) = -2x e^{-(x^2 + y^2)}$

$\quad h_y(x,\ y) = -2y e^{-(x^2 + y^2)}$

12. $g(x,\ y) = \ln\sqrt{x^2 + y^2} = \dfrac{1}{2}\ln(x^2 + y^2)$

$\quad g_x(x,\ y) = \dfrac{1}{2}\dfrac{2x}{x^2 + y^2} = \dfrac{x}{x^2 + y^2}$

$\quad g_y(x,\ y) = \dfrac{1}{2}\dfrac{2y}{x^2 + y^2} = \dfrac{y}{x^2 + y^2}$

13. $f(x,\ y) = \sqrt{x^2 + y^2}$

$\quad f_x(x,\ y) = \dfrac{1}{2}(x^2 + y^2)^{-1/2}(2x) = \dfrac{x}{\sqrt{x^2 + y^2}}$

$\quad f_y(x,\ y) = \dfrac{1}{2}(x^2 + y^2)^{-1/2}(2y) = \dfrac{y}{\sqrt{x^2 + y^2}}$

14. $f(x,\ y) = \dfrac{xy}{x^2 + y^2}$

$\quad f_x(x,\ y) = \dfrac{(x^2 + y^2)(y) - (xy)(2x)}{(x^2 + y^2)^2} = \dfrac{y^3 - x^2 y}{(x^2 + y^2)^2}$

$\quad f_y(x,\ y) = \dfrac{(x^2 + y^2)(x) - (xy)(2y)}{(x^2 + y^2)^2} = \dfrac{x^3 - xy^2}{(x^2 + y^2)^2}$

15. $z = \tan(2x - y)$

$\quad \dfrac{\partial z}{\partial x} = 2\sec^2(2x - y)$

$\quad \dfrac{\partial z}{\partial y} = -\sec^2(2x - y)$

16. $z = \sin 3x \cos 3y$

$\quad \dfrac{\partial z}{\partial x} = 3\cos 3x \cos 3y$

$\quad \dfrac{\partial z}{\partial y} = -3\sin 3x \sin 3y$

17. $z = e^y \sin xy$

$\dfrac{\partial z}{\partial x} = ye^y \cos xy$

$\dfrac{\partial z}{\partial y} = e^y \sin xy + xe^y \cos xy$

$= e^y(x \cos xy + \sin xy)$

18. $z = \cos(x^2 + y^2)$

$\dfrac{\partial z}{\partial x} = -2x \sin(x^2 + y^2)$

$\dfrac{\partial z}{\partial y} = -2y \sin(x^2 + y^2)$

19. $f(x,\ y) = \displaystyle\int_x^y (t^2 - 1)\, dt = \left[\dfrac{t^3}{3} - t\right]_x^y = \left(\dfrac{y^3}{3} - y\right) - \left(\dfrac{x^3}{3} - x\right)$

$f_x(x,\ y) = -x^2 + 1 = 1 - x^2$

$f_y(x,\ y) = y^2 - 1$

20. $f(x,\ y) = \displaystyle\int_x^y (2t + 1)\, dt + \int_y^x (2t - 1)\, dt = \int_x^y (2t + 1)\, dt - \int_x^y (2t - 1)\, dt = \int_x^y 2\, dt = \left[2t\right]_x^y = 2y - 2x$

$f_x(x,\ y) = -2$

$f_y(x,\ y) = 2$

21. $f(x,\ y) = 2x + 3y$

$\dfrac{\partial f}{\partial x} = \lim_{\Delta x \to 0} \dfrac{f(x + \Delta x,\ y) - f(x,\ y)}{\Delta x} = \lim_{\Delta x \to 0} \dfrac{2(x + \Delta x) + 3y - 2x - 3y}{\Delta x} = \lim_{\Delta x \to 0} \dfrac{2\Delta x}{\Delta x} = 2$

$\dfrac{\partial f}{\partial y} = \lim_{\Delta y \to 0} \dfrac{f(x,\ y + \Delta y) - f(x,\ y)}{\Delta y} = \lim_{\Delta y \to 0} \dfrac{2x + 3(y + \Delta y) - 2x - 3y}{\Delta y} = \lim_{\Delta y \to 0} \dfrac{3\Delta y}{\Delta y} = 3$

22. $f(x,\ y) = \dfrac{1}{x + y}$

$\dfrac{\partial f}{\partial x} = \lim_{\Delta x \to 0} \dfrac{f(x + \Delta x,\ y) - f(x,\ y)}{\Delta x} = \lim_{\Delta x \to 0} \dfrac{\dfrac{1}{x + \Delta x + y} - \dfrac{1}{x + y}}{\Delta x} = \lim_{\Delta x \to 0} \dfrac{-1}{(x + \Delta x + y)(x + y)} = \dfrac{-1}{(x + y)^2}$

$\dfrac{\partial f}{\partial y} = \lim_{\Delta y \to 0} \dfrac{f(x,\ y + \Delta y) - f(x,\ y)}{\Delta y} = \lim_{\Delta y \to 0} \dfrac{\dfrac{1}{x + y + \Delta y} - \dfrac{1}{x + y}}{\Delta y} = \lim_{\Delta y \to 0} \dfrac{-1}{(x + y + \Delta y)(x + y)} = \dfrac{-1}{(x + y)^2}$

23. $f(x,\ y) = \sqrt{x + y}$

$\dfrac{\partial f}{\partial x} = \lim_{\Delta x \to 0} \dfrac{f(x + \Delta x,\ y) - f(x,\ y)}{\Delta x} = \lim_{\Delta x \to 0} \dfrac{\sqrt{x + \Delta x + y} - \sqrt{x + y}}{\Delta x}$

$= \lim_{\Delta x \to 0} \dfrac{\left(\sqrt{x + \Delta x + y} - \sqrt{x + y}\right)\left(\sqrt{x + \Delta x + y} + \sqrt{x + y}\right)}{\Delta x\left(\sqrt{x + \Delta x + y} + \sqrt{x + y}\right)}$

$= \lim_{\Delta x \to 0} \dfrac{1}{\sqrt{x + \Delta x + y} + \sqrt{x + y}} = \dfrac{1}{2\sqrt{x + y}}$

$\dfrac{\partial f}{\partial y} = \lim_{\Delta y \to 0} \dfrac{f(x,\ y + \Delta y) - f(x,\ y)}{\Delta y} = \lim_{\Delta y \to 0} \dfrac{\sqrt{x + y + \Delta y} - \sqrt{x + y}}{\Delta y}$

$= \lim_{\Delta y \to 0} \dfrac{\left(\sqrt{x + y + \Delta y} - \sqrt{x + y}\right)\left(\sqrt{x + y + \Delta y} + \sqrt{x + y}\right)}{\Delta y\left(\sqrt{x + y + \Delta y} + \sqrt{x + y}\right)}$

$= \lim_{\Delta y \to 0} \dfrac{1}{\sqrt{x + y + \Delta y} + \sqrt{x + y}} = \dfrac{1}{2\sqrt{x + y}}$

24. $f(x, y) = x^2 - 2xy + y^2 = (x - y)^2$

$$\frac{\partial f}{\partial x} = \lim_{\Delta x \to 0} \frac{f(x + \Delta x, y) - f(x, y)}{\Delta x}$$

$$= \lim_{\Delta x \to 0} \frac{(x + \Delta x)^2 - 2(x + \Delta x)y + y^2 - x^2 + 2xy - y^2}{\Delta x} = \lim_{\Delta x \to 0} (2x + \Delta x - 2y) = 2(x - y)$$

$$\frac{\partial f}{\partial y} = \lim_{\Delta y \to 0} \frac{f(x, y + \Delta y) - f(x, y)}{\Delta y}$$

$$= \lim_{\Delta y \to 0} \frac{x^2 - 2x(y + \Delta y) + (y + \Delta y)^2 - x^2 + 2xy - y^2}{\Delta y} = \lim_{\Delta y \to 0} (-2x + 2y + \Delta y) = 2(y - x)$$

25. $g(x, y) = 4 - x^2 - y^2$

$g_x(x, y) = -2x$

At $(1, 1)$: $g_x(1, 1) = -2$

$g_y(x, y) = -2y$

At $(1, 1)$: $g_y(1, 1) = -2$

26. $h(x, y) = x^2 - y^2$

$h_x(x, y) = 2x$

At $(-2, 1)$: $h_x(-2, 1) = -4$

$h_y(x, y) = -2y$

At $(-2, 1)$: $h_y(-2, 1) = -2$

27. $z = e^{-x} \cos y$

$$\frac{\partial z}{\partial x} = -e^{-x} \cos y$$

At $(0, 0)$: $\dfrac{\partial z}{\partial x} = -1$

$$\frac{\partial z}{\partial y} = -e^{-x} \sin y$$

At $(0, 0)$: $\dfrac{\partial z}{\partial y} = 0$

28. $z = \dfrac{1}{2} \sin(2x - y)$

$$\frac{\partial z}{\partial x} = \cos(2x - y)$$

At $\left(\dfrac{\pi}{2}, \dfrac{\pi}{3}\right)$: $\dfrac{\partial z}{\partial x} = -\dfrac{1}{2}$

$$\frac{\partial z}{\partial y} = -\frac{1}{2} \cos(2x - y)$$

At $\left(\dfrac{\pi}{2}, \dfrac{\pi}{3}\right)$: $\dfrac{\partial z}{\partial y} = \dfrac{1}{4}$

29. $f(x, y) = \arctan \dfrac{y}{x}$

$$f_x(x, y) = \frac{1}{1 + (y^2/x^2)}\left(-\frac{y}{x^2}\right) = \frac{-y}{x^2 + y^2}$$

At $(2, -2)$: $f_x(2, -2) = \dfrac{1}{4}$

$$f_y(x, y) = \frac{1}{1 + (y^2/x^2)}\left(\frac{1}{x}\right) = \frac{x}{x^2 + y^2}$$

At $(2, -2)$: $f_y(2, -2) = \dfrac{1}{4}$

30. $f(x, y) = \arcsin xy$

$$f_x(x, y) = \frac{y}{\sqrt{1 - (xy)^2}}$$

At $(1, 0)$: $f_x(1, 0) = 0$

$$f_y(x, y) = \frac{x}{\sqrt{1 - (xy)^2}}$$

At $(1, 0)$: $f_y(1, 0) = 1$

31. $f(x, y) = \dfrac{xy}{x - y}$

$$f_x(x, y) = \frac{y(x - y) - xy}{(x - y)^2} = \frac{-y^2}{(x - y)^2}$$

At $(2, -2)$: $f_x(2, -2) = -\dfrac{1}{4}$

$$f_y(x, y) = \frac{x(x - y) + xy}{(x - y)^2} = \frac{x^2}{(x - y)^2}$$

At $(2, -2)$: $f_y(2, -2) = \dfrac{1}{4}$

32. $f(x, y) = \dfrac{4xy}{\sqrt{x^2 + y^2}}$

$$f_x(x, y) = \frac{4y^3}{(x^2 + y^2)^{3/2}}$$

At $(1, 0)$: $f_x(1, 0) = 0$

$$f_y(x, y) = \frac{4x^3}{(x^2 + y^2)^{3/2}}$$

At $(1, 0)$: $f_y(1, 0) = 4$

33. $z = \sqrt{49 - x^2 - y^2}$, $x = 2$, $(2, 3, 6)$

Intersecting curve: $z = \sqrt{45 - y^2}$

$$\frac{\partial z}{\partial y} = \frac{-y}{\sqrt{45 - y^2}}$$

At $(2, 3, 6)$: $\dfrac{\partial z}{\partial y} = \dfrac{-3}{\sqrt{45 - 9}} = -\dfrac{1}{2}$

34. $z = x^2 + 4y^2$, $y = 1$, $(2, 1, 8)$

Intersecting curve: $z = x^2 + 4$

$$\frac{\partial z}{\partial x} = 2x$$

At $(2, 1, 8)$: $\dfrac{\partial z}{\partial x} = 2(2) = 4$

35. $z = 9x^2 - y^2$, $y = 3$, $(1, 3, 0)$

Intersecting curve: $z = 9x^2 - 9$

$$\frac{\partial z}{\partial x} = 18x$$

At $(1, 3, 0)$: $\dfrac{\partial z}{\partial x} = 18(1) = 18$

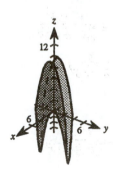

36. $z = 9x^2 - y^2$, $x = 1$, $(1, 3, 0)$

Intersecting curve: $z = 9 - y^2$

$$\frac{\partial z}{\partial y} = -2y$$

At $(1, 3, 0)$: $\dfrac{\partial z}{\partial y} = -2(3) = -6$

37. $z = x^2 - 2xy + 3y^2$

$$\frac{\partial z}{\partial x} = 2x - 2y$$

$$\frac{\partial^2 z}{\partial x^2} = 2$$

$$\frac{\partial^2 z}{\partial y \partial x} = -2$$

$$\frac{\partial z}{\partial y} = -2x + 6y$$

$$\frac{\partial^2 z}{\partial y^2} = 6$$

$$\frac{\partial^2 z}{\partial x \partial y} = -2$$

38. $z = x^4 - 3x^2 y^2 + y^4$

$$\frac{\partial z}{\partial x} = 4x^3 - 6xy^2$$

$$\frac{\partial^2 z}{\partial x^2} = 12x^2 - 6y^2$$

$$\frac{\partial^2 z}{\partial y \partial x} = -12xy$$

$$\frac{\partial z}{\partial y} = -6x^2 y + 4y^3$$

$$\frac{\partial^2 z}{\partial y^2} = -6x^2 + 12y^2$$

$$\frac{\partial^2 z}{\partial x \partial y} = -12xy$$

39. $z = \sqrt{x^2 + y^2}$

$$\frac{\partial z}{\partial x} = \frac{x}{\sqrt{x^2 + y^2}}$$

$$\frac{\partial^2 z}{\partial x^2} = \frac{y^2}{(x^2 + y^2)^{3/2}}$$

$$\frac{\partial^2 z}{\partial y \partial x} = \frac{-xy}{(x^2 + y^2)^{3/2}}$$

$$\frac{\partial z}{\partial y} = \frac{y}{\sqrt{x^2 + y^2}}$$

$$\frac{\partial^2 z}{\partial y^2} = \frac{x^2}{(x^2 + y^2)^{3/2}}$$

$$\frac{\partial^2 z}{\partial x \partial y} = \frac{-xy}{(x^2 + y^2)^{3/2}}$$

40. $z = \ln(x - y)$

$$\frac{\partial z}{\partial x} = \frac{1}{x - y}$$

$$\frac{\partial^2 z}{\partial x^2} = -\frac{1}{(x - y)^2}$$

$$\frac{\partial^2 z}{\partial y \partial x} = \frac{1}{(x - y)^2}$$

$$\frac{\partial z}{\partial y} = \frac{-1}{x - y}$$

$$\frac{\partial^2 z}{\partial y^2} = -\frac{1}{(x - y)^2}$$

$$\frac{\partial^2 z}{\partial x \partial y} = \frac{1}{(x - y)^2}$$

Therefore, $\dfrac{\partial^2 z}{\partial y \partial x} = \dfrac{\partial^2 z}{\partial x \partial y}$.

41. $z = e^x \tan y$

$$\frac{\partial z}{\partial x} = e^x \tan y$$

$$\frac{\partial^2 z}{\partial x^2} = e^x \tan y$$

$$\frac{\partial^2 z}{\partial y \partial x} = e^x \sec^2 y$$

$$\frac{\partial z}{\partial y} = e^x \sec^2 y$$

$$\frac{\partial^2 z}{\partial y^2} = 2e^x \sec^2 y \tan y$$

$$\frac{\partial^2 z}{\partial x \partial y} = e^x \sec^2 y$$

42. $z = xe^y + ye^x$

$$\frac{\partial z}{\partial x} = e^y + ye^x$$

$$\frac{\partial^2 z}{\partial x^2} = ye^x$$

$$\frac{\partial^2 z}{\partial y \partial x} = e^y + e^x$$

$$\frac{\partial z}{\partial y} = xe^y + e^x$$

$$\frac{\partial^2 z}{\partial y^2} = xe^y$$

$$\frac{\partial^2 z}{\partial x \partial y} = e^y + e^x$$

Therefore, $\dfrac{\partial^2 z}{\partial y \partial x} = \dfrac{\partial^2 z}{\partial x \partial y}$.

43. $z = \arctan \dfrac{y}{x}$

$$\frac{\partial z}{\partial x} = \frac{1}{1 + (y^2/x^2)}\left(-\frac{y}{x^2}\right) = \frac{-y}{x^2 + y^2}$$

$$\frac{\partial^2 z}{\partial x^2} = \frac{2xy}{(x^2 + y^2)^2}$$

$$\frac{\partial^2 z}{\partial y \partial x} = \frac{-(x^2 + y^2) + y(2y)}{(x^2 + y^2)^2} = \frac{y^2 - x^2}{(x^2 + y^2)^2}$$

$$\frac{\partial z}{\partial y} = \frac{1}{1 + (y^2/x^2)}\left(\frac{1}{x}\right) = \frac{x}{x^2 + y^2}$$

$$\frac{\partial^2 z}{\partial y^2} = \frac{-2xy}{(x^2 + y^2)^2}$$

$$\frac{\partial^2 z}{\partial x \partial y} = \frac{(x^2 + y^2) - x(2x)}{(x^2 + y^2)^2} = \frac{y^2 - x^2}{(x^2 + y^2)^2}$$

44. $z = \sin(x - 2y)$

$$\frac{\partial z}{\partial x} = \cos(x - 2y)$$

$$\frac{\partial^2 z}{\partial x^2} = -\sin(x - 2y)$$

$$\frac{\partial^2 z}{\partial y \partial x} = 2\sin(x - 2y)$$

$$\frac{\partial z}{\partial y} = -2\cos(x - 2y)$$

$$\frac{\partial^2 z}{\partial y^2} = -4\sin(x - 2y)$$

$$\frac{\partial^2 z}{\partial x \partial y} = 2\sin(x - 2y)$$

45. $z = x \sec y$

$$\frac{\partial z}{\partial x} = \sec y$$

$$\frac{\partial^2 z}{\partial x^2} = 0$$

$$\frac{\partial^2 z}{\partial y \partial x} = \sec y \tan y$$

$$\frac{\partial z}{\partial y} = x \sec y \tan y$$

$$\frac{\partial^2 z}{\partial y^2} = x \sec y(\sec^2 y + \tan^2 y)$$

$$\frac{\partial^2 z}{\partial x \partial y} = \sec y \tan y$$

Therefore, $\dfrac{\partial^2 z}{\partial y \partial x} = \dfrac{\partial^2 z}{\partial x \partial y}$.

46. $z = \sqrt{9 - x^2 - y^2}$

$$\frac{\partial z}{\partial x} = \frac{-x}{\sqrt{9 - x^2 - y^2}}$$

$$\frac{\partial^2 z}{\partial x^2} = \frac{y^2 - 9}{(9 - x^2 - y^2)^{3/2}}$$

$$\frac{\partial^2 z}{\partial y \partial x} = \frac{-xy}{(9 - x^2 - y^2)^{3/2}}$$

$$\frac{\partial z}{\partial y} = \frac{-y}{\sqrt{9 - x^2 - y^2}}$$

$$\frac{\partial^2 z}{\partial y^2} = \frac{x^2 - 9}{(9 - x^2 - y^2)^{3/2}}$$

$$\frac{\partial^2 z}{\partial x \partial y} = \frac{-xy}{(9 - x^2 - y^2)^{3/2}}$$

Therefore, $\dfrac{\partial^2 z}{\partial y \partial x} = \dfrac{\partial^2 z}{\partial x \partial y}$.

47.

$$z = \ln\left(\frac{x}{x^2 + y^2}\right) = \ln x - \ln(x^2 + y^2)$$

$$\frac{\partial z}{\partial x} = \frac{1}{x} - \frac{2x}{x^2 + y^2} = \frac{y^2 - x^2}{x(x^2 + y^2)}$$

$$\frac{\partial^2 z}{\partial x^2} = \frac{x^4 - 4x^2 y^2 - y^4}{x^2(x^2 + y^2)^2}$$

$$\frac{\partial^2 z}{\partial y \partial x} = \frac{4xy}{(x^2 + y^2)^2}$$

$$\frac{\partial z}{\partial y} = -\frac{2y}{x^2 + y^2}$$

$$\frac{\partial^2 z}{\partial y^2} = \frac{2(y^2 - x^2)}{(x^2 + y^2)^2}$$

$$\frac{\partial^2 z}{\partial x \partial y} = \frac{4xy}{(x^2 + y^2)^2}$$

48.

$$z = \frac{xy}{x - y}$$

$$\frac{\partial z}{\partial x} = \frac{y(x - y) - xy}{(x - y)^2} = \frac{-y^2}{(x - y)^2}$$

$$\frac{\partial^2 z}{\partial x^2} = \frac{2y^2}{(x - y)^3}$$

$$\frac{\partial^2 z}{\partial y \partial x} = \frac{(x - y)^2(-2y) + y^2(2)(x - y)(-1)}{(x - y)^4}$$

$$= \frac{-2xy}{(x - y)^3}$$

$$\frac{\partial z}{\partial y} = \frac{x(x - y) + xy}{(x - y)^2} = \frac{x^2}{(x - y)^2}$$

$$\frac{\partial^2 z}{\partial y^2} = \frac{2x^2}{(x - y)^3}$$

$$\frac{\partial^2 z}{\partial x \partial y} = \frac{(x - y)^2(2x) - x^2(2)(x - y)}{(x - y)^4}$$

$$= \frac{-2xy}{(x - y)^3}$$

49. $w = \sqrt{x^2 + y^2 + z^2}$

$$\frac{\partial w}{\partial x} = \frac{x}{\sqrt{x^2 + y^2 + z^2}}$$

$$\frac{\partial w}{\partial y} = \frac{y}{\sqrt{x^2 + y^2 + z^2}}$$

$$\frac{\partial w}{\partial z} = \frac{z}{\sqrt{x^2 + y^2 + z^2}}$$

50. $w = \frac{xy}{x + y + z}$

$$\frac{\partial w}{\partial x} = \frac{y(x + y + z) - xy}{(x + y + z)^2} = \frac{y(y + z)}{(x + y + z)^2}$$

$$\frac{\partial w}{\partial y} = \frac{x(x + y + z) - xy}{(x + y + z)^2} = \frac{x(x + z)}{(x + y + z)^2}$$

$$\frac{\partial w}{\partial z} = \frac{-xy}{(x + y + z)^2}$$

51. $F(x, \ y, \ z) = \ln\sqrt{x^2 + y^2 + z^2}$

$$= \frac{1}{2}\ln(x^2 + y^2 + z^2)$$

$$F_x(x, \ y, \ z) = \frac{x}{x^2 + y^2 + z^2}$$

$$F_y(x, \ y, \ z) = \frac{y}{x^2 + y^2 + z^2}$$

$$F_z(x, \ y, \ z) = \frac{z}{x^2 + y^2 + z^2}$$

52. $G(x, \ y, \ z) = \dfrac{1}{\sqrt{1 - x^2 - y^2 - z^2}}$

$$G_x(x, \ y, \ z) = \frac{x}{(1 - x^2 - y^2 - z^2)^{3/2}}$$

$$G_y(x, \ y, \ z) = \frac{y}{(1 - x^2 - y^2 - z^2)^{3/2}}$$

$$G_z(x, \ y, \ z) = \frac{z}{(1 - x^2 - y^2 - z^2)^{3/2}}$$

53. $H(x, \ y, \ z) = \sin(x + 2y + 3z)$

$H_x(x, \ y, \ z) = \cos(x + 2y + 3z)$

$H_y(x, \ y, \ z) = 2\cos(x + 2y + 3z)$

$H_z(x, \ y, \ z) = 3\cos(x + 2y + 3z)$

54. $f(x, \ y, \ z) = 3x^2 y - 5xyz + 10yz^2$

$f_x(x, \ y, \ z) = 6xy - 5yz$

$f_y(x, \ y, \ z) = 3x^2 - 5xz + 10z^2$

$f_z(x, \ y, \ z) = -5xy + 20yz$

55. $f(x, y, z) = xyz$

$f_x(x, y, z) = yz$

$f_y(x, y, z) = xz$

$f_{yy}(x, y, z) = 0$

$f_{xy}(x, y, z) = z$

$f_{yx}(x, y, z) = z$

$f_{yyx}(x, y, z) = 0$

$f_{xyy}(x, y, z) = 0$

$f_{yxy}(x, y, z) = 0$

Therefore, $f_{xyy} = f_{yxy} = f_{yyx} = 0$.

56. $f(x, y, z) = x^2 - 3xy + 4yz + z^3$

$f_x(x, y, z) = 2x - 3y$

$f_y(x, y, z) = -3x + 4z$

$f_{yy}(x, y, z) = 0$

$f_{xy}(x, y, z) = -3$

$f_{yx}(x, y, z) = -3$

$f_{yyx}(x, y, z) = 0$

$f_{xyy}(x, y, z) = 0$

$f_{yxy}(x, y, z) = 0$

Therefore, $f_{xyy} = f_{yxy} = f_{yyx} = 0$.

57. $f(x, y, z) = e^{-x} \sin yz$

$f_x(x, y, z) = -e^{-x} \sin yz$

$f_y(x, y, z) = ze^{-x} \cos yz$

$f_{yy}(x, y, z) = -z^2 e^{-x} \sin yz$

$f_{xy}(x, y, z) = -ze^{-x} \cos yz$

$f_{yx}(x, y, z) = -ze^{-x} \cos yz$

$f_{yyx}(x, y, z) = z^2 e^{-x} \sin yz$

$f_{xyy}(x, y, z) = z^2 e^{-x} \sin yz$

$f_{yxy}(x, y, z) = z^2 e^{-x} \sin yz$

Therefore, $f_{xyy} = f_{yxy} = f_{yyx}$.

58. $f(x, y, z) = \dfrac{x}{y+z}$

$f_x(x, y, z) = \dfrac{1}{y+z}$

$f_y(x, y, z) = -\dfrac{x}{(y+z)^2}$

$f_{yy}(x, y, z) = \dfrac{2x}{(y+z)^3}$

$f_{xy}(x, y, z) = -\dfrac{1}{(y+z)^2}$

$f_{yx}(x, y, z) = -\dfrac{1}{(y+z)^2}$

$f_{yyx}(x, y, z) = \dfrac{2}{(y+z)^3}$

$f_{xyy}(x, y, z) = \dfrac{2}{(y+z)^3}$

$f_{yxy}(x, y, z) = \dfrac{2}{(y+z)^3}$

Therefore, $f_{xyy} = f_{yxy} = f_{yyx}$.

59. $z = 5xy$

$\dfrac{\partial z}{\partial x} = 5y$

$\dfrac{\partial^2 z}{\partial x^2} = 0$

$\dfrac{\partial z}{\partial y} = 5x$

$\dfrac{\partial^2 z}{\partial y^2} = 0$

Therefore,

$\dfrac{\partial^2 z}{\partial x^2} + \dfrac{\partial^2 z}{\partial y^2} = 0 + 0 = 0.$

60. $z = \sin x \left(\dfrac{e^y - e^{-y}}{2} \right)$

$\dfrac{\partial z}{\partial x} = \cos x \left(\dfrac{e^y - e^{-y}}{2} \right)$

$\dfrac{\partial^2 z}{\partial x^2} = -\sin x \left(\dfrac{e^y - e^{-y}}{2} \right)$

$\dfrac{\partial z}{\partial y} = \sin x \left(\dfrac{e^y + e^{-y}}{2} \right)$

$\dfrac{\partial^2 z}{\partial y^2} = \sin x \left(\dfrac{e^y - e^{-y}}{2} \right)$

Therefore, $\dfrac{\partial^2 z}{\partial x^2} + \dfrac{\partial^2 z}{\partial y^2} = -\sin x \left(\dfrac{e^y - e^{-y}}{2} \right) + \sin x \left(\dfrac{e^y - e^{-y}}{2} \right) = 0.$

61. $z = e^x \sin y$

$$\frac{\partial z}{\partial x} = e^x \sin y$$

$$\frac{\partial^2 z}{\partial x^2} = e^x \sin y$$

$$\frac{\partial z}{\partial y} = e^x \cos y$$

$$\frac{\partial^2 z}{\partial y^2} = -e^x \sin y$$

Therefore, $\dfrac{\partial^2 z}{\partial x^2} + \dfrac{\partial^2 z}{\partial y^2} = e^x \sin y - e^x \sin y = 0.$

62. $z = \arctan \dfrac{y}{x}$

From Exercise 43, we have

$$\frac{\partial^2 z}{\partial x^2} + \frac{\partial^2 z}{\partial y^2} = \frac{2xy}{(x^2 + y^2)^2} + \frac{-2xy}{(x^2 + y^2)^2} = 0.$$

63. $z = \sin(x - ct)$

$$\frac{\partial z}{\partial t} = -c \cos(x - ct)$$

$$\frac{\partial^2 z}{\partial t^2} = -c^2 \sin(x - ct)$$

$$\frac{\partial z}{\partial x} = \cos(x - ct)$$

$$\frac{\partial^2 z}{\partial x^2} = -\sin(x - ct)$$

Therefore, $\dfrac{\partial^2 z}{\partial t^2} = c^2 \dfrac{\partial^2 z}{\partial x^2}.$

64. $z = \sin(wct) \sin(wx)$

$$\frac{\partial z}{\partial t} = wc \cos(wct) \sin(wx)$$

$$\frac{\partial^2 z}{\partial t^2} = -w^2 c^2 \sin(wct) \sin(wx)$$

$$\frac{\partial z}{\partial x} = w \sin(wct) \cos(wx)$$

$$\frac{\partial^2 z}{\partial x^2} = -w^2 \sin(wct) \sin(wx)$$

Therefore, $\dfrac{\partial^2 z}{\partial t^2} = c^2 \dfrac{\partial^2 z}{\partial x^2}.$

65. $z = e^{-t} \cos \dfrac{x}{c}$

$$\frac{\partial z}{\partial t} = -e^{-t} \cos \frac{x}{c}$$

$$\frac{\partial z}{\partial x} = -\frac{1}{c} e^{-t} \sin \frac{x}{c}$$

$$\frac{\partial^2 z}{\partial x^2} = -\frac{1}{c^2} e^{-t} \cos \frac{x}{c}$$

Therefore, $\dfrac{\partial z}{\partial t} = c^2 \dfrac{\partial^2 z}{\partial x^2}.$

66. $z = e^{-t} \sin \dfrac{x}{c}$

$$\frac{\partial z}{\partial t} = -e^{-t} \sin \frac{x}{c}$$

$$\frac{\partial z}{\partial x} = \frac{1}{c} e^{-t} \cos \frac{x}{c}$$

$$\frac{\partial^2 z}{\partial x^2} = -\frac{1}{c^2} e^{-t} \sin \frac{x}{c}$$

Therefore, $\dfrac{\partial z}{\partial t} = c^2 \dfrac{\partial^2 z}{\partial x^2}.$

67. $C = 32\sqrt{xy} + 175x + 205y + 1050$

$$\frac{\partial C}{\partial x} = 16\sqrt{\frac{y}{x}} + 175$$

$$\left. \frac{\partial C}{\partial x} \right]_{(80,20)} = 16\sqrt{\frac{1}{4}} + 175 = 183$$

$$\frac{\partial C}{\partial y} = 16\sqrt{\frac{x}{y}} + 205$$

$$\left. \frac{\partial C}{\partial y} \right]_{(80,20)} = 16\sqrt{4} + 205 = 237$$

68. $f(x,\ y) = 100x^{0.6}y^{0.4}$

a. $\dfrac{\partial f}{\partial x} = 60x^{-0.4}y^{0.4}$

$$= 60\left(\frac{y}{x}\right)^{0.4} = 60\left(\frac{500}{1000}\right)^{0.4} \approx 45.47$$

b. $\dfrac{\partial f}{\partial y} = 40x^{0.6}y^{-0.6}$

$$= 40\left(\frac{x}{y}\right)^{0.6} = 40\left(\frac{1000}{500}\right)^{0.6} \approx 60.63$$

69.   An increase in either price will cause a decrease in demand.

70.

$$V(I, R) = 1000 \left[\frac{1 + 0.10(1 - R)}{1 + I} \right]^{10}$$

$$V_I(I, R) = 10,000 \left[\frac{1 + 0.10(1 - R)}{1 + I} \right]^9 \left[-\frac{1 + 0.10(1 - R)}{(1 + I)^2} \right] = -10,000 \frac{[1 + 0.10(1 - R)]^{10}}{(1 + I)^{11}}$$

$$V_I(0.03, 0.28) \approx -14,478.99$$

$$V_R(I, R) = 10,000 \left[\frac{1 + 0.10(1 - R)}{1 + I} \right]^9 \left[\frac{-0.10}{1 + I} \right] = -1000 \frac{[1 + 0.10(1 - R)]^9}{(1 + I)^{10}}$$

$$V_R(0.03, 0.28) \approx -1391.17$$

The rate of inflation has the greater negative influence on the growth of the investment. (See Exercise 61 in Section 13.1.)

71. $T = 500 - 0.6x^2 - 1.5y^2$

$$\frac{\partial T}{\partial x} = -1.2x = -2.4°/\text{ft}$$

$$\frac{\partial T}{\partial y} = -3y = -9°/\text{ft}$$

72. $PV = kT$

a. $P = \dfrac{kT}{V}, \quad \dfrac{\partial P}{\partial T} = \dfrac{k}{V}$

b. $V = \dfrac{kT}{P}, \quad \dfrac{\partial V}{\partial P} = -\dfrac{kT}{P^2}$

73. $R = \dfrac{1}{32}v_0^2 \sin 2\theta$

$$\frac{\partial R}{\partial v_0} = \frac{1}{16}v_0 \sin 2\theta$$

$$\left. \frac{\partial R}{\partial v_0} \right]_{(2000,5)} = \frac{2000}{16} \sin(10°)$$

$$\approx 21.7$$

$$\frac{\partial R}{\partial \theta} = \frac{1}{16}v_0^2 \cos 2\theta$$

$$\left. \frac{\partial R}{\partial \theta} \right]_{(2000,5)} = \frac{(2000)^2}{16} \cos(10°)$$

$$\approx 246{,}201.9$$

74. $f(x, y) = \begin{cases} \dfrac{xy(x^2 - y^2)}{x^2 + y^2}, & (x, y) \neq (0, 0) \\ 0, & (x, y) = (0, 0) \end{cases}$

a. $f_x(x, y) = \dfrac{(x^2 + y^2)(3x^2y - y^3) - (x^3y - xy^3)(2x)}{(x^2 + y^2)^2} = \dfrac{y(x^4 + 4x^2y^2 - y^4)}{(x^2 + y^2)^2}$

$f_y(x, y) = \dfrac{(x^2 + y^2)(x^3 - 3xy^2) - (x^3y - xy^3)(2y)}{(x^2 + y^2)^2} = \dfrac{x(x^4 - 4x^2y^2 - y^4)}{(x^2 + y^2)^2}$

b. $f_x(0, 0) = \lim\limits_{\Delta x \to 0} \dfrac{f(\Delta x, 0) - f(0, 0)}{\Delta x} = \lim\limits_{\Delta x \to 0} \dfrac{0/[(\Delta x)^2] - 0}{\Delta x} = 0$

$f_y(0, 0) = \lim\limits_{\Delta y \to 0} \dfrac{f(0, \Delta y) - f(0, 0)}{\Delta y} = \lim\limits_{\Delta y \to 0} \dfrac{0/[(\Delta y)^2] - 0}{\Delta y} = 0$

c. $f_{xy}(0, 0) = \dfrac{\partial}{\partial y} \left(\dfrac{\partial f}{\partial x} \right) \bigg|_{(0,0)} = \lim\limits_{\Delta y \to 0} \dfrac{f_x(0, \Delta y) - f_x(0, 0)}{\Delta y} = \lim\limits_{\Delta y \to 0} \dfrac{\Delta y(-(\Delta y)^4)}{((\Delta y)^2)^2(\Delta y)} = \lim\limits_{\Delta y \to 0} (-1) = -1$

$f_{yx}(0, 0) = \dfrac{\partial}{\partial x} \left(\dfrac{\partial f}{\partial y} \right) \bigg|_{(0,0)} = \lim\limits_{\Delta x \to 0} \dfrac{f_y(\Delta x, 0) - f_y(0, 0)}{\Delta x} = \lim\limits_{\Delta x \to 0} \dfrac{\Delta x((\Delta x)^4)}{((\Delta x)^2)^2(\Delta x)} = \lim\limits_{\Delta x \to 0} 1 = 1$

d. f_{yx} or f_{xy} is not continuous at $(0, 0)$.

75. False
Let $z = x + y + 1$.

76. True

77. True

78. True

79.
$$PV = mRT$$

$$T = \frac{PV}{mR} \Rightarrow \frac{\partial T}{\partial P} = \frac{V}{mR}$$

$$P = \frac{mRT}{V} \Rightarrow \frac{\partial P}{\partial V} = -\frac{mRT}{V^2}$$

$$V = \frac{mRT}{P} \Rightarrow \frac{\partial V}{\partial T} = \frac{mR}{P}$$

$$\frac{\partial T}{\partial P} \cdot \frac{\partial P}{\partial V} \cdot \frac{\partial V}{\partial T} = \left(\frac{V}{mR}\right)\left(-\frac{mRT}{V^2}\right)\left(\frac{mR}{P}\right)$$

$$= -\frac{mRT}{VP} = -\frac{mRT}{mRT} = -1$$

80. $f(x, y) = \int_x^y \sqrt{1+t^3}\, dt$

By the Second Fundamental Theorem of Calculus,

$$\frac{\partial f}{\partial x} = \frac{d}{dx}\int_x^y \sqrt{1+t^3}\, dt$$

$$= -\frac{d}{dx}\int_y^x \sqrt{1+t^3}\, dt = -\sqrt{1+x^3}$$

$$\frac{\partial f}{\partial y} = \frac{d}{dy}\int_x^y \sqrt{1+t^3}\, dt = \sqrt{1+y^3}.$$

81. Essay

Section 13.4   Differentials

1. $z = 3x^2 y^3$

$dz = 6xy^3\, dx + 9x^2 y^2\, dy$

2. $z = \dfrac{x^2}{y}$

$dz = \dfrac{2x}{y}\, dx - \dfrac{x^2}{y^2}\, dy$

3. $z = \dfrac{-1}{x^2 + y^2}$

$dz = \dfrac{2x}{(x^2+y^2)^2}\, dx + \dfrac{2y}{(x^2+y^2)^2}\, dy$

$= \dfrac{2}{(x^2+y^2)^2}(x\, dx + y\, dy)$

4. $z = e^x \sin y$

$dz = (e^x \sin y)\, dx + (e^x \cos y)\, dy$

5. $z = x \cos y - y \cos x$

$dz = (\cos y + y \sin x)\, dx + (-x \sin y - \cos x)\, dy = (\cos y + y \sin x)\, dx - (x \sin y + \cos x)\, dy$

6. $z = \left(\dfrac{1}{2}\right)(e^{x^2+y^2} - e^{-x^2-y^2})$

$dz = 2x\left(\dfrac{e^{x^2+y^2} + e^{-x^2-y^2}}{2}\right)dx + 2y\left(\dfrac{e^{x^2+y^2} + e^{-x^2-y^2}}{2}\right)dy = (e^{x^2+y^2} + e^{-x^2-y^2})(x\, dx + y\, dy)$

7. $w = 2z^3 y \sin x$

$dw = 2z^3 y \cos x\, dx + 2z^3 \sin x\, dy + 6z^2 y \sin x\, dz$

8. $w = e^x \cos y + z$

$dw = e^x \cos y\, dx - e^x \sin y\, dy + dz$

9. $w = \dfrac{x+y}{z-2y}$

$dw = \dfrac{1}{z-2y}\, dx + \dfrac{z+2x}{(z-2y)^2}\, dy - \dfrac{x+y}{(z-2y)^2}\, dz$

10. $w = x^2 yz^2 + \sin yz$

$dw = 2xyz^2\, dx + (x^2 z^2 + z \cos yz)\, dy + (2x^2 yz + y \cos yz)\, dz$

11. a. $f(1, 2) = 4$

  $f(1.05, 2.1) = 3.4875$

  $\Delta z = f(1.05, 2.1) - f(1, 2) = -0.5125$

  b. $dz = -2x\,dx - 2y\,dy$

  $= -2(0.05) - 4(0.1) = -0.5$

12. a. $f(1, 2) = \sqrt{5} \approx 2.2361$

  $f(1.05, 2.1) = \sqrt{5.5125} \approx 2.3479$

  $\Delta z = 0.11180$

  b. $dz = \dfrac{x}{\sqrt{x^2+y^2}}\,dx + \dfrac{y}{\sqrt{x^2+y^2}}\,dy$

  $= \dfrac{x\,dx + y\,dy}{\sqrt{x^2+y^2}} = \dfrac{0.05 + 2(0.1)}{\sqrt{5}} \approx 0.11180$

13. a. $f(1, 2) = \sin 2$

  $f(1.05, 2.1) = 1.05 \sin 2.1$

  $\Delta z = f(1.05, 2.1) - f(1, 2) \approx -0.00293$

  b. $dz = \sin y\,dx + x \cos y\,dy$

  $= (\sin 2)(0.05) + (\cos 2)(0.1) \approx 0.00385$

14. a. $f(1, 2) = 2$

  $f(1.05, 2.1) = 2.205$

  $\Delta z = 0.205$

  b. $dz = y\,dx + x\,dy$

  $= 2(0.05) + (0.1) = 0.2$

15. a. $f(1, 2) = -5$

  $f(1.05, 2.1) = -5.25$

  $\Delta z = -0.25$

  b. $dz = 3\,dx - 4\,dy$

  $= 3(0.05) - 4(0.1) = -0.25$

16. a. $f(1, 2) = \dfrac{1}{2} = 0.5$

  $f(1.05, 2.1) = \dfrac{1.05}{2.1} = 0.5$

  $\Delta z = 0$

  b. $dz = \dfrac{1}{y}\,dx - \dfrac{x}{y^2}\,dy$

  $= \dfrac{1}{2}(0.05) - \dfrac{1}{4}(0.1) = 0$

17. Let $z = \sqrt{x^2 + y^2}$, $x = 5$, $y = 3$, $dx = 0.05$, $dy = 0.1$. Then:

$$dz = \frac{x}{\sqrt{x^2+y^2}}\,dx + \frac{y}{\sqrt{x^2+y^2}}\,dy$$

$$\sqrt{(5.05)^2 + (3.1)^2} - \sqrt{5^2 + 3^2} \approx \frac{5}{\sqrt{5^2+3^2}}(0.05) + \frac{3}{\sqrt{5^2+3^2}}(0.1) = \frac{0.55}{\sqrt{34}} \approx 0.094$$

18. Let $z = x^2(1 + y)^3$, $x = 2$, $y = 9$, $dx = 0.03$, $dy = -0.1$. Then:

$$dz = 2x(1 + y)^3\,dx + 3x^2(1 + y)^2\,dy$$

$$(2.03)^2(1 + 8.9)^3 - 2^2(1 + 9)^3 \approx 2(2)(1 + 9)^3(0.03) + 3(2)^2(1 + 9)^2(-0.1) = 0$$

19. Let $z = (1 - x^2)/y^2$, $x = 3$, $y = 6$, $dx = 0.05$, $dy = -0.05$. Then:

$$dz = -\frac{2x}{y^2}\,dx + \frac{-2(1 - x^2)}{y^3}\,dy$$

$$\frac{1 - (3.05)^2}{(5.95)^2} - \frac{1 - 3^2}{6^2} \approx -\frac{2(3)}{6^2}(0.05) - \frac{2(1 - 3^2)}{6^3}(-0.05) \approx -0.012$$

20. Let $z = \sin(x^2 + y^2)$, $x = y = 1$, $dx = 0.05$, $dy = -0.05$. Then:

$$dz = 2x \cos(x^2 + y^2)\,dx + 2y \cos(x^2 + y^2)\,dy$$

$$\sin[(1.05)^2 + (0.95)^2] - \sin 2 \approx 2(1) \cos(1^2 + 1^2)(0.05) + 2(1) \cos(1^2 + 1^2)(-0.05) = 0$$

21. $A = lh$

$dA = l\,dh + h\,dl$

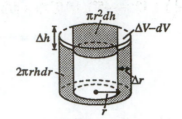

22. $V = \pi r^2 h$

$dV = 2\pi r h\,dr + \pi r^2\,dh$

23. $dV = (2\pi r h)\,dr + (\pi r^2)\,dh$

$\dfrac{dV}{V} = 2\dfrac{dr}{r} + \dfrac{dh}{h}$

$= 2(0.04) + (0.02)$

$= 0.10 = 10\%$

24. $V = \dfrac{\pi r^2 h}{3}$

$r = 3$

$h = 6$

$dV = \dfrac{2\pi r h}{3}\,dr + \dfrac{\pi r^2}{3}\,dh = \dfrac{\pi r}{3}(2h\,dr + r\,dh)$

Δr	Δh	dV	ΔV	$\Delta V - dV$
0.1	0.1	4.7124	4.8391	0.1267
0.1	−0.1	2.8274	2.8264	−0.0010
0.001	0.002	0.0566	0.0566	0.0000
−0.0001	0.0002	−0.0019	−0.0019	0.0000

25. a. $V = \dfrac{1}{2}bhl$

$= \left(18\sin\dfrac{\theta}{2}\right)\left(18\cos\dfrac{\theta}{2}\right)(16)(12)$

$= 31{,}104\sin\theta$ in.^{3}

$= 18\sin\theta$ ft^3

V is maximum when $\sin\theta = 1$ or $\theta = \pi/2$.

b. $V = \dfrac{s^2}{2}(\sin\theta)l$

$dV = s(\sin\theta)l\,ds + \dfrac{s^2}{2}l(\cos\theta)\,d\theta + \dfrac{s^2}{2}(\sin\theta)\,dl$

$= 18\left(\sin\dfrac{\pi}{2}\right)(16)(12)\left(\dfrac{1}{2}\right) + \dfrac{18^2}{2}(16)(12)\left(\cos\dfrac{\pi}{2}\right)\left(\dfrac{\pi}{90}\right) + \dfrac{18^2}{2}\left(\sin\dfrac{\pi}{2}\right)\left(\dfrac{1}{2}\right)$

$= 1809$ in$^3 \approx 1.047$ ft^3

26. a. Using the Law of Cosines:

$a^2 = b^2 + c^2 - 2bc\cos A$

$= 330^2 + 420^2 - 2(330)(420)\cos 9°$

$a \approx 107.3$ ft

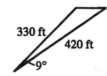

b. $a = \sqrt{b^2 + 420^2 - 2b(420)\cos\theta}$

$da = \dfrac{1}{2}\left[b^2 + 420^2 - 840b\cos\theta\right]^{-1/2}\left[(2b - 840\cos\theta)\,db + 840b\sin\theta\,d\theta\right]$

$= \dfrac{1}{2}\left[330^2 + 420^2 - 840(330)(\cos 9°)\right]^{-1/2}\left[(2(330) - 840(\cos 9°))(6) + 840(330)(\sin 9°)\left(\dfrac{\pi}{180}\right)\right]$

≈ -1.2 ft

27. $A = \frac{1}{2}ab\sin C$

$dA = \frac{1}{2}[(b\sin C)\,da + (a\sin C)\,db + (ab\cos C)\,dC]$

$\quad = \frac{1}{2}\left[4(\sin 45°)\left(\pm\frac{1}{16}\right) + 3(\sin 45°)\left(\pm\frac{1}{16}\right) + 12(\cos 45°)(\pm 0.02)\right] \approx \pm 0.24 \text{ in.}^2$

28. $a = \dfrac{v^2}{r}$

$da = \dfrac{2v}{r}\,dv - \dfrac{v^2}{r^2}\,dr$

$\dfrac{da}{a} = 2\dfrac{dv}{v} - \dfrac{dr}{r} = 2(0.02) - (-0.01) = 0.05 = 5\%$

Note: The maximum error will occur when dv and dr differ in signs.

29. $P = \dfrac{E^2}{R}$

$dP = \dfrac{2E}{R}\,dE - \dfrac{E^2}{R^2}\,dR$

$\dfrac{dP}{P} = 2\dfrac{dE}{E} - \dfrac{dR}{R} = 2(0.02) - (-0.03) = 0.07 = 7\%$

30. $\dfrac{1}{R} = \dfrac{1}{R_1} + \dfrac{1}{R_2}$

$R = \dfrac{R_1 R_2}{R_1 + R_2}$

$dR_1 = \Delta R_1 = 0.5$

$dR_2 = \Delta R_2 = -2$

$\Delta R \approx dR = \dfrac{\partial R}{\partial R_1}\,dR_1 + \dfrac{\partial R}{\partial R_2}\,dR_2 = \dfrac{R_2{}^2}{(R_1 + R_2)^2}\,\Delta R_1 + \dfrac{R_1{}^2}{(R_1 + R_2)^2}\,\Delta R_2$

When $R_1 = 10$ and $R_2 = 15$, we have $\Delta R \approx \dfrac{15^2}{(10+15)^2}(0.5) + \dfrac{10^2}{(10+15)^2}(-2) = -0.14$ ohm.

31. $L = 0.00021\left(\ln\dfrac{2h}{r} - 0.75\right)$

$dL = 0.00021\left[\dfrac{dh}{h} - \dfrac{dr}{r}\right] = 0.00021\left[\dfrac{(\pm 1/100)}{100} - \dfrac{(\pm 1/16)}{2}\right] \approx (\pm 6.5) \times 10^{-6}$

$L = 0.00021(\ln 100 - 0.75) \approx 8.096 \times 10^{-4} \pm dL = 8.096 \times 10^{-4} \pm 6.6 \times 10^{-6}$ micro-henrys

32. $T = 2\pi\sqrt{\dfrac{L}{g}}$

$dg = \Delta g = 32.24 - 32.09 = 0.15$

$dL = \Delta L = 2.48 - 2.5 = -0.02$

$\Delta T \approx dT = \dfrac{\partial T}{\partial g}\,dg + \dfrac{\partial T}{\partial L}\,dL = -\dfrac{\pi}{g}\sqrt{\dfrac{L}{g}}\,\Delta g + \dfrac{\pi}{\sqrt{Lg}}\,\Delta L$

When $g = 32.09$ and $L = 2.5$, we have $\Delta T \approx -\dfrac{\pi}{32.09}\sqrt{\dfrac{2.5}{32.09}}(0.15) + \dfrac{\pi}{\sqrt{(2.5)(32.09)}}(-0.02) \approx -0.0111$ sec.

33. $z = f(x, \ y) = x^2 - 2x + y$

$\Delta z = f(x + \Delta x, \ y + \Delta y) - f(x, \ y)$

$\quad = (x^2 + 2x(\Delta x) + (\Delta x)^2 - 2x - 2(\Delta x) + y + (\Delta y)) - (x^2 - 2x + y)$

$\quad = 2x(\Delta x) + (\Delta x)^2 - 2(\Delta x) + (\Delta y)$

$\quad = (2x - 2)\, \Delta x + \Delta y + \Delta x(\Delta x) + 0(\Delta y)$

$\quad = f_x(x, \ y)\, \Delta x + f_y(x, \ y)\, \Delta y + \epsilon_1\, \Delta x + \epsilon_2\, \Delta y$ where $\epsilon_1 = \Delta x$ and $\epsilon_2 = 0$.

As $(\Delta x, \ \Delta y) \to (0, \ 0), \epsilon_1 \to 0$ and $\epsilon_2 \to 0$.

34. $z = f(x, \ y) = x^2 + y^2$

$\Delta z = f(x + \Delta x, \ y + \Delta y) - f(x, \ y)$

$\quad = x^2 + 2x(\Delta x) + (\Delta x)^2 + y^2 + 2y(\Delta y) + (\Delta y)^2 - (x^2 + y^2)$

$\quad = 2x(\Delta x) + 2y(\Delta y) + \Delta x(\Delta x) + \Delta y(\Delta y)$

$\quad = f_x(x, \ y)\, \Delta x + f_y(x, \ y)\, \Delta y + \epsilon_1\, \Delta x + \epsilon_2\, \Delta y$ where $\epsilon_1 = \Delta x$ and $\epsilon_2 = \Delta y$.

As $(\Delta x, \ \Delta y) \to (0, \ 0), \ \epsilon_1 \to 0$ and $\epsilon_2 \to 0$.

35. $z = f(x, \ y) = x^2 y$

$\Delta z = f(x + \Delta x, \ y + \Delta y) - f(x, \ y)$

$\quad = (x^2 + 2x(\Delta x) + (\Delta x)^2)(y + \Delta y) - x^2 y$

$\quad = 2xy(\Delta x) + y(\Delta x)^2 + x^2\, \Delta y + 2x(\Delta x)(\Delta y) + (\Delta x)^2\, \Delta y$

$\quad = 2xy(\Delta x) + x^2\, \Delta y + (y\Delta x)\, \Delta x + [2x\, \Delta x + (\Delta x)^2]\, \Delta y$

$\quad = f_x(x, \ y)\, \Delta x + f_y(x, \ y)\, \Delta y + \epsilon_1\, \Delta x + \epsilon_2\, \Delta y$ where $\epsilon_1 = y(\Delta x)$ and $\epsilon_2 = 2x\, \Delta x + (\Delta x)^2$.

As $(\Delta x, \ \Delta y) \to (0, \ 0), \ \epsilon_1 \to 0$ and $\epsilon_2 \to 0$.

36. $z = f(x, \ y) = 5x - 10y + y^3$

$\Delta z = f(x + \Delta x, \ y + \Delta y) - f(x, \ y)$

$\quad = 5x + 5\, \Delta x - 10y - 10\, \Delta y + y^3 + 3y^2(\Delta y) + 3y(\Delta y)^2 + (\Delta y)^3 - (5x - 10y + y^3)$

$\quad = 5(\Delta x) + (3y^2 - 10)(\Delta y) + 0(\Delta x) + (3y(\Delta y) + (\Delta y)^2)\, \Delta y$

$\quad = f_x(x, \ y)\, \Delta x + f_y(x, \ y)\, \Delta y + \epsilon_1\, \Delta x + \epsilon_2\, \Delta y$ where $\epsilon_1 = 0$ and $\epsilon_2 = 3y(\Delta y) + (\Delta y)^2$.

As $(\Delta x, \ \Delta y) \to (0, \ 0), \ \epsilon_1 \to 0$ and $\epsilon_2 \to 0$.

37. $f(x, y) = \begin{cases} \dfrac{3x^2 y}{x^4 + y^2}, & (x, y) \neq (0, 0) \\ 0, & (x, y) = (0, 0) \end{cases}$

a. $f_x(0, 0) = \lim\limits_{\Delta x \to 0} \dfrac{f(\Delta x, 0) - f(0, 0)}{\Delta x} = \lim\limits_{\Delta x \to 0} \dfrac{0/[(\Delta x)^4] - 0}{\Delta x} = 0$

$f_y(0, 0) = \lim\limits_{\Delta y \to 0} \dfrac{f(0, \Delta y) - f(0, 0)}{\Delta y} = \lim\limits_{\Delta y \to 0} \dfrac{0/[(\Delta y)^2] - 0}{\Delta y} = 0$

Thus, the partial derivatives exist at $(0, 0)$.

b. Along the line $y = x$: $\lim\limits_{(x,y) \to (0,0)} f(x, y) = \lim\limits_{x \to 0} \dfrac{3x^3}{x^4 + x^2} = \lim\limits_{x \to 0} \dfrac{3x}{x^2 + 1} = 0$

Along the curve $y = x^2$: $\lim\limits_{(x,y) \to (0,0)} f(x, y) = \dfrac{3x^4}{2x^4} = \dfrac{3}{2}$

f is not continuous at $(0, 0)$. Therefore, f is not differentiable at $(0, 0)$.

38. $f(x, y) = \begin{cases} \dfrac{2x^2 y^2}{x^4 + y^4}, & (x, y) \neq (0, 0) \\ 0, & (x, y) = (0, 0) \end{cases}$

a. $f_x(0, 0) = \lim\limits_{\Delta x \to 0} \dfrac{f(\Delta x, 0) - f(0, 0)}{\Delta x} = \lim\limits_{\Delta x \to 0} \dfrac{0 - 0}{\Delta x} = 0$

$f_y(0, 0) = \lim\limits_{\Delta y \to 0} \dfrac{f(0, \Delta y) - f(0, 0)}{\Delta y} = \lim\limits_{\Delta y \to 0} \dfrac{0 - 0}{\Delta y} = 0$

Thus, the partial derivatives exist at $(0, 0)$.

b. Along the line $y = x$: $\lim\limits_{(x,y) \to (0,0)} f(x, y) = \lim\limits_{x \to 0} \dfrac{2x^4}{2x^4} = 1$

Along the curve $y = x^2$: $\lim\limits_{(x,y) \to (0,0)} f(x, y) = \lim\limits_{x \to 0} \dfrac{2x^6}{x^4 + x^8} = \lim\limits_{x \to 0} \dfrac{2x^2}{1 + x^4} = 0$

f is not continuous at $(0, 0)$. Therefore, f is not differentiable at $(0, 0)$.

Section 13.5   Chain Rules for Functions of Several Variables

1. $w = x^2 + y^2$

$x = e^t$

$y = e^{-t}$

$\dfrac{dw}{dt} = 2xe^t + 2y(-e^{-t}) = 2(e^{2t} - e^{-2t})$

2. $w = \sqrt{x^2 + y^2}$

$x = \sin t$

$y = e^t$

$\dfrac{dw}{dt} = \dfrac{x}{\sqrt{x^2 + y^2}} \cos t + \dfrac{y}{\sqrt{x^2 + y^2}} e^t$

$= \dfrac{x \cos t + y e^t}{\sqrt{x^2 + y^2}} = \dfrac{\sin t \cos t + e^{2t}}{\sqrt{\sin^2 t + e^{2t}}}$

3. $w = x \sec y$

$x = e^t$

$y = \pi - t$

$\dfrac{dw}{dt} = (\sec y)(e^t) + (x \sec y \tan y)(-1)$

$= e^t \sec(\pi - t)[1 - \tan(\pi - t)]$

4. $w = \ln \dfrac{y}{x}$

$x = \cos t$

$y = \sin t$

$\dfrac{dw}{dt} = \left(\dfrac{-1}{x}\right)(-\sin t) + \left(\dfrac{1}{y}\right)(\cos t)$

$= \tan t + \cot t$

5. $w = x^2 + y^2$

$x = s + t$

$y = s - t$

$\dfrac{\partial w}{\partial s} = 2x + 2y = 2(x + y) = 4s$

$\dfrac{\partial w}{\partial t} = 2x + 2y(-1) = 2(x - y) = 4t$

When $s = 2$ and $t = -1$,

$\dfrac{\partial w}{\partial s} = 8$ and $\dfrac{\partial w}{\partial t} = -4$.

6. $w = y^3 - 3x^2 y$

$x = e^s$

$y = e^t$

$\dfrac{\partial w}{\partial s} = -6xy(e^s) + (3y^2 - 3x^2)(0) = -6e^{2s+t}$

$\dfrac{\partial w}{\partial t} = -6xy(0) + (3y^2 - 3x^2)(e^t)$

$= 3e^t(e^{2t} - e^{2s})$

When $s = 0$ and $t = 1$,

$\dfrac{\partial w}{\partial s} = -6e$ and $\dfrac{\partial w}{\partial t} = 3e(e^2 - 1)$.

7. $w = x^2 - y^2$

$x = s \cos t$

$y = s \sin t$

$\dfrac{\partial w}{\partial s} = 2x \cos t - 2y \sin t = 2s \cos 2t$

$\dfrac{\partial w}{\partial t} = 2x(-s \sin t) - 2y(s \cos t) = -2s^2 \sin 2t$

When $s = 3$ and $t = \dfrac{\pi}{4}$, $\dfrac{\partial w}{\partial s} = 0$ and $\dfrac{\partial w}{\partial t} = -18$.

8. $w = \sin(2x + 3y)$

$x = s + t$

$y = s - t$

$\dfrac{\partial w}{\partial s} = 2 \cos(2x + 3y) + 3 \cos(2x + 3y)$

$= 5 \cos(2x + 3y) = 5 \cos(5s - t)$

$\dfrac{\partial w}{\partial t} = 2 \cos(2x + 3y) - 3 \cos(2x + 3y)$

$= -\cos(2x + 3y) = -\cos(5s - t)$

When $s = 0$ and $t = \dfrac{\pi}{2}$, $\dfrac{\partial w}{\partial s} = 0$ and $\dfrac{\partial w}{\partial t} = 0$.

9. $w = xy, \quad x = 2 \sin t, \quad y = \cos t$

a. $\dfrac{dw}{dt} = 2y \cos t + x(-\sin t) = 2y \cos t - x \sin t$

$= 2(\cos^2 t - \sin^2 t) = 2 \cos 2t$

b. $w = 2 \sin t \cos t = \sin 2t, \quad \dfrac{dw}{dt} = 2 \cos 2t$

10. $w = \cos(x - y), \quad x = t^2, \quad y = 1$

a. $\dfrac{dw}{dt} = -\sin(x - y)(2t) + \sin(x - y)(0)$

$= -2t \sin(x - y) = -2t \sin(t^2 - 1)$

b. $w = \cos(t^2 - 1), \quad \dfrac{dw}{dt} = -2t \sin(t^2 - 1)$

11. $w = x^2 + y^2 + z^2$

$x = e^t \cos t$

$y = e^t \sin t$

$z = e^t$

a. $\dfrac{dw}{dt} = 2x(-e^t \sin t + e^t \cos t) + 2y(e^t \cos t + e^t \sin t) + 2ze^t = 4e^{2t}$

b. $w = 2e^{2t}, \quad \dfrac{dw}{dt} = 4e^{2t}$

12. $w = xy \cos z$

$x = t$

$y = t^2$

$z = \arccos t$

a. $\dfrac{dw}{dt} = (y \cos z)(1) + (x \cos z)(2t) + (-xy \sin z)\left(-\dfrac{1}{\sqrt{1-t^2}}\right)$

$= t^2(t) + t(t)(2t) - t(t^2)\sqrt{1-t^2}\left(\dfrac{-1}{\sqrt{1-t^2}}\right) = t^3 + 2t^3 + t^3 = 4t^3$

b. $w = t^4, \quad \dfrac{dw}{dt} = 4t^3$

13. $w = xy + xz + yz, \quad x = t - 1, \quad y = t^2 - 1, \quad z = t$

a. $\dfrac{dw}{dt} = (y + z) + (x + z)(2t) + (x + y) = (t^2 - 1 + t) + (t - 1 + t)(2t) + (t - 1 + t^2 - 1) = 3(2t^2 - 1)$

b. $\quad w = (t - 1)(t^2 - 1) + (t - 1)t + (t^2 - 1)t$

$\dfrac{dw}{dt} = 2t(t - 1) + (t^2 - 1) + 2t - 1 + 3t^2 - 1 = 3(2t^2 - 1)$

14. $w = xyz, \quad x = t^2, \quad y = 2t, \quad z = e^{-t}$

a. $\dfrac{dw}{dt} = yz(2t) + xz(2) + (xy)(-e^{-t})$

$= (2t)(e^{-t})(2t) + (t^2)(e^{-t})(2) + (t^2)(2t)(-e^{-t}) = 2t^2 e^{-t}(2 + 1 - t) = 2t^2 e^{-t}(3 - t)$

b. $\quad w = (t^2)(2t)(e^{-t}) = 2t^3 e^{-t}$

$\dfrac{dw}{dt} = (2t^3)(-e^{-t}) + (e^{-t})(6t^2) = 2t^2 e^{-t}(-t + 3)$

15. $w = x^2 - 2xy + y^2, \quad x = r + \theta, \quad y = r - \theta$

a. $\dfrac{\partial w}{\partial r} = (2x - 2y)(1) + (-2x + 2y)(1) = 0$

$\dfrac{\partial w}{\partial \theta} = (2x - 2y)(1) + (-2x + 2y)(-1) = 4x - 4y = 4(x - y) = 4[(r + \theta) - (r - \theta)] = 8\theta$

b. $\quad w = (r + \theta)^2 - 2(r + \theta)(r - \theta) + (r - \theta)^2 = (r^2 + 2r\theta + \theta^2) - 2(r^2 - \theta^2) + (r^2 - 2r\theta + \theta^2) = 4\theta^2$

$\dfrac{\partial w}{\partial r} = 0$

$\dfrac{\partial w}{\partial \theta} = 8\theta$

16. $w = \sqrt{4 - 2x^2 - 2y^2}, \quad x = r\cos\theta, \quad y = r\sin\theta$

a. $\dfrac{\partial w}{\partial r} = \dfrac{-2x}{\sqrt{4 - 2x^2 - 2y^2}}\cos\theta + \dfrac{-2y}{\sqrt{4 - 2x^2 - 2y^2}}\sin\theta = \dfrac{-2r\cos^2\theta - 2r\sin^2\theta}{\sqrt{4 - 2r^2}} = \dfrac{-2r}{\sqrt{4 - 2r^2}}$

$\dfrac{\partial w}{\partial\theta} = \dfrac{-2x}{\sqrt{4 - 2x^2 - 2y^2}}(-r\sin\theta) + \dfrac{-2y}{\sqrt{4 - 2x^2 - 2y^2}}(r\cos\theta) = \dfrac{2r\sin\theta\cos\theta - 2r\sin\theta\cos\theta}{\sqrt{4 - 2r^2}} = 0$

b. $w = \sqrt{4 - 2r^2\cos^2\theta - 2r^2\sin^2\theta} = \sqrt{4 - 2r^2}$

$\dfrac{\partial w}{\partial r} = \dfrac{-2r}{\sqrt{4 - 2r^2}}$

$\dfrac{\partial w}{\partial\theta} = 0$

17. $w = \arctan\dfrac{y}{x}, \quad x = r\cos\theta, \quad y = r\sin\theta$

a. $\dfrac{\partial w}{\partial r} = \dfrac{-y}{x^2 + y^2}\cos\theta + \dfrac{x}{x^2 + y^2}\sin\theta = \dfrac{-r\sin\theta\cos\theta}{r^2} + \dfrac{r\cos\theta\sin\theta}{r^2} = 0$

$\dfrac{\partial w}{\partial\theta} = \dfrac{-y}{x^2 + y^2}(-r\sin\theta) + \dfrac{x}{x^2 + y^2}(r\cos\theta) = \dfrac{-(r\sin\theta)(-r\sin\theta)}{r^2} + \dfrac{(r\cos\theta)(r\cos\theta)}{r^2} = 1$

b. $w = \arctan\dfrac{r\sin\theta}{r\cos\theta} = \arctan(\tan\theta) = \theta$

$\dfrac{\partial w}{\partial r} = 0$

$\dfrac{\partial w}{\partial\theta} = 1$

18. $w = \dfrac{yx}{z}, \quad x = r + \theta, \quad y = r - \theta, \quad z = \theta^2$

a. $\dfrac{\partial w}{\partial r} = \dfrac{y}{z} + \dfrac{x}{z} = \dfrac{r - \theta}{\theta^2} + \dfrac{r + \theta}{\theta^2} = \dfrac{2r}{\theta^2}$

b. $\dfrac{\partial w}{\partial\theta} = \dfrac{y}{z} - \dfrac{x}{z} - \dfrac{xy}{z^2}(2\theta) = \dfrac{r - \theta}{\theta^2} - \dfrac{r + \theta}{\theta^2} - \dfrac{(r + \theta)(r - \theta)}{\theta^4}(2\theta) = \dfrac{-2r^2}{\theta^3}$

19. $F(x, y, z) = x^2 + y^2 + z^2 - 25$

$F_x = 2x$

$F_y = 2y$

$F_z = 2z$

$\dfrac{\partial z}{\partial x} = -\dfrac{F_x}{F_z} = -\dfrac{x}{z}$

$\dfrac{\partial z}{\partial y} = -\dfrac{F_y}{F_z} = -\dfrac{y}{z}$

20. $F(x, y, z) = xz + yz + xy$

$F_x = z + y$

$F_y = z + x$

$F_z = x + y$

$\dfrac{\partial z}{\partial x} = -\dfrac{F_x}{F_z} = -\dfrac{y + z}{x + y}$

$\dfrac{\partial z}{\partial y} = -\dfrac{F_y}{F_z} = -\dfrac{x + z}{x + y}$

21. $F(x, \ y, \ z) = \tan(x + y) + \tan(y + z) - 1$

$F_x = \sec^2(x + y)$

$F_y = \sec^2(x + y) + \sec^2(y + z)$

$F_z = \sec^2(y + z)$

$\dfrac{\partial z}{\partial x} = -\dfrac{F_x}{F_z} = -\dfrac{\sec^2(x + y)}{\sec^2(y + z)}$

$\dfrac{\partial z}{\partial y} = -\dfrac{F_y}{F_z} = -\dfrac{\sec^2(x + y) + \sec^2(y + z)}{\sec^2(y + z)}$

$\qquad = -\left(\dfrac{\sec^2(x + y)}{\sec^2(y + z)} + 1 \right)$

22. $F(x, \ y, \ z) = e^x \sin(y + z) - z$

$F_x = e^x \sin(y + z)$

$F_y = e^x \cos(y + z)$

$F_z = e^x \cos(y + z) - 1$

$\dfrac{\partial z}{\partial x} = -\dfrac{F_x}{F_z} = \dfrac{e^x \sin(y + z)}{1 - e^x \cos(y + z)}$

$\dfrac{\partial z}{\partial y} = -\dfrac{F_y}{F_z} = \dfrac{e^x \cos(y + z)}{1 - e^x \cos(y + z)}$

23. $x^2 + 2yz + z^2 - 1 = 0$

(i) $2x + 2y\dfrac{\partial z}{\partial x} + 2z\dfrac{\partial z}{\partial x} = 0$ implies $\dfrac{\partial z}{\partial x} = -\dfrac{x}{y + z}$.

(ii) $2y\dfrac{\partial z}{\partial y} + 2z + 2z\dfrac{\partial z}{\partial y} = 0$ implies $\dfrac{\partial z}{\partial y} = -\dfrac{z}{y + z}$.

Differentiate (i) with respect to x.

$2 + 2y\dfrac{\partial^2 z}{\partial x^2} + 2z\dfrac{\partial^2 z}{\partial x^2} + 2\left(\dfrac{\partial z}{\partial x}\right)^2 = 0$

$\dfrac{\partial^2 z}{\partial x^2} = \dfrac{-1 - (\partial z/\partial x)^2}{y + z} = -\dfrac{(y + z)^2 + x^2}{(y + z)^3}$

Differentiate (i) with respect to y.

$2y\dfrac{\partial^2 z}{\partial y \partial x} + 2\dfrac{\partial z}{\partial x} + 2z\dfrac{\partial^2 z}{\partial y \partial x} + 2\left(\dfrac{\partial z}{\partial x}\right)\left(\dfrac{\partial z}{\partial y}\right) = 0$

$\dfrac{\partial^2 z}{\partial y \partial z} = -\dfrac{\partial z/\partial x + (\partial z/\partial x)(\partial z/\partial y)}{y + z} = \dfrac{x(y + z) - xz}{(y + z)^3} = \dfrac{xy}{(y + z)^3} = \dfrac{\partial^2 z}{\partial x \partial y}$

Differentiate (ii) with respect to y.

$2y\dfrac{\partial^2 z}{\partial y^2} + 2\dfrac{\partial z}{\partial y} + 2\dfrac{\partial z}{\partial y} + 2z\dfrac{\partial^2 z}{\partial y^2} + 2\left(\dfrac{\partial z}{\partial y}\right)^2 = 0$

$\dfrac{\partial^2 z}{\partial y^2} = \dfrac{-4(\partial z/\partial y) - 2(\partial z/\partial y)^2}{2y + 2z} = \dfrac{2z(y + z) - z}{(y + z)^3} = \dfrac{z(2y + z)}{(y + z)^3}$

24. $x + \sin(y + z) = 0$

(i) $1 + \dfrac{\partial z}{\partial x} \cos(y + z) = 0$ implies $\dfrac{\partial z}{\partial x} = -\dfrac{1}{\cos(y + z)} = -\sec(y + z)$.

(ii) $\left(1 + \dfrac{\partial z}{\partial y}\right) \cos(y + z) = 0$ implies $\dfrac{\partial z}{\partial y} = -1$.

Differentiate (i) with respect to x.

$$\dfrac{\partial^2 z}{\partial x^2} \cos(y + z) - \left(\dfrac{\partial z}{\partial x}\right)^2 \sin(y + z) = 0$$

$$\dfrac{\partial^2 z}{\partial x^2} = \dfrac{\sin(y + z) \sec^2(y + z)}{\cos(y + z)} = \tan(y + z) \sec^2(y + z)$$

Differentiate (ii) with respect to x.

$$\dfrac{\partial^2 z}{\partial x \partial y} \cos(y + z) = 0 \text{ implies } \dfrac{\partial^2 z}{\partial x \partial y} = 0 = \dfrac{\partial^2 z}{\partial y \partial x}.$$

Differentiate (ii) with respect to y.

$$\dfrac{\partial^2 z}{\partial y^2} \cos(y + z) - \dfrac{\partial z}{\partial y}\left(1 + \dfrac{\partial z}{\partial y}\right) \sin(y + z) = 0$$

$$\dfrac{\partial^2 z}{\partial y^2} = 0$$

25. $F(x, y, z, w) = xyz + xzw - yzw + w^2 - 5$

$F_x = yz + zw$

$F_y = xz - zw$

$F_z = xy + xw - yw$

$F_w = xz - yz + 2w$

$\dfrac{\partial w}{\partial x} = -\dfrac{F_x}{F_w} = -\dfrac{z(y + w)}{xz - yz + 2w}$

$\dfrac{\partial w}{\partial y} = -\dfrac{F_y}{F_w} = -\dfrac{z(x - w)}{xz - yz + 2w}$

$\dfrac{\partial w}{\partial z} = -\dfrac{F_z}{F_w} = -\dfrac{xy + xw - yw}{xz - yz + 2w}$

26. $F(x, y, z, w) = x^2 + y^2 + z^2 + 6xw - 8w^2 - 5$

$F_x = 2x + 6w$

$F_y = 2y$

$F_z = 2z$

$F_w = 6x - 16w$

$\dfrac{\partial w}{\partial x} = -\dfrac{F_x}{F_w} = -\dfrac{x + 3w}{3x - 8w}$

$\dfrac{\partial w}{\partial y} = -\dfrac{F_y}{F_w} = -\dfrac{y}{3x - 8w}$

$\dfrac{\partial w}{\partial z} = -\dfrac{F_z}{F_w} = -\dfrac{z}{3x - 8w}$

27. $A = \left(x \sin \dfrac{\theta}{2}\right)\left(x \cos \dfrac{\theta}{2}\right) = \dfrac{x^2}{2} \sin \theta$

$$\dfrac{dA}{dt} = x \sin \theta \dfrac{dx}{dt} + \dfrac{x^2}{2} \cos \theta \dfrac{d\theta}{dt}$$

$$= 6\left(\sin \dfrac{\pi}{4}\right)\left(\dfrac{1}{2}\right) + \dfrac{6^2}{2}\left(\cos \dfrac{\pi}{4}\right)\left(\dfrac{\pi}{90}\right) = \dfrac{3\sqrt{2}}{2} + \dfrac{\pi\sqrt{2}}{10}$$

28. a. $V = \pi r^2 h$

$$\dfrac{dV}{dt} = \pi\left(2rh\dfrac{dr}{dt} + r^2\dfrac{dh}{dt}\right) = \pi r\left(2h\dfrac{dr}{dt} + r\dfrac{dh}{dt}\right) = \pi(12)[2(36)(6) + 12(-4)] = 4608\pi \text{ in.}^3/\text{min}$$

b. $S = 2\pi r(r + h)$

$$\dfrac{dS}{dt} = 2\pi\left[(2r + h)\dfrac{dr}{dt} + r\dfrac{dh}{dt}\right] = 2\pi[(24 + 36)(6) + 12(-4)] = 624\pi \text{ in.}^2/\text{min}$$

29. a. $V = \frac{1}{3}\pi r^2 h$

$$\frac{dV}{dt} = \frac{1}{3}\pi\left(2rh\frac{dr}{dt} + r^2\frac{dh}{dt}\right) = \frac{1}{3}\pi[2(12)(36)(6) + (12)^2(-4)] = 1536\pi \text{ in.}^3/\text{min}$$

b. $S = \pi r\sqrt{r^2 + h^2} + \pi r^2$ (Surface area includes base.)

$$\frac{dS}{dt} = \pi\left[\left(\sqrt{r^2 + h^2} + \frac{r^2}{\sqrt{r^2 + h^2}} + 2r\right)\frac{dr}{dt} + \frac{rh}{\sqrt{r^2 + h^2}}\frac{dh}{dt}\right]$$

$$= \pi\left[\left(\sqrt{12^2 + 36^2} + \frac{144}{\sqrt{12^2 + 36^2}} + 2(12)\right)(6) + \frac{36(12)}{\sqrt{12^2 + 36^2}}(-4)\right]$$

$$= \pi\left[\left(12\sqrt{10} + \frac{12}{\sqrt{10}}\right)(6) + 144 + \frac{36}{\sqrt{10}}(-4)\right]$$

$$= \frac{648\pi}{\sqrt{10}} + 144\pi \text{ in.}^2/\text{min} = \frac{36\pi}{5}(20 + 9\sqrt{10}) \text{ in.}^2/\text{min}$$

30. a. $V = \frac{\pi}{3}(r^2 + rR + R^2)h$

$$\frac{dV}{dt} = \frac{\pi}{3}\left[(2r + R)h\frac{dr}{dt} + (r + 2R)h\frac{dR}{dt} + (r^2 + rR + R^2)\frac{dh}{dt}\right]$$

$$= \frac{\pi}{3}\left[[2(15) + 25](10)(4) + [15 + 2(25)](10)(4) + [(15)^2 + (15)(25) + (25)^2](12)\right]$$

$$= \frac{\pi}{3}(19,500) = 6500\pi \text{ cm}^3/\text{min}$$

b. $S = \pi(R + r)\sqrt{(R - r)^2 + h^2}$

$$\frac{dS}{dt} = \pi\left[\left[\sqrt{(R - r)^2 + h^2} - (R + r)\frac{(R - r)}{\sqrt{(R - r)^2 + h^2}}\right]\frac{dr}{dt}\right.$$

$$\left. + \left[\sqrt{(R - r)^2 + h^2} + (R + r)\frac{(R - r)}{\sqrt{(R - r)^2 + h^2}}\right]\frac{dR}{dt} + (R + r)\frac{h}{\sqrt{(R - r)^2 + h^2}}\frac{dh}{dt}\right]$$

$$= \pi\left[\left[\sqrt{(25 - 15)^2 + 10^2} - (25 + 15)\frac{25 - 15}{\sqrt{(25 - 15)^2 + 10^2}}\right](4)\right.$$

$$\left. + \left[\sqrt{(25 - 15)^2 + 10^2} + (25 + 15)\frac{25 - 10}{\sqrt{(25 - 15)^2 + 10^2}}\right](4) + (25 + 15)\frac{10}{\sqrt{(25 - 15)^2 + 10^2}}(12)\right]$$

$$= 320\sqrt{2}\,\pi\,\text{cm}^2/\text{min}$$

31. $I = \frac{1}{2}m(r_1^2 + r_2^2)$

$$\frac{dI}{dt} = \frac{1}{2}m\left[2r_1\frac{dr_1}{dt} + 2r_2\frac{dr_2}{dt}\right]$$

$$= m[(6)(2) + (8)(2)] = 28m \text{ cm}^2/\text{sec}$$

32. $pV = RT$

$$T = \frac{1}{R}(pV)$$

$$\frac{dT}{dt} = \frac{1}{R}\left[V\frac{dp}{dt} + p\frac{dV}{dt}\right]$$

33. a. $x = 64(\cos 45°)t = 32\sqrt{2}t$

$y = 64(\sin 45°)t - 16t^2 = 32\sqrt{2}t - 16t^2$

b. $\tan\alpha = \dfrac{y}{x + 50}$

$\alpha = \arctan\left(\dfrac{y}{x + 50}\right) = \arctan\left(\dfrac{32\sqrt{2}t - 16t^2}{32\sqrt{2}t + 50}\right)$

c. $\dfrac{d\alpha}{dt} = \dfrac{1}{1 + \left(\dfrac{32\sqrt{2}t - 16t^2}{32\sqrt{2}t + 50}\right)^2} \cdot \dfrac{-64\left(8\sqrt{2}t^2 + 25t - 25\sqrt{2}\right)}{\left(32\sqrt{2}t + 50\right)^2}$

$= \dfrac{-16\left(8\sqrt{2}t^2 + 25t - 25\sqrt{2}\right)}{64t^4 - 256\sqrt{2}t^3 + 1024t^2 + 800\sqrt{2}t + 625}$

e. $\dfrac{d\alpha}{dt} = 0$ when

$8\sqrt{2}t^2 + 25t - 25\sqrt{2} = 0$

$t = \dfrac{-25 + \sqrt{25^2 - 4(8\sqrt{2})(-25\sqrt{2})}}{2(8\sqrt{2})} \approx 0.98$ second.

The projectile is at its maximum height when $dy/dt = 32\sqrt{2} - 32t = 0$ or $t = \sqrt{2} \approx 1.41$ seconds.

d.

No. The rate of change of α is greatest when the projectile is closest to the camera.

34. a. $d = \sqrt{x^2 + y^2} = \sqrt{(32\sqrt{2}t)^2 + (32\sqrt{2}t - 16t^2)^2}$

$= \sqrt{4096t^2 - 1024\sqrt{2}t^3 + 256t^4}$

$= 16t\sqrt{t^2 - 4\sqrt{2}t + 16}$

b. $\dfrac{dd}{dt} = \dfrac{32(t^2 - 3\sqrt{2}t + 8)}{\sqrt{t^2 - 4\sqrt{2}t + 16}}$

c. When $t = 2$:

$\dfrac{dd}{dt} = \dfrac{32(12 - 6\sqrt{2})}{\sqrt{20 - 8\sqrt{2}}} \approx 38.16$ ft/sec

d. $\dfrac{d^2d}{dt^2} = \dfrac{32(t^3 - 6\sqrt{2}t^2 + 36t - 32\sqrt{2})}{(t^2 - 4\sqrt{2}t + 16)^{3/2}} = 0$ when $t \approx 1.52$ seconds. The projectile is at its maximum height when $t = \sqrt{2} \approx 1.41$ seconds.

35. $f(x,\ y) = x^3 - 3xy^2 + y^3$

$f(tx,\ ty) = (tx)^3 - 3(tx)(ty)^2 + (ty)^3 = t^3(x^3 - 3xy^2 + y^3) = t^3 f(x,\ y)$

Degree: 3

$xf_x(x,\ y) + yf_y(x,\ y) = x(3x^2 - 3y^2) + y(-6xy + 3y^2) = 3x^2 - 9xy^2 + 3y^3 = 3f(x,\ y)$

36. $f(x, y) = \dfrac{xy}{\sqrt{x^2 + y^2}}$

$f(tx, ty) = \dfrac{(tx)(ty)}{\sqrt{(tx)^2 + (ty)^2}} = t\left(\dfrac{xy}{\sqrt{x^2 + y^2}}\right) = tf(x, y)$

Degree: 1

$xf_x(x, y) + yf_y(x, y) = x\left(\dfrac{y^3}{(x^2 + y^2)^{3/2}}\right) + y\left(\dfrac{x^3}{(x^2 + y^2)^{3/2}}\right) = \dfrac{xy}{\sqrt{x^2 + y^2}} = 1f(x, y)$

37. $f(x, y) = e^{x/y}$

$f(tx, ty) = e^{tx/ty} = e^{x/y} = f(x, y)$

Degree: 0

$xf_x(x, y) + yf_y(x, y) = x\left(\dfrac{1}{y}e^{x/y}\right) + y\left(-\dfrac{x}{y^2}e^{x/y}\right) = 0$

38. $f(x, y) = \dfrac{x^2}{\sqrt{x^2 + y^2}}$

$f(tx, ty) = \dfrac{(tx)^2}{\sqrt{(tx)^2 + (ty)^2}} = t\left(\dfrac{x^2}{\sqrt{x^2 + y^2}}\right) = tf(x, y)$

Degree: 1

$xf_x(x, y) + yf_y(x, y) = x\left[\dfrac{x^3 + 2xy^2}{(x^2 + y^2)^{3/2}}\right] + y\left[\dfrac{-x^2y}{(x^2 + y^2)^{3/2}}\right]$

$\qquad = \dfrac{x^4 + x^2y^2}{(x^2 + y^2)^{3/2}} = \dfrac{x^2(x^2 + y^2)}{(x^2 + y^2)^{3/2}} = \dfrac{x^2}{\sqrt{x^2 + y^2}} = f(x, y)$

39. $g(t) = f(xt, yt) = t^n f(x, y)$

Let $u = xt$, $v = yt$, then

$g'(t) = \dfrac{\partial f}{\partial u} \cdot \dfrac{du}{dt} + \dfrac{\partial f}{\partial v} \cdot \dfrac{dv}{dt} = \dfrac{\partial f}{\partial u}x + \dfrac{\partial f}{\partial v}y$

and $g'(t) = nt^{n-1}f(x, y)$. Now, let $t = 1$ and we have $u = x$, $v = y$. Thus,

$\dfrac{\partial f}{\partial x}x + \dfrac{\partial f}{\partial y}y = nf(x, y)$.

40. $w = f(x, y)$

$x = u - v$

$y = v - u$

$\dfrac{\partial w}{\partial u} = \dfrac{\partial w}{\partial x} - \dfrac{\partial w}{\partial y}$

$\dfrac{\partial w}{\partial v} = -\dfrac{\partial w}{\partial x} + \dfrac{\partial w}{\partial y}$

$\dfrac{\partial w}{\partial u} + \dfrac{\partial w}{\partial v} = 0$

41. $w = (x - y)\sin(y - x)$

$\dfrac{\partial w}{\partial x} = -(x - y)\cos(y - x) + \sin(y - x)$

$\dfrac{\partial w}{\partial y} = (x - y)\cos(y - x) - \sin(y - x)$

$\dfrac{\partial w}{\partial x} + \dfrac{\partial w}{\partial y} = 0$

42. $w = f(x, \ y), \quad x = r\cos\theta, \quad y = r\sin\theta$

$$\frac{\partial w}{\partial r} = \frac{\partial w}{\partial x}\cos\theta + \frac{\partial w}{\partial y}\sin\theta$$

$$\frac{\partial w}{\partial \theta} = \frac{\partial w}{\partial x}(-r\sin\theta) + \frac{\partial w}{\partial y}(r\cos\theta)$$

a.

$$r\cos\theta\frac{\partial w}{\partial r} = \frac{\partial w}{\partial x}r\cos^2\theta + \frac{\partial w}{\partial y}r\sin\theta\cos\theta$$

$$-\sin\theta\frac{\partial w}{\partial \theta} = \frac{\partial w}{\partial x}(r\sin^2\theta) - \frac{\partial w}{\partial x}r\sin\theta\cos\theta$$

$$r\cos\theta\frac{\partial w}{\partial r} - \sin\theta\frac{\partial w}{\partial \theta} = \frac{\partial w}{\partial x}(r\cos^2\theta + r\sin^2\theta)$$

$$r\frac{\partial w}{\partial x} = \frac{\partial w}{\partial r}(r\cos\theta) - \frac{\partial w}{\partial \theta}\sin\theta$$

$$\frac{\partial w}{\partial x} = \frac{\partial w}{\partial r}\cos\theta - \frac{\partial w}{\partial \theta}\frac{\sin\theta}{r}$$

$$r\sin\theta\frac{\partial w}{\partial r} = \frac{\partial w}{\partial x}r\sin\theta\cos\theta + \frac{\partial w}{\partial y}r\sin^2\theta$$

$$\cos\theta\frac{\partial w}{\partial \theta} = \frac{\partial w}{\partial x}(-r\sin\theta\cos\theta) + \frac{\partial w}{\partial y}(r\cos^2\theta)$$

$$r\sin\theta\frac{\partial w}{\partial r} + \cos\theta\frac{\partial w}{\partial \theta} = \frac{\partial w}{\partial y}(r\sin^2\theta + r\cos^2\theta)$$

$$r\frac{\partial w}{\partial y} = \frac{\partial w}{\partial r}r\sin\theta + \frac{\partial w}{\partial \theta}\cos\theta$$

$$\frac{\partial w}{\partial y} = \frac{\partial w}{\partial r}\sin\theta + \frac{\partial w}{\partial \theta}\frac{\cos\theta}{r}$$

b. $\left(\dfrac{\partial w}{\partial r}\right)^2 + \dfrac{1}{r^2}\left(\dfrac{\partial w}{\partial \theta}\right)^2 = \left(\dfrac{\partial w}{\partial x}\right)^2\cos^2\theta + 2\dfrac{\partial w}{\partial x}\dfrac{\partial w}{\partial y}\sin\theta\cos\theta + \left(\dfrac{\partial w}{\partial y}\right)^2\sin^2\theta + \left(\dfrac{\partial w}{\partial x}\right)^2\sin^2\theta$

$$- 2\frac{\partial w}{\partial x}\frac{\partial w}{\partial y}\sin\theta\cos\theta + \left(\frac{\partial w}{\partial y}\right)^2\cos^2\theta = \left(\frac{\partial w}{\partial x}\right)^2 + \left(\frac{\partial w}{\partial y}\right)^2$$

43. $w = \arctan\dfrac{y}{x}, \quad x = r\cos\theta, \quad y = r\sin\theta$

$$= \arctan\left(\frac{r\sin\theta}{r\cos\theta}\right) = \arctan(\tan\theta) = \theta \ \text{ for } \ -\frac{\pi}{2} < \theta < \frac{\pi}{2}$$

$$\frac{\partial w}{\partial x} = \frac{-y}{x^2 + y^2}, \quad \frac{\partial w}{\partial y} = \frac{x}{x^2 + y^2}, \quad \frac{\partial w}{\partial r} = 0, \quad \frac{\partial w}{\partial \theta} = 1$$

$$\left(\frac{\partial w}{\partial x}\right)^2 + \left(\frac{\partial w}{\partial y}\right)^2 = \frac{y^2}{(x^2 + y^2)^2} + \frac{x^2}{(x^2 + y^2)^2} = \frac{1}{x^2 + y^2} = \frac{1}{r^2}$$

$$\left(\frac{\partial w}{\partial r}\right)^2 + \left(\frac{1}{r^2}\right)\left(\frac{\partial w}{\partial \theta}\right)^2 = 0 + \frac{1}{r^2}(1) = \frac{1}{r^2}$$

Therefore, $\left(\dfrac{\partial w}{\partial x}\right)^2 + \left(\dfrac{\partial w}{\partial y}\right)^2 = \left(\dfrac{\partial w}{\partial r}\right)^2 + \dfrac{1}{r^2}\left(\dfrac{\partial w}{\partial \theta}\right)^2.$

44. Given $\dfrac{\partial u}{\partial x} = \dfrac{\partial v}{\partial y}$ and $\dfrac{\partial u}{\partial y} = -\dfrac{\partial v}{\partial x}$, $x = r\cos\theta$ and $y = r\sin\theta$.

$$\frac{\partial u}{\partial r} = \frac{\partial u}{\partial x}\cos\theta + \frac{\partial u}{\partial y}\sin\theta = \frac{\partial v}{\partial y}\cos\theta - \frac{\partial v}{\partial x}\sin\theta$$

$$\frac{\partial v}{\partial \theta} = \frac{\partial v}{\partial x}(-r\sin\theta) + \frac{\partial v}{\partial y}(r\cos\theta) = r\left[\frac{\partial v}{\partial y}\cos\theta - \frac{\partial v}{\partial x}\sin\theta\right]$$

Therefore, $\dfrac{\partial u}{\partial r} = \dfrac{1}{r}\dfrac{\partial v}{\partial \theta}$.

$$\frac{\partial v}{\partial r} = \frac{\partial v}{\partial x}\cos\theta + \frac{\partial v}{\partial y}\sin\theta = -\frac{\partial u}{\partial y}\cos\theta + \frac{\partial u}{\partial x}\sin\theta$$

$$\frac{\partial u}{\partial \theta} = \frac{\partial u}{\partial x}(-r\sin\theta) + \frac{\partial u}{\partial y}(r\cos\theta) = -r\left[-\frac{\partial u}{\partial y}\cos\theta + \frac{\partial u}{\partial x}\sin\theta\right]$$

Therefore, $\dfrac{\partial v}{\partial r} = -\dfrac{1}{r}\dfrac{\partial u}{\partial \theta}$.

45. $u = \ln\sqrt{x^2 + y^2}$, $v = \arctan\dfrac{y}{x}$

$$\frac{\partial u}{\partial r} = \frac{x}{x^2 + y^2}\cos\theta + \frac{y}{x^2 + y^2}\sin\theta = \frac{r\cos^2\theta + r\sin^2\theta}{r^2} = \frac{1}{r}$$

$$\frac{\partial v}{\partial \theta} = \frac{-y}{x^2 + y^2}(-r\sin\theta) + \frac{x}{x^2 + y^2}(r\cos\theta) = \frac{r^2\sin^2\theta + r^2\cos^2\theta}{r^2} = 1$$

Thus, $\dfrac{\partial u}{\partial r} = \dfrac{1}{r}\dfrac{\partial v}{\partial \theta}$.

$$\frac{\partial v}{\partial r} = \frac{-y}{x^2 + y^2}\cos\theta + \frac{x}{x^2 + y^2}\sin\theta = \frac{-r\sin\theta\cos\theta + r\sin\theta\cos\theta}{r^2} = 0$$

$$\frac{\partial u}{\partial \theta} = \frac{x}{x^2 + y^2}(-r\sin\theta) + \frac{y}{x^2 + y^2}(r\cos\theta) = \frac{-r^2\sin\theta\cos\theta + r^2\sin\theta\cos\theta}{r^2} = 0$$

Thus, $\dfrac{\partial v}{\partial r} = -\dfrac{1}{r}\dfrac{\partial u}{\partial \theta}$.

46. $u(x,\ t) = \dfrac{1}{2}[f(x - ct) + f(x + ct)]$

Let $r = x - ct$ and $s = x + ct$. Then $u(r,\ s) = \dfrac{1}{2}[f(r) + f(s)]$.

$$\frac{\partial u}{\partial t} = \frac{\partial u}{\partial r}\frac{\partial r}{\partial t} + \frac{\partial u}{\partial s}\frac{\partial s}{\partial t} = \frac{1}{2}\frac{df}{dr}(-c) + \frac{1}{2}\frac{df}{ds}(c)$$

$$\frac{\partial^2 u}{\partial t^2} = \frac{1}{2}\frac{d^2 f}{dr^2}(-c)^2 + \frac{1}{2}\frac{d^2 f}{ds^2}(c)^2 = \frac{c^2}{2}\left[\frac{d^2 f}{dr^2} + \frac{d^2 f}{ds^2}\right]$$

$$\frac{\partial u}{\partial x} = \frac{\partial u}{\partial r}\frac{\partial r}{\partial x} + \frac{\partial u}{\partial s}\frac{\partial s}{\partial x} = \frac{1}{2}\frac{df}{dr}(1) + \frac{1}{2}\frac{df}{ds}(1)$$

$$\frac{\partial^2 u}{\partial x^2} = \frac{1}{2}\frac{d^2 f}{dr^2}(1)^2 + \frac{1}{2}\frac{d^2 f}{ds^2}(1)^2 = \frac{1}{2}\left[\frac{d^2 f}{dr^2} + \frac{d^2 f}{ds^2}\right]$$

Thus, $\dfrac{\partial^2 u}{\partial t^2} = c^2 \dfrac{\partial^2 u}{\partial x^2}$.

Section 13.6   Directional Derivatives and Gradients

1. $f(x, y) = 3x - 4xy + 5y$

$$v = \frac{1}{2}(\mathbf{i} + \sqrt{3}\,\mathbf{j})$$

$\nabla f(x, y) = (3 - 4y)\mathbf{i} + (-4x + 5)\mathbf{j}$

$\nabla f(1, 2) = -5\mathbf{i} + \mathbf{j}$

$$\mathbf{u} = \frac{\mathbf{v}}{\|\mathbf{v}\|} = \frac{1}{2}\mathbf{i} + \frac{\sqrt{3}}{2}\mathbf{j}$$

$D_\mathbf{u}f(1, 2) = \nabla f(1, 2) \cdot \mathbf{u} = \frac{1}{2}(-5 + \sqrt{3})$

2. $f(x, y) = x^2 - y^2$

$$v = \frac{\sqrt{2}}{2}(\mathbf{i} + \mathbf{j})$$

$\nabla f(x, y) = 2x\mathbf{i} - 2y\mathbf{j}$

$\nabla f(4, 3) = 8\mathbf{i} - 6\mathbf{j}$

$$\mathbf{u} = \frac{\mathbf{v}}{\|\mathbf{v}\|} = \frac{\sqrt{2}}{2}\mathbf{i} + \frac{\sqrt{2}}{2}\mathbf{j}$$

$D_\mathbf{u}f(4, 3) = \nabla f(4, 3) \cdot \mathbf{u} = \sqrt{2}$

3. $f(x, y) = xy$

$$v = \mathbf{i} + \mathbf{j}$$

$\nabla f(x, y) = y\mathbf{i} + x\mathbf{j}$

$\nabla f(2, 3) = 3\mathbf{i} + 2\mathbf{j}$

$$\mathbf{u} = \frac{\mathbf{v}}{\|\mathbf{v}\|} = \frac{\sqrt{2}}{2}\mathbf{i} + \frac{\sqrt{2}}{2}\mathbf{j}$$

$D_\mathbf{u}f(2, 3) = \nabla f(2, 3) \cdot \mathbf{u} = \dfrac{5\sqrt{2}}{2}$

4. $f(x, y) = \dfrac{x}{y}$

$$v = -\mathbf{j}$$

$\nabla f(x, y) = \dfrac{1}{y}\mathbf{i} - \dfrac{x}{y^2}\mathbf{j}$

$\nabla f(1, 1) = \mathbf{i} - \mathbf{j}$

$$\mathbf{u} = \frac{\mathbf{v}}{\|\mathbf{v}\|} = -\mathbf{j}$$

$D_\mathbf{u}f(1, 1) = \nabla f(1, 1) \cdot \mathbf{u} = 1$

5. $g(x, y) = \sqrt{x^2 + y^2}$

$$v = 3\mathbf{i} - 4\mathbf{j}$$

$\nabla g = \dfrac{x}{\sqrt{x^2 + y^2}}\mathbf{i} + \dfrac{y}{\sqrt{x^2 + y^2}}\mathbf{j}$

$\nabla g(3, 4) = \dfrac{3}{5}\mathbf{i} + \dfrac{4}{5}\mathbf{j}$

$$\mathbf{u} = \frac{\mathbf{v}}{\|\mathbf{v}\|} = \frac{3}{5}\mathbf{i} - \frac{4}{5}\mathbf{j}$$

$D_\mathbf{u}g(3, 4) = \nabla g(3, 4) \cdot \mathbf{u} = -\dfrac{7}{25}$

6. $g(x, y) = \arcsin xy$

$$v = \mathbf{i} + 5\mathbf{j}$$

$\nabla g(x, y) = \dfrac{y}{\sqrt{1 - (xy)^2}}\mathbf{i} + \dfrac{x}{\sqrt{1 - (xy)^2}}\mathbf{j}$

$\nabla g(1, 0) = \mathbf{j}$

$$\mathbf{u} = \frac{\mathbf{v}}{\|\mathbf{v}\|} = \frac{1}{\sqrt{26}}\mathbf{i} + \frac{5}{\sqrt{26}}\mathbf{j}$$

$D_\mathbf{u}g(1, 0) = \nabla g(1, 0) \cdot \mathbf{u} = \dfrac{5\sqrt{26}}{26}$

7. $h(x, y) = e^x \sin y$

$$v = -\mathbf{i}$$

$\nabla h = e^x \sin y\,\mathbf{i} + e^x \cos y\,\mathbf{j}$

$h\left(1, \dfrac{\pi}{2}\right) = e\mathbf{i}$

$$\mathbf{u} = \frac{\mathbf{v}}{\|\mathbf{v}\|} = -\mathbf{i}$$

$D_\mathbf{u}h\left(1, \dfrac{\pi}{2}\right) = \nabla h\left(1, \dfrac{\pi}{2}\right) \cdot \mathbf{u} = -e$

8. $h(x, y) = e^{-(x^2 + y^2)}$

$$v = \mathbf{i} + \mathbf{j}$$

$\nabla h = -2xe^{-(x^2 + y^2)}\mathbf{i} - 2ye^{-(x^2 + y^2)}\mathbf{j}$

$\nabla h(0, 0) = \mathbf{0}$

$D_\mathbf{u}h(0, 0) = \nabla h(0, 0) \cdot \mathbf{u} = 0$

9. $f(x, y, z) = xy + yz + xz$

$$\mathbf{v} = 2\mathbf{i} + \mathbf{j} - \mathbf{k}$$

$$\nabla f(x, y, z) = (y + z)\mathbf{i} + (x + z)\mathbf{j} + (x + y)\mathbf{k}$$

$$\nabla f(1, 1, 1) = 2\mathbf{i} + 2\mathbf{j} + 2\mathbf{k}$$

$$\mathbf{u} = \frac{\mathbf{v}}{\|\mathbf{v}\|} = \frac{\sqrt{6}}{3}\mathbf{i} + \frac{\sqrt{6}}{6}\mathbf{j} - \frac{\sqrt{6}}{6}\mathbf{k}$$

$$D_{\mathbf{u}}f(1, 1, 1) = \nabla f(1, 1, 1) \cdot \mathbf{u} = \frac{2\sqrt{6}}{3}$$

10. $f(x, y, z) = x^2 + y^2 + z^2$

$$\mathbf{v} = \mathbf{i} - 2\mathbf{j} + 3\mathbf{k}$$

$$\nabla f = 2x\mathbf{i} + 2y\mathbf{j} + 2z\mathbf{k}$$

$$\nabla f(1, 2, -1) = 2\mathbf{i} + 4\mathbf{j} - 2\mathbf{k}$$

$$\mathbf{u} = \frac{\mathbf{v}}{\|\mathbf{v}\|} = \frac{1}{\sqrt{14}}\mathbf{i} - \frac{2}{\sqrt{14}}\mathbf{j} + \frac{3}{\sqrt{14}}\mathbf{k}$$

$$D_{\mathbf{u}}f(1, 2, -1) = \nabla f(1, 2, -1) \cdot \mathbf{u} = -\frac{6}{7}\sqrt{14}$$

11. $h(x, y, z) = x \arctan yz$

$$\mathbf{v} = \langle 1, 2, -1 \rangle$$

$$\nabla h(x, y, z) = \arctan yz\,\mathbf{i} + \frac{xz}{1 + (yz)^2}\mathbf{j} + \frac{xy}{1 + (yz)^2}\mathbf{k}$$

$$\nabla h(4, 1, 1) = \frac{\pi}{4}\mathbf{i} + 2\mathbf{j} + 2\mathbf{k}$$

$$\mathbf{u} = \frac{\mathbf{v}}{\|\mathbf{v}\|} = \left\langle \frac{1}{\sqrt{6}}, \frac{2}{\sqrt{6}}, -\frac{1}{\sqrt{6}} \right\rangle$$

$$D_{\mathbf{u}}h(4, 1, 1) = \nabla h(4, 1, 1) \cdot \mathbf{u} = \frac{\pi + 8}{4\sqrt{6}} = \frac{(\pi + 8)\sqrt{6}}{24}$$

12. $h(x, y, z) = xyz$

$$\mathbf{v} = \langle 2, 1, 2 \rangle$$

$$\nabla h = yz\mathbf{i} + xz\mathbf{j} + xy\mathbf{k}$$

$$\nabla h(2, 1, 1) = \mathbf{i} + 2\mathbf{j} + 2\mathbf{k}$$

$$\mathbf{u} = \frac{\mathbf{v}}{\|\mathbf{v}\|} = \frac{2}{3}\mathbf{i} + \frac{1}{3}\mathbf{j} + \frac{2}{3}\mathbf{k}$$

$$D_{\mathbf{u}}h(2, 1, 1) = \nabla h(2, 1, 1) \cdot \mathbf{u} = \frac{8}{3}$$

13. $f(x, y) = x^2 + y^2$

$$\mathbf{u} = \frac{1}{\sqrt{2}}\mathbf{i} + \frac{1}{\sqrt{2}}\mathbf{j}$$

$$\nabla f = 2x\mathbf{i} + 2y\mathbf{j}$$

$$D_{\mathbf{u}}f = \nabla f \cdot \mathbf{u} = \frac{2}{\sqrt{2}}x + \frac{2}{\sqrt{2}}y = \sqrt{2}(x + y)$$

14. $f(x, y) = \dfrac{y}{x + y}$

$$\mathbf{u} = \frac{\sqrt{3}}{2}\mathbf{i} - \frac{1}{2}\mathbf{j}$$

$$\nabla f = -\frac{y}{(x + y)^2}\mathbf{i} + \frac{x}{(x + y)^2}\mathbf{j}$$

$$D_{\mathbf{u}}f = \nabla f \cdot \mathbf{u} = -\frac{\sqrt{3}\,y}{2(x + y)^2} - \frac{x}{2(x + y)^2}$$

$$= -\frac{1}{2(x + y)^2}(\sqrt{3}\,y + x)$$

15. $f(x, \ y) = \sin(2x - y)$

$$\mathbf{u} = \frac{1}{2}\mathbf{i} - \frac{\sqrt{3}}{2}\mathbf{j}$$

$$\nabla f = 2\cos(2x - y)\mathbf{i} - \cos(2x - y)\mathbf{j}$$

$$D_{\mathbf{u}}f = \nabla f \cdot \mathbf{u} = \cos(2x - y) + \frac{\sqrt{3}}{2}\cos(2x - y)$$

$$= \left(\frac{2 + \sqrt{3}}{2}\right)\cos(2x - y)$$

16. $g(x, \ y) = xe^y$

$$\mathbf{u} = -\frac{1}{2}\mathbf{i} + \frac{\sqrt{3}}{2}\mathbf{j}$$

$$\nabla g = e^y\mathbf{i} + xe^y\mathbf{j}$$

$$D_{\mathbf{u}}g = -\frac{1}{2}e^y + \frac{\sqrt{3}}{2}xe^y = \frac{e^y}{2}(\sqrt{3}x - 1)$$

17. $f(x, \ y) = x^2 + 4y^2$

$$\mathbf{v} = -2\mathbf{i} - 2\mathbf{j}$$

$$\nabla f = 2x\mathbf{i} + 8y\mathbf{j}$$

$$\mathbf{u} = \frac{\mathbf{v}}{\|\mathbf{v}\|} = -\frac{1}{\sqrt{2}}\mathbf{i} - \frac{1}{\sqrt{2}}\mathbf{j}$$

$$D_{\mathbf{u}}f = -\frac{2}{\sqrt{2}}x - \frac{8}{\sqrt{2}}y = -\sqrt{2}(x + 4y)$$

At $P = (3, \ 1), \quad D_{\mathbf{u}}f = -7\sqrt{2}.$

18. $f(x, \ y) = \cos(x + y)$

$$\mathbf{v} = \frac{\pi}{2}\mathbf{i} - \pi\mathbf{j}$$

$$\nabla f = -\sin(x + y)\mathbf{i} - \sin(x + y)\mathbf{j}$$

$$\mathbf{u} = \frac{\mathbf{v}}{\|\mathbf{v}\|} = \frac{1}{\sqrt{5}}\mathbf{i} - \frac{2}{\sqrt{5}}\mathbf{j}$$

$$D_{\mathbf{u}}f = -\frac{1}{\sqrt{5}}\sin(x + y) + \frac{2}{\sqrt{5}}\sin(x + y)$$

$$= \frac{1}{\sqrt{5}}\sin(x + y) = \frac{\sqrt{5}}{5}\sin(x + y)$$

At $(0, \ \pi), \quad D_{\mathbf{u}}f = 0.$

19. $h(x, \ y, \ z) = \ln(x + y + z)$

$$\mathbf{v} = 3\mathbf{i} + 3\mathbf{j} + \mathbf{k}$$

$$\nabla h = \frac{1}{x + y + z}(\mathbf{i} + \mathbf{j} + \mathbf{k})$$

At $(1, \ 0, \ 0), \quad \nabla h = \mathbf{i} + \mathbf{j} + \mathbf{k}.$

$$\mathbf{u} = \frac{\mathbf{v}}{\|\mathbf{v}\|} = \frac{1}{\sqrt{19}}(3\mathbf{i} + 3\mathbf{j} + \mathbf{k})$$

$$D_{\mathbf{u}}h = \nabla h \cdot \mathbf{u} = \frac{7}{\sqrt{19}} = \frac{7\sqrt{19}}{19}$$

20. $g(x, \ y, \ z) = xye^z$

$$\mathbf{v} = -2\mathbf{i} - 4\mathbf{j}$$

$$\nabla g = ye^z\mathbf{i} + xe^z\mathbf{j} + xye^z\mathbf{k}$$

At $(2, \ 4, \ 0), \quad \nabla g = 4\mathbf{i} + 2\mathbf{j} + 8\mathbf{k}.$

$$\mathbf{u} = \frac{\mathbf{v}}{\|\mathbf{v}\|} = -\frac{1}{\sqrt{5}}\mathbf{i} - \frac{2}{\sqrt{5}}\mathbf{j}$$

$$D_{\mathbf{u}}g = \nabla g \cdot \mathbf{u} = -\frac{4}{\sqrt{5}} - \frac{4}{\sqrt{5}} = -\frac{8}{\sqrt{5}}$$

21. $f(x, \ y) = \frac{1}{10}(x^2 - 3xy + y^2)$

$$\nabla f(x, \ y) = \frac{1}{10}(2x - 3y)\mathbf{i} + \frac{1}{10}(2y - 3x)\mathbf{j}$$

$$\nabla f(1, \ 2) = -\frac{2}{5}\mathbf{i} + \frac{1}{10}\mathbf{j}$$

22. $f(x, \ y) = \frac{1}{2}y\sqrt{x}$

$$\nabla f(x, \ y) = \frac{y}{4\sqrt{x}}\mathbf{i} + \frac{\sqrt{x}}{2}\mathbf{j}$$

$$\nabla f(1, \ 2) = \frac{1}{2}\mathbf{i} + \frac{1}{2}\mathbf{j}$$

23. $h(x, \ y) = x\tan y$

$$\nabla h(x, \ y) = \tan y\mathbf{i} + x\sec^2 y\mathbf{j}$$

$$\nabla h\left(2, \ \frac{\pi}{4}\right) = \mathbf{i} + 4\mathbf{j}$$

$$\left\|\nabla h\left(2, \ \frac{\pi}{4}\right)\right\| = \sqrt{17}$$

24. $h(x, y) = y \cos(x - y)$

$\nabla h(x, y) = -y \sin(x - y)\mathbf{i} + [\cos(x - y) + y \sin(x - y)]\mathbf{j}$

$\nabla h\left(0, \dfrac{\pi}{3}\right) = \dfrac{\sqrt{3}\,\pi}{6}\mathbf{i} + \left(\dfrac{3 - \sqrt{3}\,\pi}{6}\right)\mathbf{j}$

$\left\|\nabla h\left(0, \dfrac{\pi}{3}\right)\right\| = \sqrt{\dfrac{3\pi^2}{36} + \dfrac{9 - 6\sqrt{3}\,\pi + 3\pi^2}{36}} = \dfrac{\sqrt{3(2\pi^2 - 2\sqrt{3}\,\pi + 3)}}{6}$

25. $g(x, y) = \ln \sqrt[3]{x^2 + y^2} = \dfrac{1}{3}\ln(x^2 + y^2)$

$\nabla g(x, y) = \dfrac{1}{3}\left[\dfrac{2x}{x^2 + y^2}\mathbf{i} + \dfrac{2y}{x^2 + y^2}\mathbf{j}\right]$

$\nabla g(1, 2) = \dfrac{1}{3}\left(\dfrac{2}{5}\mathbf{i} + \dfrac{4}{5}\mathbf{j}\right) = \dfrac{2}{15}(\mathbf{i} + 2\mathbf{j})$

$\|\nabla g(1, 2)\| = \dfrac{2\sqrt{5}}{15}$

26. $g(x, y) = ye^{-x^2}$

$\nabla g(x, y) = -2xye^{-x^2}\mathbf{i} + e^{-x^2}\mathbf{j}$

$\nabla g(0, 5) = \mathbf{j}$

$\|\nabla g(0, 5)\| = 1$

27. $f(x, y, z) = \sqrt{x^2 + y^2 + z^2}$

$\nabla f(x, y, z) = \dfrac{1}{\sqrt{x^2 + y^2 + z^2}}(x\mathbf{i} + y\mathbf{j} + z\mathbf{k})$

$\nabla f(1, 4, 2) = \dfrac{1}{\sqrt{21}}(\mathbf{i} + 4\mathbf{j} + 2\mathbf{k})$

$\|\nabla f(1, 4, 2)\| = 1$

28. $f(x, y, z) = xe^{yz}$

$\nabla f(x, y, z) = e^{yz}\mathbf{i} + xze^{yz}\mathbf{j} + xye^{yz}\mathbf{k}$

$\nabla f(2, 0, -4) = \mathbf{i} - 8\mathbf{j}$

$\|\nabla f(2, 0, -4)\| = \sqrt{65}$

29. $w = \dfrac{1}{\sqrt{1 - x^2 - y^2 - z^2}}$

$\nabla w = \dfrac{1}{\left(\sqrt{1 - x^2 - y^2 - z^2}\right)^3}(x\mathbf{i} + y\mathbf{j} + z\mathbf{k})$

$\nabla w(0, 0, 0) = \mathbf{0}$

$\|\nabla w(0, 0, 0)\| = 0$

30. $w = xy^2z^2$

$\nabla w = y^2z^2\mathbf{i} + 2xyz^2\mathbf{j} + 2xy^2z\mathbf{k}$

$\nabla w(2, 1, 1) = \mathbf{i} + 4\mathbf{j} + 4\mathbf{k}$

$\|\nabla w(2, 1, 1)\| = \sqrt{33}$

31. $f(x, y) = 3 - \dfrac{x}{3} - \dfrac{y}{2}$

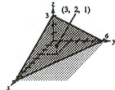

For Exercises 32–38, $f(x, y) = 3 - \dfrac{x}{3} - \dfrac{y}{2}$ and $D_\theta f(x, y) = -\left(\dfrac{1}{3}\right)\cos\theta - \left(\dfrac{1}{2}\right)\sin\theta$.

32. a. $D_{\pi/4}\, f(3, 2) = -\left(\dfrac{1}{3}\right)\dfrac{\sqrt{2}}{2} - \left(\dfrac{1}{2}\right)\dfrac{\sqrt{2}}{2} = -\dfrac{5\sqrt{2}}{12}$

b. $D_{2\pi/3}\, f(3, 2) = -\left(\dfrac{1}{3}\right)\left(-\dfrac{1}{2}\right) - \left(\dfrac{1}{2}\right)\dfrac{\sqrt{3}}{2} = \dfrac{2 - 3\sqrt{3}}{12}$

33. a. $D_{4\pi/3} f(3, 2) = -\left(\frac{1}{3}\right)\left(-\frac{1}{2}\right) - \left(\frac{1}{2}\right)\left(-\frac{\sqrt{3}}{2}\right) = \frac{2+3\sqrt{3}}{12}$

b. $D_{-\pi/6} f(3, 2) = -\left(\frac{1}{3}\right)\left(\frac{\sqrt{3}}{2}\right) - \left(\frac{1}{2}\right)\left(-\frac{1}{2}\right) = \frac{3-2\sqrt{3}}{12}$

34. a. $\mathbf{u} = \left(\frac{1}{\sqrt{2}}\right)(\mathbf{i} + \mathbf{j})$

$D_{\mathbf{u}}f = \nabla f \cdot \mathbf{u}$

$= -\left(\frac{1}{3}\right)\frac{1}{\sqrt{2}} - \left(\frac{1}{2}\right)\frac{1}{\sqrt{2}} = -\frac{5\sqrt{2}}{12}$

b. $\mathbf{v} = -3\mathbf{i} - 4\mathbf{j}$

$\|\mathbf{v}\| = \sqrt{9 + 16} = 5$

$\mathbf{u} = -\frac{3}{5}\mathbf{i} - \frac{4}{5}\mathbf{j}$

$D_{\mathbf{u}}f = \nabla f \cdot \mathbf{u} = \frac{1}{5} + \frac{2}{5} = \frac{3}{5}$

35. a. $\mathbf{v} = -3\mathbf{i} + 4\mathbf{j}$

$\|\mathbf{v}\| = \sqrt{9 + 16} = 5$

$\mathbf{u} = -\frac{3}{5}\mathbf{i} + \frac{4}{5}\mathbf{j}$

$D_{\mathbf{u}}f = \nabla f \cdot \mathbf{u}$

$= \frac{1}{5} - \frac{2}{5} = -\frac{1}{5}$

b. $\mathbf{v} = \mathbf{i} + 3\mathbf{j}$

$\|\mathbf{v}\| = \sqrt{10}$

$\mathbf{u} = \frac{1}{\sqrt{10}}\mathbf{i} + \frac{3}{\sqrt{10}}\mathbf{j}$

$D_{\mathbf{u}}f = \nabla f \cdot \mathbf{u}$

$= \frac{-11}{6\sqrt{10}} = -\frac{11\sqrt{10}}{60}$

36. $\nabla f = -\left(\frac{1}{3}\right)\mathbf{i} - \left(\frac{1}{2}\right)\mathbf{j}$

37. $\|\nabla f\| = \sqrt{\frac{1}{9} + \frac{1}{4}} = \frac{1}{6}\sqrt{13}$

38. $\nabla f = -\frac{1}{3}\mathbf{i} - \frac{1}{2}\mathbf{j}$

$\frac{\nabla f}{\|\nabla f\|} = \frac{1}{\sqrt{13}}(-2\mathbf{i} - 3\mathbf{j})$

Therefore, $\mathbf{u} = (1/\sqrt{13})(3\mathbf{i} - 2\mathbf{j})$ and $D_{\mathbf{u}}f(3, 2) = \nabla f \cdot \mathbf{u} = 0$. ∇f is the direction of greatest rate of change of f. Hence, in a direction orthogonal to ∇f, the rate of change of f is 0.

39. $f(x, y) = 9 - x^2 - y^2$

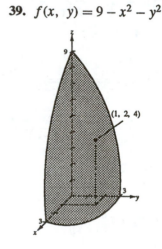

(1, 2, 4)

For Exercises 40–42, $f(x, y) = 9 - x^2 - y^2$ and $D_\theta f(x, y) = -2x\cos\theta - 2y\sin\theta = -2(x\cos\theta + y\sin\theta)$.

40. a. $D_{-\pi/4} f(1, 2) = -2\left(\frac{\sqrt{2}}{2} - \sqrt{2}\right) = \sqrt{2}$

b. $D_{\pi/3} f(1, 2) = -2\left(\frac{1}{2} + \sqrt{3}\right) = -(1 + 2\sqrt{3})$

41. $\nabla f(1, 2) = -2\mathbf{i} - 4\mathbf{j}$

$\|\nabla f(1, 2)\| = \sqrt{4 + 16} = \sqrt{20} = 2\sqrt{5}$

42. $\nabla f(1, 2) = -2\mathbf{i} - 4\mathbf{j}$

$\frac{\nabla f(1, 2)}{\|\nabla f(1, 2)\|} = \frac{1}{\sqrt{5}}(-\mathbf{i} - 2\mathbf{j})$

Therefore, $\mathbf{u} = (1/\sqrt{5})(-2\mathbf{i} + \mathbf{j})$ and $D_{\mathbf{u}}f(1, 2) = \nabla f(1, 2) \cdot \mathbf{u} = 0$.

43. $f(x, y) = x^2 + y^2$

$c = 25, \quad P = (3, 4)$

$\nabla f(x, y) = 2x\mathbf{i} + 2y\mathbf{j}$

$x^2 + y^2 = 25$

$\nabla f(3, 4) = 6\mathbf{i} + 8\mathbf{j}$

44. $f(x, y) = 6 - 2x - 3y$

$c = 6, \quad P = (0, 0)$

$\nabla f(x, y) = -2\mathbf{i} - 3\mathbf{j}$

$6 - 2x - 3y = 6$

$0 = 2x + 3y$

$\nabla f(0, 0) = -2\mathbf{i} - 3\mathbf{j}$

45. $f(x, y) = \dfrac{x}{x^2 + y^2}$

$c = \dfrac{1}{2}, \quad P = (1, 1)$

$\nabla f(x, y) = \dfrac{y^2 - x^2}{(x^2 + y^2)^2}\mathbf{i} - \dfrac{2xy}{(x^2 + y^2)^2}\mathbf{j}$

$\dfrac{x}{x^2 + y^2} = \dfrac{1}{2}$

$x^2 + y^2 - 2x = 0$

$\nabla f(1, 1) = -\dfrac{1}{2}\mathbf{j}$

46. $f(x, y) = xy$

$c = -3, \quad P = (-1, 3)$

$\nabla f(x, y) = y\mathbf{i} + x\mathbf{j}$

$xy = -3$

$\nabla f(-1, 3) = 3\mathbf{i} - \mathbf{j}$

47. $4x^2 - y = 6$

$f(x, y) = 4x^2 - y$

$\nabla f(x, y) = 8x\mathbf{i} - \mathbf{j}$

$\nabla f(2, 10) = 16\mathbf{i} - \mathbf{j}$

$\dfrac{\nabla f(2, 10)}{\|\nabla f(2, 10)\|} = \dfrac{1}{\sqrt{257}}(16\mathbf{i} - \mathbf{j}) = \dfrac{\sqrt{257}}{257}(16\mathbf{i} - \mathbf{j})$

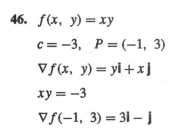

48. $3x^2 - 2y^2 = 1$

$f(x, y) = 3x^2 - 2y^2$

$\nabla f(x, y) = 6x\mathbf{i} - 4y\mathbf{j}$

$\nabla f(1, 1) = 6\mathbf{i} - 4\mathbf{j}$

$\dfrac{\nabla f(1, 1)}{\|\nabla f(1, 1)\|} = \dfrac{1}{\sqrt{13}}(3\mathbf{i} - 2\mathbf{j}) = \dfrac{\sqrt{13}}{13}(3\mathbf{i} - 2\mathbf{j})$

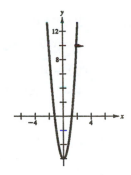

49. $9x^2 + 4y^2 = 40$

$f(x, y) = 9x^2 + 4y^2$

$\nabla f(x, y) = 18x\mathbf{i} + 8y\mathbf{j}$

$\nabla f(2, -1) = 36\mathbf{i} - 8\mathbf{j}$

$\dfrac{\nabla f(2, -1)}{\|\nabla f(2, -1)\|} = \dfrac{1}{\sqrt{85}}(9\mathbf{i} - 2\mathbf{j}) = \dfrac{\sqrt{85}}{85}(9\mathbf{i} - 2\mathbf{j})$

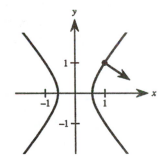

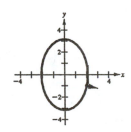

50. $xe^y - y = 5$

$f(x, y) = xe^y - y$

$\nabla f(x, y) = e^y \mathbf{i} + (xe^y - 1)\mathbf{j}$

$\nabla f(5, 0) = \mathbf{i} + 4\mathbf{j}$

$\dfrac{\nabla f(5, 0)}{\|\nabla f(5, 0)\|} = \dfrac{1}{\sqrt{17}}(\mathbf{i} + 4\mathbf{j}) = \dfrac{\sqrt{17}}{17}(\mathbf{i} + 4\mathbf{j})$

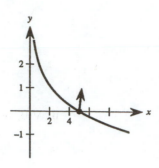

51. $T = \dfrac{x}{x^2 + y^2}$

$\nabla T = \dfrac{y^2 - x^2}{(x^2 + y^2)^2}\mathbf{i} - \dfrac{2xy}{(x^2 + y^2)^2}\mathbf{j}$

$\nabla T(3, 4) = \dfrac{7}{625}\mathbf{i} - \dfrac{24}{625}\mathbf{j} = \dfrac{1}{625}(7\mathbf{i} - 24\mathbf{j})$

52. $h(x, y) = 4000 - 0.001x^2 - 0.004y^2$

$\nabla h = -0.002x\mathbf{i} - 0.008y\mathbf{j}$

$\nabla h(500, 300) = -\mathbf{i} - 2.4\mathbf{j}$ or

$5\nabla h = -(5\mathbf{i} + 12\mathbf{j})$

53. $T(x, y) = 400 - 2x^2 - y^2,$ $P = (10, 10)$

$\dfrac{dx}{dt} = -4x$ $\dfrac{dy}{dt} = -2y$

$x(t) = C_1 e^{-4t}$ $y(t) = C_2 e^{-2t}$

$10 = x(0) = C_1$ $10 = y(0) = C_2$

$x(t) = 10e^{-4t}$ $y(t) = 10e^{-2t}$

$x = \dfrac{y^2}{10}$ $y^2(t) = 100e^{-4x}$

$y^2 = 10x$

54. $T(x, y) = 50 - x^2 - 2y^2,$ $P = (0, 0)$

$\dfrac{dx}{dt} = -2x$ $\dfrac{dy}{dt} = -4y$

$x(t) = C_1 e^{-2t}$ $y(t) = C_2 e^{-4t}$

$0 = x(0) = C_1$ $0 = y(0) = C_2$

$x(t) = 0$ $y(t) = 0$

$x = y$

55. The wind speed is greatest at A.

56. $\dfrac{x^2}{a^2} + \dfrac{y^2}{b^2} = 1$

$f(x, y) = d_1 + d_2 = 2a$

$\nabla f(x, y) = \mathbf{0}$

Therefore, $\mathbf{T} \cdot \nabla f(x, y) = 0$. $f(x, y)$ is the plane $z = 2a$ which is parallel to the xy-plane, and $D_{\mathbf{u}}f(x, y) = 0$ for all $\mathbf{u}$.

57. True

58. False

$D_{\mathbf{u}}f(x, y) = \sqrt{2} > 1$ when

$\mathbf{u} = \left(\cos\dfrac{\pi}{4}\right)\mathbf{i} + \left(\sin\dfrac{\pi}{4}\right)\mathbf{j}.$

59. True

60. True

61. $f(x, y) = \dfrac{-4y}{1 + x^2 + y^2}$

$$\nabla f(x, y) = \dfrac{8xy}{(1 + x^2 + y^2)^2}\mathbf{i} + \dfrac{4(y^2 - x^2 - 1)}{(1 + x^2 + y^2)^2}\mathbf{j}$$

a. $\nabla(0, 1) = \mathbf{0}$

b. $\nabla(0, -1) = \mathbf{0}$

c. $\nabla(2, 0) = -\dfrac{4}{5}\mathbf{j}$

d. $\nabla(-2, 0) = -\dfrac{4}{5}\mathbf{j}$

e. $\nabla(0, 0) = -4\mathbf{j}$

f. $\nabla(1, 1) = \dfrac{8}{9}\mathbf{i} - \dfrac{4}{9}\mathbf{j}$

62. We cannot use Theorem 13.9 since f is not a differentiable function of x and y. Hence, we use the definition of directional derivatives.

$$\begin{aligned}D_{\mathbf{u}}f(x, y) &= \lim_{t \to 0} \frac{f(x + t\cos\theta, \ y + t\sin\theta) - f(x, y)}{t} \\[2mm]
&= \lim_{t \to 0} \frac{f(0 + t/\sqrt{2}, \ 0 + t/\sqrt{2}) - f(0, 0)}{t} \\[2mm]
&= \lim_{t \to 0}\frac{1}{t}\left[\frac{4(t/\sqrt{2})(t/\sqrt{2})}{(t^2/2) + (t^2/2)}\right] = \lim_{t \to 0}\frac{1}{t}\left[\frac{2t^2}{t^2}\right] = \lim_{t \to 0}\frac{2}{t} \text{ which does not exist.}\end{aligned}$$

If $f(0, 0) = 2$, then

$$D_{\mathbf{u}}f(x, y) = \lim_{t \to 0} \frac{f(0 + t/\sqrt{2}, \ 0 + t/\sqrt{2}) - 2}{t} = \lim_{t \to 0}\frac{1}{t}\left[\frac{2t^2}{t^2} - 2\right] = 0$$

which implies that the directional derivatives exist.

63. Let $f(x, y, z) = e^x \cos y + \dfrac{z^2}{2} + C$. Then $\nabla f(x, y, z) = e^x \cos y\,\mathbf{i} - e^x \sin y\,\mathbf{j} + z\mathbf{k}$.

Section 13.7   Tangent Planes and Normal Lines

1. $F(x, y, z) = x + y + z - 4$

$\nabla F = \mathbf{i} + \mathbf{j} + \mathbf{k}$

$\mathbf{n} = \dfrac{\nabla F}{\|\nabla F\|} = \dfrac{1}{\sqrt{3}}(\mathbf{i} + \mathbf{j} + \mathbf{k}) = \dfrac{\sqrt{3}}{3}(\mathbf{i} + \mathbf{j} + \mathbf{k})$

2. $F(x, y, z) = x^2 + y^2 + z^2 - 11$

$\nabla F(x, y, z) = 2x\mathbf{i} + 2y\mathbf{j} + 2z\mathbf{k}$

$\nabla F(3, 1, 1) = 6\mathbf{i} + 2\mathbf{j} + 2\mathbf{k}$

$\mathbf{n} = \dfrac{\nabla F}{\|\nabla F\|} = \dfrac{1}{\sqrt{44}}(6\mathbf{i} + 2\mathbf{j} + 2\mathbf{k}) = \dfrac{1}{\sqrt{11}}(3\mathbf{i} + \mathbf{j} + \mathbf{k}) = \dfrac{\sqrt{11}}{11}(3\mathbf{i} + \mathbf{j} + \mathbf{k})$

3. $F(x, y, z) = \sqrt{x^2 + y^2} - z$

$\nabla F(x, y, z) = \dfrac{x}{\sqrt{x^2 + y^2}}\mathbf{i} + \dfrac{y}{\sqrt{x^2 + y^2}}\mathbf{j} - \mathbf{k}$

$\nabla F(3, 4, 5) = \dfrac{3}{5}\mathbf{i} + \dfrac{4}{5}\mathbf{j} - \mathbf{k}$

$\mathbf{n} = \dfrac{\nabla F}{\|\nabla F\|} = \dfrac{5}{5\sqrt{2}}\left(\dfrac{3}{5}\mathbf{i} + \dfrac{4}{5}\mathbf{j} - \mathbf{k}\right) = \dfrac{1}{5\sqrt{2}}(3\mathbf{i} + 4\mathbf{j} - 5\mathbf{k}) = \dfrac{\sqrt{2}}{10}(3\mathbf{i} + 4\mathbf{j} - 5\mathbf{k})$

4. $F(x, y, z) = x^3 - z$

$\nabla F(x, y, z) = 3x^2\mathbf{i} - \mathbf{k}$

$\nabla F(2, 1, 8) = 12\mathbf{i} - \mathbf{k}$

$\mathbf{n} = \dfrac{\nabla F}{\|\nabla F\|} = \dfrac{1}{\sqrt{145}}(12\mathbf{i} - \mathbf{k}) = \dfrac{\sqrt{145}}{145}(12\mathbf{i} - \mathbf{k})$

5. $F(x, y, z) = x^2y^4 - z$

$\nabla F(x, y, z) = 2xy^4\mathbf{i} + 4x^2y^3\mathbf{j} - \mathbf{k}$

$\nabla F(1, 2, 16) = 32\mathbf{i} + 32\mathbf{j} - \mathbf{k}$

$\mathbf{n} = \dfrac{\nabla F}{\|\nabla F\|} = \dfrac{1}{\sqrt{2049}}(32\mathbf{i} + 32\mathbf{j} - \mathbf{k}) = \dfrac{\sqrt{2049}}{2049}(32\mathbf{i} + 32\mathbf{j} - \mathbf{k})$

6. $F(x, y, z) = x^2 + 3y + z^3 - 9$

$\nabla F(x, y, z) = 2x\mathbf{i} + 3\mathbf{j} + 3z^2\mathbf{k}$

$\nabla F(2, -1, 2) = 4\mathbf{i} + 3\mathbf{j} + 12\mathbf{k}$

$\mathbf{n} = \dfrac{\nabla F}{\|\nabla F\|} = \dfrac{1}{13}(4\mathbf{i} + 3\mathbf{j} + 12\mathbf{k})$

7. $F(x, y, z) = -x\sin y + z - 4$

$\nabla F(x, y, z) = -\sin y\,\mathbf{i} - x\cos y\,\mathbf{j} + \mathbf{k}$

$\nabla F\left(6, \dfrac{\pi}{6}, 7\right) = -\dfrac{1}{2}\mathbf{i} - 3\sqrt{3}\,\mathbf{j} + \mathbf{k}$

$\mathbf{n} = \dfrac{\nabla F}{\|\nabla F\|} = \dfrac{2}{\sqrt{113}}\left(-\dfrac{1}{2}\mathbf{i} - 3\sqrt{3}\,\mathbf{j} + \mathbf{k}\right) = \dfrac{1}{\sqrt{113}}(-\mathbf{i} - 6\sqrt{3}\,\mathbf{j} + 2\mathbf{k}) = \dfrac{\sqrt{113}}{113}(-\mathbf{i} - 6\sqrt{3}\,\mathbf{j} + 2\mathbf{k})$

8. $F(x, y, z) = ze^{x^2-y^2} - 3$

$\nabla F(x, y, z) = 2xze^{x^2-y^2}\mathbf{i} - 2yze^{x^2-y^2}\mathbf{j} + e^{x^2-y^2}\mathbf{k}$

$\nabla F(2, 2, 3) = 12\mathbf{i} - 12\mathbf{j} + \mathbf{k}$

$\mathbf{n} = \dfrac{\nabla F}{\|\nabla F\|} = \dfrac{1}{17}(12\mathbf{i} - 12\mathbf{j} + \mathbf{k})$

9. $F(x, y, z) = \ln\left(\dfrac{x}{y-z}\right) = \ln x - \ln(y - z)$

$\nabla F(x, y, z) = \dfrac{1}{x}\mathbf{i} - \dfrac{1}{y-z}\mathbf{j} + \dfrac{1}{y-z}\mathbf{k}$

$\nabla F(1, 4, 3) = \mathbf{i} - \mathbf{j} + \mathbf{k}$

$\mathbf{n} = \dfrac{\nabla F}{\|\nabla F\|} = \dfrac{1}{\sqrt{3}}(\mathbf{i} - \mathbf{j} + \mathbf{k}) = \dfrac{\sqrt{3}}{3}(\mathbf{i} - \mathbf{j} + \mathbf{k})$

10. $F(x, y, z) = \sin(x - y) - z - 2$

$\nabla F(x, y, z) = \cos(x - y)\mathbf{i} - \cos(x - y)\mathbf{j} - \mathbf{k}$

$\nabla F\left(\dfrac{\pi}{3}, \dfrac{\pi}{6}, -\dfrac{3}{2}\right) = \dfrac{\sqrt{3}}{2}\mathbf{i} - \dfrac{\sqrt{3}}{2}\mathbf{j} - \mathbf{k}$

$\mathbf{n} = \dfrac{\nabla F}{\|\nabla F\|} = \dfrac{2}{\sqrt{10}}\left(\dfrac{\sqrt{3}}{2}\mathbf{i} - \dfrac{\sqrt{3}}{2}\mathbf{j} - \mathbf{k}\right) = \dfrac{1}{\sqrt{10}}(\sqrt{3}\,\mathbf{i} - \sqrt{3}\,\mathbf{j} - 2\mathbf{k}) = \dfrac{\sqrt{10}}{10}(\sqrt{3}\,\mathbf{i} - \sqrt{3}\,\mathbf{j} - 2\mathbf{k})$

11. $f(x, \ y) = 25 - x^2 - y^2, \quad (3, \ 1, \ 15)$

$F(x, \ y, \ z) = 25 - x^2 - y^2 - z$

$F_x(x, \ y, \ z) = -2x \qquad F_y(x, \ y, \ z) = -2y \qquad F_z(x, \ y, \ z) = -1$

$F_x(3, \ 1, \ 15) = -6 \qquad F_y(3, \ 1, \ 15) = -2 \qquad F_z(3, \ 1, \ 15) = -1$

$-6(x - 3) - 2(y - 1) - (z - 15) = 0$

$$0 = 6x + 2y + z - 35$$

$$6x + 2y + z = 35$$

12. $f(x, \ y) = \sqrt{x^2 + y^2}, \quad (3, \ 4, \ 5)$

$F(x, \ y, \ z) = \sqrt{x^2 + y^2} - z$

$F_x(x, \ y, \ z) = \dfrac{x}{\sqrt{x^2 + y^2}} \qquad F_y(x, \ y, \ z) = \dfrac{y}{\sqrt{x^2 + y^2}} \qquad F_z(x, \ y, \ z) = -1$

$F_x(3, \ 4, \ 5) = \dfrac{3}{5} \qquad\qquad F_y(3, \ 4, \ 5) = \dfrac{4}{5} \qquad\qquad F_z(3, \ 4, \ 5) = -1$

$\dfrac{3}{5}(x - 3) + \dfrac{4}{5}(y - 4) - (z - 5) = 0$

$3(x - 3) + 4(y - 4) - 5(z - 5) = 0$

$$3x + 4y - 5z = 0$$

13. $f(x, \ y) = \dfrac{y}{x}, \quad (1, \ 2, \ 2)$

$F(x, \ y, \ z) = \dfrac{y}{x} - z$

$F_x(x, \ y, \ z) = -\dfrac{y}{x^2} \qquad F_y(x, \ y, \ z) = \dfrac{1}{x} \qquad F_z(x, \ y, \ z) = -1$

$F_x(1, \ 2, \ 2) = -2 \qquad\quad F_y(1, \ 2, \ 2) = 1 \qquad F_z(1, \ 2, \ 2) = -1$

$-2(x - 1) + (y - 2) - (z - 2) = 0$

$$-2x + y - z + 2 = 0$$

$$2x - y + z = 2$$

14. $g(x, \ y) = \arctan\dfrac{y}{x}, \quad (1, \ 0, \ 0)$

$G(x, \ y, \ z) = \arctan\dfrac{y}{x} - z$

$G_x(x, \ y, \ z) = \dfrac{-(y/x^2)}{1 + (y^2/x^2)} = \dfrac{-y}{x^2 + y^2} \qquad G_y(x, \ y, \ z) = \dfrac{1/x}{1 + (y^2/x^2)} = \dfrac{x}{x^2 + y^2} \qquad G_z(x, \ y, \ z) = -1$

$G_x(1, \ 0, \ 0) = 0 \qquad\qquad\qquad\qquad G_y(1, \ 0, \ 0) = 1 \qquad\qquad\qquad\qquad G_z(1, \ 0, \ 0) = -1$

$y - z = 0$

15. $g(x, y) = x^2 - y^2$, (5, 4, 9)

$G(x, y, z) = x^2 - y^2 - z$

$G_x(x, y, z) = 2x$ $\qquad G_y(x, y, z) = -2y$ $\qquad G_z(x, y, z) = -1$

$G_x(5, 4, 9) = 10$ $\qquad G_y(5, 4, 9) = -8$ $\qquad G_z(5, 4, 9) = -1$

$10(x - 5) - 8(y - 4) - (z - 9) = 0$

$$10x - 8y - z = 9$$

16. $f(x, y) = 2 - \frac{2}{3}x - y$, (3, -1, 1)

$F(x, y, z) = 2 - \frac{2}{3}x - y - z$

$F_x(x, y, z) = -\frac{2}{3}$, $\qquad F_y(x, y, z) = -1$, $\qquad F_z(x, y, z) = -1$

$-\frac{2}{3}(x - 3) - (y + 1) - (z - 1) = 0$

$-\frac{2}{3}x - y - z + 2 = 0$

$$2x + 3y + 3z = 6$$

17. $z = e^x(\sin y + 1)$, $\left(0, \dfrac{\pi}{2}, 2\right)$

$F(x, y, z) = e^x(\sin y + 1) - z$

$F_x(x, y, z) = e^x(\sin y + 1)$ $\qquad F_y(x, y, z) = e^x \cos y$ $\qquad F_z(x, y, z) = -1$

$F_x\left(0, \dfrac{\pi}{2}, 2\right) = 2$ $\qquad F_y\left(0, \dfrac{\pi}{2}, 2\right) = 0$ $\qquad F_z\left(0, \dfrac{\pi}{2}, 2\right) = -1$

$2x - z = -2$

18. $z = x^3 - 3xy + y^3$, (1, 2, 3)

$F(x, y, z) = x^3 - 3xy + y^3 - z$

$F_x(x, y, z) = 3x^2 - 3y$ $\qquad F_y(x, y, z) = -3x + 3y^2$ $\qquad F_z(x, y, z) = -1$

$F_x(1, 2, 3) = -3$ $\qquad F_y(1, 2, 3) = 9$ $\qquad F_z(1, 2, 3) = -1$

$-3(x - 1) + 9(y - 2) - (z - 3) = 0$

$-3x + 9y - z - 12 = 0$

$$3x - 9y + z = -12$$

19. $h(x, y) = \ln\sqrt{x^2 + y^2}$, (3, 4, $\ln 5$)

$H(x, y, z) = \ln\sqrt{x^2 + y^2} - z = \dfrac{1}{2}\ln(x^2 + y^2) - z$

$H_x(x, y, z) = \dfrac{x}{x^2 + y^2}$ $\qquad H_y(x, y, z) = \dfrac{y}{x^2 + y^2}$ $\qquad H_z(x, y, z) = -1$

$H_x(3, 4, \ln 5) = \dfrac{3}{25}$ $\qquad H_y(3, 4, \ln 5) = \dfrac{4}{25}$ $\qquad H_z(3, 4, \ln 5) = -1$

$\dfrac{3}{25}(x - 3) + \dfrac{4}{25}(y - 4) - (z - \ln 5) = 0$

$3(x - 3) + 4(y - 4) - 25(z - \ln 5) = 0$

$$3x + 4y - 25z = 25(1 - \ln 5)$$

20. $h(x, y) = \cos y, \quad \left(5, \dfrac{\pi}{4}, \dfrac{\sqrt{2}}{2}\right)$

$H(x, y, z) = \cos y - z$

$$H_x(x, y, z) = 0 \qquad H_y(x, y, z) = -\sin y \qquad H_z(x, y, z) = -1$$

$$H_x\left(5, \dfrac{\pi}{4}, \dfrac{\sqrt{2}}{2}\right) = 0 \qquad H_y\left(5, \dfrac{\pi}{4}, \dfrac{\sqrt{2}}{2}\right) = -\dfrac{\sqrt{2}}{2} \qquad H_z\left(5, \dfrac{\pi}{4}, \dfrac{\sqrt{2}}{2}\right) = -1$$

$$-\dfrac{\sqrt{2}}{2}\left(y - \dfrac{\pi}{4}\right) - \left(z - \dfrac{\sqrt{2}}{2}\right) = 0$$

$$-\dfrac{\sqrt{2}}{2}y - z + \dfrac{\sqrt{2}\pi}{8} + \dfrac{\sqrt{2}}{2} = 0$$

$$4\sqrt{2}\,y + 8z = \sqrt{2}(\pi + 4)$$

21. $x^2 + 4y^2 + z^2 = 36, \quad (2, -2, 4)$

$F(x, y, z) = x^2 + 4y^2 + z^2 - 36$

$$F_x(x, y, z) = 2x \qquad F_y(x, y, z) = 8y \qquad F_z(x, y, z) = 2z$$

$$F_x(2, -2, 4) = 4 \qquad F_y(2, -2, 4) = -16 \qquad F_z(2, -2, 4) = 8$$

$$4(x - 2) - 16(y + 2) + 8(z - 4) = 0$$

$$(x - 2) - 4(y + 2) + 2(z - 4) = 0$$

$$x - 4y + 2z = 18$$

22. $x^2 + 2z^2 = y^2, \quad (1, 3, -2)$

$F(x, y, z) = x^2 - y^2 + 2z^2$

$$F_x(x, y, z) = 2x \qquad F_y(x, y, z) = -2y \qquad F_z(x, y, z) = 4z$$

$$F_x(1, 3, -2) = 2 \qquad F_y(1, 3, -2) = -6 \qquad F_z(1, 3, -2) = -8$$

$$2(x - 1) - 6(y - 3) - 8(z + 2) = 0$$

$$(x - 1) - 3(y - 3) - 4(z + 2) = 0$$

$$x - 3y - 4z = 0$$

23. $xy^2 + 3x - z^2 = 4, \quad (2, 1, -2)$

$F(x, y, z) = xy^2 + 3x - z^2 - 4$

$$F_x(x, y, z) = y^2 + 3 \qquad F_y(x, y, z) = 2xy \qquad F_z(x, y, z) = -2z$$

$$F_x(2, 1, -2) = 4 \qquad F_y(2, 1, -2) = 4 \qquad F_z(2, 1, -2) = 4$$

$$4(x - 2) + 4(y - 1) + 4(z + 2) = 0$$

$$x + y + z = 1$$

24. $y = x(2z - 1), \quad (4, 4, 1)$

$F(x, y, z) = x(2z - 1) - y$

$$F_x(x, y, z) = 2z - 1 \qquad F_y(x, y, z) = -1 \qquad F_z(x, y, z) = 2x$$

$$F_x(4, 4, 1) = 1 \qquad F_y(4, 4, 1) = -1 \qquad F_z(4, 4, 1) = 8$$

$$(x - 4) - (y - 4) + 8(z - 1) = 0$$

$$x - y + 8z = 8$$

25. $x^2 + y^2 + z = 9$, (1, 2, 4)

$F(x, y, z) = x^2 + y^2 + z - 9$

$F_x(x, y, z) = 2x$ $\qquad F_y(x, y, z) = 2y$ $\qquad F_z(x, y, z) = 1$

$F_x(1, 2, 4) = 2$ $\qquad F_y(1, 2, 4) = 4$ $\qquad F_z(1, 2, 4) = 1$

Plane: $2(x - 1) + 4(y - 2) + (z - 4) = 0$, $\quad 2x + 4y + z = 14$

Line: $\dfrac{x - 1}{2} = \dfrac{y - 2}{4} = \dfrac{z - 4}{1}$

26. $x^2 + y^2 + z^2 = 9$, (1, 2, 2)

$F(x, y, z) = x^2 + y^2 + z^2 - 9$

$F_x(x, y, z) = 2x$ $\qquad F_y(x, y, z) = 2y$ $\qquad F_z(x, y, z) = 2z$

$F_x(1, 2, 2) = 2$ $\qquad F_y(1, 2, 2) = 4$ $\qquad F_z(1, 2, 2) = 4$

Direction numbers: 1, 2, 2

Plane: $(x - 1) + 2(y - 2) + 2(z - 2) = 0$, $\quad x + 2y + 2z = 9$

Line: $\dfrac{x - 1}{1} = \dfrac{y - 2}{2} = \dfrac{z - 2}{2}$

27. $xy - z = 0$, (−2, −3, 6)

$F(x, y, z) = xy - z$

$F_x(x, y, z) = y$ $\qquad F_y(x, y, z) = x$ $\qquad F_z(x, y, z) = -1$

$F_x(-2, -3, 6) = -3$ $\qquad F_y(-2, -3, 6) = -2$ $\qquad F_z(-2, -3, 6) = -1$

Direction numbers: 3, 2, 1

Plane: $3(x + 2) + 2(x + 3) + (z - 6) = 0$, $\quad 3x + 2y + z = -6$

Line: $\dfrac{x + 2}{3} = \dfrac{y + 3}{2} = \dfrac{z - 6}{1}$

28. $x^2 + y^2 - z^2 = 0$, (5, 12, 13)

$F(x, y, z) = x^2 + y^2 - z^2$

$F_x(x, y, z) = 2x$ $\qquad F_y(x, y, z) = 2y$ $\qquad F_z(x, y, z) = -2z$

$F_x(5, 12, 13) = 10$ $\qquad F_y(5, 12, 13) = 24$ $\qquad F_z(5, 12, 13) = -26$

Direction numbers: 5, 12, −13

Plane: $5(x - 5) + 12(y - 12) - 13(z - 13) = 0$, $\quad 5x + 12y - 13z = 0$

Line: $\dfrac{x - 5}{5} = \dfrac{y - 12}{12} = \dfrac{z - 13}{-13}$

29. $z = \arctan\dfrac{y}{x}$, $\left(1,\ 1,\ \dfrac{\pi}{4}\right)$

$F(x,\ y,\ z) = \arctan\dfrac{y}{x} - z$

$F_x(x,\ y,\ z) = \dfrac{-y}{x^2 + y^2}$ $F_y(x,\ y,\ z) = \dfrac{x}{x^2 + y^2}$ $F_z(x,\ y,\ z) = -1$

$F_x\left(1,\ 1,\ \dfrac{\pi}{4}\right) = -\dfrac{1}{2}$ $F_y\left(1,\ 1,\ \dfrac{\pi}{4}\right) = \dfrac{1}{2}$ $F_z\left(1,\ 1,\ \dfrac{\pi}{4}\right) = -1$

Direction numbers: 1, −1, 2

Plane: $(x - 1) - (y - 1) + 2\left(z - \dfrac{\pi}{4}\right) = 0$, $x - y + 2z = \dfrac{\pi}{2}$

Line: $\dfrac{x - 1}{1} = \dfrac{y - 1}{-1} = \dfrac{z - (\pi/4)}{2}$

30. $xyz = 10$, $(1,\ 2,\ 5)$

$F(x,\ y,\ z) = xyz - 10$

$F_x(x,\ y,\ z) = yz$ $F_y(x,\ y,\ z) = xz$ $F_z(x,\ y,\ z) = xy$

$F_x(1,\ 2,\ 5) = 10$ $F_y(1,\ 2,\ 5) = 5$ $F_z(1,\ 2,\ 5) = 2$

Direction numbers: 10, 5, 2

Plane: $10(x - 1) + 5(y - 2) + 2(z - 5) = 0$, $10x + 5y + 2z = 30$

Line: $\dfrac{x - 1}{10} = \dfrac{y - 2}{5} = \dfrac{z - 5}{2}$

31. $F(x,\ y,\ z) = x^2 + y^2 - 5$ $G(x,\ y,\ z) = x - z$

$\nabla F(x,\ y,\ z) = 2x\mathbf{i} + 2y\mathbf{j}$ $\nabla G(x,\ y,\ z) = \mathbf{i} - \mathbf{k}$

$\nabla F(2,\ 1,\ 2) = 4\mathbf{i} + 2\mathbf{j}$ $\nabla G(2,\ 1,\ 2) = \mathbf{i} - \mathbf{k}$

a. $\nabla F \times \nabla G = \begin{vmatrix} \mathbf{i} & \mathbf{j} & \mathbf{k} \\ 4 & 2 & 0 \\ 1 & 0 & -1 \end{vmatrix} = -2\mathbf{i} + 4\mathbf{j} - 2\mathbf{k} = -2(\mathbf{i} - 2\mathbf{j} + \mathbf{k})$

Direction numbers: 1, −2, 1, $\dfrac{x - 2}{1} = \dfrac{y - 1}{-2} = \dfrac{z - 2}{1}$

b. $\cos\theta = \dfrac{|\nabla F \cdot \nabla G|}{\|\nabla F\|\,\|\nabla G\|} = \dfrac{4}{\sqrt{20}\sqrt{2}} = \dfrac{2}{\sqrt{10}} = \dfrac{\sqrt{10}}{5}$; not orthogonal

32. $F(x,\ y,\ z) = x^2 + y^2 - z$ $G(x,\ y,\ z) = 4 - y - z$

$\nabla F(x,\ y,\ z) = 2x\mathbf{i} + 2y\mathbf{j} - \mathbf{k}$ $\nabla G(x,\ y,\ z) = -\mathbf{j} - \mathbf{k}$

$\nabla F(2,\ -1,\ 5) = 4\mathbf{i} - 2\mathbf{j} - \mathbf{k}$ $\nabla G(2,\ -1,\ 5) = -\mathbf{j} - \mathbf{k}$

a. $\nabla F \times \nabla G = \begin{vmatrix} \mathbf{i} & \mathbf{j} & \mathbf{k} \\ 4 & -2 & -1 \\ 0 & -1 & -1 \end{vmatrix} = \mathbf{i} + 4\mathbf{j} - 4\mathbf{k}$

Direction numbers: 1, 4, −4, $\dfrac{x - 2}{1} = \dfrac{y + 1}{4} = \dfrac{z - 5}{-4}$

b. $\cos\theta = \dfrac{|\nabla F \cdot \nabla G|}{\|\nabla F\|\,\|\nabla G\|} = \dfrac{3}{\sqrt{21}\sqrt{2}} = \dfrac{3}{\sqrt{42}} = \dfrac{\sqrt{42}}{14}$; not orthogonal

33. $F(x, y, z) = x^2 + z^2 - 25$ $G(x, y, z) = y^2 + z^2 - 25$

$\nabla F = 2x\mathbf{i} + 2z\mathbf{k}$ $\nabla G = 2y\mathbf{j} + 2z\mathbf{k}$

$\nabla F(3, 3, 4) = 6\mathbf{i} + 8\mathbf{k}$ $\nabla G(3, 3, 4) = 6\mathbf{j} + 8\mathbf{k}$

a. $\nabla F \times \nabla G = \begin{vmatrix} \mathbf{i} & \mathbf{j} & \mathbf{k} \\ 6 & 0 & 8 \\ 0 & 6 & 8 \end{vmatrix} = -48\mathbf{i} - 48\mathbf{j} + 36\mathbf{k} = -12(4\mathbf{i} + 4\mathbf{j} - 3\mathbf{k})$

Direction numbers: 4, 4, −3, $\dfrac{x-3}{4} = \dfrac{y-3}{4} = \dfrac{z-4}{-3}$

b. $\cos\theta = \dfrac{|\nabla F \cdot \nabla G|}{\|\nabla F\|\,\|\nabla G\|} = \dfrac{64}{(10)(10)} = \dfrac{16}{25};$ not orthogonal

34. $F(x, y, z) = \sqrt{x^2 + y^2} - z$ $G(x, y, z) = 2x + y + 2z - 20$

$\nabla F(x, y, z) = \dfrac{x}{\sqrt{x^2+y^2}}\mathbf{i} + \dfrac{y}{\sqrt{x^2+y^2}}\mathbf{j} - \mathbf{k}$ $\nabla G(x, y, z) = 2\mathbf{i} + \mathbf{j} + 2\mathbf{k}$

$\nabla F(3, 4, 5) = \dfrac{3}{5}\mathbf{i} + \dfrac{4}{5}\mathbf{j} - \mathbf{k}$ $\nabla G(3, 4, 5) = 2\mathbf{i} + \mathbf{j} + 2\mathbf{k}$

a. $\nabla F \times \nabla G = \begin{vmatrix} \mathbf{i} & \mathbf{j} & \mathbf{k} \\ \frac{3}{5} & \frac{4}{5} & -1 \\ 2 & 1 & 2 \end{vmatrix} = \dfrac{13}{5}\mathbf{i} - \dfrac{16}{5}\mathbf{j} - \mathbf{k} = \dfrac{1}{5}(13\mathbf{i} - 16\mathbf{j} - 5\mathbf{k})$

Direction numbers: 13, −16, −5, $\dfrac{x-3}{13} = \dfrac{y-4}{-16} = \dfrac{z-5}{-5}$

b. $\cos\theta = \dfrac{|\nabla F \cdot \nabla G|}{\|\nabla F\|\,\|\nabla G\|} = 0;$ orthogonal

35. $F(x, y, z) = x^2 + y^2 + z^2 - 6$ $G(x, y, z) = x - y - z$

$\nabla F(x, y, z) = 2x\mathbf{i} + 2y\mathbf{j} + 2z\mathbf{k}$ $\nabla G(x, y, z) = \mathbf{i} - \mathbf{j} - \mathbf{k}$

$\nabla F(2, 1, 1) = 4\mathbf{i} + 2\mathbf{j} + 2\mathbf{k}$ $\nabla G(2, 1, 1) = \mathbf{i} - \mathbf{j} - \mathbf{k}$

a. $\nabla F \times \nabla G = \begin{vmatrix} \mathbf{i} & \mathbf{j} & \mathbf{k} \\ 4 & 2 & 2 \\ 1 & -1 & -1 \end{vmatrix} = 6\mathbf{j} - 6\mathbf{k} = 6(\mathbf{j} - \mathbf{k})$

Direction numbers: 0, 1, −1, $x = 2,$ $\dfrac{y-1}{1} = \dfrac{z-1}{-1}$

b. $\cos\theta = \dfrac{|\nabla F \cdot \nabla G|}{\|\nabla F\|\,\|\nabla G\|} = 0;$ orthogonal

36. $F(x, y, z) = x^2 + y^2 - z$ $G(x, y, z) = x + y + 6z - 33$

$\nabla F(x, y, z) = 2x\mathbf{i} + 2y\mathbf{j} - \mathbf{k}$ $\nabla G(x, y, z) = \mathbf{i} + \mathbf{j} + 6\mathbf{k}$

$\nabla F(1, 2, 5) = 2\mathbf{i} + 4\mathbf{j} - \mathbf{k}$ $\nabla G(1, 2, 5) = \mathbf{i} + \mathbf{j} + 6\mathbf{k}$

a. $\nabla F \times \nabla G = \begin{vmatrix} \mathbf{i} & \mathbf{j} & \mathbf{k} \\ 2 & 4 & -1 \\ 1 & 1 & 6 \end{vmatrix} = 25\mathbf{i} - 13\mathbf{j} - 2\mathbf{k}$

Direction numbers: 25, −13, −2, $\dfrac{x-1}{25} = \dfrac{y-2}{-13} = \dfrac{z-5}{-2}$

b. $\cos\theta = \dfrac{|\nabla F \cdot \nabla G|}{\|\nabla F\|\,\|\nabla G\|} = 0;$ orthogonal

37. $F(x, y, z) = 3x^2 + 2y^2 - z - 15$,   (2, 2, 5)

$\nabla F(x, y, z) = 6x\mathbf{i} + 4y\mathbf{j} - \mathbf{k}$

$\nabla F(2, 2, 5) = 12\mathbf{i} + 8\mathbf{j} - \mathbf{k}$

$\cos\theta = \dfrac{|\nabla F(2, 2, 5) \cdot \mathbf{k}|}{\|\nabla F(2, 2, 5)\|} = \dfrac{1}{\sqrt{209}}$

$\theta = \arccos\left(\dfrac{1}{\sqrt{209}}\right) \approx 86°$

38. $F(x, y, z) = xy - z^2$,   (2, 2, 2)

$\nabla F(x, y, z) = y\mathbf{i} + x\mathbf{j} - 2z\mathbf{k}$

$\nabla F(2, 2, 2) = 2\mathbf{i} + 2\mathbf{j} - 4\mathbf{k}$

$\cos\theta = \dfrac{|\nabla F(2, 2, 2) \cdot \mathbf{k}|}{\|\nabla F(2, 2, 2)\|} = \dfrac{4}{\sqrt{24}} = \dfrac{2}{\sqrt{6}}$

$\theta = \arccos\dfrac{2}{\sqrt{6}} \approx 35.26°$

39. $F(x, y, z) = x^2 - y^2 + z$,   (1, 2, 3)

$\nabla F(x, y, z) = 2x\mathbf{i} - 2y\mathbf{j} + \mathbf{k}$

$\nabla F(1, 2, 3) = 2\mathbf{i} - 4\mathbf{j} + \mathbf{k}$

$\cos\theta = \dfrac{|\nabla F(1, 2, 3) \cdot \mathbf{k}|}{\|\nabla F(1, 2, 3)\|} = \dfrac{1}{\sqrt{21}}$

$\theta = \arccos\dfrac{1}{\sqrt{21}} \approx 77.40°$

40. $F(x, y, z) = x^2 + y^2 - 5$,   (2, 1, 3)

$\nabla F(x, y, z) = 2x\mathbf{i} + 2y\mathbf{j}$

$\nabla F(2, 1, 3) = 4\mathbf{i} + 2\mathbf{j}$

$\cos\theta = \dfrac{|\nabla F(2, 1, 3) \cdot \mathbf{k}|}{\|\nabla F(2, 1, 3)\|} = 0$

$\theta = \arccos 0 = 90°$

41. $F(x, y, z) = 3 - x^2 - y^2 + 6y - z$

$\nabla F(x, y, z) = -2x\mathbf{i} + (-2y + 6)\mathbf{j} - \mathbf{k}$

$-2x = 0, \quad x = 0$

$-2y + 6 = 0, \quad y = 3$

$z = 3 - 0^2 - 3^2 + 6(3) = 12$

$(0, 3, 12)$

42. $F(x, y, z) = 3x^2 + 2y^2 - 3x + 4y - z - 5$

$\nabla F(x, y, z) = (6x - 3)\mathbf{i} + (4y + 4)\mathbf{j} - \mathbf{k}$

$6x - 3 = 0, \quad x = \frac{1}{2}$

$4y + 4 = 0, \quad y = -1$

$z = 3\left(\frac{1}{2}\right)^2 + 2(-1)^2 - 3\left(\frac{1}{2}\right) + 4(-1) - 5 = -\frac{31}{4}$

$\left(\frac{1}{2}, -1, -\frac{31}{4}\right)$

43. $T(x, y, z) = 400 - 2x^2 - y^2 - 4z^2$,   (4, 3, 10)

$\dfrac{dx}{dt} = -4x$	$\dfrac{dy}{dt} = -2y$	$\dfrac{dz}{dt} = -8z$
$x(t) = C_1 e^{-4t}$	$y(t) = C_2 e^{-2t}$	$z(t) = C_3 e^{-8t}$
$x(0) = C_1 = 4$	$y(0) = C_2 = 3$	$z(0) = C_3 = 10$
$x = 4e^{-4t}$	$y = 3e^{-2t}$	$z = 10e^{-8t}$

44. $T(x, y, z) = 100 - 3x - y - z^2$,   (2, 2, 5)

$\dfrac{dx}{dt} = -3$	$\dfrac{dy}{dt} = -1$	$\dfrac{dz}{dt} = -2z$
$x(t) = -3t + C_1$	$y(t) = -t + C_2$	$z(t) = C_3 e^{-2t}$
$x(0) = C_1 = 2$	$y(0) = C_2 = 2$	$z(0) = C_3 = 5$
$x = -3t + 2$	$y = -t + 2$	$z = 5e^{-2t}$

45. $F(x, \ y, \ z) = \dfrac{x^2}{a^2} + \dfrac{y^2}{b^2} + \dfrac{z^2}{c^2} - 1$

$F_x(x, \ y, z) = \dfrac{2x}{a^2}$

$F_y(x, \ y, \ z) = \dfrac{2y}{b^2}$

$F_z(x, \ y, \ z) = \dfrac{2z}{c^2}$

Plane:

$\dfrac{2x_0}{a^2}(x - x_0) + \dfrac{2y_0}{b^2}(y - y_0) + \dfrac{2z_0}{c^2}(z - z_0) = 0$

$\dfrac{x_0 x}{a^2} + \dfrac{y_0 y}{b^2} + \dfrac{z_0 z}{c^2} = \dfrac{x_0^2}{a^2} + \dfrac{y_0^2}{b^2} + \dfrac{z_0^2}{c^2} = 1$

46. $F(x, \ y, \ z) = \dfrac{x^2}{a^2} + \dfrac{y^2}{b^2} - \dfrac{z^2}{c^2} - 1$

$F_x(x, \ y, \ z) = \dfrac{2x}{a^2}$

$F_y(x, \ y, \ z) = \dfrac{2y}{b^2}$

$F_z(x, \ y, \ z) = \dfrac{-2z}{c^2}$

Plane:

$\dfrac{2x_0}{a^2}(x - x_0) + \dfrac{2y_0}{b^2}(y - y_0) - \dfrac{2z_0}{c^2}(z - z_0) = 0$

$\dfrac{x_0 x}{a^2} + \dfrac{y_0 y}{b^2} - \dfrac{z_0 z}{c^2} = \dfrac{x_0^2}{a^2} + \dfrac{y_0^2}{b^2} - \dfrac{z_0^2}{c^2} = 1$

47. $F(x, \ y, \ z) = a^2 x^2 + b^2 y^2 - z^2$

$F_x(x, \ y, \ z) = 2a^2 x$

$F_y(x, \ y, \ z) = 2b^2 y$

$F_z(x, \ y, \ z) = -2z$

Plane:

$2a^2 x_0 (x - x_0) + 2b^2 y_0 (y - y_0) - 2z_0 (z - z_0) = 0$

$a^2 x_0 x + b^2 y_0 y - z_0 z = a^2 x_0^2 + b^2 y_0^2 - z_0^2 = 0$

Therefore, the plane passes through the origin.

48. Consider a sphere of radius r and center $(h, \ k, \ l)$:

$F(x, \ y, \ z) = (x - h)^2 + (y - k)^2 + (z - l)^2 - r^2$

$F_x = 2(x - h)$

$F_y = 2(y - k)$

$F_z = 2(z - l)$

Line: $\dfrac{x - x_0}{x_0 - h} = \dfrac{y - y_0}{y_0 - k} = \dfrac{z - z_0}{z_0 - l}$

At the center $(h, \ k, \ l)$: $-1 = -1 = -1$

Therefore, the line passes through the center.

49. Given $w = F(x, \ y, \ z)$ where F is differentiable at $(x_0, \ y_0, \ z_0)$ and $\nabla F(x_0, \ y_0, \ z_0) \neq \mathbf{0}$, the level surface of F at $(x_0, \ y_0, \ z_0)$ is of the form $F(x, \ y, \ z) = C$ for some constant C. Let $G(x, \ y, \ z) = F(x, \ y, \ z) - C = 0$. Then $\nabla G(x_0, \ y_0, \ z_0) = \nabla F(x_0, \ y_0, \ z_0)$ where $\nabla G(x_0, \ y_0, \ z_0)$ is normal to $F(x_0, \ y_0, \ z_0) - C = 0$. Therefore, $\nabla F(x_0, \ y_0, \ z_0)$ is normal to $F(x_0, \ y_0, \ z_0) = C$.

50. Given $z = f(x, \ y)$, then:

$F(x, \ y, \ z) = f(x, \ y) - z = 0$

$\nabla F(x_0, \ y_0, \ z_0) = f_x(x_0, \ y_0)\mathbf{i} + f_y(x_0, \ y_0)\mathbf{j} - \mathbf{k}$

$\cos \theta = \dfrac{|\nabla F(x_0, \ y_0, \ z_0) \cdot \mathbf{k}|}{\|\nabla F(x_0, \ y_0, \ z_0)\| \, \|\mathbf{k}\|}$

$= \dfrac{|-1|}{\sqrt{[f_x(x_0, \ y_0)]^2 + [f_y(x_0, \ y_0)]^2 + (-1)^2}} = \dfrac{1}{\sqrt{[f_x(x_0, \ y_0)]^2 + [f_y(x_0, \ y_0)]^2 + 1}}$

51. $f(x, y) = e^{x-y}$

$f_x(x, y) = e^{x-y}$, $f_y(x, y) = -e^{x-y}$

$f_{xx}(x, y) = e^{x-y}$, $f_{yy}(x, y) = e^{x-y}$, $f_{xy}(x, y) = -e^{x-y}$

a. $P_1(x, y) \approx f(0, 0) + f_x(0, 0)x + f_y(0, 0)y = 1 + x - y$

b. $P_2(x, y) \approx f(0, 0) + f_x(0, 0)x + f_y(0, 0)y + \frac{1}{2}f_{xx}(0, 0)x^2 + f_{xy}(0, 0)xy + \frac{1}{2}f_{yy}(0, 0)y^2$

$\qquad = 1 + x - y + \frac{1}{2}x^2 - xy + \frac{1}{2}y^2$

c. If $x = 0$, $P_2(0, y) = 1 - y + \frac{1}{2}y^2$. This is the second-degree Taylor polynomial for e^{-y}.

   If $y = 0$, $P_2(x, 0) = 1 + x + \frac{1}{2}x^2$. This is the second-degree Taylor polynomial for e^x.

d.

x	y	$f(x, y)$	$P_1(x, y)$	$P_2(x, y)$
0	0	1	1	1
0	0.1	0.9048	0.9000	0.9050
0.2	0.1	1.1052	1.1000	1.1050
0.2	0.5	0.7408	0.7000	0.7450
1	0.5	1.6487	1.5000	1.6250

e.

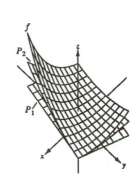

52. $f(x, y) = \cos(x + y)$

$f_x(x, y) = -\sin(x + y)$, $f_y(x, y) = -\sin(x + y)$

$f_{xx}(x, y) = -\cos(x + y)$, $f_{yy}(x, y) = -\cos(x + y)$, $f_{xy}(x, y) = -\cos(x + y)$

a. $P_1(x, y) \approx f(0, 0) + f_x(0, 0)x + f_y(0, 0)y = 1$

b. $P_2(x, y) \approx f(0, 0) + f_x(0, 0)x + f_y(0, 0)y + \frac{1}{2}f_{xx}(0, 0)x^2 + f_{xy}(0, 0)xy + \frac{1}{2}f_{yy}(0, 0)y^2$

$\qquad = 1 - \frac{1}{2}x^2 - xy - \frac{1}{2}y^2$

c. If $x = 0$, $P_2(0, y) = 1 - \frac{1}{2}y^2$. This is the second-degree Taylor polynomial for $\cos y$.

   If $y = 0$, $P_2(x, 0) = 1 - \frac{1}{2}x^2$. This is the second-degree Taylor polynomial for $\cos x$.

d.

x	y	$f(x, y)$	$P_1(x, y)$	$P_2(x, y)$
0	0	1	1	1
0	0.1	0.9950	1	0.9950
0.2	0.1	0.9553	1	0.9550
0.2	0.5	0.7648	1	0.7550
1	0.5	0.0707	1	-0.1250

e.

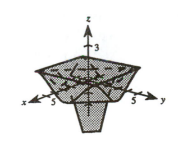

53.

$$z = xf\left(\frac{y}{x}\right)$$

$$F(x, \ y, \ z) = xf\left(\frac{y}{x}\right) - z$$

$$F_x(x, \ y, \ z) = f\left(\frac{y}{x}\right) + xf'\left(\frac{y}{x}\right)\left(-\frac{y}{x^2}\right) = f\left(\frac{y}{x}\right) - \frac{y}{x}f'\left(\frac{y}{x}\right)$$

$$F_y(x, \ y, \ z) = xf'\left(\frac{y}{x}\right)\left(\frac{1}{x}\right) = f'\left(\frac{y}{x}\right)$$

$$F_z(x, \ y, \ z) = -1$$

Tangent plane at $(x_0, \ y_0, \ z_0)$:

$$\left[f\left(\frac{y_0}{x_0}\right) - \frac{y_0}{x_0}f'\left(\frac{y_0}{x_0}\right)\right](x - x_0) + f'\left(\frac{y_0}{x_0}\right)(y - y_0) - (z - z_0) = 0$$

$$\left[f\left(\frac{y_0}{x_0}\right) - \frac{y_0}{x_0}f'\left(\frac{y_0}{x_0}\right)\right]x - x_0f\left(\frac{y_0}{x_0}\right) + y_0f'\left(\frac{y_0}{x_0}\right) + yf'\left(\frac{y_0}{x_0}\right) - y_0f'\left(\frac{y_0}{x_0}\right) - z + x_0f\left(\frac{y_0}{x_0}\right) = 0$$

$$\left[f\left(\frac{y_0}{x_0}\right) - \frac{y_0}{x_0}f'\left(\frac{y_0}{x_0}\right)\right]x + f'\left(\frac{y_0}{x_0}\right)y - z = 0$$

Therefore, the plane passes through the origin $(x, \ y, \ z) = (0, \ 0, \ 0)$.

54. a. $H = -\displaystyle\sum_{i=1}^{n} p_i \log_2 p_i$

May: $H = 3\left(-\dfrac{5}{16}\right)\log_2\dfrac{5}{16} - \dfrac{1}{16}\log_2\dfrac{1}{16} \approx 1.8232$

June: $H = 4\left(-\dfrac{1}{4}\right)\log_2\dfrac{1}{4} = 2.0000$

August: $H = 2\left(-\dfrac{1}{4}\right)\log_2\dfrac{1}{4} - \dfrac{1}{2}\log_2\dfrac{1}{2} = 1.5000$

September: $H = 0 + 1\log_2 1 = 0$

September had no diversity of wildflowers. The greatest diversity occurred in June.

b. $H = 10\left(-\dfrac{1}{10}\right)\log_2\dfrac{1}{10} \approx 3.3219 > 2.0000$

The diversity is greater with 10 types of wildflowers. An equal proportion of each type of wildflower would produce a maximum diversity.

c. $H_n = n\left(-\dfrac{1}{n}\right)\log_2\dfrac{1}{n} = \log_2 n$

$\displaystyle\lim_{n\to\infty} H_n \to \infty$

Section 13.8   Extrema of Functions of Two Variables

1. $f(x, \ y) = x^2 + y^2 + 2x - 6y + 6 = (x + 1)^2 + (y - 3)^2 - 4 \geq -4$
Relative minimum: $(-1, \ 3, \ -4)$
Check: $f_x = 2x + 2 = 0 \Rightarrow x = -1$

$$f_y = 2y - 6 = 0 \Rightarrow y = 3$$

$$f_{xx} = 2, \ f_{yy} = 2, \ f_{xy} = 0$$

At the critical point $(-1, \ 3)$, $f_{xx} > 0$ and $f_{xx}f_{yy} - (f_{xy})^2 > 0$. Therefore, $(-1, \ 3, \ -4)$ is a relative minimum.

2. $f(x, y) = -x^2 - y^2 + 4x + 8y - 11 = -(x - 2)^2 - (y - 4)^2 + 9 \leq 9$
Relative maximum: $(2, 4, 9)$
Check: $f_x = -2x + 4 = 0 \Rightarrow x = 2$

$\quad\quad\quad f_y = -2y + 8 = 0 \Rightarrow y = 4$

$\quad\quad\quad f_{xx} = -2, \ f_{yy} = -2, \ f_{xy} = 0$

At the critical point $(2, 4)$, $f_{xx} < 0$ and $f_{xx}f_{yy} - (f_{xy})^2 > 0$. Therefore, $(2, 4, 9)$ is a relative maximum.

3. $f(x, y) = \sqrt{25 - (x - 2)^2 - y^2} \leq 5$
Relative maximum: $(2, 0, 5)$
Check: $f_x = -\dfrac{x - 2}{\sqrt{25 - (x - 2)^2 - y^2}} = 0 \Rightarrow x = 2$

$\quad\quad\quad f_y = -\dfrac{y}{\sqrt{25 - (x - 2)^2 - y^2}} = 0 \Rightarrow y = 0$

$\quad\quad\quad f_{xx} = -\dfrac{25 - y^2}{[25 - (x - 2)^2 - y^2]^{3/2}}, \ f_{yy} = -\dfrac{25 - (x - 2)^2}{[25 - (x - 2)^2 - y^2]^{3/2}}, \ f_{xy} = -\dfrac{y(x - 2)}{[25 - (x - 2)^2 - y^2]^{3/2}}$

At the critical point $(2, 0)$, $f_{xx} < 0$ and $f_{xx}f_{yy} - (f_{xy})^2 > 0$. Therefore, $(2, 0, 5)$ is a relative maximum.

4. $f(x, y) = \sqrt{x^2 + y^2 + 1} \geq 1$
Relative minimum: $(0, 0, 1)$
Check: $f_x = \dfrac{x}{\sqrt{x^2 + y^2 + 1}} = 0 \Rightarrow x = 0$

$\quad\quad\quad f_y = \dfrac{y}{\sqrt{x^2 + y^2 + 1}} = 0 \Rightarrow y = 0$

$\quad\quad\quad f_{xx} = \dfrac{y^2 + 1}{(x^2 + y^2 + 1)^{3/2}}, \ f_{yy} = \dfrac{x^2 + 1}{(x^2 + y^2 + 1)^{3/2}}, \ f_{xy} = \dfrac{-xy}{(x^2 + y^2 + 1)^{3/2}}$

At the critical point $(0, 0)$, $f_{xx} > 0$ and $f_{xx}f_{yy} - (f_{xy})^2 > 0$. Therefore, $(0, 0, 1)$ is a relative minimum.

5. $g(x, y) = (x - 1)^2 + (y - 3)^2 \geq 0$
Relative minimum: $(1, 3, 0)$

6. $g(x, y) = 9 - (x - 3)^2 - (y + 2)^2 \leq 9$
Relative maximum: $(3, -2, 9)$

7. $f(x, y) = 2x^2 + 2xy + y^2 + 2x - 3$
$\left.\begin{array}{l} f_x = 4x + 2y + 2 = 0 \\ f_y = 2x + 2y = 0 \end{array}\right\}$ Solving simultaneously yields $x = -1$ and $y = 1$.

$f_{xx} = 4, \ f_{yy} = 2, \ f_{xy} = 2$

At the critical point $(-1, 1)$, $f_{xx} > 0$ and $f_{xx}f_{yy} - (f_{xy})^2 > 0$. Therefore, $(-1, 1, -4)$ is a relative minimum.

8. $f(x, y) = -x^2 - 5y^2 + 8x - 10y - 13$
$\left.\begin{array}{l} f_x = -2x + 8 = 0 \\ f_y = -10y - 10 = 0 \end{array}\right\}$ Solving simultaneously yields $x = 4$ and $y = -1$.

$f_{xx} = -2, \ f_{yy} = -10, \ f_{xy} = 0$

At the critical point $(4, -1)$, $f_{xx} < 0$ and $f_{xx}f_{yy} - (f_{xy})^2 > 0$. Therefore, $(4, -1, 8)$ is a relative maximum.

9. $f(x, \ y) = -5x^2 + 4xy - y^2 + 16x + 10$

$\left.\begin{array}{l} f_x = -10x + 4y + 16 = 0 \\[2mm] f_y = 4x - 2y = 0 \end{array}\right\}$ Solving simultaneously yields $x = 8$ and $y = 16$.

$f_{xx} = -10, \quad f_{yy} = -2, \quad f_{xy} = 4$

At the critical point $(8, 16)$, $f_{xx} < 0$ and $f_{xx}f_{yy} - (f_{xy})^2 > 0$. Therefore, $(8, \ 16, \ 74)$ is a relative maximum.

10. $f(x, \ y) = x^2 + 6xy + 10y^2 - 4y + 4$

$\left.\begin{array}{l} f_x = 2x + 6y = 0 \\[2mm] f_y = 6x + 20y - 4 = 0 \end{array}\right\}$ Solving simultaneously yields $x = -6$ and $y = 2$.

$f_{xx} = 2, \quad f_{yy} = 20, \quad f_{xy} = 6$

At the critical point $(-6, \ 2)$, $f_{xx} > 0$ and $f_{xx}f_{yy} - (f_{xy})^2 > 0$. Therefore, $(-6, \ 2, \ 0)$ is a relative minimum.

11. $f(x, \ y) = 2x^2 + 3y^2 - 4x - 12y + 13$

$f_x = 4x - 4 = 4(x - 1) = 0$ when $x = 1$.

$f_y = 6y - 12 = 6(y - 2) = 0$ when $y = 2$.

$f_{xx} = 4, \quad f_{yy} = 6, \quad f_{xy} = 0$

At the critical point $(1, \ 2)$, $f_{xx} > 0$ and $f_{xx}f_{yy} - (f_{xy})^2 > 0$. Therefore, $(1, \ 2, \ -1)$ is a relative minimum.

12. $f(x, \ y) = -3x^2 - 2y^2 + 3x - 4y + 5$

$f_x = -6x + 3 = 0$ when $x = \frac{1}{2}$.

$f_y = -4y - 4 = 0$ when $y = -1$.

$f_{xx} = -6, \quad f_{yy} = -4, \quad f_{xy} = 0$

At the critical point $\left(\frac{1}{2}, \ -1\right)$, $f_{xx} < 0$ and $f_{xx}f_{yy} - (f_{xy})^2 > 0$. Therefore, $\left(\frac{1}{2}, \ -1, \ \frac{31}{4}\right)$ is a relative maximum.

13. $h(x, \ y) = x^2 - y^2 - 2x - 4y - 4$

$h_x = 2x - 2 = 2(x - 1) = 0$ when $x = 1$.

$h_y = -2y - 4 = -2(y + 2) = 0$ when $y = -2$.

$h_{xx} = 2, \quad h_{yy} = -2, \quad h_{xy} = 0$

At the critical point $(1, \ -2)$, $h_{xx}h_{yy} - (h_{xy})^2 < 0$. Therefore, $(1, \ -2, \ -1)$ is a saddle point.

14. $h(x, \ y) = x^2 - 3xy - y^2$

$\left.\begin{array}{l} h_x = 2x - 3y = 0 \\[2mm] h_y = -3x - 2y = 0 \end{array}\right\}$ Solving simultaneously yields $x = 0$ and $y = 0$.

$h_{xx} = 2, \quad h_{yy} = -2, \quad h_{xy} = -3$

At the critical point $(0, 0)$, $h_{xx}h_{yy} - (h_{xy})^2 < 0$. Therefore, $(0, \ 0, \ 0)$ is a saddle point.

15. $g(x, \ y) = xy$

$\left.\begin{array}{l} g_x = y \\[2mm] g_y = x \end{array}\right\} x = 0$ and $y = 0$

$g_{xx} = 0, \quad g_{yy} = 0, \quad g_{xy} = 1$

At the critical point $(0, 0)$, $g_{xx}g_{yy} - (g_{xy})^2 < 0$. Therefore, $(0, \ 0, \ 0)$ is a saddle point.

16. $g(x, y) = 120x + 120y - xy - x^2 - y^2$

$\left.\begin{array}{l} g_x = 120 - y - 2x = 0 \\ g_y = 120 - x - 2y = 0 \end{array}\right\}$ Solving simultaneously yields $x = 40$ and $y = 40$.

$g_{xx} = -2, \quad g_{yy} = -2, \quad g_{xy} = -1$

At the critical point $(40, 40)$, $g_{xx} < 0$ and $g_{xx}g_{yy} - (g_{xy})^2 > 0$. Therefore, $(40, 40, 4800)$ is a relative maximum.

17. $f(x, y) = x^3 - 3xy + y^3$

$\left.\begin{array}{l} f_x = 3(x^2 - y) = 0 \\ f_y = 3(-x + y^2) = 0 \end{array}\right\}$ Solving by substitution yields two critical points $(0, 0)$ and $(1, 1)$.

$f_{xx} = 6x, \quad f_{yy} = 6y, \quad f_{xy} = -3$

At the critical point $(0, 0)$, $f_{xx}f_{yy} - (f_{xy})^2 < 0$. Therefore, $(0, 0, 0)$ is a saddle point. At the critical point $(1, 1)$, $f_{xx} = 6 > 0$ and $f_{xx}f_{yy} - (f_{xy})^2 > 0$. Therefore, $(1, 1, -1)$ is a relative minimum.

18. $f(x, y) = 4xy - x^4 - y^4$

$\left.\begin{array}{l} f_x = 4y - 4x^3 = 0 \\ f_y = 4x - 4y^3 = 0 \end{array}\right\}$ Solving by substitution yields three critical points $(0, 0)$, $(1, 1)$, $(-1, -1)$.

$f_{xx} = -12x^2, \quad f_{yy} = -12y^2, \quad f_{xy} = 4$

At the critical point $(0, 0)$, $f_{xx}f_{yy} - (f_{xy})^2 < 0$. Therefore, $(0, 0, 0)$ is a saddle point. At the critical point $(1, 1)$, $f_{xx} < 0$ and $f_{xx}f_{yy} - (f_{xy})^2 > 0$. Therefore, $(1, 1, 2)$ is a relative maximum. At the critical point $(-1, -1)$, $f_{xx} < 0$ and $f_{xx}f_{yy} - (f_{xy})^2 > 0$. Therefore, $(-1, -1, 2)$ is a relative maximum.

19. $f(x, y) = e^{-x} \sin y$

$\left.\begin{array}{l} f_x = -e^{-x} \sin y = 0 \\ f_y = e^{-x} \cos y = 0 \end{array}\right\}$ Since $e^{-x} > 0$ for all x and $\sin y$ and $\cos y$ are never both zero for a given value of y, there are no critical points.

20. $f(x, y) = \left(\dfrac{1}{2} - x^2 + y^2\right)e^{1 - x^2 - y^2}$

$\left.\begin{array}{l} f_x = (2x^3 - 2xy^2 - 3x)e^{1-x^2-y^2} = 0 \\ f_y = (2x^2y - 2y^3 + y)e^{1-x^2-y^2} = 0 \end{array}\right\}$ Solving yields the critical points $(0, 0)$, $\left(0, \pm\dfrac{\sqrt{2}}{2}\right)$, $\left(\pm\dfrac{\sqrt{6}}{2}, 0\right)$.

$f_{xx} = (-4x^4 + 4x^2y^2 + 12x^2 - 2y^2 - 3)e^{1-x^2-y^2}$

$f_{yy} = (4y^4 - 4x^2y^2 + 2x^2 - 8y^2 + 1)e^{1-x^2-y^2}$

$f_{xy} = (-4x^3y + 4xy^3 + 2xy)e^{1-x^2-y^2}$

At the critical point $(0, 0)$, $f_{xx}f_{yy} - (f_{xy})^2 < 0$. Therefore, $(0, 0, e/2)$ is a saddle point. At the critical points $(0, \pm\sqrt{2}/2)$, $f_{xx} < 0$ and $f_{xx}f_{yy} - (f_{xy})^2 > 0$. Therefore, $(0, \pm\sqrt{2}/2, \sqrt{e})$ are relative maxima. At the critical points $(\pm\sqrt{6}/2, 0)$, $f_{xx} > 0$ and $f_{xx}f_{yy} - (f_{xy})^2 > 0$. Therefore, $(\pm\sqrt{6}/2, 0, -\sqrt{e}/e)$ are relative minima.

21. $z = \dfrac{-4x}{x^2 + y^2 + 1}$

Relative minimum: $(1, 0, -2)$
Relative maximum: $(-1, 0, 2)$

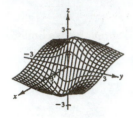

22. $f(x, y) = y^3 - 3yx^2 - 3y^2 - 3x^2 + 1$

Relative maximum: $(0, 0, 1)$
Saddle points: $(0, 2, -3)$, $(\pm\sqrt{3}, -1, -3)$

23. $z = (x^2 + 4y^2)e^{1-x^2-y^2}$

Relative minimum: $(0, 0, 0)$
Relative maxima: $(0, \pm1, 4)$
Saddle points: $(\pm1, 0, 1)$

24. $z = e^{xy}$

Saddle point: $(0, 0, 1)$

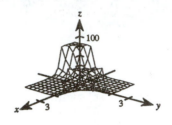

25. $f_{xx}f_{yy} - (f_{xy})^2 = (9)(4) - 6^2 = 0$
Insufficient information

26. $f_{xx} < 0$ and
$f_{xx}f_{yy} - (f_{xy})^2 = (-3)(-8) - 2^2 > 0$
f has a relative maximum at (x_0, y_0).

27. $f_{xx}f_{yy} - (f_{xy})^2 = (-9)(6) - 10^2 < 0$
f has a saddle point at (x_0, y_0).

28. $f_{xx} > 0$ and $f_{xx}f_{yy} - (f_{xy})^2 = (25)(8) - 10^2 > 0$
f has a relative minimum at (x_0, y_0).

29. $f(x, y) = 12 - 3x - 2y$ has no critical points. On the line $y = x + 1$, $0 \le x \le 1$,

$\qquad f(x, y) = f(x) = 12 - 3x - 2(x+1) = -5x + 10$

and the maximum is 10, the minimum is 5. On the line $y = -2x + 4$, $1 \le x \le 2$,

$\qquad f(x, y) = f(x) = 12 - 3x - 2(-2x + 4) = x + 4$

and the maximum is 6, the minimum is 5. On the line $y = -\frac{1}{2}x + 1$, $0 \le x \le 2$,

$\qquad f(x, y) = f(x) = 12 - 3x - 2(-\frac{1}{2}x + 1) = -2x + 10$

and the maximum is 10, the minimum is 6.
Absolute maximum: 10 at $(0, 1)$
Absolute minimum: 5 at $(1, 2)$

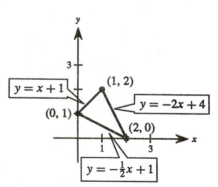

30. $f(x, y) = (2x - y)^2$

$f_x = 4(2x - y) = 0 \Rightarrow 2x = y$

$f_y = -2(2x - y) = 0 \Rightarrow 2x = y$

On the line $y = x + 1$, $0 \leq x \leq 1$,

$f(x, y) = f(x) = (2x - (x + 1))^2 = (x - 1)^2$

and the maximum is 1, the minimum is 0. On the line $y = -\frac{1}{2}x + 1$, $0 \leq x \leq 2$,

$f(x, y) = f(x) = \left(2x - \left(-\frac{1}{2}x + 1\right)\right)^2 = \left(\frac{5}{2}x - 1\right)^2$

and the maximum is 16, the minimum is 0. On the line $y = -2x + 4$, $1 \leq x \leq 2$,

$f(x, y) = f(x) = (2x - (-2x + 4))^2 = (4x - 4)^2$

and the maximum is 16, the minimum is 0.
Absolute maximum: 16 at (2, 0)
Absolute minimum: 0 at (1, 2) and elsewhere

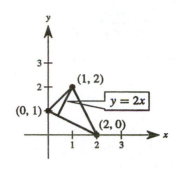

31. $f(x, y) = 3x^2 + 2y^2 - 4y$

$\left.\begin{array}{l} f_x = 6x = 0 \quad \Rightarrow x = 0 \\ f_y = 4y - 4 = 0 \Rightarrow y = 1 \end{array}\right\} f(0, 1) = -2$

On the line $y = 4$, $-2 \leq x \leq 2$,

$f(x, y) = f(x) = 3x^2 + 32 - 16 = 3x^2 + 16$

and the maximum is 28, the minimum is 16. On the curve $y = x^2$, $-2 \leq x \leq 2$,

$f(x, y) = f(x) = 3x^2 + 2(x^2)^2 - 4x^2 = 2x^4 - x^2 = x^2(2x^2 - 1)$

and the maximum is 28, the minimum is $-\frac{1}{8}$.
Absolute maximum: 28 at (± 2, 4)
Absolute minimum: -2 at (0, 1)

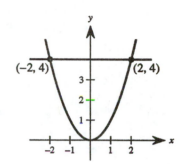

32. $f(x, y) = 2x - 2xy + y^2$

$\left.\begin{array}{l} f_x = 2 - 2y = 0 \quad \Rightarrow y = 1 \\ f_y = 2y - 2x = 0 \Rightarrow y = x \Rightarrow x = 1 \end{array}\right\} f(1, 1) = 1$

On the line $y = 1$, $-1 \leq x \leq 1$,

$f(x, y) = f(x) = 2x - 2x + 1 = 1$.

On the curve $y = x^2$, $-1 \leq x \leq 1$

$f(x, y) = f(x) = 2x - 2x(x^2) + (x^2)^2 = x^4 - 2x^3 + 2x$

and the maximum is 1, the minimum is $-\frac{11}{16}$.
Absolute maximum: 1 at (1, 1) and on $y = 1$
Absolute minimum: $-\frac{11}{16} = -0.6875$ at $\left(-\frac{1}{2}, \frac{1}{4}\right)$

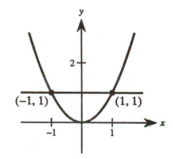

33. $f(x, y) = x^2 + xy$, $R = \{(x, y): |x| \le 2, |y| \le 1\}$

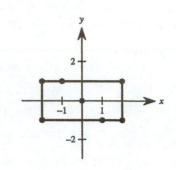

$$\left.\begin{array}{l} f_x = 2x + y = 0 \\ f_y = x = 0 \end{array}\right\} x = y = 0$$

$f(0, 0) = 0$

Along $y = 1$, $-2 \le x \le 2$, $f = x^2 + x$, $f' = 2x + 1 = 0 \Rightarrow x = -\frac{1}{2}$.

Thus, $f(-2, 1) = 2$, $f(-\frac{1}{2}, 1) = -\frac{1}{4}$ and $f(2, 1) = 6$.

Along $y = -1$, $-2 \le x \le 2$, $f = x^2 - x$, $f' = 2x - 1 = 0 \Rightarrow x = \frac{1}{2}$.

Thus, $f(-2, -1) = 6$, $f(\frac{1}{2}, -1) = -\frac{1}{4}$, $f(2, -1) = 2$.

Along $x = 2$, $-1 \le y \le 1$, $f = 4 + 2y \Rightarrow f' = 2 \ne 0$.

Along $x = -2$, $-1 \le y \le 1$, $f = 4 - 2y \Rightarrow f' = -2 \ne 0$.

Thus, the maxima are $f(2, 1) = 6$ and $f(-2, -1) = 6$ and the minima are $f(-\frac{1}{2}, 1) = -\frac{1}{4}$ and $f(\frac{1}{2}, -1) = -\frac{1}{4}$.

34. $f(x, y) = x^2 + 2xy + y^2$, $R = \{(x, y): |x| \le 2, |y| \le 1\}$

$$\left.\begin{array}{l} f_x = 2x + 2y = 0 \\ f_y = 2x + 2y = 0 \end{array}\right\} y = -x$$

$f(x, -x) = x^2 - 2x^2 + x^2 = 0$

Along $y = 1$, $-2 \le x \le 2$,

$\quad f = x^2 + 2x + 1$, $f' = 2x + 2 = 0 \Rightarrow x = -1$, $f(-2, 1) = 1$, $f(-1, 1) = 0$,
$\quad f(2, 1) = 9$.

Along $y = -1$, $-2 \le x \le 2$,

$\quad f = x^2 - 2x + 1$, $f' = 2x - 2 = 0 \Rightarrow x = 1$, $f(-2, -1) = 9$, $f(1, -1) = 0$,
$\quad f(2, -1) = 1$.

Along $x = 2$, $-1 \le y \le 1$, $f = 4 + 4y + y^2$, $f' = 2y + 4 \ne 0$.

Along $x = -2$, $-1 \le y \le 1$, $f = 4 - 4y + y^2$, $f' = 2y - 4 \ne 0$.

Thus, the maxima are $f(-2, -1) = 9$ and $f(2, 1) = 9$, and the minima are $f(x, -x) = 0$, $-1 \le x \le 1$.

35. $f(x, y) = x^2 + 2xy + y^2$, $R = \{(x, y): x^2 + y^2 \le 8\}$

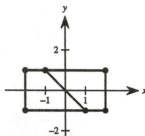

$$\left.\begin{array}{l} f_x = 2x + 2y = 0 \\ f_y = 2x + 2y = 0 \end{array}\right\} y = -x$$

$f(x, -x) = x^2 - 2x^2 + x^2 = 0$

On the boundary $x^2 + y^2 = 8$, we have $y^2 = 8 - x^2$ and $y = \pm\sqrt{8 - x^2}$. Thus,

$$f = x^2 \pm 2x\sqrt{8 - x^2} + (8 - x^2) = 8 \pm 2x\sqrt{8 - x^2}$$

$$f' = \pm(x(8 - x^2)^{-1/2}(-2x) + 2(8 - x^2)^{1/2}) = \pm\frac{16 - 4x^2}{\sqrt{8 - x^2}}.$$

Then, $f' = 0$ implies $16 = 4x^2$ or $x = \pm 2$. Thus, the maxima are $f(2, 2) = 16$ and $f(-2, -2) = 16$, and the minima are $f(x, -x)$, $|x| \le 4$.

36. $f(x, y) = x^2 - 4xy$, $R = \{(x, y): 0 \le x \le 4, 0 \le y \le \sqrt{x}\}$

$\left.\begin{array}{l} f_x = 2x - 4y = 0 \\ f_y = -4x = 0 \end{array}\right\} x = y = 0$

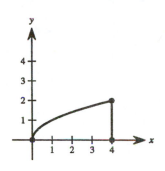

$f(0, 0) = 0$

Along $y = 0$, $0 \le x \le 4$, $f = x^2$ and $f(4, 0) = 16$.

Along $x = 4$, $0 \le y \le 2$, $f = 16 - 16y$, $f' = -16 \ne 0$ and $f(4, 2) = -16$.

Along $y = \sqrt{x}$, $0 \le x \le 4$, $f = x^2 - 4x^{3/2}$, $f' = 2x - 6x^{1/2} \ne 0$ on $[0, 4]$.

Thus, the maximum is $f(4, 0) = 16$ and the minimum is $f(4, 2) = -16$.

37. $f(x, y) = x^3 + y^3$

$\left.\begin{array}{l} f_x = 3x^2 = 0 \\ f_y = 3y^2 = 0 \end{array}\right\}$ Solving yields $x = y = 0$.

$f_{xx} = 6x$, $f_{yy} = 6y$, $f_{xy} = 0$

At $(0, 0)$, $f_{xx}f_{yy} - (f_{xy})^2 = 0$ and the test fails. $(0, 0, 0)$ is a saddle point.

38. $f(x, y) = x^3 + y^3 - 3x^2 + 6y^2 + 3x + 12y + 7$

$\left.\begin{array}{l} f_x = 3x^2 - 6x + 3 = 0 \\ f_y = 3y^2 + 12y + 12 = 0 \end{array}\right\}$ Solving yields $x = 1$ and $y = -2$.

$f_{xx} = 6x - 6$, $f_{yy} = 6y + 12$, $f_{xy} = 0$

At $(1, -2)$, $f_{xx}f_{yy} - (f_{xy})^2 = 0$ and the test fails. $(1, -2, 0)$ is a saddle point.

39. $f(x, y) = (x - 1)^2(y + 4)^2 \ge 0$

$\left.\begin{array}{l} f_x = 2(x - 1)(y + 4)^2 = 0 \\ f_y = 2(x - 1)^2(y + 4) = 0 \end{array}\right\}$ Solving yields the critical points $(1, a)$ and $(b, -4)$.

$f_{xx} = 2(y + 4)^2$, $f_{yy} = 2(x - 1)^2$, $f_{xy} = 4(x - 1)(y + 4)$

At both $(1, a)$ and $(b, -4)$, $f_{xx}f_{yy} - (f_{xy})^2 = 0$ and the test fails.

Absolute minima: $(1, a, 0)$ and $(b, -4, 0)$

40. $f(x, y) = \sqrt{(x - 1)^2 + (y + 2)^2} \ge 0$

$\left.\begin{array}{l} f_x = \dfrac{x - 1}{\sqrt{(x - 1)^2 + (y + 2)^2}} = 0 \\[3mm] f_y = \dfrac{y + 2}{\sqrt{(x - 1)^2 + (y + 2)^2}} = 0 \end{array}\right\}$ Solving yields $x = 1$ and $y = -2$.

$f_{xx} = \dfrac{(y + 2)^2}{[(x - 1)^2 + (y + 2)^2]^{3/2}}$, $f_{yy} = \dfrac{(x - 1)^2}{[(x - 1)^2 + (y + 2)^2]^{3/2}}$, $f_{xy} = \dfrac{(x - 1)(y + 2)}{[(x - 1)^2 + (y + 2)^2]^{3/2}}$

At $(1, -2)$, $f_{xx}f_{yy} - (f_{xy})^2$ is undefined and the test fails.

Absolute minimum: $(1, -2, 0)$

41. $f(x, y) = x^{2/3} + y^{2/3} \geq 0$

$\left.\begin{array}{l} f_x = \dfrac{2}{3\sqrt[3]{x}} \\[4mm] f_y = \dfrac{2}{3\sqrt[3]{y}} \end{array}\right\}$ f_x and f_y are undefined at $x = 0, \quad y = 0$. The critical point is $(0, 0)$.

$f_{xx} = -\dfrac{2}{9x\sqrt[3]{x}}, \quad f_{yy} = -\dfrac{2}{9y\sqrt[3]{y}}, \quad f_{xy} = 0$

At $(0, 0), \quad f_{xx}f_{yy} - (f_{xy})^2$ is undefined and the test fails.

Absolute minimum: $(0, 0, 0)$

42. $f(x, y) = (x^2 + y^2)^{2/3} \geq 0$

$\left.\begin{array}{l} f_x = \dfrac{4x}{3(x^2 + y^2)^{1/3}} \\[4mm] f_y = \dfrac{4y}{3(x^2 + y^2)^{1/3}} \end{array}\right\}$ f_x and f_y are undefined at $x = 0, \quad y = 0$. The critical point is $(0, 0)$.

$f_{xx} = \dfrac{4(x^2 + 3y^2)}{9(x^2 + y^2)^{4/3}}, \quad f_{yy} = \dfrac{4(3x^2 + y^2)}{9(x^2 + y^2)^{4/3}}, \quad f_{xy} = \dfrac{-8xy}{9(x^2 + y^2)^{4/3}}$

At $(0, 0), \quad f_{xx}f_{yy} - (f_{xy})^2$ is undefined and the test fails.

Absolute minimum: $(0, 0, 0)$

43. $f(x, y, z) = x^2 + (y - 3)^2 + (z + 1)^2 \geq 0$

$\left.\begin{array}{l} f_x = 2x = 0 \\[2mm] f_y = 2(y - 3) = 0 \\[2mm] f_z = 2(z + 1) = 0 \end{array}\right\}$ Solving yields the critical point $(0, 3, -1)$.

Absolute minimum: $(0, 3, -1, 0)$

44. $f(x, y, z) = 4 - [x(y - 1)(z + 2)]^2 \leq 4$

$\left.\begin{array}{l} f_x = -2x(y - 1)^2(z + 2)^2 = 0 \\[2mm] f_y = -2x^2(y - 1)(z + 2)^2 = 0 \\[2mm] f_z = -2x^2(y - 1)^2(z + 2) = 0 \end{array}\right\}$ Solving yields the critical points $(0, a, b), (c, 1, d), (e, f, -2)$. These points are all absolute maximas.

45. A and B are relative extrema. C and D are saddle points.

46. Given that f is a differentiable function such that $\nabla f(x_0, y_0) = \mathbf{0}$, then $f_x(x_0, y_0) = 0$ and $f_y(x_0, y_0) = 0$. Therefore, the tangent plane is $-(z - z_0) = 0$ or $z = z_0 = f(x_0, y_0)$ which is horizontal.

47. False

Let $f(x, y) = |1 - x - y|$.

$(0, 0, 1)$ is a relative maximum, but $f_x(0, 0)$ and $f_y(0, 0)$ do not exist.

48. True

49. False

Let $f(x, y) = x^4 - 2x^2 + y^2$.

Relative minima: $(\pm 1, 0, -1)$

Saddle point: $(0, 0, 0)$

50. True

Section 13.9   Applications of Extrema of Functions of Two Variables

1. A point on the plane is given by $(x, y, 12 - 2x - 3y)$. The square of the distance from the origin to this point is

$$S = x^2 + y^2 + (12 - 2x - 3y)^2$$

$$S_x = 2x + 2(12 - 2x - 3y)(-2)$$

$$S_y = 2y + 2(12 - 2x - 3y)(-3).$$

From the equations $S_x = 0$ and $S_y = 0$, we obtain the system

$$5x + 6y = 24$$

$$3x + 5y = 18.$$

Solving simultaneously, we have $x = \frac{12}{7}$, $y = \frac{18}{7}$, $z = 12 - \frac{24}{7} - \frac{54}{7} = \frac{6}{7}$. Therefore, the distance from the origin to $\left(\frac{12}{7}, \frac{18}{7}, \frac{6}{7}\right)$ is

$$\sqrt{\left(\frac{12}{7}\right)^2 + \left(\frac{18}{7}\right)^2 + \left(\frac{6}{7}\right)^2} = \frac{6\sqrt{14}}{7}.$$

2. A point on the plane is given by $(x, y, 12 - 2x - 3y)$. The square of the distance from $(1, 2, 3)$ to a point on the plane is given by

$$S = (x - 1)^2 + (y - 2)^2 + (9 - 2x - 3y)^2$$

$$S_x = 2(x - 1) + 2(9 - 2x - 3y)(-2)$$

$$S_y = 2(y - 2) + 2(9 - 2x - 3y)(-3).$$

From the equations $S_x = 0$ and $S_y = 0$, we obtain the system

$$5x + 6y = 19$$

$$6x + 10y = 29.$$

Solving simultaneously, we have $x = \frac{16}{14}$, $y = \frac{31}{14}$, $z = \frac{43}{14}$ and the distance is

$$\sqrt{\left(\frac{16}{14} - 1\right)^2 + \left(\frac{31}{14} - 2\right)^2 + \left(\frac{43}{14} - 3\right)^2} = \frac{1}{\sqrt{14}}.$$

3. A point on the paraboloid is given by $(x, y, x^2 + y^2)$. The square of the distance from $(5, 5, 0)$ to a point on the paraboloid is given by

$$S = (x - 5)^2 + (y - 5)^2 + (x^2 + y^2)^2$$

$$S_x = 2(x - 5) + 4x(x^2 + y^2) = 0$$

$$S_y = 2(y - 5) + 4y(x^2 + y^2) = 0.$$

From the equations $S_x = 0$ and $S_y = 0$, we obtain the system

$$2x^3 + 2xy^2 + x - 5 = 0$$

$$2y^3 + 2x^2y + y - 5 = 0.$$

Solving, we have $x = 1$, $y = 1$, $z = 2$ and the distance is $\sqrt{(1 - 5)^2 + (1 - 5)^2 + (2 - 0)^2} = 6$.

4. A point on the paraboloid is given by $(x, y, x^2 + y^2)$. The square of the distance from $(5, 0, 0)$ to a point on the paraboloid is given by

$$S = (x - 5)^2 + y^2 + (x^2 + y^2)^2$$

$$S_x = 2(x - 5) + 4x(x^2 + y^2) = 0$$

$$S_y = 2y + 4y(x^2 + y^2) = 0.$$

From the equations $S_x = 0$ and $S_y = 0$, we obtain the system

$$2x^3 + 2xy^2 + x - 5 = 0$$

$$2y^3 + 2x^2 y + y = 0.$$

Solving, we have $x \approx 1.235$, $y = 0$, $z \approx 1.525$ and the distance is

$$\sqrt{(1.235 - 5)^2 + (1.525)^2} \approx 4.06.$$

5. Let x, y and z be the numbers. Since $x + y + z = 30$, $z = 30 - x - y$.

$$P = xyz = 30xy - x^2 y - xy^2$$

$$P_x = 30y - 2xy - y^2 = y(30 - 2x - y) = 0 \ \Big\} \ 2x + y = 30$$

$$P_y = 30x - x^2 - 2xy = x(30 - x - 2y) = 0 \ \Big\} \ x + 2y = 30$$

Solving simultaneously yields $x = 10$, $y = 10$, and $z = 10$.

6. Since $x + y + z = 32$, $z = 32 - x - y$. Therefore,

$$P = xy^2 z = 32xy^2 - x^2 y^2 - xy^3$$

$$P_x = 32y^2 - 2xy^2 - y^3 = y^2(32 - 2x - y) = 0$$

$$P_y = 64xy - 2x^2 y - 3xy^2 = y(64x - 2x^2 - 3xy) = 0.$$

Ignoring the solution $y = 0$ and substituting $y = 32 - 2x$ into $P_y = 0$, we have

$$64x - 2x^2 - 3x(32 - 2x) = 0$$

$$4x(x - 8) = 0.$$

Therefore, $x = 8$, $y = 16$, and $z = 8$.

7. Let x, y, and z be the numbers and let $S = x^2 + y^2 + z^2$. Since $x + y + z = 30$, we have

$$S = x^2 + y^2 + (30 - x - y)^2$$

$$S_x = 2x + 2(30 - x - y)(-1) = 0 \ \Big\} \ 2x + y = 30$$

$$S_y = 2y + 2(30 - x - y)(-1) = 0 \ \Big\} \ x + 2y = 30.$$

Solving simultaneously yields $x = 10$, $y = 10$, and $z = 10$.

8. Let x, y, and z be the numbers and let $S = x^2 + y^2 + z^2$. Since $x + y + z = 1$, we have

$$S = x^2 + y^2 + (1 - x - y)^2$$

$$S_x = 2x - 2(1 - x - y) = 0 \ \Big\} \ 2x + y = 1$$

$$S_y = 2y - 2(1 - x - y) = 0 \ \Big\} \ x + 2y = 1.$$

Solving simultaneously yields $x = \frac{1}{3}$, $y = \frac{1}{3}$, and $z = \frac{1}{3}$.

9. Let x, y, and z be the length, width, and height, respectively. Then the sum of the length and girth is given by $x + (2y + 2z) = 108$ or $x = 108 - 2y - 2z$. The volume is given by

$$V = xyz = 108zy - 2zy^2 - 2yz^2$$

$$V_y = 108z - 4yz - 2z^2 = z(108 - 4y - 2z) = 0$$

$$V_z = 108y - 2y^2 - 4yz = y(108 - 2y - 4z) = 0.$$

Solving the system $4y + 2z = 108$ and $2y + 4z = 108$, we obtain the solution $x = 36$ inches, $y = 18$ inches, and $z = 18$ inches.

10. Let x, y, and z be the length, width, and height, respectively. Then the sum of the two perimeters of the two cross sections is given by $(2x + 2z) + (2y + 2z) = 108$ or $x = 54 - y - 2z$. The volume is given by

$$V = xyz = 54yz - y^2z - 2yz^2$$

$$V_y = 54z - 2yz - 2z^2 = z(54 - 2y - 2z) = 0$$

$$V_z = 54y - y^2 - 4yz = y(54 - y - 4z) = 0.$$

Solving the system $2y + 2z = 54$ and $y + 4z = 54$, we obtain the solution $x = 18$ inches, $y = 18$ inches, and $z = 9$ inches.

11. The distance from P to Q is $\sqrt{x^2 + 4}$. The distance from Q to R is $\sqrt{(y - x)^2 + 1}$. The distance from R to S is $10 - y$.

$$C = 3k\sqrt{x^2 + 4} + 2k\sqrt{(y - x)^2 + 1} + k(10 - y)$$

$$C_x = 3k\left(\frac{x}{\sqrt{x^2 + 4}}\right) + 2k\left(\frac{-(y - x)}{\sqrt{(y - x)^2 + 1}}\right) = 0$$

$$C_y = 2k\left(\frac{y - x}{\sqrt{(y - x)^2 + 1}}\right) - k = 0 \Rightarrow \frac{y - x}{\sqrt{(y - x)^2 + 1}} = \frac{1}{2}$$

$$3k\left(\frac{x}{\sqrt{x^2 + 4}}\right) + 2k\left(-\frac{1}{2}\right) = 0$$

$$\frac{x}{\sqrt{x^2 + 4}} = \frac{1}{3}$$

$$3x = \sqrt{x^2 + 4}$$

$$9x^2 = x^2 + 4$$

$$x^2 = \frac{1}{2}$$

$$x = \frac{\sqrt{2}}{2}$$

$$2(y - x) = \sqrt{(y - x)^2 + 1}$$

$$4(y - x)^2 = (y - x)^2 + 1$$

$$(y - x)^2 = \frac{1}{3}$$

$$y = \frac{1}{\sqrt{3}} - \frac{1}{\sqrt{2}} = \frac{2\sqrt{3} + 3\sqrt{2}}{6}$$

Therefore, $x = \dfrac{\sqrt{2}}{2} \approx 0.707$ mile and $y = \dfrac{2\sqrt{3} + 3\sqrt{2}}{6} \approx 1.284$ miles.

12. Let x, y, and z be the length, width, and height, respectively. Then $C_0 = 1.5xy + 2yz + 2xz$ and $z = \dfrac{C_0 - 1.5xy}{2(x+y)}$.
The volume is given by

$$V = xyz = \frac{C_0 xy - 1.5x^2 y^2}{2(x+y)}$$

$$V_x = \frac{y^2(2C_0 - 3x^2 - 6xy)}{4(x+y)^2}$$

$$V_y = \frac{x^2(2C_0 - 3y^2 - 6xy)}{4(x+y)^2}.$$

In solving the system $V_x = 0$ and $V_y = 0$, we note by the symmetry of the equations that $y = x$. Substituting $y = x$ into $V_x = 0$ yields

$$\frac{x^2(2C_0 - 9x^2)}{16x^2} = 0, \quad 2C_0 = 9x^2, \quad x = \frac{1}{3}\sqrt{2C_0}, \quad y = \frac{1}{3}\sqrt{2C_0}, \text{ and } z = \frac{1}{4}\sqrt{2C_0}.$$

13. Let $a + b + c = k$. Then

$$V = \frac{4\pi abc}{3} = \frac{4}{3}\pi ab(k - a - b) = \frac{4}{3}\pi(kab - a^2 b - ab^2)$$

$$\left.\begin{aligned} V_a &= \frac{4\pi}{3}(kb - 2ab - b^2) = 0 \\ V_b &= \frac{4\pi}{3}(ka - a^2 - 2ab) = 0 \end{aligned}\right| \begin{aligned} kb - 2ab - b^2 &= 0 \\ ka - a^2 - 2ab &= 0. \end{aligned}$$

Solving this system simultaneously yields $a = b$ and substitution yields $b = k/3$. Therefore, the solution is $a = b = c = k/3$.

14. Consider the sphere given by $x^2 + y^2 + z^2 = r^2$ and let a vertex of the rectangular box be $\left(x,\ y,\ \sqrt{r^2 - x^2 - y^2}\right)$.
Then the volume is given by

$$V = (2x)(2y)\left(2\sqrt{r^2 - x^2 - y^2}\right) = 8xy\sqrt{r^2 - x^2 - y^2}$$

$$V_x = 8\left(xy\frac{-x}{\sqrt{r^2 - x^2 - y^2}} + y\sqrt{r^2 - x^2 - y^2}\right) = \frac{8y}{\sqrt{r^2 - x^2 - y^2}}(r^2 - 2x^2 - y^2) = 0$$

$$V_y = 8\left(xy\frac{-y}{\sqrt{r^2 - x^2 - y^2}} + x\sqrt{r^2 - x^2 - y^2}\right) = \frac{8x}{\sqrt{r^2 - x^2 - y^2}}(r^2 - x^2 - 2y^2) = 0.$$

Solving the system

$$2x^2 + y^2 = r^2$$

$$x^2 + 2y^2 = r^2$$

yields the solution $x = y = z = r/\sqrt{3}$.

15. Let x, y, and z be the length, width, and height, respectively and let V_0 be the given volume. Then $V_0 = xyz$ and $z = V_0/xy$. The surface area is

$$S = 2xy + 2yz + 2xz = 2\left(xy + \frac{V_0}{x} + \frac{V_0}{y}\right)$$

$$\left. \begin{array}{l} S_x = 2\left(y - \dfrac{V_0}{x^2}\right) = 0 \\[2ex] S_y = 2\left(x - \dfrac{V_0}{y^2}\right) = 0 \end{array} \right| \begin{array}{l} x^2 y - V_0 = 0 \\[2ex] xy^2 - V_0 = 0. \end{array}$$

Solving simultaneously yields $x = \sqrt[3]{V_0}$, $y = \sqrt[3]{V_0}$, and $z = \sqrt[3]{V_0}$.

16. $A = \dfrac{1}{2}[(10 - 2x) + (10 - 2x) + 2x \cos\theta]x \sin\theta$

$\qquad = 10x \sin\theta - 2x^2 \sin\theta + x^2 \sin\theta \cos\theta$

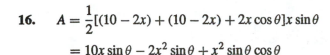

$\dfrac{\partial A}{\partial x} = 10 \sin\theta - 4x \sin\theta + 2x \sin\theta \cos\theta = 0$

$\dfrac{\partial A}{\partial \theta} = 10x \cos\theta - 2x^2 \cos\theta + x^2(2\cos^2\theta - 1) = 0$

From $\partial A/\partial x = 0$, we have

$\quad 5 - 2x + x\cos\theta = 0$

$$\cos\theta = \frac{2x - 5}{x}.$$

Substituting this expression for $\cos\theta$ in $\partial A/\partial\theta = 0$, we have

$$10x\left(\frac{2x - 5}{x}\right) - 2x^2\left(\frac{2x - 5}{x}\right) + x^2\left[2\left(\frac{2x - 5}{x}\right)^2 - 1\right] = 0$$

which simplifies to $x(3x - 10) = 0 \Rightarrow x = 0$ or $10/3$. Using $x = 10/3$, we have

$$\cos\theta = \frac{2(10/3) - 5}{10/3} = \frac{1}{2}.$$

Therefore, $\theta = 60°$, $x = 10/3$ inches.

17. We observe that the area of a trapezoidal cross section is given by

$$A = h\left[\frac{(w - 2r) + [(w - 2r) + 2x]}{2}\right] = (w - 2r + x)h$$

where $x = r\cos\theta$ and $h = r\sin\theta$. Substituting these expressions for x and h, we have

$\quad A(r, \theta) = (w - 2r + r\cos\theta)(r\sin\theta) = wr\sin\theta - 2r^2\sin\theta + r^2\sin\theta\cos\theta$

Now

$\quad A_r(r, \theta) = w\sin\theta - 4r\sin\theta + 2r\sin\theta\cos\theta = \sin\theta(w - 4r + 2r\cos\theta) = 0 \Rightarrow w = r(4 - 2\cos\theta)$

$\quad A_\theta(r, \theta) = wr\cos\theta - 2r^2\cos\theta + r^2\cos 2\theta = 0.$

Substituting the expression for w from $A_r(r, \theta) = 0$ into the equation $A_\theta(r, \theta) = 0$, we have

$\quad r^2(4 - 2\cos\theta)\cos\theta - 2r^2\cos\theta + r^2(2\cos^2\theta - 1) = 0$

$$r^2(2\cos\theta - 1) = 0 \quad \text{or} \quad \cos\theta = \frac{1}{2}.$$

Therefore, the first partial derivatives are zero when $\theta = \pi/3$ and $r = w/3$. (Ignore the solution $r = \theta = 0$.) Thus, the trapezoid of maximum area occurs when each edge of width $w/3$ is turned up 60° from the horizontal.

18. $R(x_1, x_2) = -5x_1^2 - 8x_2^2 - 2x_1x_2 + 42x_1 + 102x_2$

$$R_{x_1} = -10x_1 - 2x_2 + 42 = 0, \quad 5x_1 + x_2 = 21$$

$$R_{x_2} = -16x_2 - 2x_1 + 102 = 0, \quad x_1 + 8x_2 = 51$$

Solving this system yields $x_1 = 3$ and $x_2 = 6$.

$$R_{x_1x_1} = -10$$

$$R_{x_1x_2} = -2$$

$$R_{x_2x_2} = -16$$

$R_{x_1x_1} < 0$ and $R_{x_1x_1} R_{x_2x_2} - (R_{x_1x_2})^2 > 0$

Thus, revenue is maximized when $x_1 = 3$ and $x_2 = 6$.

19. $R(p_1, p_2) = 500p_1 + 800p_2 + 1.5p_1p_2 - 1.5p_1^2 - p_2^2$

$$R_{p_1} = 500 + 1.5p_2 - 3p_1 = 0, \quad -3p_1 + 1.5p_2 = -500$$

$$R_{p_2} = 800 + 1.5p_1 - 2p_2 = 0, \quad 3p_1 - 4.0p_2 = -1600$$

Solving this system yields $-2.5p_2 = -2100$ which implies that $p_2 = \$840$ and $p_1 \approx \$586.67$.

20. $P(x_1, x_2) = 15(x_1 + x_2) - C_1 - C_2$

$$= 15x_1 + 15x_2 - (0.02x_1^2 + 4x_1 + 500) - (0.05x_2^2 + 4x_2 + 275)$$

$$= -0.02x_1^2 - 0.05x_2^2 + 11x_1 + 11x_2 - 775$$

$$P_{x_1} = -0.04x_1 + 11 = 0, \quad x_1 = 275$$

$$P_{x_2} = -0.10x_2 + 11 = 0, \quad x_2 = 110$$

$$P_{x_1x_1} = -0.04$$

$$P_{x_1x_2} = 0$$

$$P_{x_2x_2} = -0.10$$

$P_{x_1x_1} < 0$ and $P_{x_1x_1} P_{x_2x_2} - (P_{x_1x_2})^2 > 0$

Therefore, profit is maximized when $x_1 = 275$ and $x_2 = 110$.

21. $S = d_1 + d_2 + d_3 = \sqrt{(0 - 0)^2 + (y - 0)^2} + \sqrt{(0 - 2)^2 + (y - 2)^2} + \sqrt{(0 + 2)^2 + (y - 2)^2}$

$$= y + 2\sqrt{4 + (y - 2)^2}$$

$$\frac{dS}{dy} = 1 + \frac{2(y - 2)}{\sqrt{4 + (y - 2)^2}} = 0 \text{ when } y = 2 - \frac{2\sqrt{3}}{3} = \frac{6 - 2\sqrt{3}}{3}.$$

The sum of the distances is minimized when

$$y = \frac{2(3 - \sqrt{3})}{3} \approx 0.845.$$

22. a. $S(x, y) = d_1 + d_2 + d_3$

$$= \sqrt{(x-0)^2 + (y-0)^2} + \sqrt{(x+2)^2 + (y-2)^2} + \sqrt{(x-4)^2 + (y-2)^2}$$

$$= \sqrt{x^2 + y^2} + \sqrt{(x+2)^2 + (y-2)^2} + \sqrt{(x-4)^2 + (y-2)^2}$$

From the graph we see that the surface has a minimum.

b. $S_x(x, y) = \dfrac{x}{\sqrt{x^2 + y^2}} + \dfrac{x+2}{\sqrt{(x+2)^2 + (y-2)^2}} + \dfrac{x-4}{\sqrt{(x-4)^2 + (y-2)^2}}$

$S_y(x, y) = \dfrac{y}{\sqrt{x^2 + y^2}} + \dfrac{y-2}{\sqrt{(x+2)^2 + (y-2)^2}} + \dfrac{y-2}{\sqrt{(x-4)^2 + (y-2)^2}}$

c. $-\nabla S(1, 1) = -S_x(1, 1)\mathbf{i} - S_y(1, 1)\mathbf{j} = -\dfrac{1}{\sqrt{2}}\mathbf{i} - \left(\dfrac{1}{\sqrt{2}} - \dfrac{2}{\sqrt{10}}\right)\mathbf{j}$

$\tan\theta = \dfrac{(2/\sqrt{10}) - (1/\sqrt{2})}{-1/\sqrt{2}} = 1 - \dfrac{2}{\sqrt{5}} \Rightarrow \theta \approx 186.027°$

d. $(x_2, y_2) = (x_1 - S_x(x_1, y_1)t,\ y_1 - S_y(x_1, y_1)t) = \left(1 - \dfrac{1}{\sqrt{2}}t,\ 1 + \left(\dfrac{2}{\sqrt{10}} - \dfrac{1}{\sqrt{2}}\right)t\right)$

$S\left(1 - \dfrac{1}{\sqrt{2}}t,\ 1 + \left(\dfrac{2}{\sqrt{10}} - \dfrac{1}{\sqrt{2}}\right)t\right)$

$$= \sqrt{2 + \left(\dfrac{2\sqrt{10}}{5} - 2\sqrt{2}\right)t + \left(1 - \dfrac{2\sqrt{5}}{5} + \dfrac{2}{5}\right)t^2} + \sqrt{10 - \left(\dfrac{2\sqrt{10}}{5} + 2\sqrt{2}\right)t + \left(1 - \dfrac{2\sqrt{5}}{5} + \dfrac{2}{5}\right)t^2}$$

$$+ \sqrt{10 - \left(\dfrac{2\sqrt{10}}{5} - 4\sqrt{2}\right)t + \left(1 - \dfrac{2\sqrt{5}}{5} + \dfrac{2}{5}\right)t^2}$$

Using a computer algebra system, we find that the minimum occurs when $t \approx 1.344$. Thus, $(x_2, y_2) \approx (0.05, 0.90)$.

e. $(x_3, y_3) = (x_2 - S_x(x_2, y_2)t,\ y_2 - S_y(x_2, y_2)t) \approx (0.05 + 0.03t,\ 0.90 - 0.26t)$

$S(0.05 + 0.03t,\ 0.90 - 0.26t)$

$$= \sqrt{(0.05 + 0.03t)^2 + (0.90 - 0.26t)^2} + \sqrt{(2.05 + 0.03t)^2 + (-1.10 - 0.26t)^2}$$

$$+ \sqrt{(-3.95 + 0.03t)^2 + (-1.10 - 0.26t)^2}$$

Using a computer algebra system, we find that the minimum occurs when $t \approx 1.737$. Thus, $(x_3, y_3) \approx (0.10, 0.45)$.

$(x_4, y_4) = (x_3 - S_x(x_3, y_3)t,\ y_3 - S_y(x_3, y_3)t) \approx (0.10 - 0.09t,\ 0.45 - 0.01t)$

$S(0.10 - 0.09t,\ 0.45 - 0.01t)$

$$= \sqrt{(0.10 - 0.09t)^2 + (0.45 - 0.01t)^2} + \sqrt{(2.10 - 0.09t)^2 + (-1.55 - 0.01t)^2}$$

$$+ \sqrt{(-3.90 - 0.09t)^2 + (-1.55 - 0.01t)^2}$$

Using a computer algebra system, we find that the minimum occurs when $t \approx 0.463$. Thus, $(x_4, y_4) \approx (0.06, 0.45)$.

23. a.

x	y	xy	x^2
-2	0	0	4
0	1	0	0
2	3	6	4
$\sum x_i = 0$	$\sum y_i = 4$	$\sum x_i y_i = 6$	$\sum x_i^2 = 8$

$$a = \frac{3(6) - 0(4)}{3(8) - 0^2} = \frac{3}{4},\ b = \frac{1}{3}\left[4 - \frac{3}{4}(0)\right] = \frac{4}{3},\ y = \frac{3}{4}x + \frac{4}{3}$$

b. $S = \left(-\frac{3}{2} + \frac{4}{3} - 0\right)^2 + \left(\frac{4}{3} - 1\right)^2 + \left(\frac{3}{2} + \frac{4}{3} - 3\right)^2 = \frac{1}{6}$

24. a.

x	y	xy	x^2
-3	0	0	9
-1	1	-1	1
1	1	1	1
3	2	6	9
$\sum x_i = 0$	$\sum y_i = 4$	$\sum x_i y_i = 6$	$\sum x_i^2 = 20$

$$a = \frac{4(6) - 0(4)}{4(20) - (0)^2} = \frac{3}{10},\ b = \frac{1}{4}\left[4 - \frac{3}{10}(0)\right] = 1,\ y = \frac{3}{10}x + 1$$

b. $S = \left(\frac{1}{10} - 0\right)^2 + \left(\frac{7}{10} - 1\right)^2 + \left(\frac{13}{10} - 1\right)^2 + \left(\frac{19}{10} - 2\right)^2 = \frac{1}{5}$

25. a.

x	y	xy	x^2
0	4	0	0
1	3	3	1
1	1	1	1
2	0	0	4
$\sum x_i = 4$	$\sum y_i = 8$	$\sum x_i y_i = 4$	$\sum x_i^2 = 6$

$$a = \frac{4(4) - 4(8)}{4(6) - 4^2} = -2,\ b = \frac{1}{4}[8 + 2(4)] = 4,\ y = -2x + 4$$

b. $S = (4 - 4)^2 + (2 - 3)^2 + (2 - 1)^2 + (0 - 0)^2 = 2$

26. a.

x	y	xy	x^2
3	0	0	9
1	0	0	1
2	0	0	4
3	1	3	9
4	1	4	16
4	2	8	16
5	2	10	25
6	2	12	36
$\sum x_i = 28$	$\sum y_i = 8$	$\sum x_i y_i = 37$	$\sum x_i^2 = 116$

$$a = \frac{8(37) - (28)(8)}{8(116) - (28)^2} = \frac{72}{144} = \frac{1}{2}, \; b = \frac{1}{8}\left[8 - \frac{1}{2}(28)\right] = -\frac{3}{4}, \; y = \frac{1}{2}x - \frac{3}{4}$$

b. $S = \left(\frac{3}{4} - 0\right)^2 + \left(-\frac{1}{4} - 0\right)^2 + \left(\frac{1}{4} - 0\right)^2 + \left(\frac{3}{4} - 1\right)^2 + \left(\frac{5}{4} - 1\right)^2 + \left(\frac{5}{4} - 2\right)^2$

$$+ \left(\frac{7}{4} - 2\right)^2 + \left(\frac{9}{4} - 2\right)^2 = \frac{3}{2}$$

27. (0, 0), (1, 1), (3, 4), (4, 2), (5, 5)

$$\sum x_i = 13, \qquad \sum y_i = 12,$$
$$\sum x_i y_i = 46, \qquad \sum x_i^2 = 51$$
$$a = \frac{5(46) - 13(12)}{5(51) - (13)^2} = \frac{74}{86} = \frac{37}{43}$$
$$b = \frac{1}{5}\left[12 - \frac{37}{43}(13)\right] = \frac{7}{43}$$
$$y = \frac{37}{43}x + \frac{7}{43}$$

28. (1, 0), (3, 3), (5, 6)

$$\sum x_i = 9, \qquad \sum y_i = 9,$$
$$\sum x_i y_i = 39, \qquad \sum x_i^2 = 35$$
$$a = \frac{3(39) - 9(9)}{3(35) - (9)^2} = \frac{36}{24} = \frac{3}{2}$$
$$b = \frac{1}{3}\left[9 - \frac{3}{2}(9)\right] = -\frac{9}{6} = -\frac{3}{2}$$
$$y = \frac{3}{2}x - \frac{3}{2}$$

29. (0, 6), (4, 3), (5, 0), (8, −4), (10, −5)

$$\sum x_i = 27, \qquad \sum y_i = 0,$$
$$\sum x_i y_i = -70, \qquad \sum x_i^2 = 205$$
$$a = \frac{5(-70) - (27)(0)}{5(205) - (27)^2} = \frac{-350}{296} = -\frac{175}{148}$$
$$b = \frac{1}{5}\left[0 - \left(-\frac{175}{148}\right)(27)\right] = \frac{945}{148}$$
$$y = -\frac{175}{148}x + \frac{945}{148}$$

30. (5, 2), (0, 0), (2, 1), (7, 4), (10, 6), (12, 6)

$$\sum x_i = 36, \qquad \sum y_i = 19,$$
$$\sum x_i y_i = 172, \qquad \sum x_i^2 = 322$$
$$a = \frac{6(172) - (36)(19)}{6(322) - (36)^2} = \frac{348}{636} = \frac{29}{53}$$
$$b = \frac{1}{6}\left[19 - \frac{29}{53}(36)\right] = -\frac{37}{318}$$
$$y = \frac{29}{53}x - \frac{37}{318}$$

31. a. (1.00, 450), (1.25, 375), (1.50, 330)

$$\sum x_i = 3.75, \quad \sum y_i = 1155, \quad \sum x_i^2 = 4.8125, \quad \sum x_i y_i = 1413.75$$

$$a = \frac{3(1413.75) - (3.75)(1155)}{3(4.8125) - (3.75)^2} = -240$$

$$b = \frac{1}{3}[1155 - (-240)(3.75)] = 685$$

$$y = -240x + 685$$

b. When $x = 1.40$, $y = -240(1.40) + 685 = 349$.

32. (2.67, 135.2), (1.35, 118.5), (3.93, 167.3), (5.14, 197.6), (7.43, 204.7)

$$\sum x_i = 20.52, \quad \sum y_i = 823.3, \quad \sum x_i y_i = 3715.033, \quad \sum x_i^2 = 106.0208$$

$$a = \frac{5(3715.033) - (20.52)(823.3)}{5(106.0208) - (20.52)^2} \approx 15.4177$$

$$b = \frac{1}{5}[823.3 - 15.4177(20.52)] \approx 101.3857$$

$$y = 15.4177x + 101.3857$$

33. (1.0, 32), (1.5, 41), (2.0, 48), (2.5, 53)

$$\sum x_i = 7, \quad \sum y_i = 174, \quad \sum x_i y_i = 322, \quad \sum x_i^2 = 13.5$$

$a = 14$, $b = 19$, $y = 14x + 19$

When $x = 1.6$, $y = 41.4$ bushels per acre.

34. a. (24.3, 12.8), (29.0, 18.4), (32.5, 23.3), (37.2, 31.6), (42.0, 45.6), (44.5, 54.9), (44.8, 56.2), (45.3, 56.6)

$$\sum x_i = 299.6, \quad \sum y_i = 299.4, \quad \sum x_i y_i = 12,217.4, \quad \sum x_i^2 = 11,674.96$$

$a \approx 2.21$, $b \approx -45.29$, $y = 2.21x - 45.29$

b. According to this model, 2.21 million women enter the labor force for each one point increase in the percentage of women in the labor force.

35. $S(a, b, c) = \sum_{i=1}^{n} (y_i - ax_i^2 - bx_i - c)^2$

$$\frac{\partial S}{\partial a} = \sum_{i=1}^{n} -2x_i^2(y_i - ax_i^2 - bx_i - c) = 0$$

$$\frac{\partial S}{\partial b} = \sum_{i=1}^{n} -2x_i(y_i - ax_i^2 - bx_i - c) = 0$$

$$\frac{\partial S}{\partial c} = -2\sum_{i=1}^{n} (y_i - ax_i^2 - bx_i - c) = 0$$

$$a\sum_{i=1}^{n} x_i^4 + b\sum_{i=1}^{n} x_i^3 + c\sum_{i=1}^{n} x_i^2 = \sum_{i=1}^{n} x_i^2 y_i$$

$$a\sum_{i=1}^{n} x_i^3 + b\sum_{i=1}^{n} x_i^2 + c\sum_{i=1}^{n} x_i = \sum_{i=1}^{n} x_i y_i$$

$$a\sum_{i=1}^{n} x_i^2 + b\sum_{i=1}^{n} x_i + cn = \sum_{i=1}^{n} y_i$$

36. $S(a, b) = \sum_{i=1}^{n}(ax_i + b - y_i)^2$

$$S_a(a, b) = 2a\sum_{i=1}^{n} x_i^2 + 2b\sum_{i=1}^{n} x_i - 2\sum_{i=1}^{n} x_iy_i$$

$$S_b(a, b) = 2a\sum_{i=1}^{n} x_i + 2nb - 2\sum_{i=1}^{n} y_i$$

$$S_{aa}(a, b) = 2\sum_{i=1}^{n} x_i^2$$

$$S_{bb}(a, b) = 2n$$

$$S_{ab}(a, b) = 2\sum_{i=1}^{n} x_i$$

$S_{aa}(a, b) > 0$ as long as $x_i \neq 0$ for all i. (**Note:** If $x_i = 0$ for all i, then $x = 0$ is the least squares regression line.)

$$d = S_{aa}S_{bb} - S_{ab}^2 = 4n\sum_{i=1}^{n} x_i^2 - 4\left(\sum_{i=1}^{n} x_i\right)^2 = 4\left[n\sum_{i=1}^{n} x_i^2 - \left(\sum_{i=1}^{n} x_i\right)^2\right] \geq 0 \text{ since } n\sum_{i=1}^{n} x_i^2 \geq \left(\sum_{i=1}^{n} x_i\right)^2.$$

As long as $d \neq 0$, the given values for a and b yield a minimum.

37. $(-2, 0)$, $(-1, 0)$, $(0, 1)$, $(1, 2)$, $(2, 5)$

$\sum x_i = 0$

$\sum y_i = 8$

$\sum x_i^2 = 10$

$\sum x_i^3 = 0$

$\sum x_i^4 = 34$

$\sum x_iy_i = 12$

$\sum x_i^2y_i = 22$

$34a + 10c = 22$, $10b = 12$, $10a + 5c = 8$

$a = \frac{3}{7}$, $b = \frac{6}{5}$, $c = \frac{26}{35}$, $y = \frac{3}{7}x^2 + \frac{6}{5}x + \frac{26}{35}$

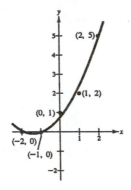

38. $(-4, 5)$, $(-2, 6)$, $(2, 6)$, $(4, 2)$

$\sum x_i = 0$

$\sum y_i = 19$

$\sum x_i^2 = 40$

$\sum x_i^3 = 0$

$\sum x_i^4 = 544$

$\sum x_iy_i = -12$

$\sum x_i^2y_i = 160$

$544a + 40c = 160$, $40b = -12$, $40a + 4c = 19$

$a = -\frac{5}{24}$, $b = -\frac{3}{10}$, $c = \frac{41}{6}$, $y = -\frac{5}{24}x^2 - \frac{3}{10}x + \frac{41}{6}$

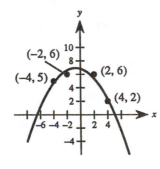

39. $(0, 0), (2, 2), (3, 6), (4, 12)$

$$\sum x_i = 9$$

$$\sum y_i = 20$$

$$\sum x_i^2 = 29$$

$$\sum x_i^3 = 99$$

$$\sum x_i^4 = 353$$

$$\sum x_i y_i = 70$$

$$\sum x_i^2 y_i = 254$$

$353a + 99b + 29c = 254$

$99a + 29b + 9c = 70$

$29a + 9b + 4c = 20$

$a = 1, \quad b = -1, \quad c = 0, \quad y = x^2 - x$

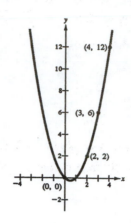

40. $(0, 10), (1, 9), (2, 6), (3, 0)$

$$\sum x_i = 6$$

$$\sum y_i = 25$$

$$\sum x_i^2 = 14$$

$$\sum x_i^3 = 36$$

$$\sum x_i^4 = 98$$

$$\sum x_i y_i = 21$$

$$\sum x_i^2 y_i = 33$$

$98a + 36b + 14c = 33$

$36a + 14b + 6c = 21$

$14a + 6b + 4c = 25$

$a = -\frac{5}{4}, \quad b = \frac{9}{20}, \quad c = \frac{199}{20}, \quad y = -\frac{5}{4}x^2 + \frac{9}{20}x + \frac{199}{20}$

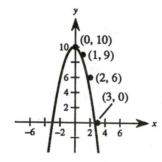

41. $(0, 0), (2, 15), (4, 30), (6, 50), (8, 65), (10, 70)$

$$\sum x_i = 30, \qquad \sum y_i = 230, \qquad \sum x_i^2 = 220, \qquad \sum x_i^3 = 1800,$$

$$\sum x_i^4 = 15{,}664, \qquad \sum x_i y_i = 1670, \qquad \sum x_i^2 y_i = 13{,}500$$

$15{,}664a + 1800b + 220c = 13{,}500$

$1800a + 220b + 30c = 1670$

$220a + 30b + 6c = 230$

$y = -\frac{25}{112}x^2 + \frac{541}{56}x - \frac{25}{14}$

42. a. $(-10, 3.7)$, $(-5, 4.1)$, $(0, 4.5)$, $(5, 4.8)$, $(10, 5.3)$

$$\sum x_i = 0, \qquad \sum y_i = 22.4, \qquad \sum x_i^2 = 250, \qquad \sum x_i^3 = 0,$$

$$\sum x_i^4 = 21{,}250, \qquad \sum x_i y_i = 19.5, \qquad \sum x_i^2 y_i = 1122.5$$

$$21{,}250a + \qquad\quad 250c = 1122.5$$

$$250b \qquad\quad = \quad 19.5$$

$$250a \qquad + \quad 5c = \quad 22.4$$

$a = \frac{1}{3500}$, $b = \frac{39}{500}$, $c = \frac{1563}{350}$, $y = \frac{1}{3500}(x^2 + 273x + 15{,}630)$

b. For the year 2010, use $x = 30$, then we have $y \approx 7.1$.

43. Use ∇S rather than $-\nabla S$.

Section 13.10 Lagrange Multipliers

1. Maximize $f(x, y) = xy$.
Constraint: $x + y = 10$
$\nabla f = \lambda \nabla g$
$y\mathbf{i} + x\mathbf{j} = \lambda(\mathbf{i} + \mathbf{j})$

$\left.\begin{array}{l} y = \lambda \\ x = \lambda \end{array}\right\} x = y$

$x + y = 10 \Rightarrow x = y = 5$

$f(5, 5) = 25$

2. Maximize $f(x, y) = xy$.
Constraint: $2x + y = 4$
$\nabla f = \lambda \nabla g$
$y\mathbf{i} + x\mathbf{j} = 2\lambda\mathbf{i} + \lambda\mathbf{j}$

$y = 2\lambda$

$x = \lambda$

$2x + y = 4 \Rightarrow 4\lambda = 4$

$\lambda = 1$, $x = 1$, $y = 2$

$f(1, 2) = 2$

3. Minimize $f(x, y) = x^2 + y^2$.
Constraint: $x + y = 4$
$\nabla f = \lambda \nabla g$
$2x\mathbf{i} + 2y\mathbf{j} = \lambda\mathbf{i} + \lambda\mathbf{j}$

$\left.\begin{array}{l} 2x = \lambda \\ 2y = \lambda \end{array}\right\} x = y$

$x + y = 4 \Rightarrow x = y = 2$

$f(2, 2) = 8$

4. Minimize $f(x, y) = x^2 + y^2$.
Constraint: $2x + 4y = 5$
$\nabla f = \lambda \nabla g$
$2x\mathbf{i} + 2y\mathbf{j} = 2\lambda\mathbf{i} + 4\lambda\mathbf{j}$

$2x = 2\lambda \Rightarrow x = \lambda$

$2y = 4\lambda \Rightarrow y = 2\lambda$

$2x + 4y = 5 \Rightarrow 10\lambda = 5$

$\lambda = \frac{1}{2}$, $x = \frac{1}{2}$, $y = 1$

$f\left(\frac{1}{2}, 1\right) = \frac{5}{4}$

5. Minimize $f(x, y) = x^2 - y^2$.

Constraint: $x - 2y = -6$

$\nabla f = \lambda \nabla g$

$2x\mathbf{i} - 2y\mathbf{j} = \lambda\mathbf{i} - 2\lambda\mathbf{j}$

$$2x = \lambda \Rightarrow x = \frac{\lambda}{2}$$

$$-2y = -2\lambda \Rightarrow y = \lambda$$

$$x - 2y = -6 \Rightarrow -\frac{3}{2}\lambda = -6$$

$$\lambda = 4, \ x = 2, \ y = 4$$

$f(2, 4) = -12$

6. Maximize $f(x, y) = x^2 - y^2$.

Constraint: $y - x^2 = 0$

$\nabla f = \lambda \nabla g$

$2x\mathbf{i} - 2y\mathbf{j} = -2x\lambda\mathbf{i} + \lambda\mathbf{j}$

$2x = -2x\lambda \Rightarrow x = 0$ or $\lambda = -1$

If $x = 0$, then $y = 0$ and $f(0, 0)$ does not yield a maximum.

$$-2y = \lambda \Rightarrow y = -\frac{\lambda}{2} = \frac{1}{2}$$

$$y - x^2 = 0 \Rightarrow x^2 = y$$

$$x = \frac{\sqrt{2}}{2}$$

$$f\left(\frac{\sqrt{2}}{2}, \frac{1}{2}\right) = \frac{1}{4}$$

7. Maximize $f(x, y) = 2x + 2xy + y$.

Constraint: $2x + y = 100$

$\nabla f = \lambda \nabla g$

$(2 + 2y)\mathbf{i} + (2x + 1)\mathbf{j} = 2\lambda\mathbf{i} + \lambda\mathbf{j}$

$$\left.\begin{array}{l} 2 + 2y = 2\lambda \Rightarrow y = \lambda - 1 \\ 2x + 1 = \lambda \ \Rightarrow x = \dfrac{\lambda - 1}{2} \end{array}\right\} y = 2x$$

$$2x + y = 100 \Rightarrow 4x = 100$$

$$x = 25, \ y = 50$$

$f(25, 50) = 2600$

8. Minimize $f(x, y) = 3x + y + 10$.

Constraint: $x^2 y = 6$

$\nabla f = \lambda \nabla g$

$3\mathbf{i} + \mathbf{j} = 2xy\lambda\mathbf{i} + x^2\lambda\mathbf{j}$

$$\left.\begin{array}{l} 3 = 2xy\lambda \Rightarrow \lambda = \dfrac{3}{2xy} \\ 1 = x^2\lambda \ \Rightarrow \lambda = \dfrac{1}{x^2} \end{array}\right\} 3x^2 = 2xy \Rightarrow y = \frac{3x}{2}$$

$(x \neq 0)$

$$x^2 y = 6 \Rightarrow x^2\left(\frac{3x}{2}\right) = 6$$

$$x^3 = 4$$

$$x = \sqrt[3]{4}, \ y = \frac{3\sqrt[3]{4}}{2}$$

$$f\left(\sqrt[3]{4}, \frac{3\sqrt[3]{4}}{2}\right) = \frac{9\sqrt[3]{4} + 20}{2}$$

9. Note: $f(x, y) = \sqrt{6 - x^2 - y^2}$ is maximum when $g(x, y)$ is maximum.

Maximize $g(x, y) = 6 - x^2 - y^2$.

Constraint: $x + y = 2$

$$\left.\begin{array}{l} -2x = \lambda \\ -2y = \lambda \end{array}\right\} x = y$$

$$x + y = 2 \Rightarrow x = y = 1$$

$$f(1, 1) = \sqrt{g(1, 1)} = 2$$

10. Note: $f(x, y) = \sqrt{x^2 + y^2}$ is minimum when $g(x, y)$ is minimum.

Minimize $g(x, y) = x^2 + y^2$.

Constraint: $2x + 4y = 15$

$$\left.\begin{array}{l} 2x = 2\lambda \\ 2y = 4\lambda \end{array}\right\} y = 2x$$

$$2x + 4y = 15 \Rightarrow 10x = 15$$

$$x = \frac{3}{2}, \ y = 3$$

$$f\left(\frac{3}{2}, 3\right) = \sqrt{g\left(\frac{3}{2}, 3\right)} = \frac{3\sqrt{5}}{2}$$

11. Maximize $f(x, y) = e^{xy}$.
Constraint: $x^2 + y^2 = 8$

$$\left. \begin{array}{r} ye^{xy} = 2x\lambda \\ xe^{xy} = 2y\lambda \end{array} \right\} x = y$$

$x^2 + y^2 = 8 \Rightarrow 2x^2 = 8$

$$x = y = 2$$

$f(2, 2) = e^4$

12. Minimize $f(x, y) = 2x + y$.
Constraint: $xy = 32$

$$\left. \begin{array}{r} 2 = y\lambda \\ 1 = x\lambda \end{array} \right\} y = 2x$$

$xy = 32 \Rightarrow 2x^2 = 32$

$$x = 4, \ y = 8$$

$f(4, 8) = 16$

13. Minimize $f(x, y, z) = x^2 + y^2 + z^2$.
Constraint: $x + y + z = 6$

$$\left. \begin{array}{r} 2x = \lambda \\ 2y = \lambda \\ 2z = \lambda \end{array} \right\} x = y = z$$

$x + y + z = 6 \Rightarrow x = y = z = 2$
$f(2, 2, 2) = 12$

14. Maximize $f(x, y, z) = xyz$.
Constraint: $x + y + z = 6$

$$\left. \begin{array}{r} yz = \lambda \\ xz = \lambda \\ xy = \lambda \end{array} \right\} x = y = z$$

$x + y + z = 6 \Rightarrow x = y = z = 2$
$f(2, 2, 2) = 8$

15. Minimize $f(x, y, z) = x^2 + y^2 + z^2$.
Constraint: $x + y + z = 1$

$$\left. \begin{array}{r} 2x = \lambda \\ 2y = \lambda \\ 2z = \lambda \end{array} \right\} x = y = z$$

$x + y + z = 1 \Rightarrow x = y = z = \frac{1}{3}$
$f\left(\frac{1}{3}, \frac{1}{3}, \frac{1}{3}\right) = \frac{1}{3}$

16. Minimize $f(x, y, z) = 2x + y + 2z$.
Constraint: $x^2 + y^2 + z^2 = 4$

$$\left. \begin{array}{l} 2 = 2x\lambda \Rightarrow x = \dfrac{1}{\lambda} \\ 1 = 2y\lambda \Rightarrow y = \dfrac{1}{2\lambda} \\ 2 = 2z\lambda \ \ \Rightarrow z = \dfrac{1}{\lambda} \end{array} \right\} y = \frac{x}{2} \text{ and } z = x$$

$x^2 + y^2 + z^2 = 4 \Rightarrow \dfrac{9x^2}{4} = 4$

$$x = \frac{4}{3}, \ y = \frac{2}{3}, \ z = \frac{4}{3}$$

$f\left(\dfrac{4}{3}, \dfrac{2}{3}, \dfrac{4}{3}\right) = 6$

17. Maximize $f(x, y, z) = xyz$.
Constraints: $x + y + z = 32$
$\qquad\qquad\quad x - y + z = 0$

$\nabla f = \lambda \nabla g + \mu \nabla h$

$yz\mathbf{i} + xz\mathbf{j} + xy\mathbf{k} = \lambda(\mathbf{i} + \mathbf{j} + \mathbf{k}) + \mu(\mathbf{i} - \mathbf{j} + \mathbf{k})$

$$\left. \begin{array}{r} yz = \lambda + \mu \\ xz = \lambda - \mu \\ xy = \lambda + \mu \end{array} \right\} yz = xy \Rightarrow x = z$$

$$\left. \begin{array}{r} x + y + z = 32 \\ x - y + z = 0 \end{array} \right\} 2x + 2z = 32 \Rightarrow x = z = 8, \ y = 16$$

$f(8, 16, 8) = 1024$

18. Minimize $f(x, y, z) = x^2 + y^2 + z^2$.

Constraints: $x + 2z = 4$

$\qquad\qquad x + y = 8$

$\nabla f = \lambda \nabla g + \mu \nabla h$

$2x\mathbf{i} + 2y\mathbf{j} + 2z\mathbf{k} = \lambda(\mathbf{i} + 2\mathbf{k}) + \mu(\mathbf{i} + \mathbf{j})$

$\left.\begin{array}{l} 2x = \lambda + \mu \\[2mm] 2y = \mu \\[2mm] 2z = 2\lambda \end{array}\right\} \; 2x = 2y + z$

$x + 2z = 4 \Rightarrow z = 2 - \dfrac{x}{2}$

$x + y = 8 \Rightarrow y = 8 - x$

$\qquad\qquad 2x = 2(8 - x) + 2 - \dfrac{x}{2}$

$\qquad\qquad x = 4, \; y = 4, \; z = 0$

$f(4, 4, 0) = 32$

19. Maximize $f(x, y, z) = xy + yz$.

Constraints: $x + 2y = 6$

$\qquad\qquad x - 3z = 0$

$\nabla f = \lambda \nabla g + \mu \nabla h$

$y\mathbf{i} + (x + z)\mathbf{j} + y\mathbf{k} = \lambda(\mathbf{i} + 2\mathbf{j}) + \mu(\mathbf{i} - 3\mathbf{k})$

$\left.\begin{array}{l} y = \lambda + \mu \\[2mm] x + z = 2\lambda \\[2mm] y = -3\mu \end{array}\right\} \; y = \dfrac{3}{4}\lambda \Rightarrow x + z = \dfrac{8}{3}y$

$x + 2y = 6 \Rightarrow y = 3 - \dfrac{x}{2}$

$x - 3z = 0 \Rightarrow z = \dfrac{x}{3}$

$\qquad\qquad x + \dfrac{x}{3} = \dfrac{8}{3}\left(3 - \dfrac{x}{2}\right)$

$\qquad\qquad x = 3, \; y = \dfrac{3}{2}, \; z = 1$

$f\left(3, \dfrac{3}{2}, 1\right) = 6$

20. Maximize $f(x, y, z) = xyz$.

Constraints: $x^2 + z^2 = 5$

$\qquad\qquad x - 2y = 0$

$\nabla f = \lambda \nabla g + \mu \nabla h$

$yz\mathbf{i} + xz\mathbf{j} + xy\mathbf{k} = \lambda(2x\mathbf{i} + 2z\mathbf{k}) + \mu(\mathbf{i} - 2\mathbf{j})$

$yz = 2x\lambda + \mu$

$xz = -2\mu \qquad \Rightarrow \mu = -\dfrac{xz}{2}$

$xy = 2z\lambda \qquad \Rightarrow \lambda = \dfrac{xy}{2z}$

$x^2 + z^2 = 5 \qquad \Rightarrow z = \sqrt{5 - x^2}$

$x - 2y = 0 \qquad \Rightarrow y = \dfrac{x}{2}$

$\qquad\qquad yz = 2x\left(\dfrac{xy}{2z}\right) - \dfrac{xz}{2}$

$\dfrac{x\sqrt{5 - x^2}}{2} = \dfrac{x^3}{2\sqrt{5 - x^2}} - \dfrac{x\sqrt{5 - x^2}}{2}$

$x\sqrt{5 - x^2} = \dfrac{x^3}{2\sqrt{5 - x^2}}$

$2x(5 - x^2) = x^3$

$\qquad\qquad 0 = 3x^3 - 10x = x(3x^2 - 10)$

$\qquad\qquad x = 0 \text{ or } x = \sqrt{\dfrac{10}{3}}, \; y = \dfrac{1}{2}\sqrt{\dfrac{10}{3}}, \; z = \sqrt{\dfrac{5}{3}}$

$f\left(\sqrt{\dfrac{10}{3}}, \dfrac{1}{2}\sqrt{\dfrac{10}{3}}, \sqrt{\dfrac{5}{3}}\right) = \dfrac{5\sqrt{15}}{9}$

Note: $f(0, 0, \sqrt{5}) = 0$ does not yield a **maximum**.

21. Maximize or minimize $f(x, y) = x^2 + 3xy + y^2$.

Constraint: $x^2 + y^2 \leq 1$

Case 1: On the circle $x^2 + y^2 = 1$

$$\left. \begin{array}{l} 2x + 3y = 2x\lambda \\ 3x + 2y = 2y\lambda \end{array} \right\} x^2 = y^2$$

$$x^2 + y^2 = 1 \Rightarrow x = \pm\frac{\sqrt{2}}{2}, \ y = \pm\frac{\sqrt{2}}{2}$$

Maxima: $f\left(\pm\frac{\sqrt{2}}{2}, \pm\frac{\sqrt{2}}{2}\right) = \frac{5}{2}$

Minima: $f\left(\pm\frac{\sqrt{2}}{2}, \mp\frac{\sqrt{2}}{2}\right) = -\frac{1}{2}$

Case 2: Inside the circle

$$\left. \begin{array}{l} f_x = 2x + 3y = 0 \\ f_y = 3x + 2y = 0 \end{array} \right\} x = y = 0$$

$f_{xx} = 2, \ f_{yy} = 2, \ f_{xy} = 3, \ f_{xx}f_{yy} - (f_{xy})^2 < 0$

Saddle point: $f(0, 0) = 0$

By combining these two cases, we have a maximum of 5/2 at $\left(\pm\sqrt{2}/2, \pm\sqrt{2}/2\right)$ and a minimum of $-1/2$ at $\left(\pm\sqrt{2}/2, \mp\sqrt{2}/2\right)$.

22. Maximize or minimize $f(x, y) = e^{-xy}$.

Constraint: $x^2 + y^2 \leq 1$

Case 1: On the circle $x^2 + y^2 = 1$

$$\left. \begin{array}{l} -ye^{-xy} = 2x\lambda \\ -xe^{-xy} = 2y\lambda \end{array} \right\} x^2 = y^2$$

$$x^2 + y^2 = 1 \Rightarrow x = \pm\frac{\sqrt{2}}{2}$$

Maxima: $f\left(\pm\frac{\sqrt{2}}{2}, \mp\frac{\sqrt{2}}{2}\right) = e^{1/2} \approx 1.649$

Minima: $f\left(\pm\frac{\sqrt{2}}{2}, \pm\frac{\sqrt{2}}{2}\right) = e^{-1/2} \approx 0.607$

Case 2: Inside the circle

$$\left. \begin{array}{l} f_x = -ye^{-xy} = 0 \\ f_y = -xe^{-xy} = 0 \end{array} \right\} x = y = 0$$

$f_{xx} = y^2 e^{-xy}, \ f_{yy} = x^2 e^{xy}, \ f_{xy} = e^{-xy}(xy - 1)$

At $(0, 0)$, $f_{xx}f_{yy} - (f_{xy})^2 < 0$.

Saddle point: $f(0, 0) = 1$

By combining these two cases, we have a maximum of $e^{1/2}$ at $\left(\pm\sqrt{2}/2, \mp\sqrt{2}/2\right)$ and a minimum of $e^{-1/2}$ at $\left(\pm\sqrt{2}/2, \pm\sqrt{2}/2\right)$.

23. Minimize the square of the distance $f(x, \ y) = x^2 + y^2$ subject to the constraint $2x + 3y = -1$.

$$\left.\begin{array}{r} 2x = 2\lambda \\ 2y = 3\lambda \end{array}\right\} y = \frac{3x}{2}$$

$$2x + 3y = -1 \Rightarrow x = -\frac{2}{13}, \ y = -\frac{3}{13}$$

The point on the line is $\left(-\frac{2}{13}, \ -\frac{3}{13}\right)$ and the desired distance is

$$d = \sqrt{\left(-\frac{2}{13}\right)^2 + \left(-\frac{3}{13}\right)^2} = \frac{\sqrt{13}}{13}.$$

24. Minimize the square of the distance $f(x, \ y) = x^2 + (y - 10)^2$ subject to the constraint $(x - 4)^2 + y^2 = 4$.

$$\left.\begin{array}{r} 2x = 2(x - 4)\lambda \\ 2(y - 10) = 2y\lambda \end{array}\right\} \frac{x}{x - 4} = \frac{y - 10}{y} \Rightarrow y = -\frac{5}{2}x + 10$$

$$(x - 4)^2 + y^2 = 4 \Rightarrow (x^2 - 8x + 16) + \left(\frac{25}{4}x^2 - 50x + 100\right) = 4$$

$$\frac{29}{4}x^2 - 58x + 112 = 0$$

Using a graphing utility, you obtain $x \approx 3.2572$ and $x \approx 4.7428$ or, by the Quadratic Formula,

$$x = \frac{58 \pm \sqrt{58^2 - 4(29/4)(112)}}{2(29/4)} = \frac{58 \pm 2\sqrt{29}}{29/2} = 4 \pm \frac{4\sqrt{29}}{29}.$$

Using the smaller value, we have

$$x = 4\left(1 - \frac{\sqrt{29}}{29}\right) \quad \text{and} \quad y = \frac{10\sqrt{29}}{29} \approx 1.8570.$$

The point on the circle is

$$\left(4\left(1 - \frac{\sqrt{29}}{29}\right), \ \frac{10\sqrt{29}}{29}\right)$$

and the desired distance is

$$d = \sqrt{16\left(1 - \frac{\sqrt{29}}{29}\right)^2 + \left(\frac{10\sqrt{29}}{29} - 10\right)^2} \approx 8.77.$$

The larger x-value does not yield a **minimum**.

25. Minimize the square of the distance $f(x, \ y, \ z) = (x - 2)^2 + (y - 1)^2 + (z - 1)^2$ subject to the constraint $x + y + z = 1$.

$$\left.\begin{array}{r} 2(x - 2) = \lambda \\ 2(y - 1) = \lambda \\ 2(z - 1) = \lambda \end{array}\right\} y = z \text{ and } y = x - 1$$

$$x + y + z = 1 \Rightarrow x + 2(x - 1) = 1$$

$$x = 1, \ y = z = 0$$

The point on the plane is $(1, 0, 0)$ and the desired distance is $d = \sqrt{(1 - 2)^2 + (0 - 1)^2 + (0 - 1)^2} = \sqrt{3}$.

26. Minimize the square of the distance $f(x, y, z) = (x - 4)^2 + y^2 + z^2$ subject to the constraint $\sqrt{x^2 + y^2} - z = 0$.

$$\left. \begin{array}{l} 2(x - 4) = \dfrac{x}{\sqrt{x^2 + y^2}} \lambda = \dfrac{x}{z} \lambda \\[3mm] 2y = \dfrac{y}{\sqrt{x^2 + y^2}} \lambda = \dfrac{y}{z} \lambda \\[3mm] 2z = -\lambda \end{array} \right\} \quad \begin{array}{l} 2(x - 4) = -2x \\[3mm] 2y = -2y \end{array}$$

$\sqrt{x^2 + y^2} - z = 0, \ x = 2, \ y = 0, \ z = 2$

The point on the cone is $(2, 0, 2)$ and the desired distance is $d = \sqrt{(2 - 4)^2 + 0^2 + 2^2} = 2\sqrt{2}$.

27. Maximize $f(x, y, z) = z$ subject to the constraints $x^2 + y^2 + z^2 = 36$ and $2x + y - z = 2$.

$$\left. \begin{array}{l} 0 = 2x\lambda + 2\mu \\[2mm] 0 = 2y\lambda + \mu \\[2mm] 1 = 2z\lambda - \mu \end{array} \right\} x = 2y$$

$x^2 + y^2 + z^2 = 36$

$2x + y - z = 2 \Rightarrow z = 2x + y - 2 = 5y - 2$

$$(2y)^2 + y^2 + (5y - 2)^2 = 36$$

$$30y^2 - 20y - 32 = 0$$

$$15y^2 - 10y - 16 = 0$$

$$y = \frac{5 \pm \sqrt{265}}{15}$$

Choosing the positive value for y we have the point

$$\left(\frac{10 + 2\sqrt{265}}{15}, \ \frac{5 + \sqrt{265}}{15}, \ \frac{-1 + \sqrt{265}}{3} \right).$$

28. Maximize $f(x, y, z) = z$ subject to the constraints $x^2 + y^2 - z^2 = 0$ and $x + 2z = 4$.

$$0 = 2x\lambda + \mu$$

$$0 = 2y\lambda \Rightarrow y = 0$$

$$1 = -2z\lambda + 2\mu$$

$$x^2 + y^2 - z^2 = 0$$

$$x + 2z = 4 \Rightarrow \qquad x = 4 - 2z$$

$$(4 - 2z)^2 + 0^2 - z^2 = 0$$

$$3z^2 - 16z + 16 = 0$$

$$(3z - 4)(z - 4) = 0$$

$$z = \tfrac{4}{3} \text{ or } z = 4$$

The maximum value of f occurs when $z = 4$ at the point $(-4, 0, 4)$.

29. Maximize $V(x, y, z) = xyz$ subject to the constraint $x + 2y + 2z = 108$.

$$\left. \begin{array}{l} yz = \lambda \\ xz = 2\lambda \\ xy = 2\lambda \end{array} \right\} \; y = z \text{ and } x = 2y$$

$x + 2y + 2z = 108 \Rightarrow 6y = 108, \; y = 18$

$$x = 36, \; y = z = 18$$

Volume is maximum when the dimensions are $36 \times 18 \times 18$ inches.

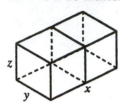

30. Maximize $V(x, y, z) = xyz$ subject to the constraint $1.5xy + 2xz + 2yz = C$.

$$\left. \begin{array}{l} yz = (1.5y + 2z)\lambda \\ xz = (1.5x + 2z)\lambda \\ xy = (2x + 2y)\lambda \end{array} \right\} \; x = y \text{ and } z = \frac{3}{4}x$$

$1.5xy + 2xz + 2yz = C \Rightarrow 1.5x^2 + \frac{3}{2}x^2 + \frac{3}{2}x^2 = C$

$$x = \frac{\sqrt{2C}}{3}$$

Volume is maximum when $x = y = \sqrt{2C}/3$ and $z = \sqrt{2C}/4$.

31. Minimize $C(x, y, z) = 5xy + 3(2xz + 2yz + xy)$ subject to the constraint $xyz = 480$.

$$\left. \begin{array}{l} 8y + 6z = yz\lambda \\ 8x + 6z = xz\lambda \\ 6x + 6y = xy\lambda \end{array} \right\} \; x = y, \; 4y = 3z$$

$xyz = 480 \Rightarrow \frac{4}{3}y^3 = 480$

$$x = y = \sqrt[3]{360}, \; z = \frac{4}{3}\sqrt[3]{360}$$

Dimensions: $\sqrt[3]{360} \times \sqrt[3]{360} \times \frac{4}{3}\sqrt[3]{360}$ feet

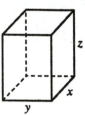

32. Minimize $A(\pi, r) = 2\pi rh + 2\pi r^2$ subject to the constraint $\pi r^2 h = V_0$.

$$\left. \begin{array}{l} 2\pi h + 4\pi r = 2\pi rh\lambda \\ 2\pi r = \pi r^2\lambda \end{array} \right\} \; h = 2r$$

$\pi r^2 h = V_0 \Rightarrow 2\pi r^3 = V_0$

Dimensions: $r = \sqrt[3]{V_0/2\pi}$ and $h = 2\sqrt[3]{V_0/2\pi}$

33. Maximize $V(x, y, z) = (2x)(2y)(2z) = 8xyz$ subject to the constraint $\dfrac{x^2}{a^2} + \dfrac{y^2}{b^2} + \dfrac{z^2}{c^2} = 1$.

$$\left. \begin{array}{l} 8yz = \dfrac{2x}{a^2}\lambda \\[2mm] 8xz = \dfrac{2y}{b^2}\lambda \\[2mm] 8xy = \dfrac{2z}{c^2}\lambda \end{array} \right\} \; \dfrac{x^2}{a^2} = \dfrac{y^2}{b^2} = \dfrac{z^2}{c^2}$$

$\dfrac{x^2}{a^2} + \dfrac{y^2}{b^2} + \dfrac{z^2}{c^2} = 1 \Rightarrow \dfrac{3x^2}{a^2} = 1, \; \dfrac{3y^2}{b^2} = 1, \; \dfrac{3z^2}{c^2} = 1$

$$x = \frac{a}{\sqrt{3}}, \; y = \frac{b}{\sqrt{3}}, \; z = \frac{c}{\sqrt{3}}$$

Therefore, the dimensions of the box are $\dfrac{2\sqrt{3}a}{3} \times \dfrac{2\sqrt{3}b}{3} \times \dfrac{2\sqrt{3}c}{3}$.

34. a. Maximize $P(x, y, z) = xyz$ subject to the constraint $x + y + z = S$.

$$\left.\begin{array}{l} yz = \lambda \\ xz = \lambda \\ xy = \lambda \end{array}\right\} x = y = z$$

$$x + y + z = S \Rightarrow x = y = z = \frac{S}{3}$$

Therefore,

$$xyz \le \left(\frac{S}{3}\right)\left(\frac{S}{3}\right)\left(\frac{S}{3}\right), \quad x, y, z > 0$$

$$xyz \le \frac{S^3}{27}$$

$$\sqrt[3]{xyz} \le \frac{S}{3}$$

$$\sqrt[3]{xyz} \le \frac{x + y + z}{3}.$$

b. Maximize $P = x_1 x_2 x_3 \ldots x_n$ subject to the constraint $\sum\limits_{i=1}^{n} x_i = S$.

$$\left.\begin{array}{l} x_2 x_3 \ldots x_n = \lambda \\ x_1 x_3 \ldots x_n = \lambda \\ x_1 x_2 \ldots x_n = \lambda \\ \quad\vdots \\ x_1 x_2 x_3 \ldots x_{n-1} = \lambda \end{array}\right\} x_1 = x_2 = x_3 = \cdots = x_n$$

$$\sum_{i=1}^{n} x_i = S \Rightarrow x_1 = x_2 = x_3 = \cdots = x_n = \frac{S}{n}$$

Therefore,

$$x_1 x_2 x_3 \ldots x_n \le \left(\frac{S}{n}\right)\left(\frac{S}{n}\right)\left(\frac{S}{n}\right)\cdots\left(\frac{S}{n}\right), \quad x_i \ge 0$$

$$x_1 x_2 x_3 \ldots x_n \le \left(\frac{S}{n}\right)^n$$

$$\sqrt[n]{x_1 x_2 x_3 \ldots x_n} \le \frac{S}{n}$$

$$\sqrt[n]{x_1 x_2 x_3 \ldots x_n} \le \frac{x_1 + x_2 + x_3 + \cdots + x_n}{n}.$$

35. Using the formula Time = Distance/Rate, minimize

$$T(x, \ y) = \frac{\sqrt{d_1^2 + x^2}}{v_1} + \frac{\sqrt{d_2^2 + y^2}}{v_2}$$

subject to the constraint $x + y = a$.

$$\left.\begin{array}{c} \dfrac{x}{v_1\sqrt{d_1^2 + x^2}} = \lambda \\[4mm] \dfrac{y}{v_2\sqrt{d_2^2 + x^2}} = \lambda \end{array}\right\} \quad \dfrac{x}{v_1\sqrt{d_1^2 + x^2}} = \dfrac{y}{v_2\sqrt{d_2^2 + x^2}}$$

$x + y = a$

Since

$$\sin \theta_1 = \frac{x}{\sqrt{d_1^2 + x^2}} \quad \text{and} \quad \sin \theta_2 = \frac{y}{\sqrt{d_2^2 + y^2}}$$

we have

$$\frac{x/\sqrt{d_1^2 + x^2}}{v_1} = \frac{y/\sqrt{d_2^2 + y^2}}{v_2} \quad \text{or} \quad \frac{\sin \theta_1}{v_1} = \frac{\sin \theta_2}{v_2}.$$

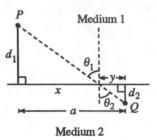

36. *Case 1:* Minimize $P(l, \ h) = 2h + l + (\pi l/2)$ subject to the constraint $lh + (\pi l^2/8) = A$.

$$1 + \frac{\pi}{2} = \left(h + \frac{\pi l}{4}\right)\lambda$$

$$2 = l\lambda \Rightarrow \lambda = \frac{2}{l}, \quad 1 + \frac{\pi}{2} = \frac{2h}{l} + \frac{\pi}{2}$$

$$l = 2h$$

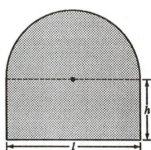

Case 2: Maximize $A(l, \ h) = lh + (\pi l^2/8)$ subject to the constraint $2h + l + (\pi l/2) = P$.

$$h + \frac{\pi l}{4} = \left(1 + \frac{\pi}{2}\right)\lambda$$

$$l = 2\lambda \Rightarrow \lambda = \frac{l}{2}, \quad h + \frac{\pi l}{4} = \frac{l}{2} + \frac{\pi l}{4}$$

$$h = \frac{l}{2} \text{ or } l = 2h$$

37. a. Maximize $g(\alpha, \ \beta, \ \gamma) = \cos \alpha \cos \beta \cos \gamma$ subject to the constraint $\alpha + \beta + \gamma = \pi$.

$$\left.\begin{array}{c} -\sin \alpha \cos \beta \cos \gamma = \lambda \\ -\cos \alpha \sin \beta \cos \gamma = \lambda \\ -\cos \alpha \cos \beta \sin \gamma = \lambda \end{array}\right\} \tan \alpha = \tan \beta = \tan \gamma \Rightarrow \alpha = \beta = \gamma$$

$$\alpha + \beta + \gamma = \pi \Rightarrow \alpha = \beta = \gamma = \frac{\pi}{3}$$

$$g\left(\frac{\pi}{3}, \ \frac{\pi}{3}, \ \frac{\pi}{3}\right) = \frac{1}{8}$$

b. $\alpha + \beta + \gamma = \pi \Rightarrow \gamma = \pi - (\alpha + \beta)$

$$g(\alpha, \ \beta) = \cos \alpha \cos \beta \cos(\pi - (\alpha + \beta))$$

$$= \cos \alpha \cos \beta [\cos \pi \cos(\alpha + \beta) + \sin \pi \sin(\alpha + \beta)]$$

$$= -\cos \alpha \cos \beta \cos(\alpha + \beta)$$

38. Maximize $T(x, y, z) = 100 + x^2 + y^2$ subject to the constraints $x^2 + y^2 + z^2 = 50$ and $x - z = 0$.

$$\left. \begin{array}{l} 2x = 2x\lambda + \mu \\ 2y = 2y\lambda \\ 0 = 2z\lambda - \mu \end{array} \right\} \lambda = 1, \ \mu = 0, \ z = 0$$

$$x^2 + y^2 + z^2 = 50$$

$$x - z = 0 \Rightarrow x = z = 0, \ y = \sqrt{50}$$

Therefore, $x = 0, \ y = \sqrt{50}, \ z = 0$, and $T(0, \ \sqrt{50}, \ 0) = 150$.

39. Maximize $P(x, y) = 100x^{0.25}y^{0.75}$ subject to the constraint $48x + 36y = 100,000$.

$$25x^{-0.75}y^{0.75} = 48\lambda \Rightarrow \left(\frac{y}{x}\right)^{0.75} = \frac{48\lambda}{25}$$

$$75x^{0.25}y^{-0.25} = 36\lambda \Rightarrow \left(\frac{x}{y}\right)^{0.25} = \frac{36\lambda}{75}$$

$$\left(\frac{y}{x}\right)^{0.75} \left(\frac{y}{x}\right)^{0.25} = \left(\frac{48\lambda}{25}\right)\left(\frac{75\lambda}{36}\right)$$

$$\frac{y}{x} = 4$$

$$y = 4x$$

$$48x + 36y = 100,000 \Rightarrow 192x = 100,000$$

$$x = \frac{3125}{6}, \ y = \frac{6250}{3}$$

Therefore, $P\left(\frac{3125}{6}, \ \frac{6250}{3}\right) \approx 147,314$.

40. Maximize $P(x, y) = 100x^{0.6}y^{0.4}$ subject to the constraint $48x + 36y = 100,000$.

$$60x^{-0.4}y^{0.4} = 48\lambda \Rightarrow \left(\frac{y}{x}\right)^{0.4} = \frac{48\lambda}{60}$$

$$40x^{0.6}y^{-0.6} = 36\lambda \Rightarrow \left(\frac{x}{y}\right)^{0.6} = \frac{36\lambda}{40}$$

$$\left(\frac{y}{x}\right)^{0.4} \left(\frac{y}{x}\right)^{0.6} = \left(\frac{48\lambda}{60}\right)\left(\frac{40}{36\lambda}\right)$$

$$\frac{y}{x} = \frac{8}{9} \Rightarrow y = \frac{8}{9}x$$

$$48x + 36y = 100,000 \Rightarrow 80x = 100,000$$

$$x = 1250, \ y = \frac{10,000}{9}$$

Therefore, $P\left(1250, \ \frac{10,000}{9}\right) \approx 119,247$.

41. Minimize $C(x,\ y) = 48x + 36y$ subject to the constraint $100x^{0.25}y^{0.75} = 20{,}000$.

$$48 = 25x^{-0.75}y^{0.75}\lambda \Rightarrow \left(\frac{y}{x}\right)^{0.75} = \frac{48}{25\lambda}$$

$$36 = 75x^{0.25}y^{-0.25}\lambda \Rightarrow \left(\frac{x}{y}\right)^{0.25} = \frac{36}{75\lambda}$$

$$\left(\frac{y}{x}\right)^{0.75}\left(\frac{y}{x}\right)^{0.25} = \left(\frac{48}{25\lambda}\right)\left(\frac{75\lambda}{36}\right)$$

$$\frac{y}{x} = 4 \Rightarrow y = 4x$$

$$100x^{0.25}y^{0.75} = 20{,}000 \Rightarrow x^{0.25}(4x)^{0.75} = 200$$

$$x = \frac{200}{4^{0.75}} = \frac{200}{2\sqrt{2}} = 50\sqrt{2}$$

$$y = 4x = 200\sqrt{2}$$

Therefore, $C\left(50\sqrt{2},\ 200\sqrt{2}\right) \approx \$13{,}576.45$.

42. Minimize $C(x,\ y) = 48x + 36y$ subject to the constraint $100x^{0.6}y^{0.4} = 20{,}000$.

$$48 = 60x^{-0.4}y^{0.4}\lambda \Rightarrow \left(\frac{y}{x}\right)^{0.4} = \frac{48}{60\lambda}$$

$$36 = 40x^{0.6}y^{-0.6}\lambda \Rightarrow \left(\frac{x}{y}\right)^{0.6} = \frac{36}{40\lambda}$$

$$\left(\frac{y}{x}\right)^{0.4}\left(\frac{y}{x}\right)^{0.6} = \left(\frac{48}{60\lambda}\right)\left(\frac{40\lambda}{36}\right)$$

$$\frac{y}{x} = \frac{8}{9} \Rightarrow y = \frac{8}{9}x$$

$$100x^{0.6}y^{0.4} = 20{,}000 \Rightarrow x^{0.6}\left(\frac{8}{9}x\right)^{0.4} = 200$$

$$x = \frac{200}{(8/9)^{0.4}} \approx 209.65$$

$$y = \frac{8}{9}\left[\frac{200}{(8/9)^{0.4}}\right] \approx 186.35$$

Therefore $C(209.65,\ 186.35) = \$16{,}771.94$.

43. a. Maximize $\displaystyle\sum_{i=1}^{n} x_i y_i$ subject to $\displaystyle\sum_{i=1}^{n} x_i{}^2 = 1$ and $\displaystyle\sum_{i=1}^{n} y_i{}^2 = 1$. Let

$$f(x_1, \ldots, x_n, y_1, \ldots, y_n) = \sum_{i=1}^{n} x_i y_i$$

$$g(x_1, \ldots, x_n, y_1, \ldots, y_n) = \sum_{i=1}^{n} x_i{}^2 = 1$$

$$h(x_1, \ldots, x_n, y_1, \ldots, y_n) = \sum_{i=1}^{n} y_i{}^2 = 1.$$

$\nabla f = \lambda \nabla g + \mu \nabla h \Rightarrow$

$$
\begin{array}{lll}
y_1 = 2x_1\lambda \Rightarrow & y_1{}^2 = 4x_1{}^2\lambda^2 \\
y_2 = 2x_2\lambda \Rightarrow & y_2{}^2 = 4x_2{}^2\lambda^2 \\
\vdots & \vdots \\
y_n = 2x_n\lambda \Rightarrow & \underline{y_n{}^2 = 4x_n{}^2\lambda^2} \\
& \displaystyle\sum_{i=1}^{n} y_i{}^2 = 4\lambda^2 \sum_{i=1}^{n} x_i{}^2 \\
& 1 = 4\lambda^2 \\
& \lambda = \dfrac{1}{2}
\end{array}
$$

and

$$
\begin{array}{lll}
x_1 = 2y_1\mu \Rightarrow & x_1{}^2 = 4y_1{}^2\mu^2 \\
x_2 = 2y_2\mu \Rightarrow & x_2{}^2 = 4y_2{}^2\mu^2 \\
\vdots & \vdots \\
x_n = 2y_n\mu \Rightarrow & \underline{x_n{}^2 = 4y_n{}^2\mu^2} \\
& \displaystyle\sum_{i=1}^{n} x_i{}^2 = 4\mu^2 \sum_{i=1}^{n} y_i{}^2 \\
& 1 = 4\mu^2 \\
& \mu = \dfrac{1}{2}
\end{array}
$$

Thus, $\displaystyle\sum_{i=1}^{n} x_i y_i$ is maximum when $x_i = y_i$. Therefore, $\displaystyle\sum_{i=1}^{n} x_i y_i = \sum_{i=1}^{n} x_i{}^2 = 1$.

b. Let $x_i = \dfrac{a_i}{\sqrt{\displaystyle\sum_{i=1}^{n} a_i{}^2}}$ and $y_i = \dfrac{b_i}{\sqrt{\displaystyle\sum_{i=1}^{n} b_i{}^2}}$.

$$x_i y_i = \dfrac{a_i b_i}{\sqrt{\displaystyle\sum_{i=1}^{n} a_i{}^2}\sqrt{\displaystyle\sum_{i=1}^{n} b_i{}^2}}$$

$$\sum_{i=1}^{n} x_i y_i = \sum_{i=1}^{n} \dfrac{a_i b_i}{\sqrt{\displaystyle\sum_{i=1}^{n} a_i{}^2}\sqrt{\displaystyle\sum_{i=1}^{n} b_i{}^2}} \leq 1, \text{ from a.}$$

Thus, $\displaystyle\sum_{i=1}^{n} a_i b_i \leq \left(\sum_{i=1}^{n} a_i{}^2\right)^{1/2} \left(\sum_{i=1}^{n} b_i{}^2\right)^{1/2}$.

Chapter 13   Review Exercises

1. $f(x, \ y) = e^{x^2 + y^2}$

The level curves are of the form

$$c = e^{x^2 + y^2}$$

$$\ln c = x^2 + y^2.$$

The level curves are circles centered at the origin.

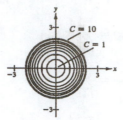

2. $f(x, \ y) = \ln xy$

The level curves are of the form

$$c = \ln xy$$

$$e^c = xy.$$

The level curves are hyperbolas.

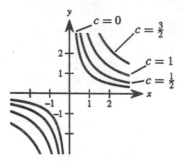

3. $f(x, \ y) = x^2 - y^2$

The level curves are of the form

$$c = x^2 - y^2$$

$$1 = \frac{x^2}{c} - \frac{y^2}{c}.$$

The level curves are hyperbolas.

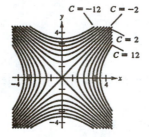

4. $f(x, \ y) = \dfrac{x}{x + y}$

The level curves are of the form

$$c = \frac{x}{x + y}$$

$$y = \left(\frac{1 - c}{c} \right) x.$$

The level curves are lines passing through the origin with slope $(1 - c)/c$.

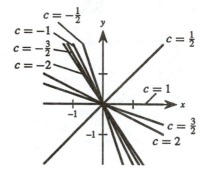

5. $f(x, \ y) = e^{-(x^2 + y^2)}$

6. $g(x, \ y) = |y|^{1 + |x|}$

7. $\displaystyle\lim_{(x,y)\to(1,1)} \frac{xy}{x^2+y^2} = \frac{1}{2}$

Continuous except at (0, 0).

8. $\displaystyle\lim_{(x,y)\to(1,1)} \frac{xy}{x^2-y^2}$

Does not exist

Continuous except when $y = \pm x$.

9. $\displaystyle\lim_{(x,y)\to(0,0)} \frac{-4x^2 y}{x^4+y^2} = -4\lim_{r\to 0} \frac{r^3 \cos^2\theta \sin\theta}{r^4 \cos^4\theta + r^2 \sin^2\theta}$

$$= -4\lim_{r\to 0} \frac{r\cos^2\theta \sin\theta}{r^2 \cos^4\theta + \sin^2\theta} = -4\lim_{r\to 0} \frac{\cos^2\theta \sin\theta}{2r\cos^4\theta} = -4\lim_{r\to 0} \frac{\sin\theta}{2r\cos^2\theta} = \frac{-4\sin\theta}{0}$$

Does not exist

Continuous except at (0, 0).

10. $\displaystyle\lim_{(x,y)\to(0,0)} \frac{y+xe^{-y^2}}{1+x^2} = \frac{0+0}{1+0}$

$$= 0$$

Continuous everywhere

11. $f(x, y) = e^x \cos y$

$f_x = e^x \cos y$

$f_y = -e^x \sin y$

12. $f(x, y) = \dfrac{xy}{x+y}$

$f_x = \dfrac{y(x+y) - xy}{(x+y)^2}$

$\quad = \dfrac{y^2}{(x+y)^2}$

$f_y = \dfrac{x^2}{(x+y)^2}$

13. $z = xe^y + ye^x$

$\dfrac{\partial z}{\partial x} = e^y + ye^x$

$\dfrac{\partial z}{\partial y} = xe^y + e^x$

14. $z = \ln(x^2 + y^2 + 1)$

$\dfrac{\partial z}{\partial x} = \dfrac{2x}{x^2+y^2+1}$

$\dfrac{\partial z}{\partial y} = \dfrac{2y}{x^2+y^2+1}$

15. $g(x, y) = \dfrac{xy}{x^2+y^2}$

$g_x = \dfrac{y(x^2+y^2) - xy(2x)}{(x^2+y^2)^2} = \dfrac{y(y^2-x^2)}{(x^2+y^2)^2}$

$g_y = \dfrac{x(x^2-y^2)}{(x^2+y^2)^2}$

16. $w = \sqrt{x^2+y^2+z^2}$

$\dfrac{\partial w}{\partial x} = \dfrac{1}{2}(x^2+y^2+z^2)^{-1/2}(2x) = \dfrac{x}{\sqrt{x^2+y^2+z^2}}$

$\dfrac{\partial w}{\partial y} = \dfrac{y}{\sqrt{x^2+y^2+z^2}}$

$\dfrac{\partial w}{\partial z} = \dfrac{z}{\sqrt{x^2+y^2+z^2}}$

17. $f(x, y, z) = z \arctan \dfrac{y}{x}$

$f_x = \dfrac{z}{1+(y^2/x^2)}\left(-\dfrac{y}{x^2}\right) = \dfrac{-yz}{x^2+y^2}$

$f_y = \dfrac{z}{1+(y^2/x^2)}\left(\dfrac{1}{x}\right) = \dfrac{xz}{x^2+y^2}$

$f_z = \arctan \dfrac{y}{x}$

18. $f(x, y, z) = \dfrac{1}{\sqrt{1-x^2-y^2-z^2}}$

$f_x = -\dfrac{1}{2}(1-x^2-y^2-z^2)^{-3/2}(-2x)$

$\quad = \dfrac{x}{(1-x^2-y^2-z^2)^{3/2}}$

$f_y = \dfrac{y}{(1-x^2-y^2-z^2)^{3/2}}$

$f_z = \dfrac{z}{(1-x^2-y^2-z^2)^{3/2}}$

19. $u(x, t) = ce^{-n^2 t} \sin(nx)$

$$\frac{\partial u}{\partial x} = cne^{-n^2 t} \cos(nx)$$

$$\frac{\partial u}{\partial t} = -cn^2 e^{-n^2 t} \sin(nx)$$

20. $u(x, t) = c(\sin akx) \cos kt$

$$\frac{\partial u}{\partial x} = akc(\cos akx) \cos kt$$

$$\frac{\partial u}{\partial t} = -kc(\sin akx) \sin kt$$

21.
$$x^2 y - 2xyz - xz - z^2 = 0$$

$$2xy - 2xy\frac{\partial z}{\partial x} - 2yz - x\frac{\partial z}{\partial x} - z - 2z\frac{\partial z}{\partial x} = 0$$

$$\frac{\partial z}{\partial x} = -\frac{2yz + z - 2xy}{2xy + x + 2z}$$

$$x^2 - 2xy\frac{\partial z}{\partial y} - 2xz - x\frac{\partial z}{\partial y} - 2z\frac{\partial z}{\partial y} = 0$$

$$\frac{\partial z}{\partial y} = \frac{x^2 - 2xz}{2xy + x + 2z}$$

22.
$$xz^2 - y \sin z = 0$$

$$2xz\frac{\partial z}{\partial x} + z^2 - y \cos z\frac{\partial z}{\partial x} = 0$$

$$\frac{\partial z}{\partial x} = \frac{z^2}{y \cos z - 2xz}$$

$$2xz\frac{\partial z}{\partial y} - y \cos z\frac{\partial z}{\partial y} - \sin z = 0$$

$$\frac{\partial z}{\partial y} = \frac{\sin z}{2xz - y \cos z}$$

23. $f(x, y) = 3x^2 - xy + 2y^3$

$$f_x = 6x - y$$

$$f_y = -x + 6y^2$$

$$f_{xx} = 6$$

$$f_{yy} = 12y$$

$$f_{xy} = -1$$

$$f_{yx} = -1$$

24. $h(x, y) = \dfrac{x}{x + y}$

$$h_x = \frac{y}{(x + y)^2}$$

$$h_y = \frac{-x}{(x + y)^2}$$

$$h_{xx} = \frac{-2y}{(x + y)^3}$$

$$h_{yy} = \frac{2x}{(x + y)^3}$$

$$h_{xy} = \frac{(x + y)^2 - 2y(x + y)}{(x + y)^4} = \frac{x - y}{(x + y)^3}$$

$$h_{yx} = \frac{-(x + y)^2 + 2x(x + y)}{(x + y)^4} = \frac{x - y}{(x + y)^3}$$

25. $h(x, y) = x \sin y + y \cos x$

$$h_x = \sin y - y \sin x$$

$$h_y = x \cos y + \cos x$$

$$h_{xx} = -y \cos x$$

$$h_{yy} = -x \sin y$$

$$h_{xy} = \cos y - \sin x$$

$$h_{yx} = \cos y - \sin x$$

26. $g(x, y) = \cos(x - 2y)$

$$g_x = -\sin(x - 2y)$$

$$g_y = 2\sin(x - 2y)$$

$$g_{xx} = -\cos(x - 2y)$$

$$g_{yy} = -4\cos(x - 2y)$$

$$g_{xy} = 2\cos(x - 2y)$$

$$g_{yx} = 2\cos(x - 2y)$$

27. $z = x^2 - y^2$

$$\frac{\partial z}{\partial x} = 2x$$

$$\frac{\partial^2 z}{\partial x^2} = 2$$

$$\frac{\partial z}{\partial y} = -2y$$

$$\frac{\partial^2 z}{\partial y^2} = -2$$

Therefore, $\dfrac{\partial^2 z}{\partial x^2} + \dfrac{\partial^2 z}{\partial y^2} = 0$.

28. $z = x^3 - 3xy^2$

$$\frac{\partial z}{\partial x} = 3x^2 - 3y^2$$

$$\frac{\partial^2 z}{\partial x^2} = 6x$$

$$\frac{\partial z}{\partial y} = -6xy$$

$$\frac{\partial^2 z}{\partial y^2} = -6x$$

Therefore, $\dfrac{\partial^2 z}{\partial x^2} + \dfrac{\partial^2 z}{\partial y^2} = 0$.

29. $z = \dfrac{y}{x^2 + y^2}$

$$\frac{\partial z}{\partial x} = \frac{-2xy}{(x^2 + y^2)^2}$$

$$\frac{\partial^2 z}{\partial x^2} = -2y\left[\frac{-4x^2}{(x^2 + y^2)^3} + \frac{1}{(x^2 + y^2)^2}\right]$$

$$= 2y\frac{3x^2 - y^2}{(x^2 + y^2)^3}$$

$$\frac{\partial z}{\partial y} = \frac{(x^2 + y^2) - 2y}{(x^2 + y^2)^2} = \frac{x^2 - y^2}{(x^2 + y^2)^2}$$

$$\frac{\partial^2 z}{\partial y^2} = \frac{(x^2 + y^2)^2(-2y) - 2(x^2 - y^2)(x^2 + y^2)(2y)}{(x^2 + y^2)^4}$$

$$= -2y\frac{3x^2 - y^2}{(x^2 + y^2)^3}$$

Therefore, $\dfrac{\partial^2 z}{\partial x^2} + \dfrac{\partial^2 z}{\partial y^2} = 0$.

30. $z = e^x \sin y$

$$\frac{\partial z}{\partial x} = e^x \sin y$$

$$\frac{\partial^2 z}{\partial x^2} = e^x \sin y$$

$$\frac{\partial z}{\partial y} = e^x \cos y$$

$$\frac{\partial^2 z}{\partial y^2} = -e^x \sin y$$

Therefore, $\dfrac{\partial^2 z}{\partial x^2} + \dfrac{\partial^2 z}{\partial y^2} = 0$.

31. $z = x \sin\dfrac{y}{x}$

$$dz = \frac{\partial z}{\partial x} dx + \frac{\partial z}{\partial y} dy = \left(\sin\frac{y}{x} - \frac{y}{x}\cos\frac{y}{x}\right) dx + \left(\cos\frac{y}{x}\right) dy$$

32. $z = \dfrac{xy}{\sqrt{x^2 + y^2}}$

$$dz = \frac{\partial z}{\partial x} dx + \frac{\partial z}{\partial y} dy$$

$$= \left[\frac{\sqrt{x^2 + y^2}\, y - xy\left(x/\sqrt{x^2 + y^2}\right)}{x^2 + y^2}\right] dx + \left[\frac{\sqrt{x^2 + y^2}\, x - xy\left(y/\sqrt{x^2 + y^2}\right)}{x^2 + y^2}\right] dy$$

$$= \frac{y^3}{(x^2 + y^2)^{3/2}} dx + \frac{x^3}{(x^2 + y^2)^{3/2}} dy$$

33. $z^2 = x^2 + y^2$

$2z\,dz = 2x\,dx + 2y\,dy$

$$dz = \frac{x}{z}\,dx + \frac{y}{z}\,dy = \frac{5}{13}\left(\frac{1}{16}\right) + \frac{12}{13}\left(\frac{1}{16}\right) = \frac{17}{208} \approx 0.082 \text{ in}$$

Percentage error: $\dfrac{dz}{z} = \dfrac{17/208}{13} \approx 0.0063 = 0.63\%$

34. From the accompanying figure we observe

$$\tan\theta = \frac{h}{x} \text{ or } h = x\tan\theta$$

$$dh = \frac{\partial h}{\partial x}\,dx + \frac{\partial h}{\partial \theta}\,d\theta = \tan\theta\,dx + x\sec^2\theta\,d\theta.$$

Letting $x = 100$,

$$dx = \pm\frac{1}{2}, \quad \theta = \frac{11\pi}{60}, \text{ and } d\theta = \frac{\pi}{180}.$$

(Note that we express the measurement of the angle in radians.) The maximum error is approximately

$$dh = \tan\left(\frac{11\pi}{60}\right)\left(\pm\frac{1}{2}\right) + 100\sec^2\left(\frac{11\pi}{60}\right)\left(\frac{\pi}{180}\right) \approx \pm2.8 \text{ feet.}$$

35. $V = \frac{1}{3}\pi r^2 h$

$$dV = \frac{2}{3}\pi rh\,dr + \frac{1}{3}\pi r^2\,dh = \frac{2}{3}\pi(2)(5)\left(\pm\frac{1}{8}\right) + \frac{1}{3}\pi(2)^2\left(\pm\frac{1}{8}\right) = \pm\frac{5}{6}\pi \pm \frac{1}{6}\pi = \pm\pi \text{ in}^3$$

36. $A = \pi r\sqrt{r^2 + h^2}$

$$dA = \left(\pi\sqrt{r^2 + h^2} + \frac{\pi r^2}{\sqrt{r^2 + h^2}}\right)dr + \frac{\pi rh}{\sqrt{r^2 + h^2}}\,dh$$

$$= \frac{\pi(2r^2 + h^2)}{\sqrt{r^2 + h^2}}\,dr + \frac{\pi rh}{\sqrt{r^2 + h^2}}\,dh = \frac{\pi(8 + 25)}{\sqrt{29}}\left(\pm\frac{1}{8}\right) + \frac{10\pi}{\sqrt{29}}\left(\pm\frac{1}{8}\right) = \pm\frac{43\pi}{8\sqrt{29}}$$

37. $u = x^2 + y^2 + z^2, \quad x = r\cos t, \quad y = r\sin t, \quad z = t$

Chain Rule: $\dfrac{\partial u}{\partial r} = \dfrac{\partial u}{\partial x}\dfrac{\partial x}{\partial r} + \dfrac{\partial u}{\partial y}\dfrac{\partial y}{\partial r} + \dfrac{\partial u}{\partial z}\dfrac{\partial z}{\partial r}$

$$= 2x\cos t + 2y\sin t + 2z(0) = 2(r\cos^2 t + r\sin^2 t) = 2r$$

$$\frac{\partial u}{\partial t} = \frac{\partial u}{\partial x}\frac{\partial x}{\partial t} + \frac{\partial u}{\partial y}\frac{\partial y}{\partial t} + \frac{\partial u}{\partial z}\frac{\partial z}{\partial t}$$

$$= 2x(-r\sin t) + 2y(r\cos t) + 2z = 2(-r^2\sin t\cos t + r^2\sin t\cos t) + 2t = 2t$$

Substitution: $u(r,\ t) = r^2\cos^2 t + r^2\sin^2 t + t^2 = r^2 + t^2$

$$\frac{\partial u}{\partial r} = 2r$$

$$\frac{\partial u}{\partial t} = 2t$$

38. $u = y^2 - x, \quad x = \cos t, \quad y = \sin t$

Chain Rule: $\dfrac{du}{dt} = \dfrac{\partial u}{\partial x}\dfrac{\partial x}{\partial t} + \dfrac{\partial u}{\partial y}\dfrac{\partial y}{\partial t} = -1(-\sin t) + 2y(\cos t) = \sin t + 2(\sin t)\cos t = \sin t(1 + 2\cos t)$

Substitution: $u = \sin^2 t - \cos t$

$$\frac{du}{dt} = 2\sin t\cos t + \sin t = \sin t(1 + 2\cos t)$$

39. $f(x, \ y) = x^2 y$

$$\nabla f = 2xy\mathbf{i} + x^2\mathbf{j}$$

$$\nabla f(2, \ 1) = 4\mathbf{i} + 4\mathbf{j}$$

$$\mathbf{u} = \frac{1}{\sqrt{2}}\mathbf{v} = \frac{\sqrt{2}}{2}\mathbf{i} - \frac{\sqrt{2}}{2}\mathbf{j}$$

$$D_{\mathbf{u}}f(2, \ 1) = \nabla f(2, \ 1) \cdot \mathbf{u} = 2\sqrt{2} - 2\sqrt{2} = 0$$

40. $f(x, \ y) = \frac{1}{4}y^2 - x^2$

$$\nabla f = -2x\mathbf{i} + \frac{1}{2}y\mathbf{j}$$

$$\nabla f(1, \ 4) = -2\mathbf{i} + 2\mathbf{j}$$

$$\mathbf{u} = \frac{1}{\sqrt{5}}\mathbf{v} = \frac{2\sqrt{5}}{5}\mathbf{i} + \frac{\sqrt{5}}{5}\mathbf{j}$$

$$D_{\mathbf{u}}f(1, \ 4) = \nabla f(1, \ 4) \cdot \mathbf{u}$$

$$= -\frac{4\sqrt{5}}{5} + \frac{2\sqrt{5}}{5} = -\frac{2\sqrt{5}}{5}$$

41. $w = y^2 + xz$

$$\nabla w = z\mathbf{i} + 2y\mathbf{j} + x\mathbf{k}$$

$$\nabla w(1, \ 2, \ 2) = 2\mathbf{i} + 4\mathbf{j} + \mathbf{k}$$

$$\mathbf{u} = \tfrac{1}{3}\mathbf{v} = \tfrac{2}{3}\mathbf{i} - \tfrac{1}{3}\mathbf{j} + \tfrac{2}{3}\mathbf{k}$$

$$D_{\mathbf{u}}w(1, \ 2, \ 2) = \nabla w(1, \ 2, \ 2) \cdot \mathbf{u} = \tfrac{4}{3} - \tfrac{4}{3} + \tfrac{2}{3} = \tfrac{2}{3}$$

42. $w = 6x^2 + 3xy - 4y^2 z$

$$\nabla w = (12x + 3y)\mathbf{i} + (3x - 8yz)\mathbf{j} + (-4y^2)\mathbf{k}$$

$$\nabla w(1, \ 0, \ 1) = 12\mathbf{i} + 3\mathbf{j}$$

$$\mathbf{u} = \frac{1}{\sqrt{3}}\mathbf{v} = \frac{\sqrt{3}}{3}\mathbf{i} + \frac{\sqrt{3}}{3}\mathbf{j} - \frac{\sqrt{3}}{3}\mathbf{k}$$

$$D_{\mathbf{u}}w(1, \ 0, \ 1) = \nabla w(1, \ 0, \ 1) \cdot \mathbf{u}$$

$$= 4\sqrt{3} + \sqrt{3} + 0 = 5\sqrt{3}$$

43. $z = \dfrac{y}{x^2 + y^2}$

$$\nabla z = -\frac{2xy}{(x^2 + y^2)^2}\mathbf{i} + \frac{x^2 - y^2}{(x^2 + y^2)^2}\mathbf{j}$$

$$\nabla z(1, \ 1) = -\frac{1}{2}\mathbf{i} = \left\langle -\frac{1}{2}, \ 0 \right\rangle$$

$$\|\nabla z(1, \ 1)\| = \frac{1}{2}$$

44. $z = \dfrac{x^2}{x - y}$

$$\nabla z = \frac{x^2 - 2xy}{(x - y)^2}\mathbf{i} + \frac{x^2}{(x - y)^2}\mathbf{j}$$

$$\nabla z(2, \ 1) = 4\mathbf{j}$$

$$\|\nabla z(2, \ 1)\| = 4$$

45. $z = e^{-x}\cos y$

$$\nabla z = -e^{-x}\cos y\mathbf{i} - e^{-x}\sin y\mathbf{j}$$

$$\nabla z\left(0, \ \frac{\pi}{4}\right) = -\frac{\sqrt{2}}{2}\mathbf{i} - \frac{\sqrt{2}}{2}\mathbf{j} = \left\langle -\frac{\sqrt{2}}{2}, \ -\frac{\sqrt{2}}{2} \right\rangle$$

$$\left\|\nabla z\left(0, \ \frac{\pi}{4}\right)\right\| = 1$$

46. $z = x^2 y$

$$\nabla z = 2xy\mathbf{i} + x^2\mathbf{j}$$

$$\nabla z(2, \ 1) = 4\mathbf{i} + 4\mathbf{j}$$

$$\|\nabla z(2, \ 1)\| = 4\sqrt{2}$$

47. $F(x, \ y, \ z) = x^2 y - z = 0$

$$\nabla F = 2xy\mathbf{i} + x^2\mathbf{j} - \mathbf{k}$$

$$\nabla F(2, \ 1, \ 4) = 4\mathbf{i} + 4\mathbf{j} - \mathbf{k}$$

Therefore, the equation of the tangent plane is
$4(x - 2) + 4(y - 1) - (z - 4) = 0$ or $4x + 4y - z = 8$,
and the equation of the normal line is

$$\frac{x - 2}{4} = \frac{y - 1}{4} = \frac{z - 4}{-1}.$$

48. $F(x, \ y, \ z) = y^2 + z^2 - 25 = 0$

$$\nabla F = 2y\mathbf{j} + 2z\mathbf{k}$$

$$\nabla F(2, \ 3, \ 4) = 6\mathbf{j} + 8\mathbf{k} = 2(3\mathbf{j} + 4\mathbf{k})$$

Therefore, the equation of the tangent plane is
$3(y - 3) + 4(z - 4) = 0$ or $3y + 4z = 25$, and the
equation of the normal line is

$$x = 2, \ \frac{y - 3}{3} = \frac{z - 4}{4}.$$

49. $F(x, y, z) = x^2 + y^2 - 4x + 6y + z + 9 = 0$

$\qquad \nabla F = (2x - 4)\mathbf{i} + (2y + 6)\mathbf{j} + \mathbf{k}$

$\nabla F(2, -3, 4) = \mathbf{k}$

Therefore, the equation of the tangent plane is $z - 4 = 0$ or $z = 4$ and the equation of the normal line is $x = 2$, $y = -3$, $z = 4 + t$.

50. $F(x, y, z) = x^2 + y^2 + z^2 - 9 = 0$

$\qquad \nabla F = 2x\mathbf{i} + 2y\mathbf{j} + 2z\mathbf{k}$

$\nabla F(1, 2, 2) = 2\mathbf{i} + 4\mathbf{j} + 4\mathbf{k} = 2(\mathbf{i} + 2\mathbf{j} + 2\mathbf{k})$

Therefore, the equation of the tangent plane is $(x - 1) + 2(y - 2) + 2(z - 2) = 0$ or $x + 2y + 2z = 9$, and the equation of the normal line is

$$\frac{x - 1}{1} = \frac{y - 2}{2} = \frac{z - 2}{2}.$$

51. $F(x, y, z) = x^2 - y^2 - z = 0$

$\quad G(x, y, z) = 3 - z = 0$

$\qquad \nabla F = 2x\mathbf{i} - 2y\mathbf{j} - \mathbf{k}$

$\qquad \nabla G = -\mathbf{k}$

$\nabla F(2, 1, 3) = 4\mathbf{i} - 2\mathbf{j} - \mathbf{k}$

$$\nabla F \times \nabla G = \begin{vmatrix} \mathbf{i} & \mathbf{j} & \mathbf{k} \\ 4 & -2 & -1 \\ 0 & 0 & -1 \end{vmatrix} = 2(\mathbf{i} + 2\mathbf{j})$$

Therefore, the equation of the tangent line is

$$\frac{x - 2}{1} = \frac{y - 1}{2}, \quad z = 3.$$

52. $F(x, y, z) = y^2 + z - 25 = 0$

$\quad G(x, y, z) = x - y = 0$

$\qquad \nabla F = 2y\mathbf{i} + \mathbf{k}$

$\qquad \nabla G = \mathbf{i} - \mathbf{j}$

$\nabla F(4, 4, 9) = 8\mathbf{i} + \mathbf{k}$

$$\nabla F \times \nabla G = \begin{vmatrix} \mathbf{i} & \mathbf{j} & \mathbf{k} \\ 8 & 0 & 1 \\ 1 & -1 & 0 \end{vmatrix} = \mathbf{i} + \mathbf{j} - 8\mathbf{k}$$

Therefore, the equation of the tangent line is

$$\frac{x - 4}{1} = \frac{y - 4}{1} = \frac{z - 9}{-8}.$$

53. $f(x, y) = x^3 - 3xy + y^2$

$\qquad f_x = 3x^2 - 3y = 3(x^2 - y) = 0$

$\qquad f_y = -3x + 2y = 0$

$\qquad f_{xx} = 6x$

$\qquad f_{yy} = 2$

$\qquad f_{xy} = -3$

From $f_x = 0$, we have $y = x^2$. Substituting this into $f_y = 0$, we have $-3x + 2x^2 = x(2x - 3) = 0$. Thus, $x = 0$ or $\frac{3}{2}$. At the critical point $(0, 0)$, $f_{xx} f_{yy} - (f_{xy})^2 < 0$. Therefore, $(0, 0, 0)$ is a saddle point. At the critical point $\left(\frac{3}{2}, \frac{9}{4}\right)$, $f_{xx} f_{yy} - (f_{xy})^2 > 0$ and $f_{xx} > 0$. Therefore, $\left(\frac{3}{2}, \frac{9}{4}, -\frac{27}{16}\right)$ is a relative minimum.

54. $f(x, y) = 2x^2 + 6xy + 9y^2 + 8x + 14$

$\qquad f_x = 4x + 6y + 8 = 0$

$\qquad f_y = 6x + 18y = 0, \quad x = -3y$

$\qquad 4(-3y) + 6y = -8 \Rightarrow y = \frac{4}{3}, \quad x = -4$

$\qquad f_{xx} = 4$

$\qquad f_{yy} = 18$

$\qquad f_{xy} = 6$

$f_{xx} f_{yy} - (f_{xy})^2 = 4(18) - (6)^2 = 36 > 0$

Therefore, $\left(-4, \frac{4}{3}, -2\right)$ is a relative minimum.

55. $f(x, y) = xy + \dfrac{1}{x} + \dfrac{1}{y}$

$$f_x = y - \frac{1}{x^2} = 0, \quad x^2 y = 1$$

$$f_y = x - \frac{1}{y^2} = 0, \quad xy^2 = 1$$

Thus, $x^2 y = xy^2$ or $x = y$ and substitution yields the critical point (1, 1).

$$f_{xx} = \frac{2}{x^3}$$

$$f_{xy} = 1$$

$$f_{yy} = \frac{2}{y^3}$$

At the critical point (1, 1), $f_{xx} = 2 > 0$ and $f_{xx} f_{yy} - (f_{xy})^2 = 3 > 0$. Thus, (1, 1, 3) is a relative minimum.

56. $z = 50(x + y) - (0.1x^3 + 20x + 150) - (0.05y^3 + 20.6y + 125)$

$z_x = 50 - 0.3x^2 - 20 = 0, \quad x = \pm 10$

$z_y = 50 - 0.15y^2 - 20.6 = 0, \quad y = \pm 14$

Critical points: (10, 14), (10, −14), (−10, 14), (−10, −14)

$z_{xx} = -0.6x, \quad z_{yy} = -0.3y, \quad z_{xy} = 0$

At (10, 14), $z_{xx} z_{yy} - (z_{xy})^2 = (-6)(-4.2) - 0^2 > 0, \quad z_{xx} < 0.$
(10, 14, 199.4) is a relative maximum.

At (10, −14), $z_{xx} z_{yy} - (z_{xy})^2 = (-6)(4.2) - 0^2 < 0.$
(10, −14, −349.4) is a saddle point.

At (−10, 14), $z_{xx} z_{yy} - (z_{xy})^2 = (6)(-4.2) - 0^2 < 0.$
(−10, 14, −200.6) is a saddle point.

At (−10, −14), $z_{xx} z_{yy} - (x_{xy})^2 = (6)(4.2) - 0^2 > 0, \quad z_{xx} > 0.$
(−10, −14, −749.4) is a relative minimum.

57. The level curves are hyperbolas. There is a saddle point at the critical point (0, 0), but there are no relative extrema. The gradient is normal to the level curve at any given point (x_0, y_0).

58. The level curves indicate that there is a relative extremum at A, the center of the ellipse in the second quadrant, and that there is a saddle point at B, the origin.

59. Optimize $f(x, y) = x^2 y$ subject to the constraint $x + 2y = 2$.

$$\left.\begin{array}{r} 2xy = \lambda \\ x^2 = 2\lambda \end{array}\right\} x^2 = 4xy \Rightarrow x = 0 \text{ or } x = 4y$$

$$x + 2y = 2$$

If $x = 0$, $y = 1$. If $x = 4y$, then $y = \frac{1}{3}$, $x = \frac{4}{3}$.

Maximum: $f\left(\frac{4}{3}, \frac{1}{3}\right) = \frac{16}{27}$

Minimum: $f(0, 1) = 0$

60. Optimize $f(x, y, z) = xy + yz + xz$ subject to the constraint $x + y + z = 1$.

$$\left.\begin{array}{r} y + z = \lambda \\ x + z = \lambda \\ x + y = \lambda \end{array}\right\} x = y = z$$

$$x + y + z = 1 \Rightarrow x = y = z = \frac{1}{3}$$

Maximum: $f\left(\frac{1}{3}, \frac{1}{3}, \frac{1}{3}\right) = \frac{1}{3}$

61. $P(x_1, x_2) = R - C_1 - C_2$

$$= [225 - 0.4(x_1 + x_2)](x_1 + x_2) - (0.05x_1{}^2 + 15x_1 + 5400) - (0.03x_2{}^2 + 15x_2 + 6100)$$

$$= -0.45x_1{}^2 - 0.43x_2{}^2 - 0.8x_1x_2 + 210x_1 + 210x_2 - 11500$$

$$P_{x_1} = -0.9x_1 - 0.8x_2 + 210 = 0$$

$$0.9x_1 + 0.8x_2 = 210$$

$$P_{x_2} = -0.86x_2 - 0.8x_1 + 210 = 0$$

$$0.8x_1 + 0.86x_2 = 210$$

Solving this system yields $x_1 \approx 94$ and $x_2 \approx 157$.

$$P_{x_1 x_1} = -0.9$$

$$P_{x_1 x_2} = -0.8$$

$$P_{x_2 x_2} = -0.86$$

$$P_{x_1 x_1} < 0$$

$$P_{x_1 x_1} P_{x_2 x_2} - (P_{x_1 x_2})^2 > 0$$

Therefore, profit is maximum when $x_1 \approx 94$ and $x_2 \approx 157$.

62. Minimize $C(x_1, x_2) = 0.25x_1{}^2 + 10x_1 + 0.15x_2{}^2 + 12x_2$ subject to the constraint $x_1 + x_2 = 1000$.

$$\left.\begin{array}{r} 0.50x_1 + 10 = \lambda \\ 0.30x_2 + 12 = \lambda \end{array}\right\} 5x_1 - 3x_2 = 20$$

$$x_1 + x_2 = 1000 \Rightarrow 3x_1 + 3x_2 = 3000$$

$$\begin{array}{r} 5x_1 - 3x_2 = 20 \\ \hline 8x_1 = 3020 \end{array}$$

$$x_1 = 377.5$$

$$x_2 = 622.5$$

$$C(377.5, 622.5) = 104,997.50$$

63. Maximize $f(x, y) = 4x + xy + 2y$ subject to the constraint $20x + 4y = 2000$.

$$\left. \begin{array}{l} 4 + y = 20\lambda \\ x + 2 = 4\lambda \end{array} \right\} 5x - y = -6$$

$$20x + 4y = 2000 \Rightarrow 5x + y = 500$$

$$\frac{5x - y = -6}{10x \quad = 494}$$

$$x = 49.4$$

$$y = 253$$

$$f(49.4, \ 253) = 13{,}201.8$$

64. a. $(2.4, 8.6)$, $(2.2, 9.0)$, $(2.7, 9.2)$, $(3.5, 10.4)$, $(4.0, 11.0)$, $(4.0, 11.5)$, $(4.2, 10.3)$, $(4.6, 10.6)$

$$\sum x_i = 27.6, \quad \sum y_i = 81.6, \quad \sum x_i y_i = 287.9, \quad \sum x_i^2 = 100.94$$

$a \approx 1.115$, $b \approx 6.352$, $y \approx 1.115x + 6.352$

b. For every one million trucks/buses sold, approximately 1.115 million cars are sold.

65. a. $(25, 28)$, $(50, 38)$, $(75, 54)$, $(100, 75)$, $(125, 102)$

$$\sum x_i = 375, \qquad \sum y_i = 297, \qquad \sum x_i^2 = 34{,}375, \qquad \sum x_i^3 = 3{,}515{,}625$$

$$\sum x_i^4 = 382{,}421{,}875, \qquad \sum x_i y_i = 26{,}900, \qquad \sum x_i^2 y_i = 2{,}760{,}000$$

$$382{,}421{,}875a + 3{,}515{,}625b + 34{,}375c = 2{,}760{,}000$$

$$3{,}515{,}625a + \quad 34{,}375b + \quad 375c = \quad 26{,}900$$

$$34{,}375a + \quad 375b + \quad 5c = \quad 297$$

$a \approx 0.0045$, $b \approx 0.0717$, $c \approx 23.2914$, $y \approx 0.0045x^2 + 0.0717x + 23.2914$

b. When $x = 80$ km/hr, $y \approx 57.8$ km.

66. True

67. False, $\nabla F(x_0, \ y_0, \ z_0)$ is normal to the surface.

68. True

CHAPTER 14
Multiple Integration

Section 14.1   Iterated Integrals and Area in the Plane

1. $\displaystyle\int_0^x (2x - y)\, dy = \left[2xy - \frac{1}{2}y^2\right]_0^x = \frac{3}{2}x^2$

2. $\displaystyle\int_x^{x^2} \frac{y}{x}\, dy = \left[\frac{1}{2}\frac{y^2}{x}\right]_x^{x^2} = \frac{1}{2}\left(\frac{x^4}{x} - \frac{x^2}{x}\right) = \frac{x}{2}(x^2 - 1)$

3. $\displaystyle\int_1^{2y} \frac{y}{x}\, dx = \left[y\ln x\right]_1^{2y} = y\ln 2y - 0 = y\ln 2y$

4. $\displaystyle\int_0^{\cos y} y\, dx = \left[yx\right]_0^{\cos y} = y\cos y$

5. $\displaystyle\int_0^{\sqrt{4-x^2}} x^2 y\, dy = \left[\frac{1}{2}x^2 y^2\right]_0^{\sqrt{4-x^2}} = \frac{4x^2 - x^4}{2}$

6. $\displaystyle\int_{x^2}^{\sqrt{x}} (x^2 + y^2)\, dy = \left[x^2 y + \frac{1}{3}y^3\right]_{x^2}^{\sqrt{x}} = x^2\sqrt{x} + \frac{1}{3}(\sqrt{x})^3 - x^4 - \frac{1}{3}x^6 = x^{5/2} + \frac{1}{3}x^{3/2} - x^4 - \frac{1}{3}x^6$

7. $\displaystyle\int_{e^y}^{y} \frac{y\ln x}{x}\, dx = \left[\frac{1}{2}y\ln^2 x\right]_{e^y}^{y} = \frac{y}{2}(\ln^2 y - y^2)$

8. $\displaystyle\int_{-\sqrt{1-y^2}}^{\sqrt{1-y^2}} (x^2 + y^2)\, dx = \left[\frac{1}{3}x^3 + y^2 x\right]_{-\sqrt{1-y^2}}^{\sqrt{1-y^2}} = 2\left[\frac{1}{3}(1-y^2)^{3/2} + y^2(1-y^2)^{1/2}\right] = \frac{2\sqrt{1-y^2}}{3}(1 + 2y^2)$

9. $\displaystyle\int_0^{x^3} ye^{-y/x}\, dy = \left[-xye^{-y/x}\right]_0^{x^3} + x\int_0^{x^3} e^{-y/x}\, dy = -x^4 e^{-x^2} - \left[x^2 e^{-y/x}\right]_0^{x^3} = x^2(1 - e^{-x^2} - x^2 e^{-x^2})$

$u = y, \quad du = dy, \quad dv = e^{-y/x}\, dy, \quad v = -xe^{-y/x}$

10. $\displaystyle\int_y^{\pi/2} \sin^3 x \cos y\, dx = \int_y^{\pi/2} (1 - \cos^2 x)\sin x \cos y\, dx$

$$= \left[\left(-\cos x + \frac{1}{3}\cos^3 x\right)\cos y\right]_y^{\pi/2} = \left(\cos y - \frac{1}{3}\cos^3 y\right)\cos y$$

11. $\displaystyle\int_0^1 \int_0^2 (x + y)\, dy\, dx = \int_0^1 \left[xy + \frac{1}{2}y^2\right]_0^2 dx = \int_0^1 (2x + 2)\, dx = \left[x^2 + 2x\right]_0^1 = 3$

12. $\displaystyle\int_0^1 \int_0^x \sqrt{1 - x^2}\, dy\, dx = \int_0^1 \left[y\sqrt{1 - x^2}\right]_0^x dx = \int_0^1 x\sqrt{1 - x^2}\, dx = \left[-\frac{1}{2}\left(\frac{2}{3}\right)(1 - x^2)^{3/2}\right]_0^1 = \frac{1}{3}$

13. $\displaystyle\int_1^2 \int_0^4 (x^2 - 2y^2 + 1)\, dx\, dy = \int_1^2 \left[\frac{1}{3}x^3 - 2xy^2 + x\right]_0^4 dy$

$$= \int_1^2 \left(\frac{64}{3} - 8y^2 + 4\right)\, dy = \frac{4}{3}\int_1^2 (19 - 6y^2)\, dy = \left[\frac{4}{3}(19y - 2y^3)\right]_1^2 = \frac{20}{3}$$

14. $\displaystyle\int_0^1\int_y^{2y}(1+2x^2+2y^2)\,dx\,dy=\int_0^1\left[x+\frac{2}{3}x^3+2xy^2\right]_y^{2y}dy=\int_0^1\left(y+\frac{20}{3}y^3\right)dy=\left[\frac{1}{2}y^2+\frac{5}{3}y^4\right]_0^1=\frac{13}{6}$

15. $\displaystyle\int_0^1\int_0^{\sqrt{1-y^2}}(x+y)\,dx\,dy=\int_0^1\left[\frac{1}{2}x^2+xy\right]_0^{\sqrt{1-y^2}}dy$

$\displaystyle=\int_0^1\left[\frac{1}{2}(1-y^2)+y\sqrt{1-y^2}\right]dy=\left[\frac{1}{2}y-\frac{1}{6}y^3-\frac{1}{2}\left(\frac{2}{3}\right)(1-y^2)^{3/2}\right]_0^1=\frac{2}{3}$

16. $\displaystyle\int_0^2\int_{3y^2-6y}^{2y-y^2}3y\,dx\,dy=\int_0^2\left[3xy\right]_{3y^2-6y}^{2y-y^2}dy=3\int_0^2(8y^2-4y^3)\,dy=\left[3\left(\frac{8}{3}y^3-y^4\right)\right]_0^2=16$

17. $\displaystyle\int_0^2\int_0^{\sqrt{4-y^2}}\frac{2}{\sqrt{4-y^2}}\,dx\,dy=\int_0^2\left[\frac{2x}{\sqrt{4-y^2}}\right]_0^{\sqrt{4-y^2}}dy=\int_0^2 2\,dy=\left[2y\right]_0^2=4$

18. $\displaystyle\int_0^{\pi/2}\int_0^{2\cos\theta}r\,dr\,d\theta=\int_0^{\pi/2}\left[\frac{r^2}{2}\right]_0^{2\cos\theta}d\theta=\int_0^{\pi/2}2\cos^2\theta\,d\theta=\left[\theta-\frac{1}{2}\sin2\theta\right]_0^{\pi/2}=\frac{\pi}{2}$

19. $\displaystyle\int_0^{\pi/2}\int_0^{\sin\theta}\theta r\,dr\,d\theta=\int_0^{\pi/2}\left[\theta\frac{r^2}{2}\right]_0^{\sin\theta}d\theta=\int_0^{\pi/2}\frac{1}{2}\theta\sin^2\theta\,d\theta$

$\displaystyle=\frac{1}{4}\int_0^{\pi/2}(\theta-\theta\cos2\theta)\,d\theta=\frac{1}{4}\left[\frac{\theta^2}{2}-\left(\frac{1}{4}\cos2\theta+\frac{\theta}{2}\sin2\theta\right)\right]_0^{\pi/2}=\frac{\pi^2}{32}+\frac{1}{8}$

20. $\displaystyle\int_0^{\pi/4}\int_0^{\cos\theta}3r^2\sin\theta\,dr\,d\theta=\int_0^{\pi/4}\left[r^3\sin\theta\right]_0^{\cos\theta}d\theta$

$\displaystyle=\int_0^{\pi/4}\cos^3\theta\sin\theta\,d\theta=\left[-\frac{\cos^4\theta}{4}\right]_0^{\pi/4}=-\frac{1}{4}\left[\left(\frac{1}{\sqrt{2}}\right)^4-1\right]=\frac{3}{16}$

21. $\displaystyle\int_1^\infty\int_0^{1/x}y\,dy\,dx=\int_1^\infty\left[\frac{y^2}{2}\right]_0^{1/x}dx=\frac{1}{2}\int_1^\infty\frac{1}{x^2}\,dx=\left[-\frac{1}{2x}\right]_1^\infty=0+\frac{1}{2}=\frac{1}{2}$

22. $\displaystyle\int_0^3\int_0^\infty\frac{x^2}{1+y^2}\,dy\,dx=\int_0^3\left[x^2\arctan y\right]_0^\infty dx=\int_0^3 x^2\left(\frac{\pi}{2}\right)dx=\left[\frac{\pi}{2}\cdot\frac{x^3}{3}\right]_0^3=\frac{9\pi}{2}$

23. $\displaystyle\int_0^\infty\int_0^\infty xye^{-(x^2+y^2)}\,dx\,dy=\int_0^\infty\left[-\frac{1}{2}ye^{-(x^2+y^2)}\right]_0^\infty dy=\int_0^\infty\frac{1}{2}ye^{-y^2}\,dy=\left[-\frac{1}{4}e^{-y^2}\right]_0^\infty=\frac{1}{4}$

24. $\displaystyle\int_1^\infty\int_1^\infty\frac{1}{xy}\,dx\,dy=\int_1^\infty\left[\frac{1}{y}\ln x\right]_1^\infty dy=\int_1^\infty\left[\frac{1}{y}(\infty)-\frac{1}{y}(0)\right]dy$

Diverges

25. $\displaystyle\int_0^4\int_0^y f(x,\,y)\,dx\,dy,\quad 0\le x\le y,\quad 0\le y\le 4$

$\displaystyle=\int_0^4\int_x^4 f(x,\,y)\,dy\,dx$

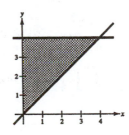

26. $\int_0^4 \int_{\sqrt{y}}^2 f(x, y)\,dx\,dy, \quad \sqrt{y} \le x \le 2, \quad 0 \le y \le 4$

$$= \int_0^2 \int_0^{x^2} f(x, y)\,dy\,dx$$

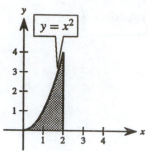

27. $\int_{-\pi/2}^{\pi/2} \int_0^{\cos x} f(x, y)\,dy\,dx, \quad 0 \le y \le \cos x, \quad -\frac{\pi}{2} \le x \le \frac{\pi}{2}$

$$= \int_0^1 \int_{-\arccos y}^{\arccos y} f(x, y)\,dx\,dy$$

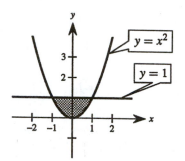

28. $\int_{-1}^1 \int_{x^2}^1 f(x, y)\,dy\,dx, \quad x^2 \le y \le 1, \quad -1 \le x \le 1$

$$= \int_0^1 \int_{-\sqrt{y}}^{\sqrt{y}} f(x, y)\,dx\,dy$$

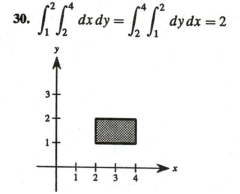

29. $\int_0^1 \int_0^2 dy\,dx = \int_0^2 \int_0^1 dx\,dy = 2$

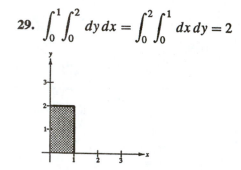

30. $\int_1^2 \int_2^4 dx\,dy = \int_2^4 \int_1^2 dy\,dx = 2$

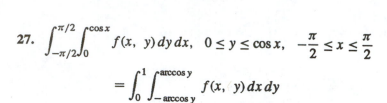

31. $\int_0^1 \int_{-\sqrt{1-y^2}}^{\sqrt{1-y^2}} dx\,dy = \int_{-1}^1 \int_0^{\sqrt{1-x^2}} dy\,dx = \frac{\pi}{2}$

32. $\int_0^2 \int_0^x dy\,dx + \int_2^4 \int_0^{4-x} dy\,dx = \int_0^2 \int_y^{4-y} dx\,dy = 4$

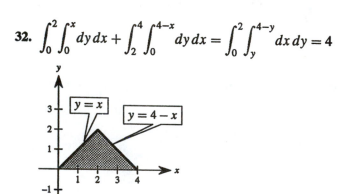

33. $\int_0^2 \int_{x/2}^1 dy\, dx = \int_0^1 \int_0^{2y} dx\, dy = 1$

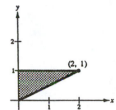
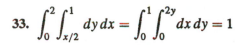

34. $\int_0^4 \int_{\sqrt{x}}^2 dy\, dx = \int_0^2 \int_0^{y^2} dx\, dy = \frac{8}{3}$

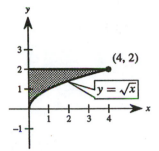

35. $\int_0^1 \int_{y^2}^{\sqrt[3]{y}} dx\, dy = \int_0^1 \int_{x^3}^{\sqrt{x}} dy\, dx = \frac{5}{12}$

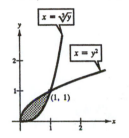

36. $\int_{-2}^2 \int_0^{4-y^2} dx\, dy = \int_0^4 \int_{-\sqrt{4-x}}^{\sqrt{4-x}} dy\, dx = \frac{32}{3}$

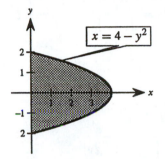

37. $A = \int_0^8 \int_0^3 dy\, dx = \int_0^8 \Big[y\Big]_0^3 dx = \int_0^8 3\, dx = \Big[3x\Big]_0^8 = 24$

$A = \int_0^3 \int_0^8 dx\, dy = \int_0^3 \Big[x\Big]_0^8 dy = \int_0^3 8\, dy = \Big[8y\Big]_0^3 = 24$

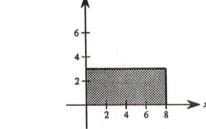

38. $A = \int_1^3 \int_1^3 dy\, dx = \int_1^3 \Big[y\Big]_1^3 dx = \int_1^3 2\, dx = \Big[2x\Big]_1^3 = 4$

$A = \int_1^3 \int_1^3 dx\, dy = \int_1^3 \Big[x\Big]_1^3 dy = \int_1^3 2\, dy = \Big[2y\Big]_1^3 = 4$

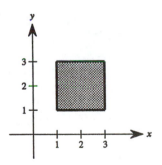

39. $A = \displaystyle\int_0^2 \int_0^{4-x^2} dy\,dx = \int_0^2 \Big[y\Big]_0^{4-x^2} dx$

$\qquad = \displaystyle\int_0^2 (4 - x^2)\,dx$

$\qquad = \Big[4x - \dfrac{x^3}{3}\Big]_0^2 = \dfrac{16}{3}$

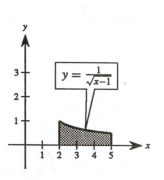

$y = 4 - x^2$

$A = \displaystyle\int_0^4 \int_0^{\sqrt{4-y}} dx\,dy$

$\qquad = \displaystyle\int_0^4 \Big[x\Big]_0^{\sqrt{4-y}} dy = \int_0^4 \sqrt{4-y}\,dy = -\int_0^4 (4-y)^{1/2}(-1)\,dy = \Big[-\dfrac{2}{3}(4-y)^{3/2}\Big]_0^4 = \dfrac{2}{3}(8) = \dfrac{16}{3}$

40. $A = \displaystyle\int_2^5 \int_0^{1/\sqrt{x-1}} dy\,dx = \int_2^5 \Big[y\Big]_0^{1/\sqrt{x-1}} dx = \int_2^5 \dfrac{1}{\sqrt{x-1}}\,dx = \Big[2\sqrt{x-1}\Big]_2^5 = 2$

$A = \displaystyle\int_0^{1/2} \int_2^5 dx\,dy + \int_{1/2}^1 \int_2^{1+(1/y^2)} dx\,dy$

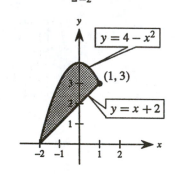

$y = \dfrac{1}{\sqrt{x-1}}$

$\qquad = \displaystyle\int_0^{1/2} \Big[x\Big]_2^5 dy + \int_{1/2}^1 \Big[x\Big]_2^{1+(1/y^2)} dy$

$\qquad = \displaystyle\int_0^{1/2} 3\,dy + \int_{1/2}^1 \Big(\dfrac{1}{y^2} - 1\Big)\,dy$

$\qquad = \Big[3y\Big]_0^{1/2} + \Big[-\dfrac{1}{y} - y\Big]_{1/2}^1 = 2$

41. $A = \displaystyle\int_{-2}^1 \int_{x+2}^{4-x^2} dy\,dx$

$\qquad = \displaystyle\int_{-2}^1 \Big[y\Big]_{x+2}^{4-x^2} dx = \int_{-2}^1 (4 - x^2 - x - 2)\,dx = \int_{-2}^1 (2 - x - x^2)\,dx = \Big[2x - \dfrac{1}{2}x^2 - \dfrac{1}{3}x^3\Big]_{-2}^1 = \dfrac{9}{2}$

$A = \displaystyle\int_0^3 \int_{-\sqrt{4-y}}^{y-2} dx\,dy + 2\int_3^4 \int_0^{\sqrt{4-y}} dx\,dy$

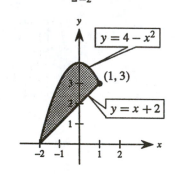

$y = 4 - x^2$

$(1, 3)$

$y = x + 2$

$\qquad = \displaystyle\int_0^3 \Big[x\Big]_{-\sqrt{4-y}}^{y-2} dy + 2\int_3^4 \Big[x\Big]_0^{\sqrt{4-y}} dy$

$\qquad = \displaystyle\int_0^3 \big(y - 2 + \sqrt{4-y}\big)\,dy + 2\int_3^4 \sqrt{4-y}\,dy$

$\qquad = \Big[\dfrac{1}{2}y^2 - 2y - \dfrac{2}{3}(4-y)^{3/2}\Big]_0^3 - \Big[\dfrac{4}{3}(4-y)^{3/2}\Big]_3^4 = \dfrac{9}{2}$

42. $A = \int_0^2 \int_0^{\sqrt{4-x^2}} dy\, dx = \int_0^2 \sqrt{4-x^2}\, dx = 4\int_0^{\pi/2} \cos^2\theta\, d\theta = 2\int_0^{\pi/2}(1+\cos 2\theta)\, d\theta = \left[2\left(\theta + \frac{1}{2}\sin 2\theta\right)\right]_0^{\pi/2} = \pi$

$(x = 2\sin\theta, \quad dx = 2\cos\theta\, d\theta, \quad \sqrt{4-x^2} = 2\cos\theta)$

$A = \int_0^2 \int_0^{\sqrt{4-y^2}} dx\, dy = \int_0^2 \sqrt{4-y^2}\, dy$

$= 4\int_0^{\pi/2}\cos^2\theta\, d\theta = 2\int_0^{\pi/2}(1+\cos 2\theta)\, d\theta$

$= \left[2\left(\theta + \frac{1}{2}\sin 2\theta\right)\right]_0^{\pi/2} = \pi$

$(y = 2\sin\theta, \quad dy = 2\cos\theta\, d\theta, \quad \sqrt{4-y^2} = 2\cos\theta)$

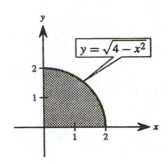

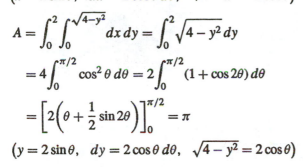

43. $\int_0^4 \int_0^{(2-\sqrt{x})^2} dy\, dx = \int_0^4 \left[y\right]_0^{(2-\sqrt{x})^2} dx$

$= \int_0^4 (4 - 4\sqrt{x} + x)\, dx$

$= \left[4x - \frac{8}{3}x\sqrt{x} + \frac{x^2}{2}\right]_0^4 = \frac{8}{3}$

$\int_0^4 \int_0^{(2-\sqrt{y})^2} dx\, dy = \frac{8}{3}$

Integration steps are similar to those above.

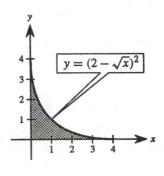

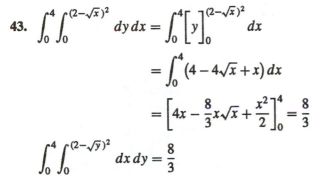

44. $A = \int_0^1 \int_{x^{3/2}}^x dy\, dx$

$= \int_0^1 \left[y\right]_{x^{3/2}}^x dx$

$= \int_0^1 (x - x^{3/2})\, dx$

$= \left[\frac{1}{2}x^2 - \frac{2}{5}x^{5/2}\right]_0^1 = \frac{1}{10}$

$A = \int_0^1 \int_y^{y^{2/3}} dx\, dy = \int_0^1 \left[x\right]_y^{y^{2/3}} dy = \int_0^1 (y^{2/3} - y)\, dy = \left[\frac{3}{5}y^{5/3} - \frac{1}{2}y^2\right]_0^1 = \frac{1}{10}$

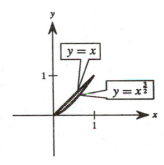

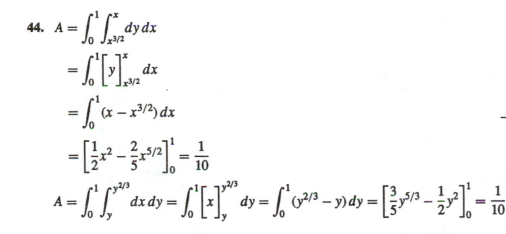

45. $A = \int_0^3 \int_0^{2x/3} dy\,dx + \int_3^5 \int_0^{5-x} dy\,dx$

$\quad = \int_0^3 \left[y\right]_0^{2x/3} dx + \int_3^5 \left[y\right]_0^{5-x} dx$

$\quad = \int_0^3 \frac{2x}{3}\,dx + \int_3^5 (5-x)\,dx$

$\quad = \left[\frac{1}{3}x^2\right]_0^3 + \left[5x - \frac{1}{2}x^2\right]_3^5 = 5$

$A = \int_0^2 \int_{3y/2}^{5-y} dx\,dy = \int_0^2 \left[x\right]_{3y/2}^{5-y} dy = \int_0^2 \left(5 - y - \frac{3y}{2}\right) dy = \int_0^2 \left(5 - \frac{5y}{2}\right) dy = \left[5y - \frac{5}{4}y^2\right]_0^2 = 5$

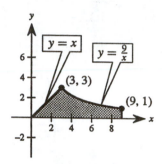

46. $A = \int_0^3 \int_0^x dy\,dx + \int_3^9 \int_0^{9/x} dy\,dx$

$\quad = \int_0^3 \left[y\right]_0^x dx + \int_3^9 \left[y\right]_0^{9/x} dx = \int_0^3 x\,dx + \int_3^9 \frac{9}{x}\,dx$

$\quad = \left[\frac{1}{2}x^2\right]_0^3 + \left[9\ln x\right]_3^9 = \frac{9}{2} + 9(\ln 9 - \ln 3) = \frac{9}{2}(1 + \ln 9)$

$A = \int_0^1 \int_y^9 dx\,dy + \int_1^3 \int_y^{9/y} dx\,dy$

$\quad = \int_0^1 \left[x\right]_y^9 dy + \int_1^3 \left[x\right]_y^{9/y} dy = \int_0^1 (9 - y)\,dy + \int_1^3 \left(\frac{9}{y} - y\right) dy$

$\quad = \left[9y - \frac{1}{2}y^2\right]_0^1 + \left[9\ln y - \frac{1}{2}y^2\right]_1^3 = \frac{9}{2}(1 + \ln 9)$

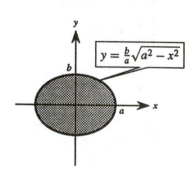

47. $\dfrac{A}{4} = \int_0^a \int_0^{(b/a)\sqrt{a^2 - x^2}} dy\,dx = \int_0^a \left[y\right]_0^{(b/a)\sqrt{a^2-x^2}} dx$

$\quad = \frac{b}{a} \int_0^a \sqrt{a^2 - x^2}\,dx = ab \int_0^{\pi/2} \cos^2\theta\,d\theta$

$\quad = \frac{ab}{2} \int_0^{\pi/2} (1 + \cos 2\theta)\,d\theta = \left[\frac{ab}{2}\left(\theta + \frac{1}{2}\sin 2\theta\right)\right]_0^{\pi/2} = \frac{\pi ab}{4}$

Therefore, $A = \pi ab$.

$(x = a\sin\theta, \quad dx = a\cos\theta\,d\theta)$

$\dfrac{A}{4} = \int_0^b \int_0^{(a/b)\sqrt{b^2 - y^2}} dx\,dy = \frac{\pi ab}{4}$

Therefore, $A = \pi ab$. Integration steps are similar to those above.

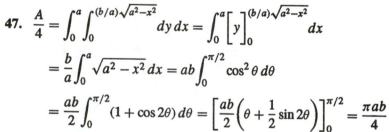

48. $A = \int_0^2 \int_{y/2}^{y} dx\, dy + \int_2^4 \int_{y/2}^{2} dx\, dy$

$\qquad = \int_0^2 \frac{y}{2}\, dy + \int_2^4 \left(2 - \frac{y}{2}\right) dy$

$\qquad = \left[\frac{y^2}{4}\right]_0^2 + \left[2y - \frac{y^2}{4}\right]_2^4$

$\qquad = 1 + (4 - 3) = 2$

$A = \int_0^2 \int_x^{2x} dy\, dx = \int_0^2 (2x - x)\, dx = \left[\frac{x^2}{2}\right]_0^2 = 2$

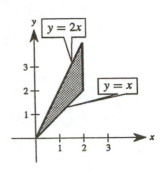

49. $\int_0^2 \int_x^2 x\sqrt{1 + y^3}\, dy\, dx = \int_0^2 \int_0^y x\sqrt{1 + y^3}\, dx\, dy = \int_0^2 \left[\sqrt{1 + y^3} \cdot \frac{x^2}{2}\right]_0^y dy$

$\qquad = \frac{1}{2} \int_0^2 \sqrt{1 + y^3}\, y^2\, dy = \left[\frac{1}{2} \cdot \frac{1}{3} \cdot \frac{2}{3}(1 + y^3)^{3/2}\right]_0^2 = \frac{1}{9}(27) - \frac{1}{9}(1) = \frac{26}{9}$

50. $\int_0^2 \int_x^2 e^{-y^2}\, dy\, dx = \int_0^2 \int_0^y e^{-y^2}\, dx\, dy$

$\qquad = \int_0^2 \left[xe^{-y^2}\right]_0^y dy = \int_0^2 ye^{-y^2}\, dy = \left[-\frac{1}{2}e^{-y^2}\right]_0^2 = -\frac{1}{2}(e^{-4}) + \frac{1}{2}e^0 = \frac{1}{2}\left(1 - \frac{1}{e^4}\right) \approx 0.4908$

51. $\int_0^1 \int_y^1 \sin(x^2)\, dx\, dy = \int_0^1 \int_0^x \sin(x^2)\, dy\, dx = \int_0^1 \left[y\sin(x^2)\right]_0^x dx$

$\qquad = \int_0^1 x\sin(x^2)\, dx = \left[-\frac{1}{2}\cos(x^2)\right]_0^1 = -\frac{1}{2}\cos 1 + \frac{1}{2}(1) = \frac{1}{2}(1 - \cos 1) \approx 0.2298$

52. $\int_0^2 \int_{y^2}^4 \sqrt{x}\sin x\, dx\, dy = \int_0^4 \int_0^{\sqrt{x}} \sqrt{x}\sin x\, dy\, dx$

$\qquad = \int_0^4 \left[y\sqrt{x}\sin x\right]_0^{\sqrt{x}} dx = \int_0^4 x\sin x\, dx = \left[\sin x - x\cos x\right]_0^4 = \sin 4 - 4\cos 4 \approx 1.858$

53. $\int_0^2 \int_{x^2}^{2x} (x^3 + 3y^2)\, dy\, dx = \frac{1664}{105} \approx 15.848$

54. $\int_0^1 \int_y^{2y} \sin(x + y)\, dx\, dy = \frac{\sin 2}{2} - \frac{\sin 3}{3} \approx 0.408$

55. $\int_0^4 \int_0^y \frac{2}{(x + 1)(y + 1)}\, dx\, dy = (\ln 5)^2 \approx 2.590$

56. $\int_0^a \int_0^{a-x} (x^2 + y^2)\, dy\, dx = \frac{a^4}{6}$

57. $\int_0^2 \int_0^{4-x^2} e^{xy}\, dy\, dx \approx 20.5648$

58. $\int_0^2 \int_x^2 \sqrt{16 - x^3 - y^3}\, dy\, dx \approx 6.8520$

59. True

60. False, let $f(x, y) = x$.

61. Ellipse

Area: πab

Matches d.

62. Asteroid

Area: $\frac{3}{8}\pi a^2$

Matches b.

63. Cardioid
Area: $6\pi a^2$
Matches f.

64. Deltoid
Area: $2\pi a^2$
Matches c.

65. Hourglass
Area: $\frac{8}{3}ab$
Matches a.

66. Teardrop
Area: $2\pi ab$
Matches e.

Section 14.2   Double Integrals and Volume

For Exercises 1–4, $\Delta x_i = \Delta y_i = 1$ and the midpoints of the squares are
$\left(\frac{1}{2}, \frac{1}{2}\right)$, $\left(\frac{3}{2}, \frac{1}{2}\right)$, $\left(\frac{5}{2}, \frac{1}{2}\right)$, $\left(\frac{7}{2}, \frac{1}{2}\right)$, $\left(\frac{1}{2}, \frac{3}{2}\right)$, $\left(\frac{3}{2}, \frac{3}{2}\right)$, $\left(\frac{5}{2}, \frac{3}{2}\right)$, $\left(\frac{7}{2}, \frac{3}{2}\right)$.

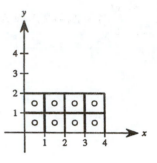

1. $f(x, y) = x + y$

$$\sum_{i=1}^{8} f(x_i, y_i)\Delta x_i \Delta y_i = 1 + 2 + 3 + 4 + 2 + 3 + 4 + 5 = 24$$

$$\int_0^4 \int_0^2 (x + y)\, dy\, dx = \int_0^4 \left[xy + \frac{y^2}{2} \right]_0^2 dx = \int_0^4 (2x + 2)\, dx = \left[x^2 + 2x \right]_0^4 = 24$$

2. $f(x, y) = xy$

$$\sum_{i=1}^{8} f(x_i, y_i)\Delta x_i \Delta y_i = \frac{1}{4} + \frac{3}{4} + \frac{5}{4} + \frac{7}{4} + \frac{3}{4} + \frac{9}{4} + \frac{15}{4} + \frac{21}{4} = 16$$

$$\int_0^4 \int_0^2 xy\, dy\, dx = \int_0^4 \left[\frac{xy^2}{2} \right]_0^2 dx = \int_0^4 2x\, dx = \left[x^2 \right]_0^4 = 16$$

3. $f(x, y) = x^2 + y^2$

$$\sum_{i=1}^{8} f(x_i, y_i)\Delta x_i \Delta y_i = \frac{2}{4} + \frac{10}{4} + \frac{26}{4} + \frac{50}{4} + \frac{10}{4} + \frac{18}{4} + \frac{34}{4} + \frac{58}{4} = 52$$

$$\int_0^4 \int_0^2 (x^2 + y^2)\, dy\, dx = \int_0^4 \left[x^2 y + \frac{y^3}{3} \right]_0^2 dx = \int_0^4 \left(2x^2 + \frac{8}{3} \right) dx = \left[\frac{2x^3}{3} + \frac{8x}{3} \right]_0^4 = \frac{160}{3}$$

4. $f(x, y) = \dfrac{1}{(x + 1)(y + 1)}$

$$\sum_{i=1}^{8} f(x_i, y_i)\Delta x_i \Delta y_i = \frac{4}{9} + \frac{4}{15} + \frac{4}{21} + \frac{4}{27} + \frac{4}{15} + \frac{4}{25} + \frac{4}{35} + \frac{4}{45} = \frac{7936}{4725} \approx 1.680$$

$$\int_0^4 \int_0^2 \frac{1}{(x + 1)(y + 1)}\, dy\, dx = \int_0^4 \left[\frac{1}{x + 1} \ln(y + 1) \right]_0^2 dx$$

$$= \int_0^4 \frac{\ln 3}{x + 1}\, dx = \left[\ln 3 \cdot \ln(x + 1) \right]_0^4 = (\ln 3)(\ln 5) \approx 1.768$$

5. $\displaystyle\iint_R f(x, y)\, dA = k \iint_R dA = kB$

6. a. $\iint_R f(x, \ y) \, dA$ represents the total volume of snow over the region R.

b. $\dfrac{\iint_R f(x, \ y) \, dA}{\iint_R dA}$ represents the average snowfall over the region R.

7.
$$\int_0^2 \int_0^1 (1 + 2x + 2y) \, dy \, dx = \int_0^2 \left[y + 2xy + y^2 \right]_0^1 dx$$

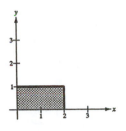

$$= \int_0^2 (2 + 2x) \, dx$$

$$= \left[2x + x^2 \right]_0^2$$

$$= 8$$

8.
$$\int_0^\pi \int_0^{\pi/2} \sin^2 x \cos^2 y \, dy \, dx = \int_0^\pi \left[\frac{1}{2} \sin^2 x \left(y + \frac{1}{2} \sin 2y \right) \right]_0^{\pi/2} dx$$

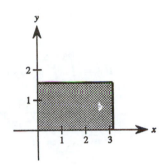

$$= \int_0^\pi \frac{1}{2} \sin^2 x \left(\frac{\pi}{2} \right) dx$$

$$= \frac{\pi}{8} \int_0^\pi (1 - \cos 2x) \, dx$$

$$= \left[\frac{\pi}{8} \left(x - \frac{1}{2} \sin 2x \right) \right]_0^\pi$$

$$= \frac{\pi^2}{8}$$

9.
$$\int_0^6 \int_{y/2}^3 (x + y) \, dx \, dy = \int_0^6 \left[\frac{1}{2} x^2 + xy \right]_{y/2}^3 dy$$

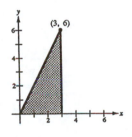

$$= \int_0^6 \left(\frac{9}{2} + 3y - \frac{5}{8} y^2 \right) dy$$

$$= \left[\frac{9}{2} y + \frac{3}{2} y^2 - \frac{5}{24} y^3 \right]_0^6$$

$$= 36$$

10.
$$\int_0^1 \int_y^{\sqrt{y}} x^2 y^2 \, dx \, dy = \int_0^1 \left[\frac{1}{3} x^3 y^2 \right]_y^{\sqrt{y}} dy$$

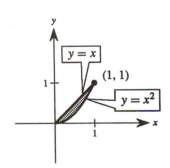

$$= \frac{1}{3} \int_0^1 (y^{7/2} - y^5) \, dy$$

$$= \left[\frac{1}{3} \left(\frac{2}{9} y^{9/2} - \frac{1}{6} y^6 \right) \right]_0^1$$

$$= \frac{1}{54}$$

11. $\displaystyle\int_{-a}^{a}\int_{-\sqrt{a^2-x^2}}^{\sqrt{a^2-x^2}}(x+y)\,dy\,dx = \int_{-a}^{a}\left[xy+\frac{1}{2}y^2\right]_{-\sqrt{a^2-x^2}}^{\sqrt{a^2-x^2}}dx$

$$= \int_{-a}^{a} 2x\sqrt{a^2-x^2}\,dx$$

$$= \left[-\frac{2}{3}(a^2-x^2)^{3/2}\right]_{-a}^{a} = 0$$

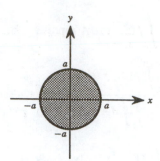

12. $\displaystyle\int_{0}^{1}\int_{y-1}^{0}e^{x+y}\,dx\,dy + \int_{0}^{1}\int_{0}^{1-y}e^{x+y}\,dx\,dy = \int_{0}^{1}\left[e^{x+y}\right]_{y-1}^{0}dy + \int_{0}^{1}\left[e^{x+y}\right]_{0}^{1-y}dy$

$$= \int_{0}^{1}(e - e^{2y-1})\,dy$$

$$= \left[ey - \frac{1}{2}e^{2y-1}\right]_{0}^{1}$$

$$= \frac{1}{2}(e + e^{-1})$$

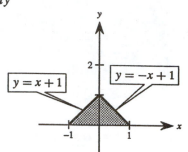

$y = x+1$ $y = -x+1$

13. $\displaystyle\int_{0}^{5}\int_{0}^{3}xy\,dx\,dy = \int_{0}^{3}\int_{0}^{5}xy\,dy\,dx$

$$= \int_{0}^{3}\left[\frac{1}{2}xy^2\right]_{0}^{5}dx$$

$$= \frac{25}{2}\int_{0}^{3}x\,dx$$

$$= \left[\frac{25}{4}x^2\right]_{0}^{3} = \frac{225}{4}$$

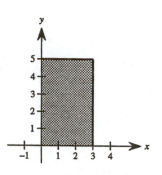

14. $\displaystyle\int_{0}^{\pi/2}\int_{-\pi}^{\pi}\sin x\sin y\,dx\,dy = \int_{-\pi}^{\pi}\int_{0}^{\pi/2}\sin x\sin y\,dy\,dx$

$$= \int_{-\pi}^{\pi}\left[-\sin x\cos y\right]_{0}^{\pi/2}dx$$

$$= \int_{-\pi}^{\pi}\sin x\,dx$$

$$= 0$$

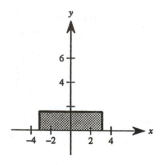

15. $\displaystyle\int_{0}^{2}\int_{y/2}^{y}\frac{y}{x^2+y^2}\,dx\,dy + \int_{2}^{4}\int_{y/2}^{2}\frac{y}{x^2+y^2}\,dx\,dy = \int_{0}^{2}\int_{x}^{2x}\frac{y}{x^2+y^2}\,dy\,dx$

$$= \frac{1}{2}\int_{0}^{2}\left[\ln(x^2+y^2)\right]_{x}^{2x}dx$$

$$= \frac{1}{2}\int_{0}^{2}(\ln 5x^2 - \ln 2x^2)\,dx$$

$$= \frac{1}{2}\ln\frac{5}{2}\int_{0}^{2}dx$$

$$= \left[\frac{1}{2}\left(\ln\frac{5}{2}\right)x\right]_{0}^{2} = \ln\frac{5}{2}$$

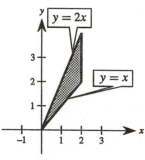

$y = 2x$ $y = x$

16. $\displaystyle\int_0^2\int_{y^2}^4 \frac{y}{1+x^2}\,dx\,dy = \int_0^4\int_0^{\sqrt{x}} \frac{y}{1+x^2}\,dy\,dx$

$\displaystyle\qquad = \frac{1}{2}\int_0^4\left[\frac{y^2}{1+x^2}\right]_0^{\sqrt{x}} dx$

$\displaystyle\qquad = \frac{1}{2}\int_0^4 \frac{x}{1+x^2}\,dx$

$\displaystyle\qquad = \left[\frac{1}{4}\ln(1+x^2)\right]_0^4 = \frac{1}{4}\ln(17)$

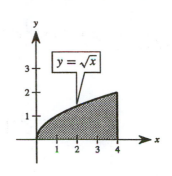

17. $\displaystyle\int_0^4\int_0^{3x/4} x\,dy\,dx + \int_4^5\int_0^{\sqrt{25-x^2}} x\,dy\,dx = \int_0^3\int_{4y/3}^{\sqrt{25-y^2}} x\,dx\,dy$

$\displaystyle\qquad = \int_0^3\left[\frac{1}{2}x^2\right]_{4y/3}^{\sqrt{25-y^2}} dy$

$\displaystyle\qquad = \frac{25}{18}\int_0^3 (9-y^2)\,dy$

$\displaystyle\qquad = \left[\frac{25}{18}\left(9y-\frac{1}{3}y^3\right)\right]_0^3 = 25$

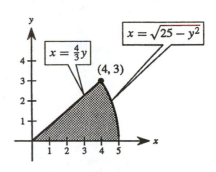

18. $\displaystyle\int_{-2}^2\int_0^{\sqrt{4-x^2}} (x^2+y^2)\,dy\,dx$

$\displaystyle\qquad = \int_0^2\int_{-\sqrt{4-y^2}}^{\sqrt{4-y^2}} (x^2+y^2)\,dx\,dy$

$\displaystyle\qquad = \int_{-2}^2\left[x^2 y+\frac{1}{3}y^3\right]_0^{\sqrt{4-x^2}} dx$

$\displaystyle\qquad = \int_{-2}^2\left[x^2\sqrt{4-x^2}+\frac{1}{3}(4-x^2)^{3/2}\right] dx$

$\displaystyle\qquad = \left[-\frac{x}{4}(4-x^2)^{3/2}+\frac{1}{2}\left(x\sqrt{4-x^2}+4\arcsin\frac{x}{2}\right)+\frac{1}{12}\left[x(4-x^2)^{3/2}+6x\sqrt{4-x^2}+24\arcsin\frac{x}{2}\right]\right]_{-2}^2 = 4\pi$

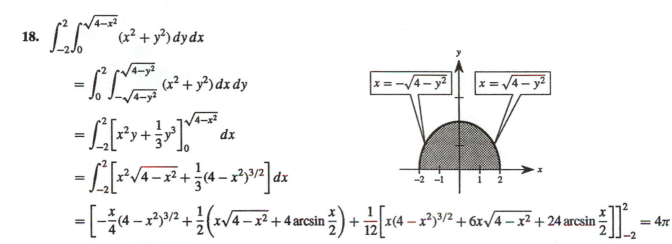

19. $\displaystyle\int_0^4\int_0^2 \frac{y}{2}\,dy\,dx = \int_0^4\left[\frac{y^2}{4}\right]_0^2 dx$

$\displaystyle\qquad = \int_0^4 dx = 4$

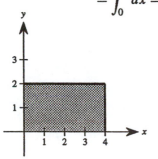

20. $\displaystyle\int_0^4\int_0^2 (6-2y)\,dy\,dx = \int_0^4\left[6y-y^2\right]_0^2 dx$

$\displaystyle\qquad = \int_0^4 8\,dx = 32$

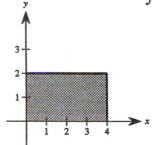

21. $\displaystyle\int_0^2\int_0^y (6-x-y)\,dx\,dy = \int_0^2\left[6x-\frac{x^2}{2}-xy\right]_0^y dy$

$$= \int_0^2\left(6y-\frac{3}{2}y^2\right)dy$$

$$= \left[3y^2-\frac{1}{2}y^3\right]_0^2$$

$$= 8$$

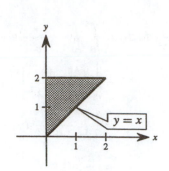

22. $\displaystyle\int_0^2\int_0^x 6\,dy\,dx = \int_0^2 6x\,dx$

$$= \left[3x^2\right]_0^2$$

$$= 12$$

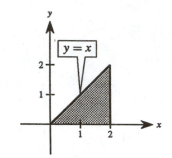

23. $\displaystyle\int_0^6\int_0^{(-2/3)x+4}\left(\frac{12-2x-3y}{4}\right)dy\,dx = \int_0^6\left[\frac{1}{4}\left(12y-2xy-\frac{3}{2}y^2\right)\right]_0^{(-2/3)x+4}dx$

$$= \int_0^6\left(\frac{1}{6}x^2-2x+6\right)dx$$

$$= \left[\frac{1}{18}x^3-x^2+6x\right]_0^6$$

$$= 12$$

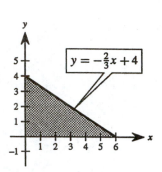

24. $\displaystyle\int_0^1\int_0^{1-x}(1-x-y)\,dy\,dx = \int_0^1\left[y-xy-\frac{y^2}{2}\right]_0^{1-x}dx$

$$= \int_0^1\frac{1}{2}(1-x)^2\,dx$$

$$= \left[-\frac{1}{6}(1-x)^3\right]_0^1$$

$$= \frac{1}{6}$$

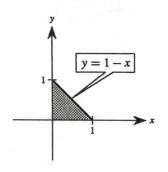

25. $\displaystyle\int_0^1\int_0^y (1-xy)\,dx\,dy = \int_0^1\left[x-\frac{x^2y}{2}\right]_0^y dy$

$$= \int_0^1\left(y-\frac{y^3}{2}\right)dy$$

$$= \left[\frac{y^2}{2}-\frac{y^4}{8}\right]_0^1$$

$$= \frac{3}{8}$$

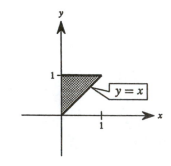

26. $\int_0^2 \int_0^y (4 - y^2) \, dx \, dy = \int_0^2 (4y - y^3) \, dy$

$$= \left[2y^2 - \frac{y^4}{4} \right]_0^2$$

$$= 4$$

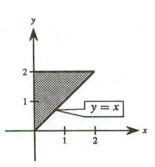

27. $\int_0^\infty \int_0^\infty \frac{1}{(x+1)^2(y+1)^2} \, dy \, dx = \int_0^\infty \left[-\frac{1}{(x+1)^2(y+1)} \right]_0^\infty dx = \int_0^\infty \frac{1}{(x+1)^2} \, dx = \left[-\frac{1}{(x+1)} \right]_0^\infty = 1$

28. $\int_0^\infty \int_0^\infty e^{-(x+y)/2} \, dy \, dx = \int_0^\infty \left[-2e^{-(x+y)/2} \right]_0^\infty dx = \int_0^\infty 2e^{-x/2} \, dx = \left[-4e^{-x/2} \right]_0^\infty = 4$

29. $4 \int_0^2 \int_0^{\sqrt{4-x^2}} (4 - x^2 - y^2) \, dy \, dx = 8\pi$

30. $\int_0^1 \int_0^x \sqrt{1 - x^2} \, dx = \frac{1}{3}$

31. $V = \int_0^1 \int_0^x xy \, dy \, dx$

$$= \int_0^1 \left[\frac{1}{2}xy^2 \right]_0^x dx = \frac{1}{2} \int_0^1 x^3 \, dx$$

$$= \left[\frac{1}{8}x^4 \right]_0^1 = \frac{1}{8}$$

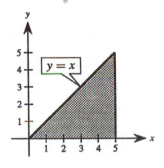

32. $V = \int_0^5 \int_0^x x \, dy \, dx$

$$= \int_0^5 \left[xy \right]_0^x dx = \int_0^5 x^2 \, dx$$

$$= \left[\frac{1}{3}x^3 \right]_0^5 = \frac{125}{3}$$

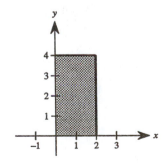

33. $V = \int_0^2 \int_0^4 x^2 \, dy \, dx$

$$= \int_0^2 \left[x^2 y \right]_0^4 dx = \int_0^2 4x^2 \, dx$$

$$= \left[\frac{4x^3}{3} \right]_0^2 = \frac{32}{3}$$

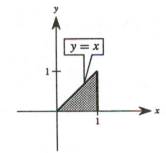

34. $V = 8\int_0^r \int_0^{\sqrt{r^2-x^2}} \sqrt{r^2 - x^2 - y^2}\, dy\, dx$

$\qquad = 4\int_0^r \left[\left[y\sqrt{r^2-x^2-y^2} + (r^2-x^2)\arcsin\frac{y}{\sqrt{r^2-x^2}} \right]\right]_0^{\sqrt{r^2-x^2}} dx$

$\qquad = 4\left(\frac{\pi}{2}\right)\int_0^r (r^2 - x^2)\, dx$

$\qquad = \left[2\pi\left(r^2 x - \frac{1}{3}x^3 \right) \right]_0^r$

$\qquad = \dfrac{4\pi r^3}{3}$

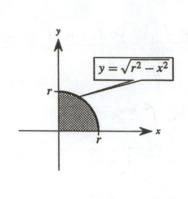

35. Divide the solid into two equal parts.

$\qquad V = 2\int_0^1 \int_0^x \sqrt{1-x^2}\, dy\, dx$

$\qquad = 2\int_0^1 \left[y\sqrt{1-x^2} \right]_0^x dx$

$\qquad = 2\int_0^1 x\sqrt{1-x^2}\, dx$

$\qquad = \left[-\frac{2}{3}(1-x^2)^{3/2} \right]_0^1 = \frac{2}{3}$

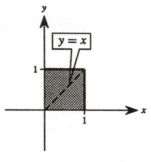

36. $V = \int_0^1 \int_0^{1-x^2} (1-x^2)\, dy\, dx$

$\qquad = \int_0^1 \left[y(1-x^2) \right]_0^{1-x^2} dx$

$\qquad = \int_0^1 (1-x^2)^2\, dx$

$\qquad = \int_0^1 (1 - 2x^2 + x^4)\, dx$

$\qquad = \left[x - \frac{2}{3}x^3 + \frac{1}{5}x^5 \right]_0^1 = \frac{8}{15}$

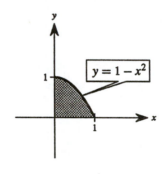

37. $V = \int_0^2 \int_0^{\sqrt{4-x^2}} (x+y)\, dy\, dx$

$\qquad = \int_0^2 \left[xy + \frac{1}{2}y^2 \right]_0^{\sqrt{4-x^2}} dx$

$\qquad = \int_0^2 \left(x\sqrt{4-x^2} + 2 - \frac{1}{2}x^2 \right) dx$

$\qquad = \left[-\frac{1}{3}(4-x^2)^{3/2} + 2x - \frac{1}{6}x^3 \right]_0^2 = \frac{16}{3}$

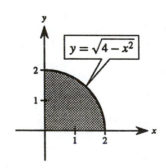

38. $V = \int_0^2 \int_0^\infty \dfrac{1}{1+y^2}\,dy\,dx$

$= \int_0^2 \Big[\arctan y\Big]_0^\infty\,dx$

$= \int_0^2 \dfrac{\pi}{2}\,dx$

$= \Big[\dfrac{\pi x}{2}\Big]_0^2 = \pi$

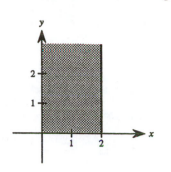

39. $V = 4\int_0^2 \int_0^{\sqrt{4-x^2}} (x^2+y^2)\,dy\,dx$

$= 4\int_0^2 \Big[x^2\sqrt{4-x^2} + \dfrac{1}{3}(4-x^2)^{3/2}\Big]\,dx, \quad x = 2\sin\theta$

$= 4\int_0^{\pi/2}\Big(16\cos^2\theta - \dfrac{32}{3}\cos^4\theta\Big)\,d\theta$

$= 4\Big[16\Big(\dfrac{\pi}{4}\Big) - \dfrac{32}{3}\Big(\dfrac{3\pi}{16}\Big)\Big]$

$= 8\pi$

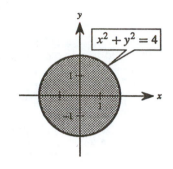

$x^2+y^2=4$

40. $V = \int_0^5 \int_0^\pi \sin^2 x\,dx\,dy$

$= \int_0^5 \dfrac{\pi}{2}\,dy$

$= \Big[\dfrac{\pi}{2}y\Big]_0^5$

$= \dfrac{5\pi}{2}$

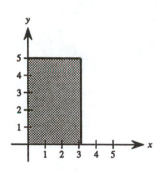

41. $V = 4\int_0^2 \int_0^{\sqrt{4-x^2}} (4-x^2-y^2)\,dy\,dx = 8\pi$

42. $V = \int_0^9 \int_0^{\sqrt{9-y}} \sqrt{9-y}\,dx\,dy = \dfrac{81}{2}$

43. $V = \int_0^2 \int_0^{-0.5x+1} \dfrac{2}{1+x^2+y^2}\,dy\,dx \approx 1.2315$

44. $V = \int_0^{16} \int_0^{4-\sqrt{y}} \ln(1+x+y)\,dx\,dy \approx 231.6494$

45. f is a continuous function such that $0 \le f(x,\ y) \le 1$ over a region R of area 1. Let $f(m,\ n) =$ the minimum value of f over R and $f(M,\ N) =$ the maximum value of f over R. Then

$$f(m,\ n)\iint_R dA \le \iint_R f(x,\ y)\,dA \le f(M,\ N)\iint_R dA.$$

Since $\iint_R dA = 1$ and $0 \le f(m,\ n) \le f(M,\ N) \le 1$, we have

$$0 \le f(m,\ n)(1) \le \iint_R f(x,\ y)\,dA \le f(M,\ N)(1) \le 1.$$

Therefore, $0 \le \iint_R f(x,\ y)\,dA \le 1.$

46. $\dfrac{x}{a} + \dfrac{y}{b} + \dfrac{z}{c} = 1$

$z = c\left(1 - \dfrac{x}{a} - \dfrac{y}{b}\right)$

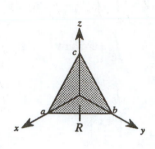

$$V = \iint\limits_{R} f(x,\ y)\,dA = \int_{0}^{a} \int_{0}^{b[1-(x/a)]} c\left(1 - \dfrac{x}{a} - \dfrac{y}{b}\right) dy\,dx$$

$$= \int_{0}^{a} \left[c\left(y - \dfrac{xy}{a} - \dfrac{y^2}{2b}\right)\right]_{0}^{b[1-(x/a)]} dx$$

$$= \int_{0}^{a} c\left[b\left(1 - \dfrac{x}{a}\right) - \dfrac{xb}{a}\left(1 - \dfrac{x}{a}\right) - \dfrac{b^2}{2b}\left(1 - \dfrac{x}{a}\right)^2\right] dx$$

$$= c\left[-\dfrac{ab}{2}\left(1 - \dfrac{x}{a}\right)^2 - \dfrac{x^2 b}{2a} + \dfrac{x^3 b}{3a^2} + \dfrac{ab}{6}\left(1 - \dfrac{x}{a}\right)^3\right]_{0}^{a}$$

$$= c\left[\left(-\dfrac{ab}{2} + \dfrac{ab}{3}\right) - \left(-\dfrac{ab}{2} + \dfrac{ab}{6}\right)\right] = \dfrac{abc}{6}$$

47. $\displaystyle\int_{0}^{1}\int_{y/2}^{1/2} e^{-x^2}\,dx\,dy = \int_{0}^{1/2}\int_{0}^{2x} e^{-x^2}\,dy\,dx$

$$= \int_{0}^{1/2} 2x e^{-x^2}\,dx$$

$$= \left[-e^{-x^2}\right]_{0}^{1/2}$$

$$= -e^{-1/4} + 1$$

$$= 1 - e^{-1/4} \approx 0.221$$

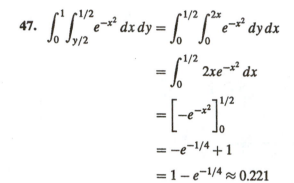

48. $\displaystyle\int_{0}^{1}\int_{0}^{\arccos y} \sin x \sqrt{1 + \sin^2 x}\,dx\,dy$

$$= \int_{0}^{\pi/2}\int_{0}^{\cos x} \sin x \sqrt{1 + \sin^2 x}\,dy\,dx$$

$$= \int_{0}^{\pi/2} (1 + \sin^2 x)^{1/2} \sin x \cos x\,dx$$

$$= \left[\dfrac{1}{2} \cdot \dfrac{2}{3}(1 + \sin^2 x)^{3/2}\right]_{0}^{\pi/2} = \dfrac{1}{3}\left[2\sqrt{2} - 1\right]$$

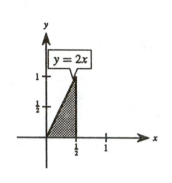

49. $\displaystyle\int_{0}^{\ln 10}\int_{e^x}^{10} \dfrac{1}{\ln y}\,dy\,dx = \int_{1}^{10}\int_{0}^{\ln y} \dfrac{1}{\ln y}\,dx\,dy$

$$= \int_{1}^{10} \left[\dfrac{x}{\ln y}\right]_{0}^{\ln y} dy$$

$$= \int_{1}^{10} dy = \left[y\right]_{1}^{10} = 9$$

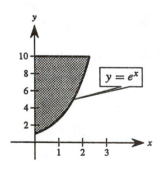

50. $\displaystyle\int_0^2 \int_{x^2}^4 \sqrt{y}\cos y\,dy\,dx = \int_0^4 \int_0^{\sqrt{y}} \sqrt{y}\cos y\,dx\,dy$

$$= \int_0^4 \left[x\sqrt{y}\cos y \right]_0^{\sqrt{y}} dy$$

$$= \int_0^4 y\cos y\,dy$$

$$= \left[y\sin y + \cos y \right]_0^4$$

$$= 4\sin 4 + \cos 4 - 1$$

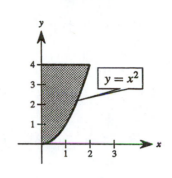

$y = x^2$

51. Average $= \dfrac{1}{8}\displaystyle\int_0^4 \int_0^2 x\,dy\,dx = \dfrac{1}{8}\int_0^4 2x\,dx = \left[\dfrac{x^2}{8}\right]_0^4 = 2$

52. Average $= \dfrac{1}{8}\displaystyle\int_0^4 \int_0^2 xy\,dy\,dx = \dfrac{1}{8}\int_0^4 2x\,dx = \left[\dfrac{x^2}{8}\right]_0^4 = 2$

53. Average $= \dfrac{1}{4}\displaystyle\int_0^2 \int_0^2 (x^2 + y^2)\,dx\,dy$

$$= \dfrac{1}{4}\int_0^2 \left[\dfrac{x^3}{3} + xy^2\right]_0^2 dy = \dfrac{1}{4}\int_0^2 \left(\dfrac{8}{3} + 2y^2\right) dy = \left[\dfrac{1}{4}\left(\dfrac{8}{3}y + \dfrac{2}{3}y^3\right)\right]_0^2 = \dfrac{8}{3}$$

54. Average $= \dfrac{1}{1/2}\displaystyle\int_0^1 \int_0^x e^{x+y}\,dy\,dx = 2\int_0^1 (e^{2x} - e^x)\,dx = 2\left[\dfrac{1}{2}e^{2x} - e^x\right]_0^1 = (e - 1)^2$

55. Average $= \dfrac{1}{1250}\displaystyle\int_{300}^{325} \int_{200}^{250} 100x^{0.6}y^{0.4}\,dx\,dy$

$$= \dfrac{1}{1250}\int_{300}^{325} \left[(100y^{0.4})\dfrac{x^{1.6}}{1.6}\right]_{200}^{250} dy = \dfrac{128{,}844.1}{1250}\int_{300}^{325} y^{0.4}\,dy = 103.0753\left[\dfrac{y^{1.4}}{1.4}\right]_{300}^{325} \approx 25{,}645.24$$

56. Average $= \dfrac{1}{150}\displaystyle\int_{45}^{60} \int_{40}^{50} [192x + 576y - x^2 - 5y^2 - 2xy - 5000]\,dx\,dy \approx 13{,}246.67$

57. $f(x, y) \geq 0$ for all (x, y) and

$$\int_{-\infty}^{\infty} \int_{-\infty}^{\infty} f(x, y)\,dA = \int_0^5 \int_0^2 \dfrac{1}{10}\,dy\,dx = \int_0^5 \dfrac{1}{5}\,dx = 1$$

$$P(0 \leq x \leq 2,\ 1 \leq y \leq 2) = \int_0^2 \int_1^2 \dfrac{1}{10}\,dy\,dx = \int_0^2 \dfrac{1}{10}\,dx = \dfrac{1}{5}.$$

58. $f(x, y) \geq 0$ for all (x, y) and

$$\int_{-\infty}^{\infty} \int_{-\infty}^{\infty} f(x, y)\,dA = \int_0^2 \int_0^2 \dfrac{1}{4}xy\,dy\,dx = \int_0^2 \dfrac{x}{2}\,dx = 1$$

$$P(0 \leq x \leq 1,\ 1 \leq y \leq 2) = \int_0^1 \int_1^2 \dfrac{1}{4}xy\,dy\,dx = \int_0^1 \dfrac{3x}{8}\,dx = \dfrac{3}{16}.$$

59. $f(x, y) \ge 0$ for all (x, y) and

$$\int_{-\infty}^{\infty} \int_{-\infty}^{\infty} f(x, y) \, dA = \int_0^3 \int_3^6 \frac{1}{27}(9 - x - y) \, dy \, dx$$

$$= \int_0^3 \frac{1}{27}\left[9y - xy - \frac{y^2}{2}\right]_3^6 dx = \int_0^3 \left(\frac{1}{2} - \frac{1}{9}x\right) dx = \left[\frac{x}{2} - \frac{x^2}{18}\right]_0^3 = 1$$

$$P(0 < x < 1, \ 4 < y < 6) = \int_0^1 \int_4^6 \frac{1}{27}(9 - x - y) \, dy \, dx = \int_0^1 \frac{2}{27}(4 - x) \, dx = \frac{7}{27}.$$

60. $f(x, y) \ge 0$ for all (x, y) and

$$\int_{-\infty}^{\infty} \int_{-\infty}^{\infty} f(x, y) \, dA = \int_0^{\infty} \int_0^{\infty} e^{-x-y} \, dy \, dx$$

$$= \int_0^{\infty} \lim_{b \to \infty}\left[-e^{-x-y}\right]_0^b dx = \int_0^{\infty} e^{-x} \, dx = \lim_{b \to \infty}\left[-e^{-x}\right]_0^b = 1$$

$$P(0 \le x \le 1, \ x \le y \le 1) = \int_0^1 \int_x^1 e^{-x-y} \, dy \, dx = \int_0^1 \left[-e^{-x-y}\right]_x^1 dx = \int_0^1 (e^{-2x} - e^{-x-1}) \, dx$$

$$= \left[-\frac{1}{2}e^{-2x} + e^{-x-1}\right]_0^1 = \frac{1}{2}e^{-2} - e^{-1} + \frac{1}{2} = \frac{1}{2}(e^{-1} - 1)^2 \approx 0.1998.$$

61. Line through $(0, 0)$ and $(2, 1)$: $y = \dfrac{x}{2}$

Line through $(0, 5)$ and $(2, 1)$: $y = -2x + 5$

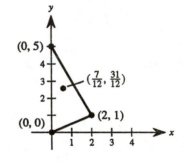

$$\iint_R y \, dA = \int_0^2 \int_{x/2}^{-2x+5} y \, dy \, dx = \int_0^2 \left[\frac{y^2}{2}\right]_{x/2}^{-2x+5} dx$$

$$= \frac{1}{2}\int_0^2 \left[(-2x+5)^2 - \frac{x^2}{4}\right] dx$$

$$= \frac{1}{2}\left[-\frac{1}{6}(-2x+5)^3 - \frac{x^3}{12}\right]_0^2 = \frac{1}{2}\left[-\frac{1}{6} - \frac{4}{6}\right] - \frac{1}{2}\left[-\frac{125}{6}\right] = 10$$

$$\iint_R xy \, dA = \int_0^2 \int_{x/2}^{-2x+5} xy \, dy \, dx$$

$$= \frac{1}{2}\int_0^2 x\left[(-2x+5)^2 - \frac{x^2}{4}\right] dx = \frac{1}{2}\int_0^2 \left[4x^3 - 20x^2 + 25x - \frac{x^3}{4}\right] dx$$

$$= \frac{1}{2}\left[x^4 - \frac{20x^3}{3} + \frac{25x^2}{2} - \frac{x^4}{16}\right]_0^2 = \frac{1}{2}\left[16 - \frac{160}{3} + 50 - 1\right] = \frac{35}{6}$$

$$\iint_R y^2 \, dA = \int_0^2 \int_{x/2}^{-2x+5} y^2 \, dy \, dx = \frac{1}{3}\int_0^2 \left[(-2x+5)^3 - \frac{x^3}{8}\right] dx$$

$$= \frac{1}{3}\left[-\frac{1}{8}(-2x+5)^4 - \frac{x^4}{32}\right]_0^2 = \frac{1}{3}\left[\left(-\frac{1}{8} - \frac{1}{2}\right) - \left(-\frac{625}{8}\right)\right] = \frac{155}{6}$$

Therefore, $x_p = \dfrac{\displaystyle\iint_R xy \, dA}{\displaystyle\iint_R y \, dA} = \dfrac{35/6}{10} = \dfrac{7}{12}$ and $y_p = \dfrac{\displaystyle\iint_R y^2 \, dA}{\displaystyle\iint_R y \, dA} = \dfrac{155/6}{10} = \dfrac{31}{12}.$

62. Line through $(0, 0)$ and $(3, 1)$: $y = \dfrac{x}{3}$

Line through $(0, 7)$ and $(3, 1)$: $y = -2x + 7$

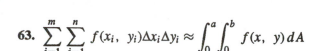

$$\iint_R y\, dA = \int_0^3 \int_{x/3}^{-2x+7} y\, dy\, dx = 28$$

$$\iint_R xy\, dA = \int_0^3 \int_{x/3}^{-2x+7} xy\, dy\, dx = \frac{189}{8}$$

$$\iint_R y^2\, dA = \int_0^3 \int_{x/3}^{-2x+7} y^2\, dy\, dx = \frac{399}{4}$$

Therefore, $x_p = \dfrac{\displaystyle\iint_R xy\, dA}{\displaystyle\iint_R y\, dA} = \dfrac{189/8}{28} = \dfrac{27}{32}$ and $y_p = \dfrac{\displaystyle\iint_R y^2\, dA}{\displaystyle\iint_R y\, dA} = \dfrac{399/4}{28} = \dfrac{57}{16}$.

63. $\displaystyle\sum_{i=1}^m \sum_{j=1}^n f(x_i,\ y_i)\Delta x_i \Delta y_i \approx \int_0^a \int_0^b f(x,\ y)\, dA$

Your computer program will vary depending on your computer system.

64. $\displaystyle\int_0^1 \int_0^2 \sin\sqrt{x+y}\, dy\, dx \approx 1.784$

65. $V \approx 125$
Matches d.

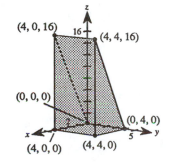

66. $V \approx 50$
Matches a.

67. False

$$V = 8\int_0^1 \int_0^{\sqrt{1-y^2}} \sqrt{1-x^2-y^2}\, dx\, dy$$

68. True

69. Essay

70. $\int_1^2 e^{-xy}\,dy = \left[-\frac{1}{x}e^{-xy}\right]_1^2 = \frac{e^{-x}-e^{-2x}}{x}$

Thus,

$$\int_0^\infty \frac{e^{-x}-e^{-2x}}{x}\,dx = \int_0^\infty \int_1^2 e^{-xy}\,dy\,dx$$

$$= \int_1^2 \int_0^\infty e^{-xy}\,dx\,dy = \int_1^2 \left[-\frac{e^{-xy}}{y}\right]_0^\infty dy = \int_1^2 \frac{1}{y}\,dy = \left[\ln y\right]_1^2 = \ln 2.$$

71. Average $= \int_0^1 f(x)\,dx = \int_0^1 \int_1^x e^{t^2}\,dt\,dx = -\int_0^1 \int_x^1 e^{t^2}\,dt\,dx$

$$= -\int_0^1 \int_0^t e^{t^2}\,dx\,dt = -\int_0^1 te^{t^2}\,dt$$

$$= \left[-\frac{1}{2}e^{t^2}\right]_0^1 = -\frac{1}{2}(e-1) = \frac{1}{2}(1-e)$$

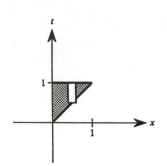

Section 14.3   Change of Variables: Polar Coordinates

1. $\displaystyle\int_0^{2\pi}\int_0^6 3r^2\sin\theta\,dr\,d\theta = \int_0^{2\pi}\left[r^3\sin\theta\right]_0^6 d\theta$

$$= \int_0^{2\pi} 216\sin\theta\,d\theta$$

$$= \left[-216\cos\theta\right]_0^{2\pi} = 0$$

2. $\displaystyle\int_0^{\pi/4}\int_0^4 r^2\sin\theta\cos\theta\,dr\,d\theta = \int_0^{\pi/4}\left[\frac{r^3}{3}\sin\theta\cos\theta\right]_0^4 d\theta$

$$= \left[\left(\frac{64}{3}\right)\frac{\sin^2\theta}{2}\right]_0^{\pi/4}$$

$$= \frac{16}{3}$$

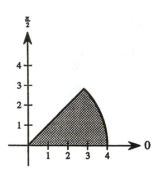

3. $\displaystyle\int_0^{\pi/2}\int_2^3 \sqrt{9-r^2}\,r\,dr\,d\theta = \int_0^{\pi/2}\left[-\frac{1}{3}(9-r^2)^{3/2}\right]_2^3 d\theta$

$$= \left[\frac{5\sqrt{5}}{3}\theta\right]_0^{\pi/2}$$

$$= \frac{5\sqrt{5}\,\pi}{6}$$

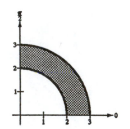

4. $\int_0^{\pi/2}\int_0^3 re^{-r^2}\,dr\,d\theta = \int_0^{\pi/2}\left[-\frac{1}{2}e^{-r^2}\right]_0^3 d\theta$

$= \left[-\frac{1}{2}(e^{-9}-1)\theta\right]_0^{\pi/2}$

$= \frac{\pi}{4}\left(1 - \frac{1}{e^9}\right)$

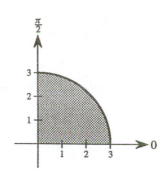

5. $\int_0^{\pi/2}\int_0^{1+\sin\theta}\theta\,dr\,d\theta = \int_0^{\pi/2}(\theta + \theta\sin\theta)\,d\theta$

$= \left[\frac{\theta^2}{2} + \sin\theta - \theta\cos\theta\right]_0^{\pi/2}$

$= \frac{\pi^2}{8} + 1$

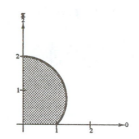

6. $\int_0^{\pi/2}\int_0^{1-\cos\theta}\sin\theta\,dr\,d\theta = \int_0^{\pi/2}(\sin\theta - \sin\theta\cos\theta)\,d\theta$

$= \left[-\cos\theta + \frac{\cos^2\theta}{2}\right]_0^{\pi/2}$

$= \frac{1}{2}$

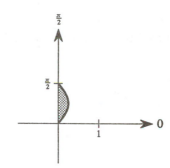

7. $A = \int_0^{\pi}\int_0^{6\cos\theta} r\,dr\,d\theta = \int_0^{\pi} 18\cos^2\theta\,d\theta = 9\int_0^{\pi}(1+\cos2\theta)\,d\theta = \left[9\left(\theta + \frac{1}{2}\sin2\theta\right)\right]_0^{\pi} = 9\pi$

8. $A = \int_0^{2\pi}\int_2^4 r\,dr\,d\theta = \int_0^{2\pi} 6\,d\theta = 12\pi$

9. $\int_0^{2\pi}\int_0^{1+\cos\theta} r\,dr\,d\theta = \frac{1}{2}\int_0^{2\pi}(1+2\cos\theta+\cos^2\theta)\,d\theta$

$= \frac{1}{2}\int_0^{2\pi}\left(1+2\cos\theta+\frac{1+\cos2\theta}{2}\right)d\theta = \frac{1}{2}\left[\theta + 2\sin\theta + \frac{1}{2}\left(\theta + \frac{1}{2}\sin2\theta\right)\right]_0^{2\pi} = \frac{3\pi}{2}$

10. $\int_0^{2\pi}\int_0^{3-2\sin\theta} r\,dr\,d\theta = \frac{1}{2}\int_0^{2\pi}(9 - 12\sin\theta + 4\sin^2\theta)\,d\theta = \frac{1}{2}\int_0^{2\pi}[9 - 12\sin\theta + 2(1-\cos2\theta)]\,d\theta$

$= \frac{1}{2}\left[9\theta + 12\cos\theta + 2\theta - \sin2\theta\right]_0^{2\pi} = \frac{1}{2}[22\pi + 12 - 12] = 11\pi$

11. $3\int_0^{\pi/3}\int_0^{2\sin3\theta} r\,dr\,d\theta = \frac{3}{2}\int_0^{\pi/3} 4\sin^2 3\theta\,d\theta = 3\int_0^{\pi/3}(1-\cos6\theta)\,d\theta = 3\left[\theta - \frac{1}{6}\sin6\theta\right]_0^{\pi/3} = \pi$

12. $8\int_0^{\pi/4}\int_0^{3\cos2\theta} r\,dr\,d\theta = 4\int_0^{\pi/4} 9\cos^2 2\theta\,d\theta = 18\int_0^{\pi/4}(1+\cos4\theta)\,d\theta = 18\left[\theta + \frac{1}{4}\sin4\theta\right]_0^{\pi/4} = \frac{9\pi}{2}$

13. $\displaystyle\int_0^a \int_0^{\sqrt{a^2-y^2}} y\,dx\,dy = \int_0^{\pi/2}\int_0^a r^2\sin\theta\,dr\,d\theta = \frac{a^3}{3}\int_0^{\pi/2}\sin\theta\,d\theta = \left[\frac{a^3}{3}(-\cos\theta)\right]_0^{\pi/2} = \frac{a^3}{3}$

14. $\displaystyle\int_0^a \int_0^{\sqrt{a^2-x^2}} x\,dy\,dx = \int_0^{\pi/2}\int_0^a r^2\cos\theta\,dr\,d\theta = \frac{a^3}{3}\int_0^{\pi/2}\cos\theta\,d\theta = \left[\frac{a^3}{3}\sin\theta\right]_0^{\pi/2} = \frac{a^3}{3}$

15. $\displaystyle\int_0^3 \int_0^{\sqrt{9-x^2}} (x^2+y^2)^{3/2}\,dy\,dx = \int_0^{\pi/2}\int_0^3 r^4\,dr\,d\theta = \frac{243}{5}\int_0^{\pi/2} d\theta = \frac{243\pi}{10}$

16. $\displaystyle\int_0^2 \int_y^{\sqrt{8-y^2}} \sqrt{x^2+y^2}\,dx\,dy = \int_0^{\pi/4}\int_0^{2\sqrt{2}} r^2\,dr\,d\theta$

$\displaystyle = \int_0^{\pi/4} \frac{(2\sqrt{2})^3}{3}\,d\theta = \left[\frac{(2\sqrt{2})^3}{3}\theta\right]_0^{\pi/4} = \frac{(2\sqrt{2})^3}{3}\cdot\frac{\pi}{4} = \frac{4\sqrt{2}\,\pi}{3}$

17. $\displaystyle\int_0^2 \int_0^{\sqrt{2x-x^2}} xy\,dy\,dx = \int_0^{\pi/2}\int_0^{2\cos\theta} r^3\cos\theta\sin\theta\,dr\,d\theta = 4\int_0^{\pi/2}\cos^5\theta\sin\theta\,d\theta = \left[-\frac{4\cos^6\theta}{6}\right]_0^{\pi/2} = \frac{2}{3}$

18. $\displaystyle\int_0^4 \int_0^{\sqrt{4y-y^2}} x^2\,dx\,dy = \int_0^{\pi/2}\int_0^{4\sin\theta} r^3\cos^2\theta\,dr\,d\theta = \int_0^{\pi/2} 64\sin^4\theta\cos^2\theta\,d\theta$

$\displaystyle = 64\int_0^{\pi/2}(\sin^4\theta - \sin^6\theta)\,d\theta = \frac{64}{6}\left[\sin^5\theta\cos\theta - \frac{\sin^3\theta\cos\theta}{4} + \frac{3}{8}(\theta - \sin\theta\cos\theta)\right]_0^{\pi/2} = 2\pi$

19. $\displaystyle\int_0^2 \int_0^x \sqrt{x^2+y^2}\,dy\,dx + \int_2^{2\sqrt{2}} \int_0^{\sqrt{8-x^2}} \sqrt{x^2+y^2}\,dy\,dx = \int_0^{\pi/4}\int_0^{2\sqrt{2}} r^2\,dr\,d\theta$

$\displaystyle = \int_0^{\pi/4} \frac{16\sqrt{2}}{3}\,d\theta$

$\displaystyle = \frac{4\sqrt{2}\,\pi}{3}$

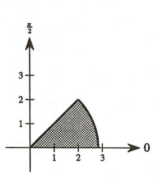

20. $\displaystyle\int_0^{(5\sqrt{2})/2} \int_0^x xy\,dy\,dx + \int_{(5\sqrt{2})/2}^5 \int_0^{\sqrt{25-x^2}} xy\,dy\,dx = \int_0^{\pi/4}\int_0^5 r^3\sin\theta\cos\theta\,dr\,d\theta$

$\displaystyle = \int_0^{\pi/4} \frac{625}{4}\sin\theta\cos\theta\,d\theta$

$\displaystyle = \left[\frac{625}{8}\sin^2\theta\right]_0^{\pi/4}$

$\displaystyle = \frac{625}{16}$

21. $\displaystyle\int_0^2 \int_0^{\sqrt{4-x^2}} (x+y)\,dy\,dx = \int_0^{\pi/2}\int_0^2 (r\cos\theta + r\sin\theta)r\,dr\,d\theta = \int_0^{\pi/2}\int_0^2 (\cos\theta+\sin\theta)r^2\,dr\,d\theta$

$\displaystyle = \frac{8}{3}\int_0^{\pi/2}(\cos\theta+\sin\theta)\,d\theta = \left[\frac{8}{3}(\sin\theta - \cos\theta)\right]_0^{\pi/2} = \frac{16}{3}$

22. $\displaystyle\int_0^2 \int_0^{\sqrt{4-x^2}} e^{-(x^2+y^2)}\,dy\,dx = \int_0^{\pi/2}\int_0^2 e^{-r^2}r\,dr\,d\theta = \frac{1}{2}\int_0^{\pi/2}(1-e^{-4})\,d\theta = \frac{\pi}{4}(1-e^{-4})$

23. $\displaystyle\int_0^{1/\sqrt{2}}\int_{\sqrt{1-y^2}}^{\sqrt{4-y^2}}\arctan\frac{y}{x}\,dx\,dy+\int_{1/\sqrt{2}}^{\sqrt{2}}\int_{y}^{\sqrt{4-y^2}}\arctan\frac{y}{x}\,dx\,dy$

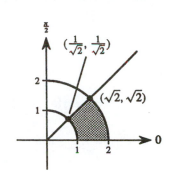

$$=\int_0^{\pi/4}\int_1^2\theta r\,dr\,d\theta$$

$$=\int_0^{\pi/4}\frac{3}{2}\theta\,d\theta=\left[\frac{3\theta^2}{4}\right]_0^{\pi/4}=\frac{3\pi^2}{64}$$

24. $\displaystyle\int_0^3\int_0^{\sqrt{9-x^2}}(9-x^2-y^2)\,dy\,dx=\int_0^{\pi/2}\int_0^3(9-r^2)r\,dr\,d\theta$

$$=\int_0^{\pi/2}\int_0^3(9r-r^3)\,dr\,d\theta=\int_0^{\pi/2}\left[\frac{9}{2}r^2-\frac{1}{4}r^4\right]_0^3d\theta=\frac{81}{4}\int_0^{\pi/2}d\theta=\frac{81\pi}{8}$$

25. $V=\displaystyle\int_0^{\pi/2}\int_0^1(r\cos\theta)(r\sin\theta)r\,dr\,d\theta$

$$=\frac{1}{2}\int_0^{\pi/2}\int_0^1 r^3\sin 2\theta\,dr\,d\theta$$

$$=\frac{1}{8}\int_0^{\pi/2}\sin 2\theta\,d\theta$$

$$=\left[-\frac{1}{16}\cos 2\theta\right]_0^{\pi/2}=\frac{1}{8}$$

26. $V=\displaystyle 4\int_0^{\pi/2}\int_0^2(r^2+1)r\,dr\,d\theta=24\int_0^{\pi/2}d\theta=12\pi$

27. $V=\displaystyle\int_0^{2\pi}\int_0^5 r^2\,dr\,d\theta=\frac{250\pi}{3}$

28. $V=\displaystyle\int_0^{2\pi}\int_2^4 r^2\,dr\,d\theta=\frac{112\pi}{3}$

29. $V=\displaystyle 2\int_0^{\pi/2}\int_0^{4\cos\theta}\sqrt{16-r^2}\,r\,dr\,d\theta=2\int_0^{\pi/2}\left[-\frac{1}{3}\left(\sqrt{16-r^2}\right)^3\right]_0^{4\cos\theta}d\theta=-\frac{2}{3}\int_0^{\pi/2}(64\sin^3\theta-64)\,d\theta$

$$=\frac{128}{3}\int_0^{\pi/2}[1-\sin\theta(1-\cos^2\theta)]\,d\theta=\frac{128}{3}\left[\theta+\cos\theta-\frac{\cos^3\theta}{3}\right]_0^{\pi/2}=\frac{64}{9}(3\pi-4)$$

30. $V=\displaystyle\int_0^{2\pi}\int_1^4\sqrt{16-r^2}\,r\,dr\,d\theta=\int_0^{2\pi}\left[-\frac{1}{3}\left(\sqrt{16-r^2}\right)^3\right]_1^4 d\theta=\int_0^{2\pi}5\sqrt{15}\,d\theta=10\sqrt{15}\,\pi$

31. $V=\displaystyle\int_0^{2\pi}\int_a^4\sqrt{16-r^2}\,r\,dr\,d\theta=\int_0^{2\pi}\left[-\frac{1}{3}\left(\sqrt{16-r^2}\right)^3\right]_a^4 d\theta=\frac{1}{3}\left(\sqrt{16-a^2}\right)^3(2\pi)$

One-half the volume of the hemisphere is $(64\pi)/3$.

$$\frac{2\pi}{3}(16-a^2)^{3/2}=\frac{64\pi}{3}$$

$$(16-a^2)^{3/2}=32$$

$$16-a^2=32^{2/3}$$

$$a^2=16-32^{2/3}=16-8\sqrt[3]{2}$$

$$a=\sqrt{4\left(4-2\sqrt[3]{2}\right)}=2\sqrt{4-2\sqrt[3]{2}}\approx 2.4332$$

32. $x^2 + y^2 + z^2 = a^2 \Rightarrow z = \sqrt{a^2 - (x^2 + y^2)} = \sqrt{a^2 - r^2}$

$$V = 8 \int_0^{\pi/2} \int_0^a \sqrt{a^2 - r^2}\, r\, dr\, d\theta \quad \text{(8 times the volume in the first octant)}$$

$$= 8 \int_0^{\pi/2} \left[-\frac{1}{2} \cdot \frac{2}{3}(a^2 - r^2)^{3/2} \right]_0^a d\theta$$

$$= 8 \int_0^{\pi/2} \frac{a^3}{3}\, d\theta = \left[\frac{8a^3}{3}\theta \right]_0^{\pi/2} = \frac{4\pi a^3}{3}$$

33. a. $I^2 = \int_{-\infty}^{\infty} \int_{-\infty}^{\infty} e^{-(x^2+y^2)/2}\, dA = 4 \int_0^{\pi/2} \int_0^{\infty} e^{-r^2/2}\, r\, dr\, d\theta = 4 \int_0^{\pi/2} \left[-e^{-r^2/2} \right]_0^{\infty} d\theta = 4 \int_0^{\pi/2} d\theta = 2\pi$

b. Therefore, $I = \sqrt{2\pi}$.

34. a. Let $u = \sqrt{2}x$, then $\int_{-\infty}^{\infty} e^{-x^2}\, dx = \int_{-\infty}^{\infty} e^{-u^2/2} \frac{1}{\sqrt{2}}\, du = \frac{1}{\sqrt{2}}(\sqrt{2\pi}) = \sqrt{\pi}$.

b. Let $u = 2x$, then $\int_{-\infty}^{\infty} e^{-4x^2}\, dx = \int_{-\infty}^{\infty} e^{-u^2} \frac{1}{2}\, du = \frac{1}{2}\sqrt{\pi}$.

35. $\int_{-7}^{7} \int_{-\sqrt{49-x^2}}^{\sqrt{49-x^2}} 4000 e^{-0.01(x^2+y^2)}\, dy\, dx = \int_0^{2\pi} \int_0^7 4000 e^{-0.01 r^2}\, r\, dr\, d\theta = \int_0^{2\pi} \left[-200{,}000 e^{-0.01 r^2} \right]_0^7 d\theta$

$$= 2\pi(-200{,}000)(e^{-0.49} - 1) = 400{,}000\pi(1 - e^{-0.49}) \approx 486{,}788$$

36. $\int_0^{\infty} \int_0^{\infty} k e^{-(x^2+y^2)}\, dy\, dx = \int_0^{\pi/2} \int_0^{\infty} k e^{-r^2}\, r\, dr\, d\theta = \int_0^{\pi/2} \left[-\frac{k}{2} e^{-r^2} \right]_0^{\infty} d\theta = \int_0^{\pi/2} \frac{k}{2}\, d\theta = \frac{k\pi}{4}$

For $f(x, y)$ to be a probability density function,

$$\frac{k\pi}{4} = 1$$

$$k = \frac{4}{\pi}.$$

37. a. $\int_2^4 \int_{y/\sqrt{3}}^{y} f\, dx\, dy$

b. $\int_{2/\sqrt{3}}^{2} \int_2^{\sqrt{3}x} f\, dy\, dx + \int_2^{4/\sqrt{3}} \int_x^{\sqrt{3}x} f\, dy\, dx + \int_{4/\sqrt{3}}^{4} \int_x^{4} f\, dy\, dx$

c. $\int_{\pi/4}^{\pi/3} \int_{2\csc\theta}^{4\csc\theta} f r\, dr\, d\theta$

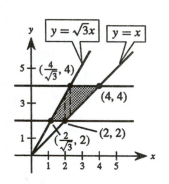

38. a. $4\int_0^2 \int_2^{2+\sqrt{4-y^2}} f\, dx\, dy$

b. $4\int_0^2 \int_0^{\sqrt{4-(x-2)^2}} f\, dy\, dx$

c. $2\int_0^{\pi/2} \int_0^{4\cos\theta} f r\, dr\, d\theta$

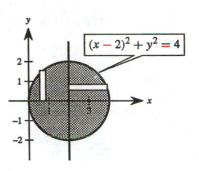

39. $A = \dfrac{\Delta\theta r_2^2}{2} - \dfrac{\Delta\theta r_1^2}{2} = \Delta\theta\left(\dfrac{r_1 + r_2}{2}\right)(r_2 - r_1) = r\,\Delta r\,\Delta\theta$

40. $\displaystyle\iint_R \ln(x^2 + y^2)\, dA = \int_0^{\pi/2} \int_1^2 (\ln r^2) r\, dr\, d\theta$

$$= 2\int_0^{\pi/2} \int_1^2 r\ln r\, dr\, d\theta$$

$$= 2\int_0^{\pi/2} \left[\frac{r^2}{4}(-1 + 2\ln r)\right]_1^2 d\theta$$

$$= 2\int_0^{\pi/2} \left(\ln 4 - \frac{3}{4}\right) d\theta$$

$$= \pi\left(\ln 4 - \frac{3}{4}\right)$$

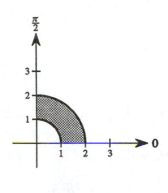

41. $16\displaystyle\int_0^a \int_0^{\sqrt{a^2-x^2}} \int_0^{\sqrt{a^2-x^2-y^2}} \int_0^{\sqrt{a^2-x^2-y^2-z^2}} dw\, dz\, dy\, dx$

$$= 16\int_0^a \int_0^{\sqrt{a^2-x^2}} \int_0^{\sqrt{a^2-x^2-y^2}} \sqrt{a^2 - x^2 - y^2 - z^2}\, dz\, dy\, dx$$

$$= 16\int_0^{\pi/2} \int_0^a \int_0^{\sqrt{a^2-r^2}} \sqrt{(a^2 - r^2) - z^2}\, dz(r\, dr\, d\theta)$$

$$= 16\int_0^{\pi/2} \int_0^a \frac{1}{2}\left[z\sqrt{(a^2 - r^2) - z^2} + (a^2 - r^2)\arcsin\frac{z}{\sqrt{a^2 - r^2}}\right]_0^{\sqrt{a^2-r^2}} r\, dr\, d\theta$$

$$= 8\int_0^{\pi/2} \int_0^a \frac{\pi}{2}(a^2 - r^2) r\, dr\, d\theta$$

$$= 4\pi\int_0^{\pi/2} \left[\frac{a^2 r^2}{2} - \frac{r^4}{4}\right]_0^a d\theta$$

$$= a^4\pi\int_0^{\pi/2} d\theta = \frac{a^4\pi^2}{2}$$

Section 14.4   Center of Mass and Moments of Inertia

1. a. $m = \int_0^a \int_0^b k\,dy\,dx = kab$

    $M_x = \int_0^a \int_0^b ky\,dy\,dx = \dfrac{kab^2}{2}$

    $M_y = \int_0^a \int_0^b kx\,dy\,dx = \dfrac{ka^2b}{2}$

    $\bar{x} = \dfrac{M_y}{m} = \dfrac{ka^2b/2}{kab} = \dfrac{a}{2}$

    $\bar{y} = \dfrac{M_x}{m} = \dfrac{kab^2/2}{kab} = \dfrac{b}{2}$

b. $m = \int_0^a \int_0^b ky\,dy\,dx = \dfrac{kab^2}{2}$

    $M_x = \int_0^a \int_0^b ky^2\,dy\,dx = \dfrac{kab^3}{3}$

    $M_y = \int_0^a \int_0^b kxy\,dy\,dx = \dfrac{ka^2b^2}{4}$

    $\bar{x} = \dfrac{M_y}{m} = \dfrac{ka^2b^2/4}{kab^2/2} = \dfrac{a}{2}$

    $\bar{y} = \dfrac{M_x}{m} = \dfrac{kab^3/3}{kab^2/2} = \dfrac{2}{3}b$

2. a. $m = \int_0^a \int_0^b kxy\,dy\,dx = \dfrac{ka^2b^2}{4}$

    $M_x = \int_0^a \int_0^b kxy^2\,dy\,dx = \dfrac{ka^2b^3}{6}$

    $M_y = \int_0^a \int_0^b kx^2y\,dy\,dx = \dfrac{ka^3b^2}{6}$

    $\bar{x} = \dfrac{M_y}{m} = \dfrac{ka^3b^2/6}{ka^2b^2/4} = \dfrac{2}{3}a$

    $\bar{y} = \dfrac{M_x}{m} = \dfrac{ka^2b^3/6}{ka^2b^2/4} = \dfrac{2}{3}b$

b. $m = \int_0^a \int_0^b k(x^2 + y^2)\,dy\,dx = \dfrac{kab}{3}(a^2 + b^2)$

    $M_x = \int_0^a \int_0^b k(x^2y + y^3)\,dy\,dx = \dfrac{kab^2}{12}(2a^2 + 3b^2)$

    $M_y = \int_0^a \int_0^b k(x^3 + xy^2)\,dy\,dx = \dfrac{ka^2b}{12}(3a^2 + 2b^2)$

    $\bar{x} = \dfrac{M_y}{m} = \dfrac{(ka^2b/12)(3a^2 + 2b^2)}{(kab/3)(a^2 + b^2)} = \dfrac{a}{4}\left(\dfrac{3a^2 + 2b^2}{a^2 + b^2}\right)$

    $\bar{y} = \dfrac{M_x}{m} = \dfrac{(kab^2/12)(2a^2 + 3b^2)}{(kab/3)(a^2 + b^2)} = \dfrac{b}{4}\left(\dfrac{2a^2 + 3b^2}{a^2 + b^2}\right)$

3. a. $m = \dfrac{k}{2}bh$

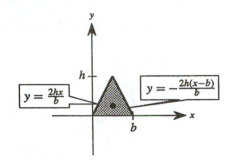

$\bar{x} = \dfrac{b}{2}$ by symmetry

$M_x = \displaystyle\int_0^{b/2} \int_0^{2hx/b} ky \, dy \, dx + \int_{b/2}^b \int_0^{-2h(x-b)/b} ky \, dy \, dx$

$\quad = \dfrac{kbh^2}{12} + \dfrac{kbh^2}{12} = \dfrac{kbh^2}{6}$

$\bar{y} = \dfrac{M_x}{m} = \dfrac{kbh^2/6}{kbh/2} = \dfrac{h}{3}$

b. $m = \displaystyle\int_0^{b/2} \int_0^{2hx/b} ky \, dy \, dx + \int_{b/2}^b \int_0^{-2h(x-b)/b} ky \, dy \, dx = \dfrac{kbh^2}{6}$

$M_x = \displaystyle\int_0^{b/2} \int_0^{2hx/b} ky^2 \, dy \, dx + \int_{b/2}^b \int_0^{-2h(x-b)/b} ky^2 \, dy \, dx = \dfrac{kbh^3}{12}$

$M_y = \displaystyle\int_0^{b/2} \int_0^{2hx/b} kxy \, dy \, dx + \int_{b/2}^b \int_0^{-2h(x-b)/b} kxy \, dy \, dx = \dfrac{kb^2h^2}{12}$

$\bar{x} = \dfrac{M_y}{m} = \dfrac{kb^2h^2/12}{kbh^2/6} = \dfrac{b}{2}$

$\bar{y} = \dfrac{M_x}{m} = \dfrac{kbh^3/12}{kbh^2/6} = \dfrac{h}{2}$

4. a. $m = \dfrac{a^2 k}{2}$

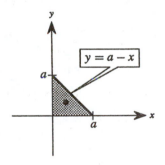

$M_x = \displaystyle\int_0^a \int_0^{a-x} ky \, dy \, dx = \dfrac{ka^3}{6}$

$M_y = M_x$ by symmetry

$\bar{x} = \bar{y} = \dfrac{M_x}{m} = \dfrac{ka^3/6}{ka^2/2} = \dfrac{a}{3}$

b. $m = \displaystyle\int_0^a \int_0^{a-x} (x^2 + y^2) \, dy \, dx$

$\quad = \displaystyle\int_0^a \left[x^2 y + \dfrac{1}{3} y^3 \right]_0^{a-x} dx = \int_0^a \left[ax^2 - x^3 + \dfrac{1}{3}(a-x)^3 \right] dx = \dfrac{a^4}{6}$

$M_y = \displaystyle\int_0^a \int_0^{a-x} (x^3 + xy^2) \, dy \, dx$

$\quad = \displaystyle\int_0^a \left(ax^3 - x^4 + \dfrac{1}{3}a^3 x - a^2 x^2 + ax^3 - \dfrac{1}{3}x^4 \right) dx = \dfrac{1}{3} \int_0^a (a^3 x - 3a^2 x^2 + 6ax^3 - 4x^4) \, dx = \dfrac{a^5}{15}$

$\bar{x} = \dfrac{M_y}{m} = \dfrac{a^5/15}{a^4/6} = \dfrac{2a}{5}$

$\bar{y} = \dfrac{2a}{5}$ by symmetry

5. a. $\bar{x} = 0$ by symmetry

$$m = \frac{\pi a^2 k}{2}$$

$$M_x = \int_{-a}^{a} \int_{0}^{\sqrt{a^2-x^2}} ky\,dy\,dx = \frac{2a^3k}{3}$$

$$\bar{y} = \frac{M_x}{m} = \frac{2a^3k}{3} \cdot \frac{2}{\pi a^2 k} = \frac{4a}{3\pi}$$

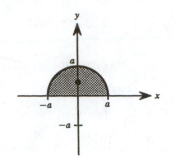

b. $m = \int_{-a}^{a} \int_{0}^{\sqrt{a^2-x^2}} k(a-y)y\,dy\,dx = \frac{a^4k}{24}(16 - 3\pi)$

$$M_x = \int_{-a}^{a} \int_{0}^{\sqrt{a^2-x^2}} k(a-y)y^2\,dy\,dx = \frac{a^5k}{120}(15\pi - 32)$$

$$M_y = \int_{-a}^{a} \int_{0}^{\sqrt{a^2-x^2}} kx(a-y)y\,dy\,dx = 0$$

$$\bar{x} = \frac{M_y}{m} = 0$$

$$\bar{y} = \frac{M_x}{m} = \frac{a}{5}\left[\frac{15\pi - 32}{16 - 3\pi}\right]$$

6. a. $m = \int_{0}^{a} \int_{0}^{\sqrt{a^2-x^2}} k\,dy\,dx = \frac{k\pi a^2}{4}$

$$M_y = \int_{0}^{a} \int_{0}^{\sqrt{a^2-x^2}} kx\,dy\,dx = k\int_{0}^{a} x\sqrt{a^2 - x^2}\,dx = \left[-\frac{k}{3}(a^2 - x^2)^{3/2}\right]_{0}^{a} = \frac{ka^3}{3}$$

$$\bar{x} = \frac{M_y}{m} = \frac{ka^3/3}{k\pi a^2/4} = \frac{4a}{3\pi}$$

$$\bar{y} = \frac{4a}{3\pi} \text{ by symmetry}$$

b. $m = \int_{0}^{a} \int_{0}^{\sqrt{a^2-x^2}} k(x^2 + y^2)\,dy\,dx = \int_{0}^{\pi/2} \int_{0}^{a} kr^3\,dr\,d\theta = \frac{ka^4\pi}{8}$

$$M_x = \int_{0}^{a} \int_{0}^{\sqrt{a^2-x^2}} k(x^2 + y^2)y\,dy\,dx = \int_{0}^{\pi/2} \int_{0}^{a} kr^4 \sin\theta\,dr\,d\theta = \frac{ka^5}{5}$$

$$M_y = M_x \text{ by symmetry}$$

$$\bar{x} = \bar{y} = \frac{M_x}{m} = \frac{ka^5}{5} \cdot \frac{8}{ka^4\pi} = \frac{8a}{5\pi}$$

7. $m = \int_{0}^{4} \int_{0}^{\sqrt{x}} kxy\,dy\,dx = \frac{32k}{3}$

$$M_x = \int_{0}^{4} \int_{0}^{\sqrt{x}} kxy^2\,dy\,dx = \frac{256k}{21}$$

$$M_y = \int_{0}^{4} \int_{0}^{\sqrt{x}} kx^2 y\,dy\,dx = 32k$$

$$\bar{x} = \frac{M_y}{m} = \frac{32k}{1} \cdot \frac{3}{32k} = 3$$

$$\bar{y} = \frac{M_x}{m} = \frac{256k}{21} \cdot \frac{3}{32k} = \frac{8}{7}$$

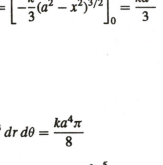

$y = \sqrt{x}$

8. $m = \displaystyle\int_0^4 \int_0^{x^2} kx \, dy \, dx = 64k$

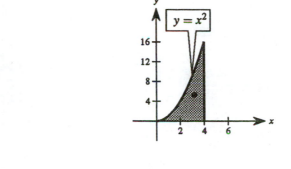

$M_x = \displaystyle\int_0^4 \int_0^{x^2} kxy \, dy \, dx = \dfrac{1024k}{3}$

$M_y = \displaystyle\int_0^4 \int_0^{x^2} kx^2 \, dy \, dx = \dfrac{1024k}{5}$

$\overline{x} = \dfrac{M_y}{m} = \dfrac{1024k}{5} \cdot \dfrac{1}{64k} = \dfrac{16}{5}$

$\overline{y} = \dfrac{M_x}{m} = \dfrac{1024k}{3} \cdot \dfrac{1}{64k} = \dfrac{16}{3}$

9. $\overline{x} = 0$ by symmetry

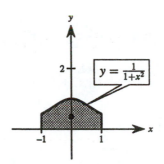

$m = \displaystyle\int_{-1}^1 \int_0^{1/(1+x^2)} k \, dy \, dx = \dfrac{k\pi}{2}$

$M_x = \displaystyle\int_{-1}^1 \int_0^{1/(1+x^2)} ky \, dy \, dx = \dfrac{k}{8}(2+\pi)$

$\overline{y} = \dfrac{M_x}{m} = \dfrac{k}{8}(2+\pi) \cdot \dfrac{2}{k\pi} = \dfrac{2+\pi}{4\pi}$

10. $m = \displaystyle\int_1^4 \int_0^{4/x} kx^2 \, dy \, dx = 30k$

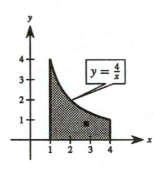

$M_x = \displaystyle\int_1^4 \int_0^{4/x} kx^2 y \, dy \, dx = 24k$

$M_y = \displaystyle\int_1^4 \int_0^{4/x} kx^3 \, dy \, dx = 84k$

$\overline{x} = \dfrac{M_y}{m} = \dfrac{84k}{30k} = \dfrac{14}{5}$

$\overline{y} = \dfrac{M_x}{m} = \dfrac{24k}{30k} = \dfrac{4}{5}$

11. $\overline{y} = 0$ by symmetry

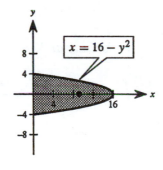

$m = \displaystyle\int_{-4}^4 \int_0^{16-y^2} kx \, dx \, dy = \dfrac{8192k}{15}$

$M_y = \displaystyle\int_{-4}^4 \int_0^{16-y^2} kx^2 \, dx \, dy = \dfrac{524{,}288k}{105}$

$\overline{x} = \dfrac{M_y}{m} = \dfrac{524{,}288k}{105} \cdot \dfrac{15}{8192k} = \dfrac{64}{7}$

12. $\bar{x} = 0$ by symmetry

$$m = \int_{-3}^{3} \int_{0}^{9-x^2} ky^2 \, dy \, dx = \frac{23{,}328k}{35}$$

$$M_x = \int_{-3}^{3} \int_{0}^{9-x^2} ky^3 \, dy \, dx = \frac{139{,}968k}{35}$$

$$\bar{y} = \frac{M_x}{m} = \frac{139{,}968k}{35} \cdot \frac{35}{23{,}328k} = 6$$

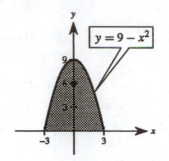
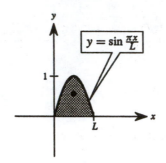

13. $\bar{x} = \dfrac{L}{2}$ by symmetry

$$m = \int_{0}^{L} \int_{0}^{\sin \pi x / L} ky \, dy \, dx = \frac{kL}{4}$$

$$M_x = \int_{0}^{L} \int_{0}^{\sin \pi x / L} ky^2 \, dy \, dx = \frac{4kL}{9\pi}$$

$$\bar{y} = \frac{M_x}{m} = \frac{4kL}{9\pi} \cdot \frac{4}{kL} = \frac{16}{9\pi}$$

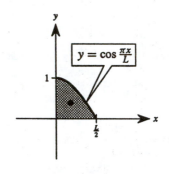

14. $$m = \int_{0}^{L/2} \int_{0}^{\cos \pi x / L} k \, dy \, dx = \frac{kL}{\pi}$$

$$M_x = \int_{0}^{L/2} \int_{0}^{\cos \pi x / L} ky \, dy \, dx = \frac{kL}{8}$$

$$M_y = \int_{0}^{L/2} \int_{0}^{\cos \pi x / L} kx \, dy \, dx = \frac{L^2(\pi - 2)k}{2\pi^2}$$

$$\bar{x} = \frac{M_y}{m} = \frac{L^2(\pi - 2)k}{2\pi^2} \cdot \frac{\pi}{kL} = \frac{L(\pi - 2)}{2\pi}$$

$$\bar{y} = \frac{M_x}{m} = \frac{kL}{8} \cdot \frac{\pi}{kL} = \frac{\pi}{8}$$

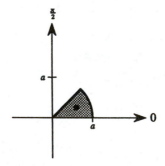

15. $$m = \frac{\pi a^2 k}{8}$$

$$M_x = \iint_{R} ky \, dA = \int_{0}^{\pi/4} \int_{0}^{a} kr^2 \sin \theta \, dr \, d\theta = \frac{ka^3(2 - \sqrt{2})}{6}$$

$$M_y = \iint_{R} kx \, dA = \int_{0}^{\pi/4} \int_{0}^{a} kr^2 \cos \theta \, dr \, d\theta = \frac{ka^3 \sqrt{2}}{6}$$

$$\bar{x} = \frac{M_y}{m} = \frac{ka^3 \sqrt{2}}{6} \cdot \frac{8}{\pi a^2 k} = \frac{4a\sqrt{2}}{3\pi}$$

$$\bar{y} = \frac{M_x}{m} = \frac{ka^3(2 - \sqrt{2})}{6} \cdot \frac{8}{\pi a^2 k} = \frac{4a(2 - \sqrt{2})}{3\pi}$$

16. $m = \iint_R k\sqrt{x^2 + y^2}\, dA = \int_0^{\pi/4} \int_0^a kr^2\, dr\, d\theta = \dfrac{ka^3\pi}{12}$

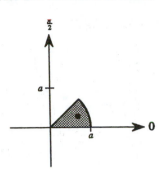

$M_x = \iint_R k\sqrt{x^2 + y^2}\, y\, dA = \int_0^{\pi/4} \int_0^a kr^3 \sin\theta\, d\theta = \dfrac{ka^4(2 - \sqrt{2})}{8}$

$M_y = \iint_R k\sqrt{x^2 + y^2}\, x\, dA = \int_0^{\pi/4} \int_0^a kr^3 \cos\theta\, d\theta = \dfrac{ka^4\sqrt{2}}{8}$

$\bar{x} = \dfrac{M_y}{m} = \dfrac{ka^4\sqrt{2}}{8} \cdot \dfrac{12}{ka^3\pi} = \dfrac{3\sqrt{2}\,a}{2\pi}$

$\bar{y} = \dfrac{M_x}{m} = \dfrac{ka^4(2 - \sqrt{2})}{8} \cdot \dfrac{12}{ka^3\pi} = \dfrac{3(2 - \sqrt{2})a}{2\pi}$

17. $m = \int_0^2 \int_0^{e^{-x}} ky\, dy\, dx = \dfrac{k}{4}(1 - e^{-4})$

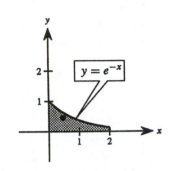

$M_x = \int_0^2 \int_0^{e^{-x}} ky^2\, dy\, dx = \dfrac{k}{9}(1 - e^{-6})$

$M_y = \int_0^2 \int_0^{e^{-x}} kxy\, dy\, dx = \dfrac{k(1 - 5e^{-4})}{8}$

$\bar{x} = \dfrac{M_y}{m} = \dfrac{k(e^4 - 5)}{8e^4} \cdot \dfrac{4e^4}{k(e^4 - 1)} = \dfrac{e^4 - 5}{2(e^4 - 1)}$

$\bar{y} = \dfrac{M_x}{m} = \dfrac{k(e^6 - 1)}{9e^6} \cdot \dfrac{4e^4}{k(e^4 - 1)} = \dfrac{4}{9}\left[\dfrac{e^6 - 1}{e^6 - e^2}\right]$

18. $m = \int_1^e \int_0^{\ln x} \dfrac{k}{x}\, dy\, dx = \dfrac{k}{2}$

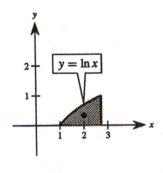

$M_x = \int_1^e \int_0^{\ln x} \dfrac{k}{x}\, y\, dy\, dx = \dfrac{k}{6}$

$M_y = \int_1^e \int_0^{\ln x} k\, dy\, dx = k$

$\bar{x} = \dfrac{M_y}{m} = \dfrac{k}{1} \cdot \dfrac{2}{k} = 2$

$\bar{y} = \dfrac{M_x}{m} = \dfrac{k}{6} \cdot \dfrac{2}{k} = \dfrac{1}{3}$

19. $\bar{y} = 0$ by symmetry

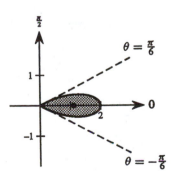

$m = \iint_R k\, dA = \int_{-\pi/6}^{\pi/6} \int_0^{2\cos 3\theta} kr\, dr\, d\theta = \dfrac{k\pi}{3}$

$M_y = \iint_R kx\, dA$

$= \int_{-\pi/6}^{\pi/6} \int_0^{2\cos 3\theta} kr^2 \cos\theta\, dr\, d\theta = \int_{-\pi/6}^{\pi/6} \dfrac{8k}{3} \cos^3 3\theta \cos\theta\, d\theta \approx 1.17k$

$\bar{x} = \dfrac{M_y}{m} \approx 1.17k\left(\dfrac{3}{\pi k}\right) \approx 1.12$

20. $\bar{y} = 0$ by symmetry

$$m = \int_R\!\!\int k\,dA = \int_0^{2\pi}\!\int_0^{1+\cos\theta} kr\,dr\,d\theta = \frac{3\pi k}{2}$$

$$M_y = \int_R\!\!\int kx\,dA = \int_0^{2\pi}\!\int_0^{1+\cos\theta} kr^2\cos\theta\,dr\,d\theta$$

$$= \frac{k}{3}\int_0^{2\pi} \cos\theta(1 + 3\cos\theta + 3\cos^2\theta + \cos^3\theta)\,d\theta$$

$$= \frac{k}{3}\int_0^{2\pi}\left[\cos\theta + \frac{3}{2}(1+\cos\theta) + 3\cos\theta(1-\sin^2\theta) + \frac{1}{4}(1+\cos 2\theta)^2\right]d\theta$$

$$= \frac{5k\pi}{4}$$

$$\bar{x} = \frac{M_y}{m} = \frac{5k\pi}{4}\cdot\frac{2}{3k\pi} = \frac{5}{6}$$

21. $m = bh$

$$I_x = \int_0^b\!\int_0^h y^2\,dy\,dx = \frac{bh^3}{3}$$

$$I_y = \int_0^b\!\int_0^h x^2\,dy\,dx = \frac{b^3h}{3}$$

$$\bar{\bar{x}} = \sqrt{\frac{I_y}{m}} = \sqrt{\frac{b^3h}{3}\cdot\frac{1}{bh}} = \sqrt{\frac{b^2}{3}} = \frac{b}{\sqrt{3}}$$

$$\bar{\bar{y}} = \sqrt{\frac{I_x}{m}} = \sqrt{\frac{bh^3}{3}\cdot\frac{1}{bh}} = \sqrt{\frac{h^2}{3}} = \frac{h}{\sqrt{3}}$$

22. $m = \int_0^b\!\int_0^{h-(hx/b)} dy\,dx = \frac{bh}{2}$

$$I_x = \int_0^b\!\int_0^{h-(hx/b)} y^2\,dy\,dx = \frac{bh^3}{12}$$

$$I_y = \int_0^b\!\int_0^{h-(hx/b)} x^2\,dy\,dx = \frac{b^3h}{12}$$

$$\bar{\bar{x}} = \sqrt{\frac{I_y}{m}} = \sqrt{\frac{b^3h/12}{bh/2}} = \frac{b}{\sqrt{6}}$$

$$\bar{\bar{y}} = \sqrt{\frac{I_x}{m}} = \sqrt{\frac{bh^3/12}{bh/2}} = \frac{h}{\sqrt{6}}$$

23. $m = \pi r^2$

$$I_x = \int_R\!\!\int y^2\,dA = \int_0^{2\pi}\!\int_0^r r^3\sin^2\theta\,dr\,d\theta = \frac{r^4\pi}{4}$$

$$I_y = \int_R\!\!\int x^2\,dA = \int_0^{2\pi}\!\int_0^r r^3\cos^2\theta\,dr\,d\theta = \frac{r^4\pi}{4}$$

$$I_0 = I_x + I_y = \frac{r^4\pi}{4} + \frac{r^4\pi}{4} = \frac{r^4\pi}{2}$$

$$\bar{\bar{x}} = \bar{\bar{y}} = \sqrt{\frac{I_x}{m}} = \sqrt{\frac{r^4\pi}{4}\cdot\frac{1}{\pi r^2}} = \frac{r}{2}$$

24. $m = \dfrac{\pi r^2}{2}$

$$I_x = \int_R\!\!\int y^2\,dA = \int_0^{\pi}\!\int_0^r r^3\sin^2\theta\,dr\,d\theta = \frac{r^4\pi}{8}$$

$$I_y = \int_R\!\!\int x^2\,dA = \int_0^{\pi}\!\int_0^r r^3\cos^2\theta\,dr\,d\theta = \frac{r^4\pi}{8}$$

$$I_0 = I_x + I_y = \frac{r^4\pi}{8} + \frac{r^4\pi}{8} = \frac{r^4\pi}{4}$$

$$\bar{\bar{x}} = \bar{\bar{y}} = \sqrt{\frac{I_x}{m}} = \sqrt{\frac{r^4\pi}{8}\cdot\frac{2}{\pi r^2}} = \frac{r}{2}$$

25. $m = \dfrac{\pi r^2}{4}$

$$I_x = \int_R\int y^2 \, dA = \int_0^{\pi/2}\int_0^r r^3 \sin^2\theta \, dr \, d\theta = \frac{\pi r^4}{16}$$

$$I_y = \int_R\int x^2 \, dA = \int_0^{\pi/2}\int_0^r r^3 \cos^2\theta \, dr \, d\theta = \frac{\pi r^4}{16}$$

$$I_0 = I_x + I_y = \frac{\pi r^4}{16} + \frac{\pi r^4}{16} = \frac{\pi r^4}{8}$$

$$\bar{\bar{x}} = \bar{\bar{y}} = \sqrt{\frac{I_x}{m}} = \sqrt{\frac{\pi r^4}{16}\cdot\frac{4}{\pi r^2}} = \frac{r}{2}$$

26. $m = \pi ab$

$$I_x = 4\int_0^a\int_0^{(b/a)\sqrt{a^2-x^2}} y^2 \, dy \, dx$$

$$= 4\int_0^a \frac{b}{3a}(a^2-x^2)^{3/2} \, dx = \frac{4b}{3a}\int_0^a \left[a^2\sqrt{a^2-x^2} - x^2\sqrt{a^2-x^2}\right] dx$$

$$= \frac{4b}{3a}\left[\frac{a^2}{2}\left(x\sqrt{a^2-x^2} + a^2\arcsin\frac{x}{a}\right) - \frac{1}{8}\left[x(2x^2-a^2)\sqrt{a^2-x^2} + a^4\arcsin\frac{x}{a}\right]\right]_0^a = \frac{a^3 b\pi}{4}$$

$$I_y = 4\int_0^b\int_0^{(a/b)\sqrt{b^2-y^2}} x^2 \, dx \, dy = \frac{ab^3\pi}{4}$$

$$I_0 = I_x + I_y = \frac{a^3 b\pi}{4} + \frac{ab^3\pi}{4} = \frac{ab\pi}{4}(a^2+b^2)$$

$$\bar{\bar{x}} = \sqrt{\frac{I_y}{m}} = \sqrt{\frac{ab^3\pi}{4}\cdot\frac{1}{\pi ab}} = \frac{b}{2}$$

$$\bar{\bar{y}} = \sqrt{\frac{I_x}{m}} = \sqrt{\frac{a^3 b\pi}{4}\cdot\frac{1}{\pi ab}} = \frac{a}{2}$$

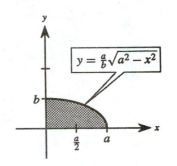

$y = \frac{a}{b}\sqrt{a^2-x^2}$

27. $\rho = ky$

$$m = k\int_0^a\int_0^b y \, dy \, dx = \frac{kab^2}{2}$$

$$I_x = k\int_0^a\int_0^b y^3 \, dy \, dx = \frac{kab^4}{4}$$

$$I_y = k\int_0^a\int_0^b x^2 y \, dy \, dx = \frac{ka^3 b^2}{6}$$

$$I_0 = I_x + I_y = \frac{3kab^4 + 2kb^2 a^3}{12}$$

$$\bar{\bar{x}} = \sqrt{\frac{I_y}{m}} = \sqrt{\frac{ka^3 b^2/6}{kab^2/2}} = \sqrt{\frac{a^2}{3}} = \frac{a}{\sqrt{3}}$$

$$\bar{\bar{y}} = \sqrt{\frac{I_x}{m}} = \sqrt{\frac{kab^4/4}{kab^2/2}} = \sqrt{\frac{b^2}{2}} = \frac{b}{\sqrt{2}}$$

28. $\rho = ky$

$$m = 2k\int_0^a\int_0^{\sqrt{a^2-x^2}} y \, dy \, dx$$

$$= k\int_0^a (a^2-x^2) \, dx = \frac{2ka^3}{3}$$

$$I_x = k\int_{-a}^a\int_0^{\sqrt{a^2-x^2}} y^3 \, dy \, dx = \frac{4ka^5}{15}$$

$$I_y = k\int_{-a}^a\int_0^{\sqrt{a^2-x^2}} x^2 y \, dy \, dx = \frac{2ka^5}{15}$$

$$I_0 = I_x + I_y = \frac{2ka^5}{5}$$

$$\bar{\bar{x}} = \sqrt{\frac{I_y}{m}} = \sqrt{\frac{2ka^5/15}{2ka^3/3}} = \sqrt{\frac{a^2}{5}} = \frac{a}{\sqrt{5}}$$

$$\bar{\bar{y}} = \sqrt{\frac{I_x}{m}} = \sqrt{\frac{4ka^5/15}{2ka^3/3}} = \sqrt{\frac{2a^2}{5}} = \frac{2a}{\sqrt{10}}$$

29. $\rho = kx$

$$m = k\int_0^2 \int_0^{4-x^2} x\,dy\,dx = 4k$$

$$I_x = k\int_0^2 \int_0^{4-x^2} xy^2\,dy\,dx = \frac{32k}{3}$$

$$I_y = k\int_0^2 \int_0^{4-x^2} x^3\,dy\,dx = \frac{16k}{3}$$

$$I_0 = I_x + I_y = 16k$$

$$\overline{\overline{x}} = \sqrt{\frac{I_y}{m}} = \sqrt{\frac{16k/3}{4k}} = \sqrt{\frac{4}{3}} = \frac{2}{\sqrt{3}}$$

$$\overline{\overline{y}} = \sqrt{\frac{I_x}{m}} = \sqrt{\frac{32k/3}{4k}} = \sqrt{\frac{8}{3}} = \frac{4}{\sqrt{6}}$$

30. $\rho = kxy$

$$m = k\int_0^1 \int_{x^2}^{x} xy\,dy\,dx = \frac{k}{2}\int_0^1 (x^3 - x^5)\,dx = \frac{k}{24}$$

$$I_x = k\int_0^1 \int_{x^2}^{x} xy^3\,dy\,dx = \frac{k}{4}\int_0^1 (x^5 - x^9)\,dx = \frac{k}{60}$$

$$I_y = k\int_0^1 \int_{x^2}^{x} x^3 y\,dy\,dx = \frac{k}{2}\int_0^1 (x^5 - x^7)\,dx = \frac{k}{48}$$

$$I_0 = I_x + I_y = \frac{9k}{240} = \frac{3k}{80}$$

$$\overline{\overline{x}} = \sqrt{\frac{I_y}{m}} = \sqrt{\frac{k/48}{k/24}} = \frac{1}{\sqrt{2}}$$

$$\overline{\overline{y}} = \sqrt{\frac{I_x}{m}} = \sqrt{\frac{k/60}{k/24}} = \sqrt{\frac{2}{5}}$$

31. $\rho = kxy$

$$m = \int_0^4 \int_0^{\sqrt{x}} kxy\,dy\,dx = \frac{32k}{3}$$

$$I_x = \int_0^4 \int_0^{\sqrt{x}} kxy^3\,dy\,dx = 16k$$

$$I_y = \int_0^4 \int_0^{\sqrt{x}} kx^3 y\,dy\,dx = \frac{512k}{5}$$

$$I_0 = I_x + I_y = \frac{592k}{5}$$

$$\overline{\overline{x}} = \sqrt{\frac{I_y}{m}} = \sqrt{\frac{512k}{5}\cdot\frac{3}{32k}} = \sqrt{\frac{48}{5}} = \frac{4\sqrt{15}}{5}$$

$$\overline{\overline{y}} = \sqrt{\frac{I_x}{m}} = \sqrt{\frac{16k}{1}\cdot\frac{3}{32k}} = \sqrt{\frac{3}{2}} = \frac{\sqrt{6}}{2}$$

32. $\rho = x^2 + y^2$

$$m = \int_0^1 \int_{x^2}^{\sqrt{x}} (x^2 + y^2)\,dy\,dx = \frac{6}{35}$$

$$I_x = \int_0^1 \int_{x^2}^{\sqrt{x}} (x^2 + y^2)y^2\,dy\,dx = \frac{158}{2079}$$

$$I_y = \int_0^1 \int_{x^2}^{\sqrt{x}} (x^2 + y^2)x^2\,dy\,dx = \frac{158}{2079}$$

$$I_0 = I_x + I_y = \frac{316}{2079}$$

$$\overline{\overline{x}} = \sqrt{\frac{I_y}{m}} = \sqrt{\frac{158}{2079}\cdot\frac{35}{6}} = \sqrt{\frac{395}{891}}$$

$$\overline{\overline{y}} = \sqrt{\frac{I_x}{m}} = \overline{\overline{x}} = \sqrt{\frac{395}{891}}$$

33. $\rho = kx$

$$m = \int_0^1 \int_{x^2}^{\sqrt{x}} kx\,dy\,dx = \frac{3k}{20}$$

$$I_x = \int_0^1 \int_{x^2}^{\sqrt{x}} kxy^2\,dy\,dx = \frac{3k}{56}$$

$$I_y = \int_0^1 \int_{x^2}^{\sqrt{x}} kx^3\,dy\,dx = \frac{k}{18}$$

$$I_0 = I_x + I_y = \frac{55k}{504}$$

$$\overline{\overline{x}} = \sqrt{\frac{I_y}{m}} = \sqrt{\frac{k}{18}\cdot\frac{20}{3k}} = \frac{\sqrt{30}}{9}$$

$$\overline{\overline{y}} = \sqrt{\frac{I_x}{m}} = \sqrt{\frac{3k}{56}\cdot\frac{20}{3k}} = \frac{\sqrt{70}}{14}$$

34. $\rho = ky$

$$m = 2\int_0^2 \int_{x^3}^{4x} ky\,dy\,dx = \frac{512k}{21}$$

$$I_x = 2\int_0^2 \int_{x^3}^{4x} ky^3\,dy\,dx = \frac{32{,}768k}{65}$$

$$I_y = 2\int_0^2 \int_{x^3}^{4x} kx^2 y\,dy\,dx = \frac{2048k}{45}$$

$$I_0 = I_x + I_y = \frac{321{,}536k}{585}$$

$$\overline{\overline{x}} = \sqrt{\frac{I_y}{m}} = \sqrt{\frac{2048k}{45}\cdot\frac{21}{512k}} = \sqrt{\frac{28}{15}} = \frac{2\sqrt{105}}{15}$$

$$\overline{\overline{y}} = \sqrt{\frac{I_x}{m}} = \sqrt{\frac{32{,}768k}{65}\cdot\frac{21}{512k}} = \frac{8\sqrt{1365}}{65}$$

35. $I = 2k \int_{-b}^{b} \int_{0}^{\sqrt{b^2-x^2}} (x-a)^2 \, dy \, dx = 2k \int_{-b}^{b} (x-a)^2 \sqrt{b^2 - x^2} \, dx$

$= 2k \left[\int_{-b}^{b} x^2 \sqrt{b^2 - x^2} \, dx - 2a \int_{-b}^{b} x\sqrt{b^2 - x^2} \, dx + a^2 \int_{-b}^{b} \sqrt{b^2 - x^2} \, dx \right]$

$= 2k \left[\dfrac{\pi b^4}{8} + 0 + \dfrac{\pi a^2 b^2}{2} \right] = \dfrac{k\pi b^2}{4}(b^2 + 4a^2)$

36. $I = \int_{0}^{4} \int_{0}^{2} k(x-6)^2 \, dy \, dx = \int_{0}^{4} 2k(x-6)^2 \, dx = \left[\dfrac{2k}{3}(x-6)^3 \right]_{0}^{4} = \dfrac{416k}{3}$

37. $I = \int_{0}^{4} \int_{0}^{\sqrt{x}} kx(x-6)^2 \, dy \, dx = \int_{0}^{4} kx\sqrt{x}(x^2 - 12x + 36) \, dx = k \left[\dfrac{2}{9}x^{9/2} - \dfrac{24}{7}x^{7/2} + \dfrac{72}{5}x^{5/2} \right]_{0}^{4} = \dfrac{42{,}752k}{315}$

38. $I = \int_{-a}^{a} \int_{0}^{\sqrt{a^2-x^2}} ky(y-a)^2 \, dy \, dx$

$= \int_{-a}^{a} k \left[\dfrac{y^4}{4} - \dfrac{2ay^3}{3} + \dfrac{a^2 y^2}{2} \right]_{0}^{\sqrt{a^2-x^2}} dx$

$= \int_{-a}^{a} k \left[\dfrac{1}{4}(a^4 - 2a^2 x^2 + x^4) - \dfrac{2a}{3}\left(a^2\sqrt{a^2-x^2} - x^2\sqrt{a^2-x^2}\right) + \dfrac{a^2}{2}(a^2 - x^2) \right] dx$

$= k \left[\dfrac{1}{4}\left(a^4 x - \dfrac{2a^2 x^3}{3} + \dfrac{a^5}{5}\right) - \dfrac{2a}{3}\left[\dfrac{a^2}{2}\left(x\sqrt{a^2-x^2} + a^2 \arcsin \dfrac{x}{a}\right) \right. \right.$

$\left. \left. - \dfrac{1}{8}\left(x(2x^2 - a^2)\sqrt{a^2-x^2} + a^4 \arcsin \dfrac{x}{a}\right)\right] + \dfrac{a^2}{2}\left(a^2 x - \dfrac{x^3}{3}\right) \right]_{-a}^{a}$

$= 2k \left[\dfrac{1}{4}\left(a^5 - \dfrac{2}{3}a^5 + \dfrac{1}{5}a^5\right) - \dfrac{2a}{3}\left(\dfrac{a^4 \pi}{4} - \dfrac{a^4 \pi}{16}\right) + \dfrac{a^2}{2}\left(a^3 - \dfrac{a^3}{3}\right) \right]$

$= 2k \left(\dfrac{7a^5}{15} - \dfrac{a^5 \pi}{8} \right) = ka^5 \left(\dfrac{56 - 15\pi}{60} \right)$

39. $I = \int_{0}^{a} \int_{0}^{\sqrt{a^2-x^2}} k(a-y)(y-a)^2 \, dy \, dx = \int_{0}^{a} \int_{0}^{\sqrt{a^2-x^2}} k(a-y)^3 \, dy \, dx = \int_{0}^{a} \left[-\dfrac{k}{4}(a-y)^4 \right]_{0}^{\sqrt{a^2-x^2}} dx$

$= -\dfrac{k}{4} \int_{0}^{a} \left[a^4 - 4a^3 y + 6a^2 y^2 - 4ay^3 + y^4 \right]_{0}^{\sqrt{a^2-x^2}} dx$

$= -\dfrac{k}{4} \int_{0}^{a} \left[a^4 - 4a^3 \sqrt{a^2-x^2} + 6a^2(a^2 - x^2) - 4a(a^2 - x^2)\sqrt{a^2-x^2} + (a^4 - 2a^2 x^2 + x^4) - a^4 \right] dx$

$= -\dfrac{k}{4} \int_{0}^{a} \left[7a^4 - 8a^2 x^2 + x^4 - 8a^3 \sqrt{a^2-x^2} + 4ax^2 \sqrt{a^2-x^2} \right] dx$

$= -\dfrac{k}{4} \left[7a^4 x - \dfrac{8a^2}{3}x^3 + \dfrac{x^5}{5} - 4a^3\left(x\sqrt{a^2-x^2} + a^2 \arcsin \dfrac{x}{a}\right) + \dfrac{a}{2}\left(x(2x^2 - a^2)\sqrt{a^2-x^2} + a^4 \arcsin \dfrac{x}{a}\right) \right]_{0}^{a}$

$= -\dfrac{k}{4}\left(7a^5 - \dfrac{8}{3}a^5 + \dfrac{1}{5}a^5 - 2a^5 \pi + \dfrac{1}{4}a^5 \pi \right) = a^5 k \left(\dfrac{7\pi}{16} - \dfrac{17}{15} \right)$

40. $I = \int_{-2}^{2}\int_{0}^{4-x^2} k(y-2)^2\, dy\, dx = \int_{-2}^{2}\left[\frac{k}{3}(y-1)^3\right]_{0}^{4-x^2} dx = \int_{-2}^{2}\frac{k}{3}[(2-x^2)^3 + 8]\, dx$

$\qquad = \frac{k}{3}\int_{-2}^{2}(16 - 12x^2 + 6x^4 - x^6)\, dx = \left[\frac{k}{3}\left(16x - 4x^3 + \frac{6}{5}x^5 - \frac{1}{7}x^7\right)\right]_{-2}^{2}$

$\qquad = \frac{2k}{3}\left(32 - 32 + \frac{192}{5} - \frac{128}{7}\right) = \frac{1408k}{105}$

41. $\bar{y} = \frac{L}{2},\ A = bL,\ h = \frac{L}{2}$

$I_{\bar{y}} = \int_{0}^{b}\int_{0}^{L}\left(y - \frac{L}{2}\right)^2 dy\, dx$

$\qquad = \int_{0}^{b}\left[\frac{[y-(L/2)]^3}{3}\right]_{0}^{L} dx = \frac{L^3 b}{12}$

$y_a = \bar{y} - \frac{I_{\bar{y}}}{hA} = \frac{L}{2} - \frac{L^3 b/12}{(L/2)(bL)} = \frac{L}{3}$

42. $\bar{y} = \frac{a}{2},\ A = ab,\ h = L - \frac{a}{2}$

$I_{\bar{y}} = \int_{0}^{b}\int_{0}^{a}\left(y - \frac{a}{2}\right)^2 dy\, dx = \frac{a^3 b}{12}$

$y_a = \frac{a}{2} - \frac{a^3 b/12}{[L-(a/2)]ab} = \frac{a(3L - 2a)}{3(2L - a)}$

43. $\bar{y} = \frac{2L}{3},\ A = \frac{bL}{2},\ h = \frac{L}{3}$

$I_{\bar{y}} = 2\int_{0}^{b/2}\int_{2Lx/b}^{L}\left(y - \frac{2L}{3}\right)^2 dy\, dx$

$\qquad = \frac{2}{3}\int_{0}^{b/2}\left[\left(y - \frac{2L}{3}\right)^3\right]_{2Lx/b}^{L} dx$

$\qquad = \frac{2}{3}\int_{0}^{b}\left[\frac{L^3}{27} - \left(\frac{2Lx}{b} - \frac{2L}{3}\right)^3\right] dx$

$\qquad = \frac{2}{3}\left[\frac{L^3 x}{27} - \frac{b}{8L}\left(\frac{2Lx}{b} - \frac{2L}{3}\right)^4\right]_{0}^{b} = \frac{L^3 b}{36}$

$y_a = \frac{2L}{3} - \frac{L^3 b/36}{L^2 b/6} = \frac{L}{2}$

44. $\bar{y} = 0,\ A = \pi a^2,\ h = L$

$I_{\bar{y}} = \int_{-a}^{a}\int_{-\sqrt{a^2-x^2}}^{\sqrt{a^2-x^2}} y^2\, dy\, dx$

$\qquad = \int_{0}^{2\pi}\int_{0}^{a} r^3 \sin^2\theta\, dr\, d\theta$

$\qquad = \int_{0}^{2\pi}\frac{a^4}{4}\sin^2\theta\, d\theta$

$\qquad = \frac{a^4\pi}{4}$

$y_a = -\frac{(a^4\pi/4)}{L\pi^2 a^2} = -\frac{a^2}{4\pi L}$

45. Orient the xy-coordinate system so that L is along the y-axis and R is in the first quadrant. Then the volume of the solid is

$V = \iint_{R} 2\pi x\, dA$

$\qquad = 2\pi \iint_{R} x\, dA$

$\qquad = 2\pi\left(\frac{\iint_{R} x\, dA}{\iint_{R} dA}\right)\iint_{R} dA$

$\qquad = 2\pi \bar{x} A.$

By our positioning, $\bar{x} = r$. Therefore, $V = 2\pi r A$.

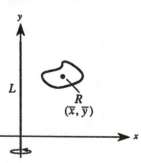

Section 14.5   Surface Area

1. $f(x, \ y) = 2x + 2y$
$R = $ triangle with vertices $(0, 0)$, $(2, 0)$, $(0, 2)$
$f_x = 2, \quad f_y = 2$

$$\sqrt{1 + (f_x)^2 + (f_y)^2} = 3$$

$$S = \int_0^2 \int_0^{2-x} 3 \, dy \, dx = 3 \int_0^2 (2 - x) \, dx = \left[3\left(2x - \frac{x^2}{2} \right) \right]_0^2 = 6$$

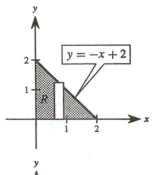

2. $f(x, \ y) = 10 + 2x - 3y$
$R = $ square with vertices $(0, 0)$, $(2, 0)$, $(0, 2)$, $(2, 2)$
$f_x = 2, \quad f_y = -3$

$$\sqrt{1 + (f_x)^2 + (f_y)^2} = \sqrt{14}$$

$$S = \int_0^2 \int_0^2 \sqrt{14} \, dy \, dx = \int_0^2 2\sqrt{14} \, dx = 4\sqrt{14}$$

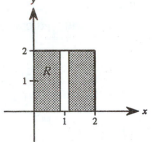

3. $f(x, \ y) = 8 + 2x + 2y$
$R = \{(x, \ y): x^2 + y^2 \le 4\}$
$f_x = 2, \quad f_y = 2$

$$\sqrt{1 + (f_x)^2 + (f_y)^2} = 3$$

$$S = \int_{-2}^2 \int_{-\sqrt{4-x^2}}^{\sqrt{4-x^2}} 3 \, dy \, dx = \int_0^{2\pi} \int_0^2 3r \, dr \, d\theta = 12\pi$$

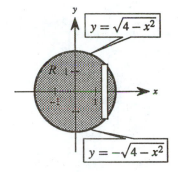

4. $f(x, \ y) = 10 + 2x - 3y$
$R = \{(x, \ y): x^2 + y^2 \le 9\}$
$f_x = 2, \quad f_y = -3$

$$\sqrt{1 + (f_x)^2 + (f_y)^2} = \sqrt{14}$$

$$S = \int_{-3}^3 \int_{-\sqrt{9-x^2}}^{\sqrt{9-x^2}} \sqrt{14} \, dy \, dx = \int_0^{2\pi} \int_0^3 \sqrt{14} \, r \, dr \, d\theta = 9\sqrt{14} \, \pi$$

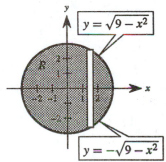

5. $f(x, \ y) = 9 - x^2$
$R = $ square with vertices $(0, 0)$, $(3, 0)$, $(0, 3)$, $(3, 3)$
$f_x = -2x, \quad f_y = 0$

$$\sqrt{1 + (f_x)^2 + (f_y)^2} = \sqrt{1 + 4x^2}$$

$$S = \int_0^3 \int_0^3 \sqrt{1 + 4x^2} \, dy \, dx = \int_0^3 3\sqrt{1 + 4x^2} \, dx$$

$$= \left[\frac{3}{4} \left(2x\sqrt{1 + 4x^2} + \ln \left| 2x + \sqrt{1 + 4x^2} \right| \right) \right]_0^3 = \frac{3}{4} \left(6\sqrt{37} + \ln \left| 6 + \sqrt{37} \right| \right)$$

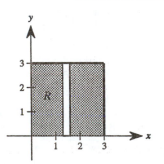

6. $f(x, y) = y^2$

R = square with vertices $(0, 0)$, $(3, 0)$, $(0, 3)$, $(3, 3)$

$f_x = 0$, $f_y = 2y$

$\sqrt{1 + (f_x)^2 + (f_y)^2} = \sqrt{1 + 4y^2}$

$S = \int_0^3 \int_0^3 \sqrt{1 + 4y^2}\, dx\, dy = \int_0^3 3\sqrt{1 + 4y^2}\, dy$

$= \left[\frac{3}{4} \left(2y\sqrt{1 + 4y^2} + \ln\left| 2y + \sqrt{1 + 4y^2} \right| \right) \right]_0^3 = \frac{3}{4}\left(6\sqrt{37} + \ln\left| 6 + \sqrt{37} \right| \right)$

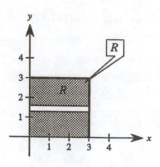

7. $f(x, y) = 2 + x^{3/2}$

R = quadrangle with vertices $(0, 0)$, $(0, 4)$, $(3, 4)$, $(3, 0)$

$f_x = \frac{3}{2}x^{1/2}$, $f_y = 0$

$\sqrt{1 + (f_x)^2 + (f_y)^2} = \sqrt{1 + \left(\frac{9}{4} \right)x} = \frac{\sqrt{4 + 9x}}{2}$

$S = \int_0^3 \int_0^4 \frac{\sqrt{4 + 9x}}{2}\, dy\, dx = \int_0^3 4\left(\frac{\sqrt{4 + 9x}}{2} \right) dx$

$= \left[\frac{4}{27}(4 + 9x)^{3/2} \right]_0^3 = \frac{4}{27}(31\sqrt{31} - 8)$

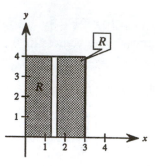

8. $f(x) = 2 + \frac{2}{3}x^{3/2}$

$R = \{(x, y): 0 \le x \le 1,\ 0 \le y \le 1 - x\}$

$f_x = x^{1/2}$, $f_y = 0$

$\sqrt{1 + (f_x)^2 + (f_y)^2} = \sqrt{1 + x}$

$S = \int_0^1 \int_0^{1-y} \sqrt{1 + x}\, dx\, dy = \int_0^1 \left[\frac{2}{3}(1 + x)^{3/2} \right]_0^{1-y} dy$

$= \frac{2}{3} \int_0^1 [(2 - y)^{3/2} - 1]\, dy = \frac{2}{3}\left[-\frac{2}{5}(2 - y)^{5/2} - y \right]_0^1$

$= \frac{2}{3}\left[\left(-\frac{2}{5} - 1 \right) - \left(-\frac{2}{5}[4\sqrt{2}] \right) \right] = \frac{2}{3}\left(\frac{8\sqrt{2} - 7}{5} \right)$

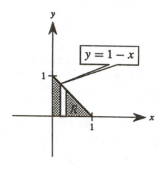

9. $f(x, y) = \ln|\sec x|$

$R = \{(x, y): 0 \le x \le \frac{\pi}{4},\ 0 \le y \le \tan x\}$

$f_x = \tan x$, $f_y = 0$

$\sqrt{1 + (f_x)^2 + (f_y)^2} = \sqrt{1 + \tan^2 x} = \sec x$

$S = \int_0^{\pi/4} \int_0^{\tan x} \sec x\, dy\, dx = \int_0^{\pi/4} \sec x \tan x\, dx = \left[\sec x \right]_0^{\pi/4} = \sqrt{2} - 1$

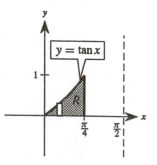

10. $f(x, \ y) = 4 + x^2 - y^2$

$R = \{(x, \ y): \ x^2 + y^2 \le 1\}$

$f_x = 2x, \quad f_y = -2y$

$\sqrt{1 + (f_x)^2 + (f_y)^2} = \sqrt{1 + 4x^2 + 4y^2}$

$S = \int_{-1}^{1} \int_{-\sqrt{1-x^2}}^{\sqrt{1-x^2}} \sqrt{1 + 4x^2 + 4y^2} \, dy \, dx$

$= \int_{0}^{2\pi} \int_{0}^{1} \sqrt{1 + 4r^2} \, r \, dr \, d\theta = \dfrac{(5\sqrt{5} - 1)\pi}{6}$

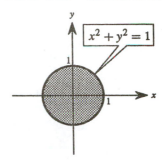

11. $f(x, \ y) = \sqrt{x^2 + y^2}$

$R = \{(x, \ y): \ 0 \le f(x, \ y) \le 1\}$

$0 \le \sqrt{x^2 + y^2} \le 1, \quad x^2 + y^2 \le 1$

$f_x = \dfrac{x}{\sqrt{x^2 + y^2}}, \quad f_y = \dfrac{y}{\sqrt{x^2 + y^2}}$

$\sqrt{1 + (f_x)^2 + (f_y)^2} = \sqrt{1 + \dfrac{x^2}{x^2 + y^2} + \dfrac{y^2}{x^2 + y^2}} = \sqrt{2}$

$S = \int_{-1}^{1} \int_{-\sqrt{1-x^2}}^{\sqrt{1-x^2}} \sqrt{2} \, dy \, dx = \int_{0}^{2\pi} \int_{0}^{1} \sqrt{2} \, r \, dr \, d\theta = \sqrt{2}\pi$

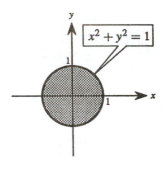

12. $f(x, \ y) = xy$

$R = \{(x, \ y): \ x^2 + y^2 \le 16\}$

$f_x = y, \quad f_y = x$

$\sqrt{1 + (f_x)^2 + (f_y)^2} = \sqrt{1 + y^2 + x^2}$

$S = \int_{-4}^{4} \int_{-\sqrt{16-x^2}}^{\sqrt{16-x^2}} \sqrt{1 + y^2 + x^2} \, dy \, dx$

$= \int_{0}^{2\pi} \int_{0}^{4} \sqrt{1 + r^2} \, r \, dr \, d\theta = \dfrac{2\pi}{3}\left(17\sqrt{17} - 1\right)$

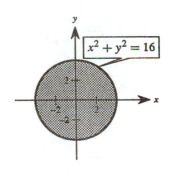

13. $f(x, \ y) = \sqrt{a^2 - x^2 - y^2}$

$R = \{(x, \ y): \ x^2 + y^2 \le b^2, \quad b < a\}$

$f_x = \dfrac{-x}{\sqrt{a^2 - x^2 - y^2}}, \quad f_y = \dfrac{-y}{\sqrt{a^2 - x^2 - y^2}}$

$\sqrt{1 + (f_x)^2 + (f_y)^2} = \sqrt{1 + \dfrac{x^2}{a^2 - x^2 - y^2} + \dfrac{y^2}{a^2 - x^2 - y^2}} = \dfrac{a}{\sqrt{a^2 - x^2 - y^2}}$

$S = \int_{-b}^{b} \int_{-\sqrt{b^2-x^2}}^{\sqrt{b^2-x^2}} \dfrac{a}{\sqrt{a^2 - x^2 - y^2}} \, dy \, dx$

$= \int_{0}^{2\pi} \int_{0}^{b} \dfrac{a}{\sqrt{a^2 - r^2}} \, r \, dr \, d\theta = 2\pi a\left(a - \sqrt{a^2 - b^2}\right)$

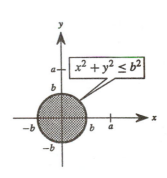

14. See Exercise 13.

$S = \int_{-a}^{a} \int_{-\sqrt{a^2-x^2}}^{\sqrt{a^2-x^2}} \dfrac{a}{\sqrt{a^2 - x^2 - y^2}} \, dy \, dx = \int_{0}^{2\pi} \int_{0}^{a} \dfrac{a}{\sqrt{a^2 - r^2}} \, r \, dr \, d\theta = 2\pi a^2$

15. $z = 24 - 3x - 2y$

$$\sqrt{1 + (f_x)^2 + (f_y)^2} = \sqrt{14}$$

$$S = \int_0^8 \int_0^{-(3/2)x+12} \sqrt{14} \, dy \, dx = 48\sqrt{14}$$

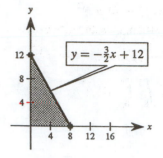

$y = -\frac{3}{2}x + 12$

16. $z = 16 - x^2 - y^2$

$$\sqrt{1 + (f_x)^2 + (f_y)^2} = \sqrt{1 + 4x^2 + 4y^2}$$

$$S = \int_0^4 \int_0^{\sqrt{16-x^2}} \sqrt{1 + 4(x^2 + y^2)} \, dy \, dx$$

$$= \int_0^{\pi/2} \int_0^4 \sqrt{1 + 4r^2} \, r \, dr \, d\theta = \frac{\pi}{24}(65\sqrt{65} - 1)$$

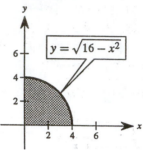

$y = \sqrt{16 - x^2}$

17. $z = \sqrt{25 - x^2 - y^2}$

$$\sqrt{1 + (f_x)^2 + (f_y)^2} = \sqrt{1 + \frac{x^2}{25 - x^2 - y^2} + \frac{y^2}{25 - x^2 - y^2}} = \frac{5}{\sqrt{25 - x^2 - y^2}}$$

$$S = 2\int_{-3}^3 \int_{-\sqrt{9-x^2}}^{\sqrt{9-x^2}} \frac{5}{\sqrt{25 - (x^2 + y^2)}} \, dy \, dx$$

$$= 2\int_0^{2\pi} \int_0^3 \frac{5}{\sqrt{25 - r^2}} r \, dr \, d\theta = 20\pi$$

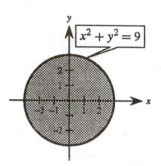

$x^2 + y^2 = 9$

18. $z = \sqrt{x^2 + y^2}$

$$\sqrt{1 + (f_x)^2 + (f_y)^2} = \sqrt{1 + \frac{x^2}{x^2 + y^2} + \frac{y^2}{x^2 + y^2}} = \sqrt{2}$$

$$S = \int_{-1}^1 \int_{-\sqrt{1-x^2}}^{\sqrt{1-x^2}} \sqrt{2} \, dy \, dx$$

$$= \int_0^{2\pi} \int_0^1 \sqrt{2} \, r \, dr \, d\theta = \sqrt{2}\pi$$

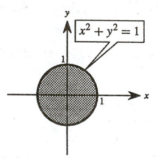

$x^2 + y^2 = 1$

19. $f(x, y) = 2y + x^2$

R = triangle with vertices $(0, 0)$, $(1, 0)$, $(1, 1)$

$$\sqrt{1 + (f_x)^2 + (f_y)^2} = \sqrt{5 + 4x^2}$$

$$S = \int_0^1 \int_0^x \sqrt{5 + 4x^2} \, dy \, dx = \frac{1}{12}(27 - 5\sqrt{5})$$

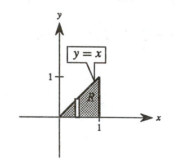

$y = x$

20. $f(x, y) = 2x + y^2$

$R = $ triangle with vertices $(0, 0), (2, 0), (2, 2)$

$$\sqrt{1 + (f_x)^2 + (f_y)^2} = \sqrt{5 + 4x^2}$$

$$S = \int_0^2 \int_0^x \sqrt{5 + 4x^2}\, dy\, dx = \frac{1}{12}(21\sqrt{21} - 5\sqrt{5})$$

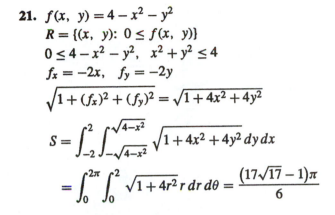

21. $f(x, y) = 4 - x^2 - y^2$

$R = \{(x, y): 0 \le f(x, y)\}$

$0 \le 4 - x^2 - y^2, \quad x^2 + y^2 \le 4$

$f_x = -2x, \quad f_y = -2y$

$$\sqrt{1 + (f_x)^2 + (f_y)^2} = \sqrt{1 + 4x^2 + 4y^2}$$

$$S = \int_{-2}^{2} \int_{-\sqrt{4-x^2}}^{\sqrt{4-x^2}} \sqrt{1 + 4x^2 + 4y^2}\, dy\, dx$$

$$= \int_0^{2\pi} \int_0^2 \sqrt{1 + 4r^2}\, r\, dr\, d\theta = \frac{(17\sqrt{17} - 1)\pi}{6}$$

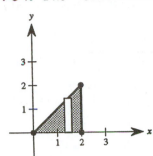

22. $f(x, y) = x^2 + y^2$

$R = \{(x, y): 0 \le f(x, y) \le 16\}$

$0 \le x^2 + y^2 \le 16$

$f_x = 2x, \quad f_y = 2y$

$$\sqrt{1 + (f_x)^2 + (f_y)^2} = \sqrt{1 + 4x^2 + 4y^2}$$

$$S = \int_{-4}^{4} \int_{-\sqrt{16-x^2}}^{\sqrt{16-x^2}} \sqrt{1 + 4x^2 + 4y^2}\, dy\, dx$$

$$= \int_0^{2\pi} \int_0^4 \sqrt{1 + 4r^2}\, r\, dr\, d\theta = \frac{(65\sqrt{65} - 1)\pi}{6}$$

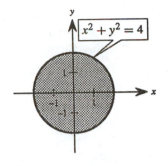

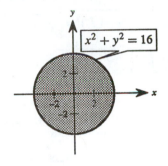

23. $f(x, y) = 4 - x^2 - y^2$

$R = \{(x, y): 0 \le x \le 1, \ 0 \le y \le 1\}$

$f_x = -2x, \quad f_y = -2y$

$$\sqrt{1 + (f_x)^2 + (f_y)^2} = \sqrt{1 + 4x^2 + 4y^2}$$

$$S = \int_0^1 \int_0^1 \sqrt{(1 + 4x^2) + 4y^2}\, dy\, dx \approx 1.8616$$

24. $f(x, y) = \frac{2}{3}x^{3/2} + \cos x$

$R = \{(x, y): 0 \le x \le 1, \ 0 \le y \le 1\}$

$f_x = x^{1/2} - \sin x, \quad f_y = 0$

$$\sqrt{1 + (f_x)^2 + (f_y)^2} = \sqrt{1 + \left(\sqrt{x} - \sin x\right)^2}$$

$$S = \int_0^1 \int_0^1 \sqrt{1 + \left(\sqrt{x} - \sin x\right)^2}\, dy\, dx \approx 1.02185$$

25. $f(x, y) = e^x$

$R = \{(x, y): 0 \le x \le 1, \ 0 \le y \le 1\}$

$f_x = e^x, \quad f_y = 0$

$\sqrt{1 + (f_x)^2 + (f_y)^2} = \sqrt{1 + e^{2x}}$

$S = \int_0^1 \int_0^1 \sqrt{1 + e^{2x}} \, dy \, dx$

$= \int_0^1 \sqrt{1 + e^{2x}} \, dx \approx 2.0035$

26. $f(x, y) = \frac{2}{5} y^{5/2}$

$R = \{(x, y): 0 \le x \le 1, \ 0 \le y \le 1\}$

$f_x = 0, \quad f_y = y^{3/2}$

$\sqrt{1 + (f_x)^2 + (f_y)^2} = \sqrt{1 + y^3}$

$S = \int_0^1 \int_0^1 \sqrt{1 + y^3} \, dx \, dy$

$= \int_0^1 \sqrt{1 + y^3} \, dy \approx 1.1114$

27. $f(x, y) = x^3 - 3xy + y^3$

$R = $ square with vertices $(1, 1), (-1, 1), (-1, -1), (1, -1)$

$f_x = 3x^2 - 3y = 3(x^2 - y), \quad f_y = -3x + 3y^2 = 3(y^2 - x)$

$S = \int_{-1}^1 \int_{-1}^1 \sqrt{1 + 9(x^2 - y)^2 + 9(y^2 - x)^2} \, dy \, dx$

28. $f(x, y) = e^{-x} \sin y$

$R = \{(x, y): 0 \le x \le 4, \ 0 \le y \le x\}$

$f_x = -e^{-x} \sin y, \quad f_y = e^{-x} \cos y$

$\sqrt{1 + (f_x)^2 + (f_y)^2} = \sqrt{1 + e^{-2x} \sin^2 y + e^{-2x} \cos^2 y} = \sqrt{1 + e^{-2x}}$

$S = \int_0^4 \int_0^x \sqrt{1 + e^{-2x}} \, dy \, dx$

29. $f(x, y) = e^{-x} \sin y$

$R = \{(x, y): x^2 + y^2 \le 4\}$

See Exercise 28.

$S = \int_{-2}^2 \int_{-\sqrt{4-x^2}}^{\sqrt{4-x^2}} \sqrt{1 + e^{-2x}} \, dy \, dx$

30. $f(x, y) = x^2 - 3xy - y^2$

$R = \{(x, y): 0 \le x \le 4, \ 0 \le y \le x\}$

$f_x = 2x - 3y, \quad f_y = -3x - 2y = -(3x + 2y)$

$\sqrt{1 + (f_x)^2 + (f_y)^2} = \sqrt{1 + (2x - 3y)^2 + (3x + 2y)^2}$

$= \sqrt{1 + 13(x^2 + y^2)}$

$S = \int_0^4 \int_0^x \sqrt{1 + 13(x^2 + y^2)} \, dy \, dx$

31. $f(x, y) = e^{xy}$

$R = \{(x, y): 0 \le x \le 4, \ 0 \le y \le 10\}$

$f_x = ye^{xy}, \quad f_y = xe^{xy}$

$\sqrt{1 + (f_x)^2 + (f_y)^2} = \sqrt{1 + y^2 e^{2xy} + x^2 e^{2xy}} = \sqrt{1 + e^{2xy}(x^2 + y^2)}$

$S = \int_0^4 \int_0^{10} \sqrt{1 + e^{2xy}(x^2 + y^2)} \, dy \, dx$

32. $f(x, y) = \cos(x^2 + y^2)$

$R = \left\{ (x, y): x^2 + y^2 \le \dfrac{\pi}{2} \right\}$

$f_x = -2x \sin(x^2 + y^2), \quad f_y = -2y \sin(x^2 + y^2)$

$\sqrt{1 + (f_x)^2 + (f_y)^2} = \sqrt{1 + 4x^2 \sin^2(x^2 + y^2) + 4y^2 \sin^2(x^2 + y^2)} = \sqrt{1 + 4[\sin^2(x^2 + y^2)](x^2 + y^2)}$

$S = \displaystyle\int_{-\sqrt{\pi/2}}^{\sqrt{\pi/2}} \int_{-\sqrt{(\pi/2)-x^2}}^{\sqrt{(\pi/2)-x^2}} \sqrt{1 + 4(x^2 + y^2)\sin^2(x^2 + y^2)} \, dy \, dx$

33. $f(x, y) = \sqrt{1 - y^2}$

$S = \displaystyle\int_R \int \sqrt{1 + (f_x)^2 + (f_y)^2} \, dA = 16 \int_0^1 \int_0^x \sqrt{1 + \dfrac{y^2}{1 - y^2}} \, dy \, dx$

$\quad = 16 \displaystyle\int_0^1 \int_0^x \dfrac{1}{\sqrt{1 - y^2}} \, dy \, dx = 16 \int_0^1 \arcsin x \, dx$

$\quad = 16 \left[x \arcsin x + \sqrt{1 - x^2} \right]_0^1 = 16 \left(\dfrac{\pi}{2} - 1 \right)$

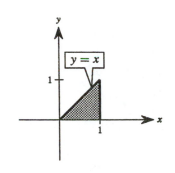

34. $f(x, y) = k\sqrt{x^2 + y^2}$

$\sqrt{1 + (f_x)^2 + (f_y)^2} = \sqrt{1 + \dfrac{k^2 x^2}{x^2 + y^2} + \dfrac{k^2 y^2}{x^2 + y^2}} = \sqrt{k^2 + 1}$

$S = \displaystyle\int_R \int \sqrt{1 + (f_x)^2 + (f_y)^2} \, dA = \int_R \int \sqrt{k^2 + 1} \, dA = \sqrt{k^2 + 1} \int_R \int dA = A\sqrt{k^2 + 1} = \pi r^2 \sqrt{k^2 + 1}$

35. a. $V = \displaystyle\int_0^{50} \int_0^{\sqrt{50^2 - x^2}} \left(20 + \dfrac{xy}{100} - \dfrac{x + y}{5} \right) dy \, dx$

$\quad = \displaystyle\int_0^{50} \left[20\sqrt{50^2 - x^2} + \dfrac{x}{200}(50^2 - x^2) - \dfrac{x}{5}\sqrt{50^2 - x^2} - \dfrac{50^2 - x^2}{10} \right] dy$

$\quad = \left[10 \left(x\sqrt{50 - x^2} + 50^2 \arcsin \dfrac{x}{50} \right) + \dfrac{25}{4}x^2 - \dfrac{x^4}{800} + \dfrac{1}{15}(50^2 - x^2)^{3/2} - 250x + \dfrac{x^3}{30} \right]_0^{50}$

$\quad \approx 30{,}415.74 \text{ ft}^3$

b. $z = 20 + \dfrac{xy}{100}$

$\quad \sqrt{1 + (f_x)^2 + (f_y)^2} = \sqrt{1 + \dfrac{y^2}{100^2} + \dfrac{x^2}{100^2}} = \dfrac{\sqrt{100^2 + x^2 + y^2}}{100}$

$\quad S = \dfrac{1}{100} \displaystyle\int_0^{50} \int_0^{\sqrt{50^2 - x^2}} \sqrt{100^2 + x^2 + y^2} \, dy \, dx$

$\quad = \dfrac{1}{100} \displaystyle\int_0^{\pi/2} \int_0^{50} \sqrt{100^2 + r^2} \, r \, dr \, d\theta \approx 2081.53 \text{ ft}^2$

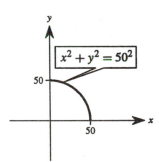

36. a. $V = \iint_R f(x, y)\, dA$

$$= 8 \iint_R \sqrt{625 - x^2 - y^2}\, dA \quad \text{where } R \text{ is the region in the first quadrant}$$

$$= 8 \int_0^{\pi/2} \int_4^{25} \sqrt{625 - r^2}\, r\, dr\, d\theta$$

$$= -4 \int_0^{\pi/2} \left[\frac{2}{3}(625 - r^2)^{3/2} \right]_4^{25} d\theta$$

$$= -\frac{8}{3}\left[0 - 609\sqrt{609} \right] \cdot \frac{\pi}{2}$$

$$= 812\pi\sqrt{609} \text{ cm}^3$$

The figure at the top right shows a quarter-annulus region R in the first quadrant, with axes labeled in increments of 4 up to 24 on both x and y.

b. $A = \iint_R \sqrt{1 + (f_x)^2 + (f_y)^2}\, dA = 8 \iint_R \sqrt{1 + \dfrac{x^2}{625 - x^2 - y^2} + \dfrac{y^2}{625 - x^2 - y^2}}\, dA$

$$= 8 \iint_R \frac{25}{\sqrt{625 - x^2 - y^2}}\, dA = 8 \int_0^{\pi/2} \int_4^{25} \frac{25}{\sqrt{625 - r^2}}\, r\, dr\, d\theta$$

$$= \lim_{b \to 25^-} \left[-200\sqrt{625 - r^2} \right]_4^{b} \cdot \frac{\pi}{2} = 100\pi\sqrt{609} \text{ cm}^2$$

37. $A = l \cdot w = \left(\dfrac{\Delta x}{\cos \theta} \right) \Delta y = \sec \theta\, \Delta x\, \Delta y$

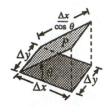

Area in xy-plane: $\Delta x \Delta y$

38. a. $V = \iint_R \dfrac{1}{\sqrt{x^2 + y^2}}\, dy\, dx$

$$= \iint_R \frac{1}{r}\, r\, dr\, d\theta$$

$$= \int_{2\arctan 0.01}^{\pi/2} \int_0^{9\csc\theta} dr\, d\theta$$

$$= \left[9 \ln |\csc\theta - \cot\theta| \right]_{2\arctan 0.01}^{\pi/2} \approx 41.45 \text{ in}^3$$

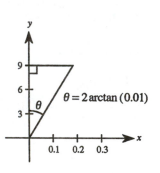

$\theta = 2\arctan(0.01)$

b. $f(x, y) = \dfrac{1}{\sqrt{x^2 + y^2}}$

$$\sqrt{1 + (f_x)^2 + (f_y)^2} = \sqrt{1 + \frac{x^2}{(x^2 + y^2)^3} + \frac{y^2}{(x^2 + y^2)^3}} = \frac{\sqrt{(x^2 + y^2)^2 + 1}}{x^2 + y^2}$$

$$S = \iint_R \frac{\sqrt{(x^2 + y^2)^2 + 1}}{x^2 + y^2}\, dy\, dx = \iint_R \frac{\sqrt{r^4 + 1}}{r^2}\, r\, dr\, d\theta = \int_{2\arctan 0.01}^{\pi/2} \int_0^{9\csc\theta} \frac{\sqrt{r^4 + 1}}{r}\, dr\, d\theta$$

$$\approx 501.69 \text{ in.}^2$$

Section 14.6   Triple Integrals and Applications

1. $\displaystyle\int_0^3\int_0^2\int_0^1 (x+y+z)\,dx\,dy\,dz = \int_0^3\int_0^2\left[\frac{1}{2}x^2+xy+xz\right]_0^1 dy\,dx$

$$= \int_0^3\int_0^2\left(\frac{1}{2}+y+z\right)dy\,dz = \int_0^3\left[\frac{1}{2}y+\frac{1}{2}y^2+yz\right]_0^2 dz = \left[3z+z^2\right]_0^3 = 18$$

2. $\displaystyle\int_{-1}^1\int_{-1}^1\int_{-1}^1 x^2y^2z^2\,dx\,dy\,dz = \frac{1}{3}\int_{-1}^1\int_{-1}^1\left[x^3y^2z^2\right]_{-1}^1 dy\,dz$

$$= \frac{2}{3}\int_{-1}^1\int_{-1}^1 y^2z^2\,dy\,dz = \frac{2}{9}\int_{-1}^1\left[y^3z^2\right]_{-1}^1 dz = \frac{4}{9}\int_{-1}^1 z^2\,dz = \left[\frac{4}{27}z^3\right]_{-1}^1 = \frac{8}{27}$$

3. $\displaystyle\int_0^1\int_0^x\int_0^{xy} x\,dz\,dy\,dx = \int_0^1\int_0^x\left[xz\right]_0^{xy} dy\,dx$

$$= \int_0^1\int_0^x x^2y\,dy\,dx = \int_0^1\left[\frac{x^2y^2}{2}\right]_0^x dx = \int_0^1 \frac{x^4}{2}\,dx = \left[\frac{x^5}{10}\right]_0^1 = \frac{1}{10}$$

4. $\displaystyle\int_0^4\int_0^\pi\int_0^{1-x} x\sin y\,dz\,dy\,dx = \int_0^4\int_0^\pi\left[(x\sin y)z\right]_0^{1-x} dy\,dx = \int_0^4\int_0^\pi x(1-x)\sin y\,dy\,dx$

$$= \int_0^4\left[-x(1-x)\cos y\right]_0^\pi dx = \int_0^4 2x(1-x)\,dx = \left[x^2-\frac{2x^3}{3}\right]_0^4 = -\frac{80}{3}$$

5. $\displaystyle\int_1^4\int_0^1\int_0^x 2ze^{-x^2}\,dy\,dx\,dz = \int_1^4\int_0^1\left[(2ze^{-x^2})y\right]_0^x dx\,dz = \int_1^4\int_0^1 2xze^{-x^2}\,dx\,dz$

$$= \int_1^4\left[-ze^{-x^2}\right]_0^1 dz = \int_1^4 z(1-e^{-1})\,dz = \left[(1-e^{-1})\frac{z^2}{2}\right]_1^4 = \frac{15}{2}\left(1-\frac{1}{e}\right)$$

6. $\displaystyle\int_1^4\int_1^{e^2}\int_0^{1/xz} \ln z\,dy\,dz\,dx = \int_1^4\int_1^{e^2}\left[(\ln z)y\right]_0^{1/xz} dz\,dx = \int_1^4\int_1^{e^2} \frac{\ln z}{xz}\,dz\,dx$

$$= \int_1^4\left[\frac{1}{x}\frac{(\ln z)^2}{2}\right]_1^{e^2} dx = \int_1^4 \frac{2}{x}\,dx = \left[2\ln|x|\right]_1^4 = 2\ln 4$$

7. $\displaystyle\int_0^9\int_0^{y/3}\int_0^{\sqrt{y^2-9x^2}} z\,dz\,dx\,dy = \frac{1}{2}\int_0^9\int_0^{y/3}(y^2-9x^2)\,dx\,dy$

$$= \frac{1}{2}\int_0^9\left[xy^2-3x^3\right]_0^{y/3} dy = \frac{2}{18}\int_0^9 y^3\,dy = \left[\frac{1}{36}y^4\right]_0^9 = \frac{729}{4}$$

8. $\displaystyle\int_0^{\pi/2}\int_0^{y/2}\int_0^{1/y} \sin y\,dz\,dx\,dy = \int_0^{\pi/2}\int_0^{y/2}\frac{\sin y}{y}\,dx\,dy = \frac{1}{2}\int_0^{\pi/2}\sin y\,dy = \left[-\frac{1}{2}\cos y\right]_0^{\pi/2} = \frac{1}{2}$

9. $\displaystyle\int_0^2\int_{-\sqrt{4-x^2}}^{\sqrt{4-x^2}}\int_0^{x^2} x\,dz\,dy\,dx = \int_0^2\int_{-\sqrt{4-x^2}}^{\sqrt{4-x^2}} x^3\,dy\,dx = \frac{128}{15}$

10. $\displaystyle\int_0^{\sqrt{2}}\int_0^{\sqrt{2-x^2}}\int_{2x^2+y^2}^{4-y^2} y\,dz\,dy\,dx = \int_0^{\sqrt{2}}\int_0^{\sqrt{2-x^2}}(4y-2x^2y-2y^3)\,dy\,dx = \frac{16\sqrt{2}}{15}$

11. $\displaystyle\int_0^2 \int_0^{\sqrt{4-x^2}} \int_1^4 \frac{x^2 \sin y}{z}\, dz\, dy\, dx = \int_0^2 \int_0^{\sqrt{4-x^2}} \Big[x^2 \sin y \ln |z| \Big]_1^4 dy\, dx$

$\displaystyle = \int_0^2 \Big[x^2 \ln 4 (-\cos y) \Big]_0^{\sqrt{4-x^2}} dx = \int_0^2 x^2 \ln 4 \Big[1 - \cos \sqrt{4-x^2} \Big] dx \approx 2.4416$

12. $\displaystyle\int_0^3 \int_0^{2-(2y/3)} \int_0^{6-2y-3z} ze^{-x^2 y^2}\, dx\, dz\, dy = \int_0^6 \int_0^{(6-x)/2} \int_0^{(6-x-2y)/3} ze^{-x^2 y^2}\, dz\, dy\, dx$

$\displaystyle = \int_0^6 \int_0^{3-(x/2)} \frac{1}{2} \left(\frac{6-x-2y}{3} \right)^2 e^{-x^2 y^2}\, dy\, dx \approx 2.188$

13. Plane: $3x + 6y + 4z = 12$

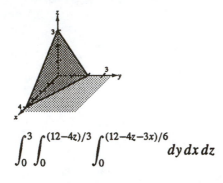

$\displaystyle\int_0^3 \int_0^{(12-4z)/3} \int_0^{(12-4z-3x)/6} dy\, dx\, dz$

14. Top plane: $x + y + z = 10$
    Side cylinder: $x^2 + y^2 = 16$

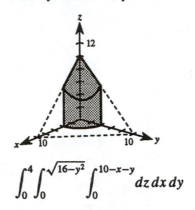

$\displaystyle\int_0^4 \int_0^{\sqrt{16-y^2}} \int_0^{10-x-y} dz\, dx\, dy$

15. Top cylinder: $y^2 + z^2 = 1$
    Side plane: $x = y$

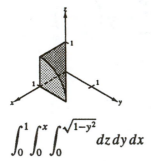

$\displaystyle\int_0^1 \int_0^x \int_0^{\sqrt{1-y^2}} dz\, dy\, dx$

16. Elliptic cone: $4x^2 + z^2 = y^2$

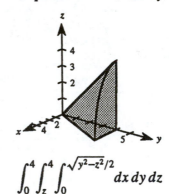

$\displaystyle\int_0^4 \int_z^4 \int_0^{\sqrt{y^2-z^2}/2} dx\, dy\, dz$

17. $Q = \{(x,\, y,\, z)\colon 0 \le x \le 1,\ \ 0 \le y \le x,\ \ 0 \le z \le 3\}$

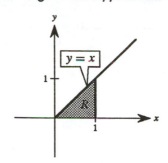

$$\iiint_{Q} xyz\, dV = \int_{0}^{3}\int_{0}^{1}\int_{y}^{1} xyz\, dx\, dy\, dz = \int_{0}^{3}\int_{0}^{1}\int_{0}^{x} xyz\, dy\, dx\, dz$$

$$= \int_{0}^{1}\int_{0}^{3}\int_{y}^{1} xyz\, dx\, dz\, dy$$

$$= \int_{0}^{1}\int_{0}^{3}\int_{0}^{x} xyz\, dy\, dz\, dx$$

$$= \int_{0}^{1}\int_{y}^{1}\int_{0}^{3} xyz\, dz\, dx\, dy$$

$$= \int_{0}^{1}\int_{0}^{x}\int_{0}^{3} xyz\, dz\, dy\, dx$$

18. $Q = \{(x,\, y,\, z)\colon 0 \le x \le 2,\ \ x^2 \le y \le 4,\ \ 0 \le z \le 2-x\}$

$$\iiint_{Q} xyz\, dV = \int_{0}^{2}\int_{x^2}^{4}\int_{0}^{2-x} xyz\, dz\, dy\, dx$$

$$= \int_{0}^{4}\int_{0}^{\sqrt{y}}\int_{0}^{2-x} xyz\, dz\, dx\, dy$$

$$= \int_{0}^{2}\int_{0}^{2-x}\int_{x^2}^{4} xyz\, dy\, dz\, dx$$

$$= \int_{0}^{2}\int_{0}^{2-z}\int_{x^2}^{4} xyz\, dy\, dx\, dz$$

$$= \int_{0}^{2}\int_{0}^{(2-z)^2}\int_{0}^{\sqrt{y}} xyz\, dx\, dy\, dz + \int_{0}^{2}\int_{(2-z)^2}^{4}\int_{0}^{2-z} xyz\, dx\, dy\, dz$$

$$= \int_{0}^{4}\int_{0}^{2-\sqrt{y}}\int_{0}^{\sqrt{y}} xyz\, dx\, dz\, dy + \int_{0}^{4}\int_{2-\sqrt{y}}^{2}\int_{0}^{2-z} dx\, dz\, dy$$

19. $\displaystyle \int_{-2}^{2}\int_{0}^{4-y^2}\int_{0}^{x} dz\, dx\, dy = \int_{-2}^{2}\int_{0}^{4-y^2} x\, dx\, dy$

$$= \frac{1}{2}\int_{-2}^{2} (4 - y^2)^2\, dy = \int_{0}^{2} (16 - 8y^2 + y^4)\, dy = \left[16y - \frac{8}{3}y^3 + \frac{1}{5}y^5\right]_{0}^{2} = \frac{256}{15}$$

20. $\displaystyle \int_{0}^{1}\int_{0}^{1}\int_{0}^{xy} dz\, dy\, dx = \int_{0}^{1}\int_{0}^{1} xy\, dy\, dx = \int_{0}^{1} \frac{x}{2}\, dx = \left[\frac{x^2}{4}\right]_{0}^{1} = \frac{1}{4}$

21. $\displaystyle 8\int_{0}^{r}\int_{0}^{\sqrt{r^2-x^2}}\int_{0}^{\sqrt{r^2-x^2-y^2}} dz\, dy\, dx = 8\int_{0}^{r}\int_{0}^{\sqrt{r^2-x^2}} \sqrt{r^2 - x^2 - y^2}\, dy\, dx$

$$= 4\int_{0}^{r}\left[y\sqrt{r^2 - x^2 - y^2} + (r^2 - x^2)\arcsin\left(\frac{y}{\sqrt{r^2 - x^2}}\right)\right]_{0}^{\sqrt{r^2-x^2}} dx$$

$$= 4\left(\frac{\pi}{2}\right)\int_{0}^{r} (r^2 - x^2)\, dx = \left[2\pi\left(r^2 x - \frac{1}{3}x^3\right)\right]_{0}^{r} = \frac{4}{3}\pi r^3$$

22. $4\displaystyle\int_0^3\int_0^{\sqrt{9-x^2}}\int_0^{9-x^2-y^2} dz\,dy\,dx = 4\int_0^3\int_0^{\sqrt{9-x^2}}(9-x^2-y^2)\,dy\,dx = 4\int_0^3\left[9y - x^2 y - \frac{1}{3}y^3\right]_0^{\sqrt{9-x^2}} dx$

$$= 4\int_0^3\left[9\sqrt{9-x^2} - x^2\sqrt{9-x^2} - \frac{1}{3}(9-x^2)^{3/2}\right]dx = \frac{8}{3}\int_0^3(9-x^2)^{3/2}\,dx$$

$$= \frac{2}{3}\left[x(9-x^2)^{3/2} + \frac{27x}{2}\sqrt{9-x^2} + \frac{243}{2}\arcsin\frac{x}{3}\right]_0^3 = \frac{81\pi}{2}$$

23. $\displaystyle\int_0^2\int_0^{4-x^2}\int_0^{4-x^2} dz\,dy\,dx = \int_0^2(4-x^2)^2\,dx = \int_0^2(16 - 8x^2 + x^4)\,dx = \left[16x - \frac{8}{3}x^3 + \frac{1}{5}x^5\right]_0^2 = \frac{256}{15}$

24. $\displaystyle\int_0^2\int_0^{2-x}\int_0^{9-x^2} dz\,dy\,dx = \int_0^2\int_0^{2-x}(9-x^2)\,dy\,dx = \int_0^2(9-x^2)(2-x)\,dx$

$$= \int_0^2(18 - 9x - 2x^2 + x^3)\,dx = \left[18x - \frac{9}{2}x^2 - \frac{2}{3}x^3 + \frac{1}{4}x^4\right]_0^2 = \frac{50}{3}$$

25. $m = k\displaystyle\int_0^6\int_0^{4-(2x/3)}\int_0^{2-(y/2)-(x/3)} dz\,dy\,dx$

$= 8k$

$M_{yz} = k\displaystyle\int_0^6\int_0^{4-(2x/3)}\int_0^{2-(y/2)-(x/3)} x\,dz\,dy\,dx$

$= 12k$

$\bar{x} = \dfrac{M_{yz}}{m} = \dfrac{12k}{8k} = \dfrac{3}{2}$

26. $m = k\displaystyle\int_0^6\int_0^{(12-2x)/3}\int_0^{(12-2x-3y)/6} y\,dz\,dy\,dx$

$= 8k$

$M_{xz} = k\displaystyle\int_0^6\int_0^{(12-2x)/3}\int_0^{(12-2x-3y)/6} y^2\,dz\,dy\,dx$

$= \dfrac{64k}{5}$

$\bar{y} = \dfrac{M_{xz}}{m} = \dfrac{8k}{5}$

27. $m = k\displaystyle\int_0^4\int_0^4\int_0^{4-x} x\,dz\,dy\,dx = k\int_0^4\int_0^4 x(4-x)\,dy\,dx = 4k\int_0^4(4x - x^2)\,dx = \dfrac{128k}{3}$

$M_{xy} = k\displaystyle\int_0^4\int_0^4\int_0^{4-x} xz\,dz\,dy\,dx = k\int_0^4\int_0^4 x\frac{(4-x)^2}{2}\,dy\,dx = 2k\int_0^4(16x - 8x^2 + x^3)\,dx = \dfrac{128k}{3}$

$\bar{z} = \dfrac{M_{xy}}{m} = 1$

28. $m = k\displaystyle\int_0^b\int_0^{a[1-(y/b)]}\int_0^{c[1-(y/b)-(x/a)]} dz\,dx\,dy = \dfrac{kabc}{6}$

$M_{xz} = k\displaystyle\int_0^b\int_0^{a[1-(y/b)]}\int_0^{c[1-(y/b)-(x/a)]} y\,dz\,dx\,dy = \dfrac{kab^2c}{24}$

$\bar{y} = \dfrac{M_{xz}}{m} = \dfrac{kab^2c/24}{kabc/6} = \dfrac{b}{4}$

29. $m = k\int_0^b \int_0^b \int_0^b xy\,dz\,dy\,dx = \dfrac{kb^5}{4}$

$M_{yz} = k\int_0^b \int_0^b \int_0^b x^2 y\,dz\,dy\,dx = \dfrac{kb^6}{6}$

$M_{xz} = k\int_0^b \int_0^b \int_0^b xy^2\,dz\,dy\,dx = \dfrac{kb^6}{6}$

$M_{xy} = k\int_0^b \int_0^b \int_0^b xyz\,dz\,dy\,dx = \dfrac{kb^6}{8}$

$\bar{x} = \dfrac{M_{yz}}{m} = \dfrac{kb^6/6}{kb^5/4} = \dfrac{2b}{3}$

$\bar{y} = \dfrac{M_{xz}}{m} = \dfrac{kb^6/6}{kb^5/4} = \dfrac{2b}{3}$

$\bar{z} = \dfrac{M_{xy}}{m} = \dfrac{kb^6/8}{kb^5/4} = \dfrac{b}{2}$

30. $m = k\int_0^a \int_0^b \int_0^c z\,dz\,dy\,dx = \dfrac{kabc^2}{2}$

$M_{xy} = k\int_0^a \int_0^b \int_0^c z^2\,dz\,dy\,dx = \dfrac{kabc^3}{3}$

$M_{yz} = k\int_0^a \int_0^b \int_0^c xz\,dz\,dy\,dx = \dfrac{ka^2bc^2}{4}$

$M_{xz} = k\int_0^a \int_0^b \int_0^c yz\,dz\,dy\,dx = \dfrac{kab^2c^2}{4}$

$\bar{x} = \dfrac{M_{yz}}{m} = \dfrac{ka^2bc^2/4}{kabc^2/2} = \dfrac{a}{2}$

$\bar{y} = \dfrac{M_{xz}}{m} = \dfrac{kab^2c^2/4}{kabc^2/2} = \dfrac{b}{2}$

$\bar{z} = \dfrac{M_{xy}}{m} = \dfrac{kabc^3/3}{kabc^2/2} = \dfrac{2c}{3}$

31. $m = \dfrac{1}{3}k\pi r^2 h$

$\bar{x} = \bar{y} = 0$ by symmetry

$M_{xy} = 4k\int_0^r \int_0^{\sqrt{r^2-x^2}} \int_{h\sqrt{x^2+y^2}/r}^{h} z\,dz\,dy\,dx$

$= \dfrac{3kh^2}{r^2}\int_0^r \int_0^{\sqrt{r^2-x^2}} (r^2 - x^2 - y^2)\,dy\,dx$

$= \dfrac{4kh^2}{3r^2}\int_0^r (r^2 - x^2)^{3/2}\,dx$

$= \dfrac{k\pi r^2 h^2}{4}$

$\bar{z} = \dfrac{M_{xy}}{m} = \dfrac{k\pi r^2 h^2/4}{k\pi r^2 h/3} = \dfrac{3h}{4}$

32. $m = 2k\int_0^2 \int_0^{\sqrt{4-x^2}} \int_0^y dz\,dy\,dx$

$= k\int_0^2 (4 - x^2)\,dx = \dfrac{16k}{3}$

$M_{yz} = 2k\int_0^2 \int_0^{\sqrt{4-x^2}} \int_0^y x\,dz\,dy\,dx = 4k$

$M_{xz} = 2k\int_0^2 \int_0^{\sqrt{4-x^2}} \int_0^y y\,dz\,dy\,dx = 2k\pi$

$M_{xy} = 2k\int_0^2 \int_0^{\sqrt{4-x^2}} \int_0^y z\,dz\,dy\,dx = k\pi$

$\bar{x} = \dfrac{M_{yz}}{m} = \dfrac{4k}{16k/3} = \dfrac{3}{4}$

$\bar{y} = \dfrac{M_{xz}}{m} = \dfrac{2k\pi}{16k/3} = \dfrac{3\pi}{8}$

$\bar{z} = \dfrac{M_{xy}}{m} = \dfrac{k\pi}{16k/3} = \dfrac{3\pi}{16}$

33. $m = \dfrac{128k\pi}{3}$

$\bar{x} = \bar{y} = 0$ by symmetry

$z = \sqrt{4^2 - x^2 - y^2}$

$M_{xy} = 4k \displaystyle\int_0^4 \int_0^{\sqrt{4^2-x^2}} \int_0^{\sqrt{4^2-x^2-y^2}} z \, dz \, dy \, dx$

$= 2k \displaystyle\int_0^4 \int_0^{\sqrt{4^2-x^2}} (4^2 - x^2 - y^2) \, dy \, dx = 2k \int_0^4 \left[16y - x^2 y - \frac{1}{3} y^3 \right]_0^{\sqrt{4^2-x^2}} dx = \frac{4k}{3} \int_0^4 (4^2 - x^2)^{3/2} \, dx$

$= \dfrac{1024k}{3} \displaystyle\int_0^{\pi/2} \cos^4 \theta \, d\theta$ (let $x = 4 \sin \theta$)

$= 64\pi k$ by Wallis's Formula

$\bar{z} = \dfrac{M_{xy}}{m} = \dfrac{64k\pi}{1} \cdot \dfrac{3}{128k\pi} = \dfrac{3}{2}$

34. $\bar{x} = 0$

$m = 2k \displaystyle\int_0^2 \int_0^1 \int_0^{1/(y^2+1)} dz \, dy \, dx = 2k \int_0^2 \int_0^1 \frac{1}{y^2 + 1} \, dy \, dx = 2k\left(\frac{\pi}{4}\right) \int_0^2 dx = k\pi$

$M_{xz} = 2k \displaystyle\int_0^2 \int_0^1 \int_0^{1/(y^2+1)} y \, dz \, dy \, dx = 2k \int_0^2 \int_0^1 \frac{y}{y^2 + 1} \, dy \, dx = k \int_0^2 (\ln 2) \, dx = k \ln 4$

$M_{xy} = 2k \displaystyle\int_0^2 \int_0^1 \int_0^{1/(y^2+1)} z \, dz \, dy \, dx$

$= k \displaystyle\int_0^2 \int_0^1 \frac{1}{(y^2 + 1)^2} \, dy \, dx = k \int_0^2 \left[\frac{y}{2(y^2 + 1)} + \frac{1}{2} \arctan y \right]_0^1 dx = k\left(\frac{1}{4} + \frac{\pi}{8}\right) \int_0^2 dx = k\left(\frac{1}{2} + \frac{\pi}{4}\right)$

$\bar{y} = \dfrac{M_{xz}}{m} = \dfrac{k \ln 4}{k\pi} = \dfrac{\ln 4}{\pi}$

$\bar{z} = \dfrac{M_{xy}}{m} = k\left(\dfrac{1}{2} + \dfrac{\pi}{4}\right) \Big/ k\pi = \dfrac{2 + \pi}{4\pi}$

35. $f(x, \ y) = \dfrac{5}{12} y$

$m = k \displaystyle\int_0^{20} \int_0^{-(3/5)x+12} \int_0^{(5/12)y} dz \, dy \, dx = 200k$

$M_{yz} = k \displaystyle\int_0^{20} \int_0^{-(3/5)x+12} \int_0^{(5/12)y} x \, dz \, dy \, dx = 1000k$

$M_{xz} = k \displaystyle\int_0^{20} \int_0^{-(3/5)x+12} \int_0^{(5/12)y} y \, dz \, dy \, dx = 1200k$

$M_{xy} = k \displaystyle\int_0^{20} \int_0^{-(3/5)x+12} \int_0^{(5/12)y} z \, dz \, dy \, dx = 250k$

$\bar{x} = \dfrac{M_{yz}}{m} = \dfrac{1000k}{200k} = 5$

$\bar{y} = \dfrac{M_{xz}}{m} = \dfrac{1200k}{200k} = 6$

$\bar{z} = \dfrac{M_{xy}}{m} = \dfrac{250k}{200k} = \dfrac{5}{4}$

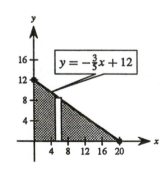

$y = -\frac{3}{5}x + 12$

36. $f(x, y) = \dfrac{1}{15}(60 - 12x - 20y)$

$$m = k \int_0^5 \int_0^{-(3/5)x+3} \int_0^{(1/15)(60-12x-20y)} dz\,dy\,dx = 10k$$

$$M_{yz} = k \int_0^5 \int_0^{-(3/5)x+3} \int_0^{(1/15)(60-12x-20y)} x\,dz\,dy\,dx = \frac{25k}{2}$$

$$M_{xz} = k \int_0^5 \int_0^{-(3/5)x+3} \int_0^{(1/15)(60-12x-20y)} y\,dz\,dy\,dx = \frac{15k}{2}$$

$$M_{xy} = k \int_0^5 \int_0^{-(3/5)x+3} \int_0^{(1/15)(60-12x-20y)} z\,dz\,dy\,dx = 10k$$

$$\overline{x} = \frac{M_{yz}}{m} = \frac{25k/2}{10k} = \frac{5}{4}$$

$$\overline{y} = \frac{M_{xz}}{m} = \frac{15k/2}{10k} = \frac{3}{4}$$

$$\overline{z} = \frac{M_{xy}}{m} = \frac{10k}{10k} = 1$$

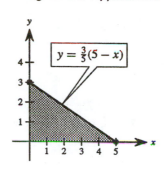

37. a. $I_x = k \displaystyle\int_0^a \int_0^a \int_0^a (y^2 + z^2)\,dx\,dy\,dz = ka \int_0^a \int_0^a (y^2 + z^2)\,dy\,dz$

$$= ka \int_0^a \left[\frac{1}{3}y^3 + z^2 y\right]_0^a dz = ka \int_0^a \left(\frac{1}{3}a^3 + az^2\right) dz = \left[ka\left(\frac{1}{3}a^3 z + \frac{1}{3}az^3\right)\right]_0^a = \frac{2ka^5}{3}$$

$$I_x = I_y = I_z = \frac{2ka^5}{3} \text{ by symmetry}$$

b. $I_x = k \displaystyle\int_0^a \int_0^a \int_0^a (y^2 + z^2)xyz\,dx\,dy\,dz = \frac{ka^2}{2} \int_0^a \int_0^a (y^3 z + yz^3)\,dy\,dz$

$$= \frac{ka^2}{2} \int_0^a \left[\frac{y^4 z}{4} + \frac{y^2 z^3}{2}\right]_0^a dz = \frac{ka^4}{8} \int_0^a (a^2 z + 2z^3)\,dz = \left[\frac{ka^4}{8}\left(\frac{a^2 z^2}{2} + \frac{2z^4}{4}\right)\right]_0^a = \frac{ka^8}{8}$$

$$I_x = I_y = I_z = \frac{ka^8}{8} \text{ by symmetry}$$

38. a. $I_{xy} = k \displaystyle\int_{-a/2}^{a/2} \int_{-a/2}^{a/2} \int_{-a/2}^{a/2} z^2\,dz\,dy\,dx = \frac{ka^5}{12}$

$$I_{xz} = I_{yz} = \frac{ka^5}{12} \text{ by symmetry}$$

$$I_x = I_y = I_z = \frac{ka^5}{12} + \frac{ka^5}{12} = \frac{ka^5}{6}$$

b. $I_{xy} = k \displaystyle\int_{-a/2}^{a/2} \int_{-a/2}^{a/2} \int_{-a/2}^{a/2} z^2(x^2 + y^2)\,dz\,dy\,dx = \frac{a^3 k}{12} \int_{-a/2}^{a/2} \int_{-a/2}^{a/2} (x^2 + y^2)\,dy\,dx = \frac{a^7 k}{72}$

$$I_{xz} = k \int_{-a/2}^{a/2} \int_{-a/2}^{a/2} \int_{-a/2}^{a/2} y^2(x^2 + y^2)\,dz\,dy\,dx = ka \int_{-a/2}^{a/2} \int_{-a/2}^{a/2} (x^2 y^2 + y^4)\,dy\,dx = \frac{7ka^7}{360}$$

$$I_{yz} = I_{xz} \text{ by symmetry}$$

$$I_x = I_{xy} + I_{xz} = \frac{a^7 k}{30}$$

$$I_y = I_{xy} + I_{yz} = \frac{a^7 k}{30}$$

$$I_z = I_{yz} + I_{xz} = \frac{7ka^7}{180}$$

39. a. $I_x = k \int_0^4 \int_0^4 \int_0^{4-x} (y^2 + z^2)\, dz\, dy\, dx = k \int_0^4 \int_0^4 \left[y^2(4-x) + \frac{1}{3}(4-x)^3 \right] dy\, dx$

$= k \int_0^4 \left[\frac{y^3}{3}(4-x) + \frac{y}{3}(4-x)^3 \right]_0^4 dx = k \int_0^4 \left[\frac{64}{3}(4-x) + \frac{4}{3}(4-x)^3 \right] dx$

$= k \left[-\frac{32}{3}(4-x)^2 - \frac{1}{3}(4-x)^4 \right]_0^4 = 256k$

$I_y = k \int_0^4 \int_0^4 \int_0^{4-x} (x^2 + z^2)\, dz\, dy\, dx = k \int_0^4 \int_0^4 \left[x^2(4-x) + \frac{1}{3}(4-x)^3 \right] dy\, dx$

$= 4k \int_0^4 \left[4x^2 - x^3 + \frac{1}{3}(4-x)^3 \right] dx = 4k \left[\frac{4}{3}x^3 - \frac{1}{4}x^4 - \frac{1}{12}(4-x)^4 \right]_0^4 = \frac{512k}{3}$

$I_z = k \int_0^4 \int_0^4 \int_0^{4-x} (x^2 + y^2)\, dz\, dy\, dx = k \int_0^4 \int_0^4 (x^2 + y^2)(4-x)\, dy\, dx$

$= k \int_0^4 \left[\left(x^2 y + \frac{y^3}{3} \right)(4-x) \right]_0^4 dx = k \int_0^4 \left(4x^2 + \frac{64}{3} \right)(4-x)\, dx$

$= k \int_0^4 \left(16x^2 - 4x^3 - \frac{64}{3}x + \frac{256}{3} \right) dx = k \left[\frac{16}{3}x^3 - x^4 - \frac{32}{3}x^2 + \frac{256}{3}x \right]_0^4 = 256k$

b. $I_x = k \int_0^4 \int_0^4 \int_0^{4-x} y(y^2 + z^2)\, dz\, dy\, dx = k \int_0^4 \int_0^4 \left[y^3(4-x) + \frac{1}{3}y(4-x)^3 \right] dy\, dx$

$= k \int_0^4 \left[\frac{y^4}{4}(4-x) + \frac{y^2}{6}(4-x)^3 \right]_0^4 dx = k \int_0^4 \left[64(4-x) + \frac{8}{3}(4-x)^3 \right] dx$

$= k \left[-32(4-x)^2 - \frac{2}{3}(4-x)^4 \right]_0^4 = \frac{2048k}{3}$

$I_y = k \int_0^4 \int_0^4 \int_0^{4-x} y(x^2 + z^2)\, dz\, dy\, dx = k \int_0^4 \int_0^4 \left[x^2 y(4-x) + \frac{1}{3}y(4-x)^3 \right] dy\, dx$

$= 8k \int_0^4 \left[4x^2 - x^3 + \frac{1}{3}(4-x)^3 \right] dx = 8k \left[\frac{4}{3}x^3 - \frac{1}{4}x^4 - \frac{1}{12}(4-x)^4 \right]_0^4 = \frac{1024k}{3}$

$I_z = k \int_0^4 \int_0^4 \int_0^{4-x} y(x^2 + y^2)\, dz\, dy\, dx = k \int_0^4 \int_0^4 (x^2 y + y^3)(4-x)\, dx$

$= k \int_0^4 \left[\left(\frac{x^2 y^2}{2} + \frac{y^4}{4} \right)(4-x) \right]_0^4 dx = k \int_0^4 (8x^2 + 64)(4-x)\, dx$

$= 8k \int_0^4 (32 - 8x + 4x^2 - x^3)\, dx = \left[8k \left(32x - 4x^2 + \frac{4}{3}x^3 - \frac{1}{4}x^4 \right) \right]_0^4 = \frac{2048k}{3}$

40. a. $I_{xy} = k \int_0^4 \int_0^2 \int_0^{4-y^2} z^3 \, dz \, dy \, dx = k \int_0^4 \int_0^2 \frac{1}{4}(4-y^2)^4 \, dy \, dx$

$\qquad = \frac{k}{4} \int_0^4 \int_0^2 (256 - 256y^2 + 96y^4 - 16y^6 + y^8) \, dy \, dx$

$\qquad = \frac{k}{4} \int_0^4 \left[256y - \frac{256y^3}{3} + \frac{96y^5}{5} - \frac{16y^7}{7} + \frac{y^9}{9} \right]_0^2 dx = k \int_0^4 \frac{49{,}152}{945} \, dx = \frac{65{,}536k}{315}$

$I_{xz} = k \int_0^4 \int_0^2 \int_0^{4-y^2} y^2 z \, dz \, dy \, dx = k \int_0^4 \int_0^2 \frac{1}{2} y^2 (4-y^2)^2 \, dy \, dx$

$\qquad = k \int_0^4 \int_0^2 \frac{1}{2}(16y^2 - 8y^4 + y^6) \, dy \, dx = \frac{k}{2} \int_0^4 \left[\frac{16y^3}{3} - \frac{8y^5}{5} + \frac{y^7}{7} \right]_0^2 dx = \frac{k}{2} \int_0^4 \frac{1024}{105} \, dx = \frac{2048k}{105}$

$I_{yz} = k \int_0^4 \int_0^2 \int_0^{4-y^2} x^2 z \, dz \, dy \, dx = k \int_0^4 \int_0^2 \frac{1}{2} x^2 (4-y^2)^2 \, dy \, dx$

$\qquad = k \int_0^4 \int_0^2 \frac{1}{2} x^2 (16 - 8y^2 + y^4) \, dy \, dx = \frac{k}{2} \int_0^4 \left[x^2 \left(16y - \frac{8y^3}{3} + \frac{y^5}{5} \right) \right]_0^2 dx = \frac{k}{2} \int_0^4 \frac{256}{15} x^2 \, dx = \frac{8192k}{45}$

$I_x = I_{xz} + I_{xy} = \frac{2048k}{9}, \quad I_y = I_{yz} + I_{xy} = \frac{8192k}{21}, \quad I_z = I_{yz} + I_{xz} = \frac{63{,}488k}{315}$

b. $I_{xy} = \int_0^4 \int_0^2 \int_0^{4-y^2} z^2 (4-z) \, dz \, dy \, dx$

$\qquad = k \int_0^4 \int_0^2 \int_0^{4-y^2} 4z^2 \, dz \, dy \, dx - k \int_0^4 \int_0^2 \int_0^{4-y^2} z^3 \, dz \, dy \, dx = \frac{32{,}768k}{105} - \frac{65{,}536k}{315} = \frac{32{,}768k}{315}$

$I_{xz} = k \int_0^4 \int_0^2 \int_0^{4-y^2} y^2 (4-z) \, dz \, dy \, dx$

$\qquad = k \int_0^4 \int_0^2 \int_0^{4-y^2} 4y^2 \, dz \, dy \, dx - k \int_0^4 \int_0^2 \int_0^{4-y^2} y^2 z \, dz \, dy \, dx = \frac{1024k}{15} - \frac{2048k}{105} = \frac{1024k}{21}$

$I_{yz} = k \int_0^4 \int_0^2 \int_0^{4-y^2} x^2 (4-z) \, dz \, dy \, dx$

$\qquad = k \int_0^4 \int_0^2 \int_0^{4-y^2} 4x^2 \, dz \, dy \, dx - k \int_0^4 \int_0^2 \int_0^{4-y^2} x^2 z \, dz \, dy \, dx = \frac{4096k}{9} - \frac{8192k}{45} = \frac{4096k}{15}$

$I_x = I_{xz} + I_{xy} = \frac{48{,}128k}{315}, \quad I_y = I_{yz} + I_{xy} = \frac{118{,}784k}{315}, \quad I_z = I_{xz} + I_{yz} = \frac{11{,}264k}{35}$

41. $I_{xy} = k \int_{-L/2}^{L/2} \int_{-a}^{a} \int_{-\sqrt{a^2-x^2}}^{\sqrt{a^2-x^2}} z^2 \, dz \, dx \, dy = k \int_{-L/2}^{L/2} \int_{-a}^{a} \frac{2}{3}(a^2 - x^2)\sqrt{a^2 - x^2} \, dx \, dy$

$$= \frac{2}{3}\int_{-L/2}^{L/2} \left[\frac{a^2}{2}\left(x\sqrt{a^2 - x^2} + a^2 \arcsin\frac{x}{a}\right) - \frac{1}{8}\left[x(2x^2 - a^2)\sqrt{x^2 - a^2} + a^4 \arcsin\frac{x}{a}\right]\right]_{-a}^{a} dy$$

$$= \frac{2k}{3}\int_{-L/2}^{L/2} 2\left(\frac{a^4\pi}{4} - \frac{a^4\pi}{16}\right) dy = \frac{a^4\pi Lk}{4}$$

Since $m = \pi a^2 Lk$, $I_{xy} = ma^2/4$.

$I_{xz} = k \int_{-L/2}^{L/2} \int_{-a}^{a} \int_{-\sqrt{a^2-x^2}}^{\sqrt{a^2-x^2}} y^2 \, dz \, dx \, dy = 2k \int_{-L/2}^{L/2} \int_{-a}^{a} y^2\sqrt{a^2 - x^2} \, dx \, dy$

$$= 2k \int_{-L/2}^{L/2} \left[\frac{y^2}{2}\left(x\sqrt{a^2 - x^2} + a^2 \arcsin\frac{x}{a}\right)\right]_{-a}^{a} dy = k\pi a^2 \int_{-L/2}^{L/2} y^2 \, dy = \frac{2k\pi a^2}{3}\left(\frac{L^3}{8}\right) = \frac{1}{12}mL^2$$

$I_{yz} = k \int_{-L/2}^{L/2} \int_{-a}^{a} \int_{-\sqrt{a^2-x^2}}^{\sqrt{a^2-x^2}} x^2 \, dz \, dx \, dy = 2k \int_{-L/2}^{L/2} \int_{-a}^{a} x^2\sqrt{a^2 - x^2} \, dx \, dy$

$$= 2k \int_{-L/2}^{L/2} \frac{1}{8}\left[x(2x^2 - a^2)\sqrt{a^2 - x^2} + a^4 \arcsin\frac{x}{a}\right]_{-a}^{a} dy = \frac{ka^4\pi}{4}\int_{-L/2}^{L/2} dy = \frac{ka^4\pi L}{4} = \frac{ma^2}{4}$$

$I_x = I_{xy} + I_{xz} = \dfrac{ma^2}{4} + \dfrac{mL^2}{12} = \dfrac{m}{12}(3a^2 + L^2)$

$I_y = I_{xy} + I_{yz} = \dfrac{ma^2}{4} + \dfrac{ma^2}{4} = \dfrac{ma^2}{2}$

$I_z = I_{xz} + I_{yz} = \dfrac{mL^2}{12} + \dfrac{ma^2}{4} = \dfrac{m}{12}(3a^2 + L^2)$

42. $I_{xy} = \int_{-c/2}^{c/2} \int_{-a/2}^{a/2} \int_{-b/2}^{b/2} z^2 \, dz \, dy \, dx = \frac{b^3}{12}\int_{-c/2}^{c/2} \int_{-a/2}^{a/2} dy \, dx = \frac{1}{12}b^2(abc) = \frac{1}{12}mb^2$

$I_{xz} = \int_{-c/2}^{c/2} \int_{-a/2}^{a/2} \int_{-b/2}^{b/2} y^2 \, dz \, dy \, dx = b\int_{-c/2}^{c/2} \int_{-a/2}^{a/2} y^2 \, dy \, dx = \frac{ba^3}{12}\int_{-c/2}^{c/2} dx = \frac{ba^3c}{12} = \frac{1}{12}a^2(abc) = \frac{1}{12}ma^2$

$I_{yz} = \int_{-c/2}^{c/2} \int_{-a/2}^{a/2} \int_{-b/2}^{b/2} x^2 \, dz \, dy \, dx = ab\int_{-c/2}^{c/2} x^2 \, dx = \frac{abc^3}{12} = \frac{1}{12}c^2(abc) = \frac{1}{12}mc^2$

$I_x = I_{xy} + I_{xz} = \dfrac{1}{12}m(a^2 + b^2)$

$I_y = I_{xy} + I_{yz} = \dfrac{1}{12}m(b^2 + c^2)$

$I_z = I_{xz} + I_{yz} = \dfrac{1}{12}m(a^2 + c^2)$

43. $\displaystyle\int_{-1}^{1} \int_{-1}^{1} \int_{0}^{1-x} (x^2 + y^2)\sqrt{x^2 + y^2 + z^2} \, dz \, dy \, dx$ **44.** $\displaystyle\int_{-1}^{1} \int_{-\sqrt{1-x^2}}^{\sqrt{1-x^2}} \int_{0}^{4-x^2-y^2} kx^2(x^2 + y^2) \, dz \, dy \, dx$

Section 14.7   Triple Integrals in Cylindrical and Spherical Coordinates

1. $\displaystyle\int_0^4\int_0^{\pi/2}\int_0^2 r\cos\theta\,dr\,d\theta\,dz = \int_0^4\int_0^{\pi/2}\left[\frac{r^2}{2}\cos\theta\right]_0^2 d\theta\,dz$

$$= \int_0^4\int_0^{\pi/2} 2\cos\theta\,d\theta\,dz = \int_0^4\left[2\sin\theta\right]_0^{\pi/2} dz = \int_0^4 2\,dz = 8$$

2. $\displaystyle\int_0^{\pi/4}\int_0^2\int_0^{2-r} rz\,dz\,dr\,d\theta = \int_0^{\pi/4}\int_0^2\left[\frac{rz^2}{2}\right]_0^{2-r} dr\,d\theta$

$$= \frac{1}{2}\int_0^{\pi/4}\int_0^2 (4r-4r^2+r^3)\,dr\,d\theta = \frac{1}{2}\int_0^{\pi/4}\left[2r^2-\frac{4r^3}{3}+\frac{r^4}{4}\right]_0^2 d\theta = \frac{2}{3}\int_0^{\pi/4} d\theta = \frac{\pi}{6}$$

3. $\displaystyle\int_0^{\pi/2}\int_0^{2\cos^2\theta}\int_0^{4-r^2} r\sin\theta\,dz\,dr\,d\theta = \int_0^{\pi/2}\int_0^{2\cos^2\theta} r(4-r^2)\sin\theta\,dr\,d\theta = \int_0^{\pi/2}\left[\left(2r^2-\frac{r^4}{4}\right)\sin\theta\right]_0^{2\cos^2\theta} d\theta$

$$= \int_0^{\pi/2} [8\cos^4\theta - 4\cos^8\theta]\sin\theta\,d\theta = \left[-\frac{8\cos^5\theta}{5}+\frac{4\cos^9\theta}{9}\right]_0^{\pi/2} = \frac{52}{45}$$

4. $\displaystyle\int_0^{\pi/2}\int_0^{\pi}\int_0^2 e^{-\rho^3}\rho^2\,d\rho\,d\theta\,d\phi = \int_0^{\pi/2}\int_0^{\pi}\left[-\frac{1}{3}e^{-\rho^3}\right]_0^2 d\theta\,d\phi = \int_0^{\pi/2}\int_0^{\pi}\frac{1}{3}(1-e^{-8})\,d\theta\,d\phi = \frac{\pi^2}{6}(1-e^{-8})$

5. $\displaystyle\int_0^{2\pi}\int_0^{\pi/4}\int_0^{\cos\phi} \rho^2\sin\phi\,d\rho\,d\phi\,d\theta = \frac{1}{3}\int_0^{2\pi}\int_0^{\pi/4}\cos^3\phi\sin\phi\,d\phi\,d\theta = -\frac{1}{12}\int_0^{2\pi}\left[\cos^4\phi\right]_0^{\pi/4} d\theta = \frac{\pi}{8}$

6. $\displaystyle\int_0^{\pi/4}\int_0^{\pi/4}\int_0^{\cos\theta}\rho^2\sin\phi\cos\phi\,d\rho\,d\theta\,d\phi = \frac{1}{3}\int_0^{\pi/4}\int_0^{\pi/4}\cos^3\theta\sin\phi\cos\phi\,d\theta\,d\phi$

$$= \frac{1}{3}\int_0^{\pi/4}\int_0^{\pi/4}\sin\phi\cos\phi[\cos\theta(1-\sin^2\theta)]\,d\theta\,d\phi$$

$$= \frac{1}{3}\int_0^{\pi/4}\sin\phi\cos\phi\left[\sin\theta-\frac{\sin^3\theta}{3}\right]_0^{\pi/4} d\phi$$

$$= \frac{5\sqrt{2}}{36}\int_0^{\pi/4}\sin\phi\cos\phi\,d\phi = \left[\frac{5\sqrt{2}}{36}\frac{\sin^2\phi}{2}\right]_0^{\pi/4} = \frac{5\sqrt{2}}{144}$$

7. $\displaystyle\int_0^4\int_0^z\int_0^{\pi/2} re^r\,d\theta\,dr\,dz = \pi(e^4+3)$

8. $\displaystyle\int_0^{\pi/2}\int_0^{\pi}\int_0^{\sin\theta}(2\cos\phi)\rho^2\,d\rho\,d\theta\,d\phi = \frac{4}{9}$

9. $\displaystyle\int_0^{\pi/2}\int_0^3\int_0^{e^{-r^2}} r\,dz\,dr\,d\theta = \int_0^{\pi/2}\int_0^3 re^{-r^2}\,dr\,d\theta$

$$= \int_0^{\pi/2}\left[-\frac{1}{2}e^{-r^2}\right]_0^3 d\theta$$

$$= \int_0^{\pi/2}\frac{1}{2}(1-e^{-9})\,d\theta$$

$$= \frac{\pi}{4}(1-e^{-9})$$

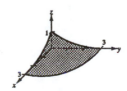

10. $\displaystyle\int_0^{2\pi}\int_0^{\sqrt{3}}\int_0^{3-r^2} r\,dz\,dr\,d\theta = \int_0^{2\pi}\int_0^{\sqrt{3}} r(3-r^2)\,dr\,d\theta$

$$= \int_0^{2\pi}\left(\frac{3r^2}{2} - \frac{r^4}{4}\right)\Bigg]_0^{\sqrt{3}} d\theta$$

$$= \int_0^{2\pi}\frac{9}{4}\,d\theta = \frac{9\pi}{2}$$

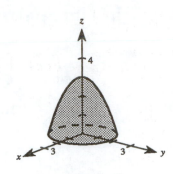

11. $\displaystyle\int_0^{2\pi}\int_{\pi/6}^{\pi/2}\int_0^{4} \rho^2\sin\phi\,d\rho\,d\phi\,d\theta = \frac{64}{3}\int_0^{2\pi}\int_{\pi/6}^{\pi/2}\sin\phi\,d\phi\,d\theta$

$$= \frac{64}{3}\int_0^{2\pi}\left[-\cos\phi\right]_{\pi/6}^{\pi/2} d\theta$$

$$= \frac{32\sqrt{3}}{3}\int_0^{2\pi} d\theta$$

$$= \frac{64\sqrt{3}\,\pi}{3}$$

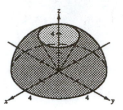

12. $\displaystyle\int_0^{2\pi}\int_0^{\pi}\int_2^{5} \rho^2\sin\phi\,d\rho\,d\phi\,d\theta = \frac{117}{3}\int_0^{2\pi}\int_0^{\pi}\sin\phi\,d\phi\,d\theta$

$$= \frac{117}{3}\int_0^{2\pi}\left[-\cos\phi\right]_0^{\pi} d\theta$$

$$= \frac{468\pi}{3}$$

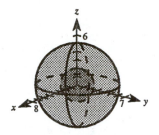

13. a. $\displaystyle\int_0^{2\pi}\int_0^{2}\int_{r^2}^{4} r^2\cos\theta\,dz\,dr\,d\theta = 0$

b. $\displaystyle\int_0^{2\pi}\int_0^{\arctan(1/2)}\int_0^{4\sec\phi} \rho^3\sin^2\phi\cos\theta\,d\rho\,d\phi\,d\theta + \int_0^{2\pi}\int_{\arctan(1/2)}^{\pi/2}\int_0^{\cot\phi\csc\phi} \rho^3\sin^2\phi\cos\theta\,d\rho\,d\phi\,d\theta = 0$

14. a. $\displaystyle\int_0^{\pi/2}\int_0^{2}\int_0^{\sqrt{16-r^2}} r^2\,dz\,dr\,d\theta = \frac{8\pi^2}{3} - 2\pi\sqrt{3}$

b. $\displaystyle\int_0^{\pi/2}\int_0^{\pi/6}\int_0^{4} \rho^3\sin^2\phi\,d\rho\,d\phi\,d\theta + \int_0^{\pi/2}\int_{\pi/6}^{\pi/2}\int_4^{2\csc\phi} \rho^3\sin^2\phi\,d\rho\,d\phi\,d\theta = \frac{8\pi^2}{3} - 2\pi\sqrt{3}$

15. a. $\displaystyle\int_0^{2\pi}\int_0^{a}\int_a^{a+\sqrt{a^2-r^2}} r^2\cos\theta\,dz\,dr\,d\theta = 0$

b. $\displaystyle\int_0^{\pi/4}\int_0^{2\pi}\int_{a\sec\phi}^{2a\cos\phi} \rho^3\sin^2\phi\cos\theta\,d\rho\,d\theta\,d\phi = 0$

16. a. $\displaystyle\int_0^{\pi/2}\int_0^{1}\int_0^{\sqrt{1-r^2}} r\sqrt{r^2+z^2}\,dz\,dr\,d\theta = \frac{\pi}{8}$

b. $\displaystyle\int_0^{\pi/2}\int_0^{\pi/2}\int_0^{1} \rho^3\sin\phi\,d\rho\,d\phi\,d\theta = \frac{\pi}{8}$

17. $z = h - \dfrac{h}{r_0}\sqrt{x^2+y^2} = \dfrac{h}{r_0}(r_0 - r)$

$$V = 4\int_0^{\pi/2}\int_0^{r_0}\int_0^{h(r_0-r)/r_0} r\,dz\,dr\,d\theta$$

$$= \frac{4h}{r_0}\int_0^{\pi/2}\int_0^{r_0}(r_0 r - r^2)\,dr\,d\theta$$

$$= \frac{4h}{r_0}\int_0^{\pi/2}\frac{r_0^3}{6}\,d\theta$$

$$= \frac{4h}{r_0}\left(\frac{r_0^3}{6}\right)\left(\frac{\pi}{2}\right) = \frac{1}{3}\pi r_0^2 h$$

18. $\bar{x} = \bar{y} = 0$ by symmetry

$m = \dfrac{1}{3}\pi r_0^2 hk$ from Exercise 17

$$M_{xy} = 4k\int_0^{\pi/2}\int_0^{r_0}\int_0^{h(r_0-r)/r_0} zr\,dz\,dr\,d\theta$$

$$= \frac{2kh^2}{r_0^2}\int_0^{\pi/2}\int_0^{r_0}(r_0^2 r - 2r_0 r^2 + r^3)\,dr\,d\theta$$

$$= \frac{2kh^2}{r_0^2}\left(\frac{r_0^4}{12}\right)\left(\frac{\pi}{2}\right) = \frac{kr_0^2 h^2\pi}{12}$$

$$\bar{z} = \frac{M_{xy}}{m} = \frac{kr_0^2 h^2\pi}{12}\left(\frac{3}{\pi r_0^2 hk}\right) = \frac{h}{4}$$

19. $\rho = k\sqrt{x^2+y^2} = kr$

$\bar{x} = \bar{y} = 0$ by symmetry

$$m = 4k\int_0^{\pi/2}\int_0^{r_0}\int_0^{h(r_0-r)/r_0} r^2\,dz\,dr\,d\theta$$

$$= \frac{1}{6}k\pi r_0^3 h$$

$$M_{xy} = 4k\int_0^{\pi/2}\int_0^{r_0}\int_0^{h(r_0-r)/r_0} r^2 z\,dz\,dr\,d\theta$$

$$= \frac{1}{30}k\pi r_0^3 h^2$$

$$\bar{z} = \frac{M_{xy}}{m} = \frac{k\pi r_0^3 h^2/30}{k\pi r_0^3 h/6} = \frac{h}{5}$$

20. $\rho = kz$

$\bar{x} = \bar{y} = 0$ by symmetry

$$m = 4k\int_0^{\pi/2}\int_0^{r_0}\int_0^{h(r_0-r)/r_0} zr\,dz\,dr\,d\theta$$

$$= \frac{1}{12}k\pi r_0^2 h^2$$

$$M_{xy} = 4k\int_0^{\pi/2}\int_0^{r_0}\int_0^{h(r_0-r)/r_0} z^2 r\,dz\,dr\,d\theta$$

$$= \frac{1}{30}k\pi r_0^2 h^3$$

$$\bar{z} = \frac{M_{xy}}{m} = \frac{k\pi r_0^2 h^3/30}{k\pi r_0^2 h^2/12} = \frac{2h}{5}$$

21. $I_z = 4k\displaystyle\int_0^{\pi/2}\int_0^{r_0}\int_0^{h(r_0-r)/r_0} r^3\,dz\,dr\,d\theta$

$$= \frac{4kh}{r_0}\int_0^{\pi/2}\int_0^{r_0}(r_0 r^3 - r^4)\,dr\,d\theta$$

$$= \frac{4kh}{r_0}\left(\frac{r_0^5}{20}\right)\left(\frac{\pi}{2}\right)$$

$$= \frac{1}{10}k\pi r_0^4 h$$

$$= \left(\frac{1}{3}k\pi r_0^2 h\right)\left(\frac{3}{10}r_0^2\right)$$

$$= \frac{3}{10}mr_0^2$$

22. $I_z = 4k\displaystyle\int_0^{\pi/2}\int_0^{r_0}\int_0^{(r_0-r)/r_0} r^5\,dz\,dr\,d\theta$

$$= \frac{4kh}{r_0}\int_0^{\pi/2}\int_0^{r_0}(r_0 r^5 - r^6)\,dr\,d\theta$$

$$= \frac{4kh}{r_0}\left(\frac{r_0^7}{42}\right)\left(\frac{\pi}{2}\right)$$

$$= \frac{kr_0^6 h\pi}{21}$$

$$= \left(\frac{1}{3}k\pi r_0^2 h\right)\left(\frac{1}{7}r_0^4\right)$$

$$= \frac{1}{7}mr_0^4$$

23. $m = k(\pi b^2 h - \pi a^2 h) = k\pi h(b^2 - a^2)$

$$I_z = 4k \int_0^{\pi/2} \int_a^b \int_0^h r^3 \, dz \, dr \, d\theta$$

$$= 4kh \int_0^{\pi/2} \int_a^b r^3 \, dr \, d\theta$$

$$= kh \int_0^{\pi/2} (b^4 - a^4) \, d\theta$$

$$= \frac{k\pi(b^4 - a^4)h}{2}$$

$$= \frac{k\pi(b^2 - a^2)(b^2 + a^2)h}{2}$$

$$= \frac{1}{2} m(a^2 + b^2)$$

24. $m = k\pi a^2 h$

$$I_z = 2k \int_0^{\pi/2} \int_0^{r_0 \sin\theta} \int_0^h r^3 \, dz \, dr \, d\theta$$

$$= \frac{3}{2} k\pi a^4 h$$

$$= \frac{3}{2} ma^2$$

25. $V = 4 \int_0^{\pi/2} \int_0^{a\cos\theta} \int_0^{\sqrt{a^2-r^2}} r \, dz \, dr \, d\theta = 4 \int_0^{\pi/2} \int_0^{a\cos\theta} r\sqrt{a^2 - r^2} \, dr \, d\theta$

$$= \frac{4}{3} a^3 \int_0^{\pi/2} (1 - \sin^3\theta) \, d\theta = \frac{4}{3} a^3 \left[\theta + \frac{1}{3}\cos\theta(\sin^2\theta + 2) \right]_0^{\pi/2} = \frac{4}{3} a^3 \left(\frac{\pi}{2} - \frac{2}{3} \right)$$

26. $V = \frac{2}{3}\pi(4)^3 + 4 \left[\int_0^{\pi/2} \int_0^{2\sqrt{2}} \int_0^r r \, dz \, dr \, d\theta + \int_0^{\pi/2} \int_{2\sqrt{2}}^4 \int_0^{\sqrt{16-r^2}} r \, dz \, dr \, d\theta \right]$

(Volume of lower hemisphere) + 4(Volume in the first octant)

$$V = \frac{128\pi}{3} + 4 \left[\int_0^{\pi/2} \int_0^{2\sqrt{2}} r^2 \, dr \, d\theta + \int_0^{\pi/2} \int_{2\sqrt{2}}^4 r\sqrt{16 - r^2} \, dr \, d\theta \right]$$

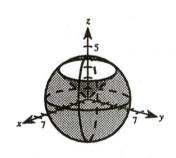

$$= \frac{128\pi}{3} + 4 \left[\frac{8\sqrt{2}\pi}{3} + \int_0^{\pi/2} \left[-\frac{1}{3}(16 - r^2)^{3/2} \right]_{2\sqrt{2}}^4 d\theta \right]$$

$$= \frac{128\pi}{3} + 4 \left[\frac{8\sqrt{2}\pi}{3} + \frac{8\sqrt{2}\pi}{3} \right]$$

$$= \frac{128\pi}{3} + \frac{64\sqrt{2}\pi}{3} = \frac{64\pi}{3}(2 + \sqrt{2})$$

27. $V = 2 \int_0^{\pi} \int_0^{a\cos\theta} \int_0^{\sqrt{a^2-r^2}} r \, dz \, dr \, d\theta$

$$= 2 \int_0^{\pi} \int_0^{a\cos\theta} r\sqrt{a^2 - r^2} \, dr \, d\theta$$

$$= 2 \int_0^{\pi} \left[-\frac{1}{3}(a^2 - r^2)^{3/2} \right]_0^{a\cos\theta} d\theta$$

$$= \frac{2a^3}{3} \int_0^{\pi} (1 - \sin^3\theta) \, d\theta$$

$$= \frac{2a^3}{3} \left[\theta + \cos\theta - \frac{\cos^3\theta}{3} \right]_0^{\pi}$$

$$= \frac{2a^3}{9}(3\pi - 4)$$

28. $V = \int_0^{2\pi} \int_0^{\sqrt{2}} \int_r^{\sqrt{4-r^2}} r \, dz \, dr \, d\theta$

$$= \int_0^{2\pi} \int_0^{\sqrt{2}} \left(r\sqrt{4 - r^2} - r^2 \right) dr \, d\theta$$

$$= \int_0^{2\pi} \left[-\frac{1}{3}(4 - r^2)^{3/2} - \frac{r^3}{3} \right]_0^{\sqrt{2}} d\theta$$

$$= \frac{8\pi}{3}(2 - \sqrt{2})$$

29. $V = \int_0^{2\pi} \int_0^\pi \int_0^{4\sin\phi} \rho^2 \sin\phi \, d\rho \, d\phi \, d\theta = 16\pi^2$

30. $V = 8\int_{\pi/4}^{\pi/2} \int_0^{\pi/2} \int_a^b \rho^2 \sin\phi \, d\rho \, d\theta \, d\phi$

$= \dfrac{8}{3}(b^3 - a^3)\int_{\pi/4}^{\pi/2} \int_0^{\pi/2} \sin\phi \, d\theta \, d\phi$

$= \dfrac{4\pi}{3}(b^3 - a^3)\int_{\pi/4}^{\pi/2} \sin\phi \, d\phi$

$= \left[\dfrac{4\pi}{3}(b^3 - a^3)(-\cos\phi)\right]_{\pi/4}^{\pi/2}$

$= \dfrac{2\sqrt{2}\,\pi}{3}(b^3 - a^3)$

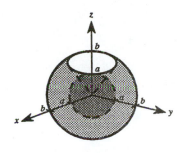

31. $m = 8k\int_0^{\pi/2} \int_0^{\pi/2} \int_0^a \rho^3 \sin\phi \, d\rho \, d\theta \, d\phi$

$= 2ka^4\int_0^{\pi/2} \int_0^{\pi/2} \sin\phi \, d\theta \, d\phi$

$= k\pi a^4\int_0^{\pi/2} \sin\phi \, d\phi$

$= \left[k\pi a^4(-\cos\phi)\right]_0^{\pi/2}$

$= k\pi a^4$

32. $m = 8k\int_0^{\pi/2} \int_0^{\pi/2} \int_0^a \rho^3 \sin^2\phi \, d\rho \, d\theta \, d\phi$

$= 2ka^4\int_0^{\pi/2} \int_0^{\pi/2} \sin^2\phi \, d\theta \, d\phi$

$= k\pi a^4\int_0^{\pi/2} \sin^2\phi \, d\phi$

$= \left[k\pi a^4\left(\dfrac{1}{2}\phi - \dfrac{1}{4}\sin 2\phi\right)\right]_0^{\pi/2}$

$= k\pi a^4\dfrac{\pi}{4} = \dfrac{1}{4}k\pi^2 a^4$

33. $m = \dfrac{2}{3}k\pi r^3$

$\overline{x} = \overline{y} = 0$ by symmetry

$M_{xy} = 4k\int_0^{\pi/2} \int_0^{\pi/2} \int_0^r \rho^3 \cos\phi \sin\phi \, d\rho \, d\theta \, d\phi$

$= \dfrac{1}{2}kr^4\int_0^{\pi/2} \int_0^{\pi/2} \sin 2\phi \, d\theta \, d\phi$

$= \dfrac{kr^4\pi}{4}\int_0^{\pi/2} \sin 2\phi \, d\phi$

$= \left[-\dfrac{1}{8}k\pi r^4 \cos 2\phi\right]_0^{\pi/2}$

$= \dfrac{1}{4}k\pi r^4$

$\overline{z} = \dfrac{M_{xy}}{m} = \dfrac{k\pi r^4/4}{2k\pi r^3/3} = \dfrac{3r}{8}$

34. $\overline{x} = \overline{y} = 0$ by symmetry

$m = k\left(\dfrac{2}{3}\pi R^3 - \dfrac{2}{3}\pi r^3\right) = \dfrac{2}{3}k\pi(R^3 - r^3)$

$M_{xy} = 4k\int_0^{\pi/2} \int_0^{\pi/2} \int_r^R \rho^3 \cos\phi \sin\phi \, d\rho \, d\theta \, d\phi$

$= \dfrac{1}{2}k(R^4 - r^4)\int_0^{\pi/2} \int_0^{\pi/2} \sin 2\phi \, d\theta \, d\phi$

$= \dfrac{1}{4}k\pi(R^4 - r^4)\int_0^{\pi/2} \sin 2\phi \, d\phi$

$= \left[-\dfrac{1}{8}k\pi(R^4 - r^4)\cos 2\phi\right]_0^{\pi/2}$

$= \dfrac{1}{4}k\pi(R^4 - r^4)$

$\overline{z} = \dfrac{M_{xy}}{m} = \dfrac{k\pi(R^4 - r^4)/4}{2k\pi(R^3 - r^3)/3} = \dfrac{3(R^4 - r^4)}{8(R^3 - r^3)}$

35. $I_z = 4k \int_{\pi/4}^{\pi/2} \int_0^{\pi/2} \int_0^{\cos\phi} \rho^4 \sin^3\phi \, d\rho \, d\theta \, d\phi$

$= \frac{4}{5} k \int_{\pi/4}^{\pi/2} \int_0^{\pi/2} \cos^5\phi \sin^3\phi \, d\theta \, d\phi$

$= \frac{2}{5} k\pi \int_{\pi/4}^{\pi/2} \cos^5\phi (1 - \cos^2\phi) \sin\phi \, d\phi$

$= \left[\frac{2}{5} k\pi \left(-\frac{1}{6} \cos^6\phi + \frac{1}{8} \cos^8\phi \right) \right]_{\pi/4}^{\pi/2}$

$= \frac{k\pi}{192}$

36. $I_z = 4k \int_0^{\pi/2} \int_0^{\pi/2} \int_r^R \rho^4 \sin^3\phi \, d\rho \, d\theta \, d\phi$

$= \frac{4k}{5} (R^5 - r^5) \int_0^{\pi/2} \int_0^{\pi/2} \sin^3\phi \, d\theta \, d\phi$

$= \frac{2k\pi}{5} (R^5 - r^5) \int_0^{\pi/2} \sin\phi (1 - \cos^2\phi) \, d\phi$

$= \left[\frac{2k\pi}{5} (R^5 - r^5) \left(-\cos\phi + \frac{\cos^3\phi}{3} \right) \right]_0^{\pi/2}$

$= \frac{4k\pi}{15} (R^5 - r^5)$

37. $\int_0^{2\pi} \int_0^\pi \int_0^{1+0.2\sin 8\theta \sin\phi} \rho^2 \sin\phi \, d\rho \, d\phi \, d\theta$

$= \frac{1}{3} \int_0^{2\pi} \int_0^\pi (1 + 0.2\sin 8\theta \sin\phi)^3 \sin\phi \, d\phi \, d\theta$

$= \frac{1}{3} \int_0^{2\pi} \int_0^\pi \left[\sin\phi + 0.6\sin 8\theta \sin^2\phi + 0.12\sin^2 8\theta \sin^3\phi + 0.008\sin^3 8\theta \sin^4\phi \right] d\phi \, d\theta$

$= \frac{1}{3} \int_0^{2\pi} \left[-\cos\phi + 0.6\sin 8\theta \left(\frac{1}{2} \right) (\phi - \sin\phi\cos\phi) + 0.12\sin^2 8\theta \left(-\frac{\sin^2\phi\cos\phi}{3} - \frac{2}{3}\cos\phi \right) \right.$

$\left. + 0.008\sin^3 8\theta \left(-\frac{\sin^3\phi\cos\phi}{4} + \frac{3}{8}(\phi - \sin\phi\cos\phi) \right) \right]_0^\pi d\theta$

$= \frac{1}{3} \int_0^{2\pi} (2 + 0.3\pi \sin 8\theta + 0.16\sin^2 8\theta + 0.003\pi \sin^3 8\theta) \, d\theta$

$= \frac{1}{3} \left[2\theta - \frac{0.3\pi}{8} \cos 8\theta + \frac{0.16}{8} \left(\frac{1}{2} \right) (8\theta - \sin 8\theta\cos 8\theta) + \frac{0.003\pi}{8} \left(-\frac{\sin^2 8\theta\cos 8\theta}{3} - \frac{2}{3}\cos 8\theta \right) \right]_0^{2\pi}$

$= \frac{104\pi}{75}$

38. $\int_0^{2\pi} \int_0^\pi \int_0^{1+0.2\sin 8\theta \sin 4\phi} \rho^2 \sin\phi \, d\rho \, d\phi \, d\theta = \frac{1}{3} \int_0^{2\pi} \int_0^\pi (1 + 0.2\sin 8\theta \sin 4\phi)^3 \sin\phi \, d\phi \, d\theta \approx 4.316$

Section 14.8   Change of Variables: Jacobians

1. $x = -\frac{1}{2}(u - v)$

$y = \frac{1}{2}(u + v)$

$\frac{\partial x}{\partial u} \frac{\partial y}{\partial v} - \frac{\partial y}{\partial u} \frac{\partial x}{\partial v} = \left(-\frac{1}{2} \right)\left(\frac{1}{2} \right) - \left(\frac{1}{2} \right)\left(\frac{1}{2} \right)$

$= -\frac{1}{2}$

2. $x = au + bv$

$y = cu + dv$

$\frac{\partial x}{\partial u} \frac{\partial y}{\partial v} - \frac{\partial y}{\partial u} \frac{\partial x}{\partial v} = ad - cb$

3. $x = u - v^2$

$y = u + v$

$$\dfrac{\partial x}{\partial u}\dfrac{\partial y}{\partial v} - \dfrac{\partial y}{\partial u}\dfrac{\partial x}{\partial v} = (1)(1) - (1)(-2v) = 1 + 2v$$

4. $x = u - uv$

$y = uv$

$$\dfrac{\partial x}{\partial u}\dfrac{\partial y}{\partial v} - \dfrac{\partial y}{\partial u}\dfrac{\partial x}{\partial v} = (1 - v)u - v(-u) = u$$

5. $x = u\cos\theta - v\sin\theta$

$y = u\sin\theta + v\cos\theta$

$$\dfrac{\partial x}{\partial u}\dfrac{\partial y}{\partial v} - \dfrac{\partial y}{\partial u}\dfrac{\partial x}{\partial v} = \cos^2\theta + \sin^2\theta = 1$$

6. $x = u + a$

$y = v + a$

$$\dfrac{\partial x}{\partial u}\dfrac{\partial y}{\partial v} - \dfrac{\partial y}{\partial u}\dfrac{\partial x}{\partial v} = (1)(1) - (0)(0) = 1$$

7. $x = e^u\sin v$

$y = e^u\cos v$

$$\dfrac{\partial x}{\partial u}\dfrac{\partial y}{\partial v} - \dfrac{\partial y}{\partial u}\dfrac{\partial x}{\partial v} = (e^u\sin v)(-e^u\sin v) - (e^u\cos v)(e^u\cos v) = -e^{2u}$$

8. $x = \dfrac{u}{v}$

$y = u + v$

$$\dfrac{\partial x}{\partial u}\dfrac{\partial y}{\partial v} - \dfrac{\partial y}{\partial u}\dfrac{\partial x}{\partial v} = \left(\dfrac{1}{v}\right)(1) - (1)\left(-\dfrac{u}{v^2}\right) = \dfrac{1}{v} + \dfrac{u}{v^2} = \dfrac{u + v}{v^2}$$

9. $x = 3u + 2v$

$y = 3v$

$v = \dfrac{y}{3}$

$u = \dfrac{x - 2v}{3} = \dfrac{x - 2(y/3)}{3} = \dfrac{x}{3} - \dfrac{2y}{9}$

$(x,\ y)$	$(u,\ v)$
$(0, 0)$	$(0, 0)$
$(3, 0)$	$(1, 0)$
$(2, 3)$	$(0, 1)$

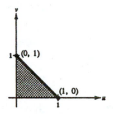

10. $x = 4u + v$

$y = u + 2v$

$u = \dfrac{2x - y}{7}$

$v = \dfrac{-x + 4y}{7}$

$(x,\ y)$	$(u,\ v)$
$(0, 0)$	$(0, 0)$
$(1, 2)$	$(0, 1)$
$(5, 3)$	$(1, 1)$
$(4, 1)$	$(1, 0)$

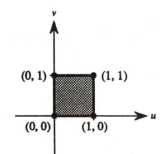

11. $x = \dfrac{1}{2}(u+v)$

$y = \dfrac{1}{2}(u-v)$

$\dfrac{\partial x}{\partial u}\dfrac{\partial y}{\partial v} - \dfrac{\partial y}{\partial u}\dfrac{\partial x}{\partial v} = \left(\dfrac{1}{2}\right)\left(-\dfrac{1}{2}\right) - \left(\dfrac{1}{2}\right)\left(\dfrac{1}{2}\right) = -\dfrac{1}{2}$

$\displaystyle\iint_R 48xy\,dx\,dy = \int_0^2\int_0^2 48\left[\dfrac{1}{2}(u+v)\right]\left[\dfrac{1}{2}(u-v)\right]\left(\dfrac{1}{2}\right)dv\,du$

$\displaystyle = 6\int_0^2\int_0^2 (u^2-v^2)\,dv\,du = 6\int_0^2\left(2u^2 - \dfrac{8}{3}\right)du = \left[6\left(\dfrac{2}{3}u^3 - \dfrac{8}{3}u\right)\right]_0^2 = 0$

12. $x = \dfrac{1}{2}(u+v)$

$y = \dfrac{1}{2}(u-v)$

$\dfrac{\partial x}{\partial u}\dfrac{\partial y}{\partial v} - \dfrac{\partial y}{\partial u}\dfrac{\partial x}{\partial v} = -\dfrac{1}{2}$ (Same as in Exercise 11)

$\displaystyle\iint_R 4(x^2+y^2)\,dx\,dy = \int_{-1}^1\int_{-1}^1 4\left[\dfrac{1}{4}(u+v)^2 + \dfrac{1}{4}(u-v)^2\right]\left(\dfrac{1}{2}\right)dv\,du$

$\displaystyle = \int_{-1}^1\int_{-1}^1 (u^2+v^2)\,dv\,du = \int_{-1}^1 2\left(u^2 + \dfrac{1}{3}\right)du = \left[2\left(\dfrac{u^3}{3} + \dfrac{u}{3}\right)\right]_{-1}^1 = \dfrac{8}{3}$

13. $x = \dfrac{1}{2}(u+v)$

$y = \dfrac{1}{2}(u-v)$

$\dfrac{\partial x}{\partial u}\dfrac{\partial y}{\partial v} - \dfrac{\partial y}{\partial u}\dfrac{\partial x}{\partial v} = -\dfrac{1}{2}$ (Same as in Exercise 11)

$\displaystyle\iint_R 4(x+y)e^{x-y}\,dy\,dx = \int_0^2\int_{u-2}^0 4ue^v\left(\dfrac{1}{2}\right)dv\,du$

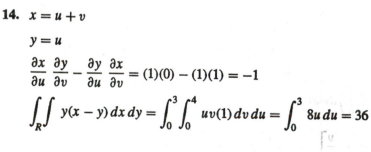

$v = u - 2$

$\displaystyle = \int_0^2 2u(1-e^{u-2})\,du = 2\left[\dfrac{u^2}{2} - ue^{u-2} + e^{u-2}\right]_0^2 = 2(1-e^{-2})$

14. $x = u+v$

$y = u$

$\dfrac{\partial x}{\partial u}\dfrac{\partial y}{\partial v} - \dfrac{\partial y}{\partial u}\dfrac{\partial x}{\partial v} = (1)(0) - (1)(1) = -1$

$\displaystyle\iint_R y(x-y)\,dx\,dy = \int_0^3\int_0^4 uv(1)\,dv\,du = \int_0^3 8u\,du = 36$

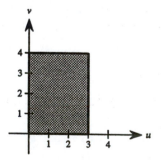

15. $x = u, \quad y = uv, \quad u = x, \quad v = \dfrac{y}{x}$

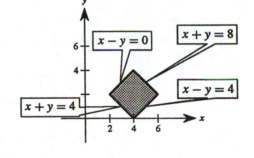

$$\frac{\partial x}{\partial u}\frac{\partial y}{\partial v} - \frac{\partial y}{\partial u}\frac{\partial x}{\partial v} = u$$

$$\iint_R \frac{\sqrt{x+y}}{x}\,dy\,dx = \int_0^1 \int_1^4 \frac{\sqrt{u+uv}}{u}\,u\,du\,dv$$

$$= \int_0^1 \int_1^4 \sqrt{u(1+v)}\,du\,dv$$

$$= \int_0^1 \left[\sqrt{1+v}\left(\frac{2}{3}u^{3/2}\right)\right]_1^4 dv = \frac{14}{3}\left[\frac{2}{3}(1+v)^{3/2}\right]_0^1 = \frac{28}{9}[2\sqrt{2}-1]$$

16. $x = \dfrac{u}{v}$

$y = v$

$$\frac{\partial x}{\partial u}\frac{\partial y}{\partial v} - \frac{\partial y}{\partial u}\frac{\partial x}{\partial v} = \frac{1}{v}$$

$$\iint_R y\sin xy\,dy\,dx = \int_1^4 \int_1^4 v(\sin u)\frac{1}{v}\,dv\,du = \int_1^4 3\sin u\,du = \Big[-3\cos u\Big]_1^4 = 3(\cos 1 - \cos 4) \approx 3.5818$$

17. $u = x+y = 4, \qquad v = x-y = 0$

$u = x+y = 8, \qquad v = x-y = 4$

$x = \dfrac{1}{2}(u+v), \qquad y = \dfrac{1}{2}(u-v)$

$\dfrac{\partial(x,\,y)}{\partial(u,\,v)} = -\dfrac{1}{2}$

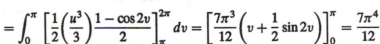

$$\iint_R (x+y)e^{x-y}\,dA = \int_4^8 \int_0^4 ue^v\left(\frac{1}{2}\right)dv\,du$$

$$= \frac{1}{2}\int_4^8 u(e^4-1)\,du = \left[\frac{1}{4}u^2(e^4-1)\right]_4^8 = 12(e^4-1)$$

18. $u = x+y = \pi, \qquad v = x-y = 0$

$u = x+y = 2\pi, \qquad v = x-y = \pi$

$x = \dfrac{1}{2}(u+v), \qquad y = \dfrac{1}{2}(u-v)$

$\dfrac{\partial(x,\,y)}{\partial(u,\,v)} = -\dfrac{1}{2}$

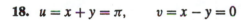

$$\iint_R (x+y)^2 \sin^2(x-y)\,dA = \int_0^\pi \int_\pi^{2\pi} u^2 \sin^2 v\left(\frac{1}{2}\right)du\,dv$$

$$= \int_0^\pi \left[\frac{1}{2}\left(\frac{u^3}{3}\right)\frac{1-\cos 2v}{2}\right]_\pi^{2\pi} dv = \left[\frac{7\pi^3}{12}\left(v + \frac{1}{2}\sin 2v\right)\right]_0^\pi = \frac{7\pi^4}{12}$$

19. $u = x + 4y = 0, \quad v = x - y = 0$

$u = x + 4y = 5, \quad v = x - y = 5$

$x = \dfrac{1}{5}(u + 4v), \quad y = \dfrac{1}{5}(u - v)$

$\dfrac{\partial x}{\partial u}\dfrac{\partial y}{\partial v} - \dfrac{\partial y}{\partial u}\dfrac{\partial x}{\partial v} = \left(\dfrac{1}{5}\right)\left(-\dfrac{1}{5}\right) - \left(\dfrac{1}{5}\right)\left(\dfrac{4}{5}\right) = -\dfrac{1}{5}$

$\displaystyle\iint_R \sqrt{(x-y)(x+4y)}\,dA = \int_0^5\int_0^5 \sqrt{uv}\left(\dfrac{1}{5}\right)du\,dv$

$\displaystyle = \int_0^5\left[\dfrac{1}{5}\left(\dfrac{2}{3}\right)u^{3/2}\sqrt{v}\right]_0^5 dv = \left[\dfrac{2\sqrt{5}}{3}\left(\dfrac{2}{3}\right)v^{3/2}\right]_0^5 = \dfrac{100}{9}$

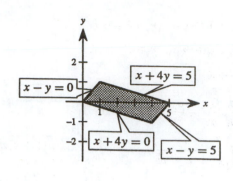

20. $u = 3x + 2y = 0, \quad v = 2y - x = 0$

$u = 3x + 2y = 16, \quad v = 2y - x = 8$

$x = \dfrac{1}{4}(u - v), \quad y = \dfrac{1}{8}(u + 3v)$

$\dfrac{\partial x}{\partial u}\dfrac{\partial y}{\partial v} - \dfrac{\partial y}{\partial u}\dfrac{\partial x}{\partial v} = \left(\dfrac{1}{4}\right)\left(\dfrac{3}{8}\right) - \left(\dfrac{1}{8}\right)\left(-\dfrac{1}{4}\right) = \dfrac{1}{8}$

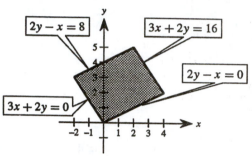

$\displaystyle\iint_R (3x+2y)^2\sqrt{2y-x}\,dA = \int_0^8\int_0^{16} u^2\sqrt{v}\left(\dfrac{1}{8}\right)du\,dv$

$\displaystyle = \int_0^8\left[\dfrac{1}{8}\left(\dfrac{u^3}{3}\right)\sqrt{v}\right]_0^{16} dv = \left[\dfrac{512}{3}\left(\dfrac{2}{3}\right)v^{3/2}\right]_0^8 = \dfrac{16{,}384\sqrt{2}}{9}$

21. $u = x + y, \quad v = x - y, \quad x = \dfrac{1}{2}(u + v), \quad y = \dfrac{1}{2}(u - v)$

$\dfrac{\partial x}{\partial u}\dfrac{\partial y}{\partial v} - \dfrac{\partial y}{\partial u}\dfrac{\partial x}{\partial v} = -\dfrac{1}{2}$

$\displaystyle\iint_R \sqrt{x+y}\,dA = \int_0^a\int_{-u}^u \sqrt{u}\left(\dfrac{1}{2}\right)dv\,du = \int_0^a u\sqrt{u}\,du = \left[\dfrac{2}{5}u^{5/2}\right]_0^a = \dfrac{2}{5}a^{5/2}$

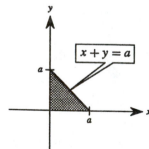

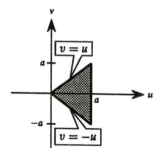

22. $u = x = 1, \quad v = xy = 1$

$u = x = 4, \quad v = xy = 4$

$x = u, \qquad y = \dfrac{v}{u}$

$\dfrac{\partial x}{\partial u}\dfrac{\partial y}{\partial v} - \dfrac{\partial y}{\partial u}\dfrac{\partial x}{\partial v} = \dfrac{1}{u}$

$\displaystyle \iint_R \frac{xy}{1+x^2y^2}\,dA = \int_1^4 \int_1^4 \frac{v}{1+v^2}\left(\frac{1}{u}\right)dv\,du$

$\displaystyle = \int_1^4 \left[\frac{1}{2}\ln(1+v^2)\right]_1^4 \frac{1}{u}\,du = \left[\frac{1}{2}[\ln 17 - \ln 2]\ln u\right]_1^4 = \frac{1}{2}\left(\ln\frac{17}{2}\right)(\ln 4)$

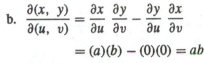

23. $\dfrac{x^2}{a^2} + \dfrac{y^2}{b^2} = 1, \quad x = au, \quad y = bv$

$\dfrac{(au)^2}{a^2} + \dfrac{(bv)^2}{b^2} = 1$

$u^2 + v^2 = 1$

a. $\dfrac{x^2}{a^2} + \dfrac{y^2}{b^2} = 1 \qquad u^2 + v^2 = 1$

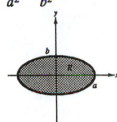 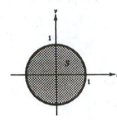

b. $\dfrac{\partial(x,\ y)}{\partial(u,\ v)} = \dfrac{\partial x}{\partial u}\dfrac{\partial y}{\partial v} - \dfrac{\partial y}{\partial u}\dfrac{\partial x}{\partial v}$

$= (a)(b) - (0)(0) = ab$

c. $A = \displaystyle\iint_S ab\,du\,dv$

$= ab(\pi(1)^2) = \pi ab$

24. a. $f(x,\ y) = 16 - x^2 - y^2$

$R: \dfrac{x^2}{16} + \dfrac{y^2}{9} \le 1$

$V = \displaystyle\iint_R f(x,\ y)\,dA$

Let $x = 4u$ and $y = 3v$.

$\displaystyle\iint_R (16 - x^2 - y^2)\,dA = \int_{-1}^{1}\int_{-\sqrt{1-u^2}}^{\sqrt{1-u^2}} (16 - 16u^2 - 9v^2)12\,dv\,du \quad \text{(Let } u = r\cos\theta, \quad v = r\sin\theta.)$

$\displaystyle = \int_0^{2\pi}\int_0^1 (16 - 16r^2\cos^2\theta - 9r^2\sin^2\theta)12r\,dr\,d\theta$

$\displaystyle = 12\int_0^{2\pi}\left[8r^2 - 4r^4\cos^2\theta - \frac{9}{4}r^4\sin^2\theta\right]_0^1 d\theta = 12\int_0^{2\pi}\left[8 - 4\cos^2\theta - \frac{9}{4}\sin^2\theta\right]d\theta$

$\displaystyle = 12\int_0^{2\pi}\left[8 - 4\left(\frac{1+\cos 2\theta}{2}\right) - \frac{9}{4}\left(\frac{1-\cos 2\theta}{2}\right)\right]d\theta = 12\int_0^{2\pi}\left[\frac{39}{8} - \frac{7}{8}\cos 2\theta\right]d\theta$

$\displaystyle = 12\left[\frac{39}{8}\theta - \frac{7}{16}\sin 2\theta\right]_0^{2\pi} = 12\left[\frac{39\pi}{4}\right] = 117\pi$

24. —CONTINUED—

b. $f(x, \ y) = A \cos\left[\dfrac{\pi}{2}\sqrt{\dfrac{x^2}{a^2} + \dfrac{y^2}{b^2}}\right]$

$R: \ \dfrac{x^2}{a^2} + \dfrac{y^2}{b^2} \le 1$

Let $x = au$ and $y = bv$.

$$\iint_R f(x, \ y)\,dA = \int_{-1}^{1}\int_{-\sqrt{1-u^2}}^{\sqrt{1-u^2}} A\cos\left[\dfrac{\pi}{2}\sqrt{u^2 + v^2}\right]ab\,dv\,du$$

Let $u = r\cos\theta, \quad v = r\sin\theta$.

$$Aab\int_0^{2\pi}\int_0^1 \cos\left[\dfrac{\pi}{2}r\right]r\,dr\,d\theta = Aab\left[\dfrac{2r}{\pi}\sin\left(\dfrac{\pi r}{2}\right) + \dfrac{4}{\pi^2}\cos\left(\dfrac{\pi r}{2}\right)\right]_0^1 (2\pi)$$

$$= 2\pi Aab\left[\left(\dfrac{2}{\pi}+0\right) - \left(0+\dfrac{4}{\pi^2}\right)\right] = \dfrac{4(\pi - 2)Aab}{\pi}$$

25. $x = u(1 - v), \quad y = uv(1 - w), \quad z = uvw$

$$\dfrac{\partial(x, \ y, \ z)}{\partial(u, \ v, \ w)} = \begin{vmatrix} 1-v & -u & 0 \\ v(1-w) & u(1-w) & -uv \\ vw & uw & uv \end{vmatrix} = (1-v)[u^2v(1-w) + u^2vw] + u[uv^2(1-w) + uv^2w]$$

$$= (1-v)(u^2v) + u(uv^2)$$

$$= u^2v$$

26. $x = 4u - v, \quad y = 4v - w, \quad z = u + w$

$$\dfrac{\partial(x, \ y, \ z)}{\partial(u, \ v, \ w)} = \begin{vmatrix} 4 & -1 & 0 \\ 0 & 4 & -1 \\ 1 & 0 & 1 \end{vmatrix} = 17$$

27. $x = \rho\sin\phi\cos\theta, \quad y = \rho\sin\phi\sin\theta, \quad z = \rho\cos\phi$

$$\dfrac{\partial(x, \ y, \ z)}{\partial(\rho, \ \theta, \ \phi)} = \begin{vmatrix} \sin\phi\cos\theta & -\rho\sin\phi\sin\theta & \rho\cos\phi\cos\theta \\ \sin\phi\sin\theta & \rho\sin\phi\cos\theta & \rho\cos\phi\sin\theta \\ \cos\phi & 0 & -\rho\sin\phi \end{vmatrix}$$

$$= \cos\phi[-\rho^2\sin\phi\cos\phi\sin^2\theta - \rho^2\sin\phi\cos\phi\cos^2\theta] - \rho\sin\phi[\rho\sin^2\phi\cos^2\theta + \rho\sin^2\phi\sin^2\theta]$$

$$= \cos\phi[-\rho^2\sin\phi\cos\phi(\sin^2\theta + \cos^2\theta)] - \rho\sin\phi[\rho\sin^2\phi(\cos^2\theta + \sin^2\theta)]$$

$$= -\rho^2\sin\phi\cos^2\phi - \rho^2\sin^3\phi$$

$$= -\rho^2\sin\phi(\cos^2\phi + \sin^2\phi)$$

$$= -\rho^2\sin\phi$$

28. $x = r\cos\theta, \quad y = r\sin\theta, \quad z = z$

$$\dfrac{\partial(x, \ y, \ z)}{\partial(r, \ \theta, \ z)} = \begin{vmatrix} \cos\theta & -r\sin\theta & 0 \\ \sin\theta & r\cos\theta & 0 \\ 0 & 0 & 1 \end{vmatrix} = 1[r\cos^2\theta + r\sin^2\theta] = r$$

Chapter 14   Review Exercises

1. $\displaystyle\int_{1}^{x^2} x \ln y \, dy = \Big[xy(-1 + \ln y) \Big]_{1}^{x^2} = x^3(-1 + \ln x^2) + x$

2. $\displaystyle\int_{y}^{2y} (x^2 + y^2) \, dx = \Big[\frac{x^3}{3} + xy^2 \Big]_{y}^{2y} = \frac{10y^3}{3}$

3. $\displaystyle\int_{0}^{1}\int_{0}^{1+x} (3x + 2y) \, dy \, dx = \int_{0}^{1}\Big[3xy + y^2 \Big]_{0}^{1+x} dx = \int_{0}^{1}(4x^2 + 5x + 1) \, dx = \Big[\frac{4}{3}x^3 + \frac{5}{2}x^2 + x \Big]_{0}^{1} = \frac{29}{6}$

4. $\displaystyle\int_{0}^{2}\int_{x^2}^{2x} (x^2 + 2y) \, dy \, dx = \int_{0}^{2}\Big[x^2 y + y^2 \Big]_{x^2}^{2x} dx = \int_{0}^{2}(4x^2 + 2x^3 - 2x^4) \, dx = \Big[\frac{4}{3}x^3 + \frac{1}{2}x^4 - \frac{2}{5}x^5 \Big]_{0}^{2} = \frac{88}{15}$

5. $\displaystyle\int_{0}^{3}\int_{0}^{\sqrt{9-x^2}} 4x \, dy \, dx = \int_{0}^{3} 4x\sqrt{9 - x^2} \, dx = \Big[-\frac{4}{3}(9 - x^2)^{3/2} \Big]_{0}^{3} = 36$

6. $\displaystyle\int_{0}^{\sqrt{3}}\int_{2-\sqrt{4-y^2}}^{2+\sqrt{4-y^2}} dx \, dy = 2\int_{0}^{\sqrt{3}} \sqrt{4 - y^2} \, dy = \Big[y\sqrt{4 - y^2} + 4\arcsin\frac{y}{2} \Big]_{0}^{\sqrt{3}} = \sqrt{3} + \frac{4\pi}{3}$

7. $\displaystyle\int_{0}^{h}\int_{0}^{x} \sqrt{x^2 + y^2} \, dy \, dx = \int_{0}^{\pi/4}\int_{0}^{h\sec\theta} r^2 \, dr \, d\theta = \frac{h^3}{3}\int_{0}^{\pi/4} \sec^3\theta \, d\theta$

$\displaystyle\qquad\qquad = \frac{h^3}{6}\Big[\sec\theta\tan\theta + \ln|\sec\theta + \tan\theta| \Big]_{0}^{\pi/4} = \frac{h^3}{6}\big[\sqrt{2} + \ln(\sqrt{2} + 1) \big]$

8. $\displaystyle\int_{0}^{4}\int_{0}^{\sqrt{16-y^2}} (x^2 + y^2) \, dx \, dy = \int_{0}^{\pi/2}\int_{0}^{4} r^3 \, dr \, d\theta = \int_{0}^{\pi/2}\Big[\frac{r^4}{4} \Big]_{0}^{4} d\theta = \int_{0}^{\pi/2} 64 \, d\theta = 32\pi$

9. $\displaystyle\int_{-3}^{3}\int_{-\sqrt{9-x^2}}^{\sqrt{9-x^2}}\int_{x^2+y^2}^{9} \sqrt{x^2 + y^2} \, dz \, dy \, dx = \int_{0}^{2\pi}\int_{0}^{3}\int_{r^2}^{9} r^2 \, dz \, dr \, d\theta = \int_{0}^{2\pi}\int_{0}^{3}(9r^2 - r^4) \, dr \, d\theta$

$\displaystyle\qquad\qquad = \int_{0}^{2\pi}\Big[3r^3 - \frac{r^5}{5} \Big]_{0}^{3} d\theta = \frac{162}{5}\int_{0}^{2\pi} d\theta = \frac{324\pi}{5}$

10. $\displaystyle\int_{-2}^{2}\int_{-\sqrt{4-x^2}}^{\sqrt{4-x^2}}\int_{0}^{(x^2+y^2)/2} (x^2 + y^2) \, dz \, dy \, dx = \int_{0}^{2\pi}\int_{0}^{2}\int_{0}^{r^2/2} r^3 \, dz \, dr \, d\theta$

$\displaystyle\qquad\qquad = \frac{1}{2}\int_{0}^{2\pi}\int_{0}^{2} r^5 \, dr \, d\theta = \frac{16}{3}\int_{0}^{2\pi} d\theta = \frac{32\pi}{3}$

11. $\displaystyle\int_{0}^{a}\int_{0}^{b}\int_{0}^{c} (x^2 + y^2 + z^2) \, dx \, dy \, dz = \int_{0}^{a}\int_{0}^{b} \Big(\frac{1}{3}c^3 + cy^2 + cz^2 \Big) dy \, dz = \int_{0}^{a}\Big(\frac{1}{3}bc^3 + \frac{1}{3}b^3 c + bcz^2 \Big) dz$

$\displaystyle\qquad\qquad = \frac{1}{3}abc^3 + \frac{1}{3}ab^3 c + \frac{1}{3}a^3 bc = \frac{1}{3}abc(a^2 + b^2 + c^2)$

12. $\displaystyle\int_0^5\int_0^{\sqrt{25-x^2}}\int_0^{\sqrt{25-x^2-y^2}}\frac{1}{1+x^2+y^2+z^2}\,dz\,dy\,dx = \int_0^{\pi/2}\int_0^{\pi/2}\int_0^5\frac{\rho^2}{1+\rho^2}\sin\phi\,d\rho\,d\phi\,d\theta$

$$= \int_0^{\pi/2}\int_0^{\pi/2}\Big[\rho-\arctan\rho\Big]_0^5\sin\phi\,d\phi\,d\theta$$

$$= \int_0^{\pi/2}\Big[(5-\arctan 5)(-\cos\phi)\Big]_0^{\pi/2}d\theta = \frac{\pi}{2}(5-\arctan 5)$$

13. $\displaystyle\int_{-2}^4\int_{y^2/4}^{(4+y)/2}(x-y)\,dx\,dy = \frac{27}{5}$

14. $\displaystyle\int_{-2}^2\int_6^{4-y^2}(8x-2y^2)\,dx\,dy = \frac{1792}{15}$

15. $\displaystyle\int_{-1}^1\int_{-\sqrt{1-x^2}}^{\sqrt{1-x^2}}\int_{-\sqrt{1-x^2-y^2}}^{\sqrt{1-x^2-y^2}}(x^2+y^2)\,dz\,dy\,dx = \int_0^{2\pi}\int_0^1\int_{-\sqrt{1-r^2}}^{\sqrt{1-r^2}}r^3\,dz\,dr\,d\theta = \frac{8\pi}{15}$

16. $\displaystyle\int_0^2\int_0^{\sqrt{4-x^2}}\int_0^{\sqrt{4-x^2-y^2}}xyz\,dz\,dy\,dx = \frac{4}{3}$

17. $\displaystyle\int_0^3\int_0^{(3-x)/3}dy\,dx = \int_0^1\int_0^{3-3y}dx\,dy$

$A = \displaystyle\int_0^1\int_0^{3-3y}dx\,dy$

$= \displaystyle\int_0^1(3-3y)\,dy = \Big[3y-\frac{3}{2}y^2\Big]_0^1 = \frac{3}{2}$

18. $\displaystyle\int_0^2\int_0^x dy\,dx + \int_2^3\int_0^{6-2x}dy\,dx = \int_0^2\int_y^{(6-y)/2}dx\,dy$

$A = \displaystyle\int_0^2\int_y^{(6-y)/2}dx\,dy$

$= \displaystyle\frac{1}{2}\int_0^2(6-3y)\,dy = \Big[\frac{1}{2}\Big(6y-\frac{3}{2}y^2\Big)\Big]_0^2 = 3$

19. $\displaystyle\int_{-5}^3\int_{-\sqrt{25-x^2}}^{\sqrt{25-x^2}}dy\,dx = \int_{-5}^{-4}\int_{-\sqrt{25-y^2}}^{\sqrt{25-y^2}}dx\,dy + \int_{-4}^4\int_{-\sqrt{25-y^2}}^3 dx\,dy + \int_4^5\int_{-\sqrt{25-y^2}}^{\sqrt{25-y^2}}dx\,dy$

$A = \displaystyle 2\int_{-5}^3\int_0^{\sqrt{25-x^2}}dy\,dx$

$= \displaystyle 2\int_{-5}^3\sqrt{25-x^2}\,dx = \Big[x\sqrt{25-x^2}+25\arcsin\frac{x}{5}\Big]_{-5}^3 = \frac{25\pi}{2}+12+25\arcsin\frac{3}{5}\approx 67.36$

20. $\displaystyle\int_0^4\int_{x^2-2x}^{6x-x^2}dy\,dx = \int_{-1}^0\int_{1-\sqrt{1+y}}^{1+\sqrt{1+y}}dx\,dy + \int_0^8\int_{3-\sqrt{9-y}}^{1+\sqrt{1+y}}dx\,dy + \int_8^9\int_{3-\sqrt{9-y}}^{3+\sqrt{9-y}}dx\,dy$

$A = \displaystyle\int_0^4\int_{x^2-2x}^{6x-x^2}dy\,dx = \int_0^4(8x-2x^2)\,dx = \Big[4x^2-\frac{2}{3}x^3\Big]_0^4 = \frac{64}{3}$

21. $A = \displaystyle 4\int_0^1\int_0^{x\sqrt{1-x^2}}dy\,dx = 4\int_0^1 x\sqrt{1-x^2}\,dx = \Big[-\frac{4}{3}(1-x^2)^{3/2}\Big]_0^1 = \frac{4}{3}$

$A = \displaystyle 4\int_0^{1/2}\int_{\sqrt{(1-\sqrt{1-4y^2})/2}}^{\sqrt{(1+\sqrt{1-4y^2})/2}}dx\,dy$

22. $A = \displaystyle\int_0^2\int_0^{y^2+1}dx\,dy = \int_0^1\int_0^2 dy\,dx + \int_1^5\int_{\sqrt{x-1}}^2 dy\,dx = \frac{14}{3}$

23. $A = \displaystyle\int_2^5\int_{x-3}^{\sqrt{x-1}}dy\,dx + 2\int_1^2\int_0^{\sqrt{x-1}}dy\,dx = \int_{-1}^2\int_{y^2+1}^{y+3}dx\,dy = \frac{9}{2}$

24. $A = \int_0^3 \int_{-y}^{2y-y^2} dx\,dy = \int_{-3}^0 \int_{-x}^{1+\sqrt{1-x}} dy\,dx + \int_0^1 \int_{1-\sqrt{1-x}}^{1+\sqrt{1-x}} dy\,dx = \dfrac{9}{2}$

25. $V = \int_0^4 \int_0^{x^2+4} (x^2 - y + 4)\,dy\,dx$

$= \int_0^4 \left[x^2 y - \dfrac{1}{2}y^2 + 4y \right]_0^{x^2+4} dx = \int_0^4 \left(\dfrac{1}{2}x^4 + 4x^2 + 8 \right) dx = \left[\dfrac{1}{10}x^5 + \dfrac{4}{3}x^3 + 8x \right]_0^4 = \dfrac{3296}{15}$

26. $V = \int_0^3 \int_0^x (x + y)\,dy\,dx = \int_0^3 \left[xy + \dfrac{1}{2}y^2 \right]_0^x dx = \dfrac{3}{2}\int_0^3 x^2\,dx = \left[\dfrac{1}{2}x^3 \right]_0^3 = \dfrac{27}{2}$

27. $V = 4\int_0^h \int_0^{\pi/2} \int_1^{\sqrt{1+z^2}} r\,dr\,d\theta\,dz = 2\int_0^h \int_0^{\pi/2} (1 + z^2 - 1)\,d\theta\,dz = \pi \int_0^h z^2\,dz = \left[\pi \left(\dfrac{1}{3}z^3 \right) \right]_0^h = \dfrac{\pi h^3}{3}$

28. $V = 8\int_0^{\pi/2} \int_b^R \sqrt{R^2 - r^2}\,r\,dr\,d\theta = -\dfrac{8}{3}\int_0^{\pi/2} \left[(R^2 - r^2)^{3/2} \right]_b^R d\theta = \dfrac{8}{3}(R^2 - b^2)^{3/2} \int_0^{\pi/2} d\theta = \dfrac{4}{3}\pi(R^2 - b^2)^{3/2}$

29. $V = 2\int_0^{\pi/2} \int_0^{2\cos\theta} \int_0^{\sqrt{4-r^2}} r\,dz\,dr\,d\theta = 2\int_0^{\pi/2} \int_0^{2\cos\theta} r\sqrt{4-r^2}\,dr\,d\theta$

$= -\int_0^{\pi/2} \left[\dfrac{2}{3}(4-r^2)^{3/2} \right]_0^{2\cos\theta} d\theta = \dfrac{16}{3}\int_0^{\pi/2} (1 - \sin^3\theta)\,d\theta = \dfrac{16}{3}\left[\theta + \cos\theta - \dfrac{1}{3}\cos^3\theta \right]_0^{\pi/2} = \dfrac{16}{3}\left(\dfrac{\pi}{2} - \dfrac{2}{3} \right)$

30. $V = 2\int_0^{\pi/2} \int_0^{2\sin\theta} \int_0^{16-r^2} r\,dz\,dr\,d\theta = 2\int_0^{\pi/2} \int_0^{2\sin\theta} r(16 - r^2)\,dr\,d\theta = 2\int_0^{\pi/2} (32\sin^2\theta - 4\sin^4\theta)\,d\theta$

$= 8\int_0^{\pi/2} (8\sin^2\theta - \sin^4\theta)\,d\theta = 8\left[4\theta - 2\sin 2\theta + \dfrac{1}{4}\sin^3\theta\cos\theta - \dfrac{3}{4}\left(\dfrac{1}{2}\theta - \dfrac{1}{4}\sin 2\theta \right) \right]_0^{\pi/2} = \dfrac{29\pi}{2}$

31. $\int_0^\infty \int_0^\infty kxye^{-(x+y)}\,dy\,dx = \int_0^\infty \left[-kxe^{-(x+y)}(y + 1) \right]_0^\infty dx = \int_0^\infty kxe^{-x}\,dx = \left[-k(x + 1)e^{-x} \right]_0^\infty = k$

Therefore, $k = 1$.

$P = \int_0^1 \int_0^1 xye^{-(x+y)}\,dy\,dx \approx 0.070$

32. $\int_0^1 \int_0^x kxy\,dy\,dx = \int_0^1 \left[\dfrac{kxy^2}{2} \right]_0^x dx = \int_0^1 \dfrac{kx^3}{2}\,dx = \left[\dfrac{kx^4}{8} \right]_0^1 = \dfrac{k}{8}$

Since $k/8 = 1$, we have $k = 8$.

$P = \int_0^{0.5} \int_0^{0.25} 8xy\,dy\,dx = 0.03125$

33. a. $m = k \displaystyle\int_0^1 \int_{2x^3}^{2x} xy\,dy\,dx = \dfrac{k}{4}$

$M_x = k \displaystyle\int_0^1 \int_{2x^3}^{2x} xy^2\,dy\,dx = \dfrac{16k}{55}$

$M_y = k \displaystyle\int_0^1 \int_{2x^3}^{2x} x^2 y\,dy\,dx = \dfrac{8k}{45}$

$\bar{x} = \dfrac{M_y}{m} = \dfrac{32}{45}$

$\bar{y} = \dfrac{M_x}{m} = \dfrac{64}{55}$

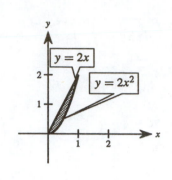

b. $m = k \displaystyle\int_0^1 \int_{2x^3}^{2x} (x^2 + y^2)\,dy\,dx = \dfrac{17k}{30}$

$M_x = k \displaystyle\int_0^1 \int_{2x^3}^{2x} y(x^2 + y^2)\,dy\,dx = \dfrac{392k}{585}$

$M_y = k \displaystyle\int_0^1 \int_{2x^3}^{2x} x(x^2 + y^2)\,dy\,dx = \dfrac{156k}{385}$

$\bar{x} = \dfrac{M_y}{m} = \dfrac{936}{1309}$

$\bar{y} = \dfrac{M_x}{m} = \dfrac{784}{663}$

34. $m = k \displaystyle\int_0^L \int_0^{(h/2)[2-(x/L)-(x^2/L^2)]} dy\,dx = \dfrac{kh}{2}\int_0^L \left(2 - \dfrac{x}{L} - \dfrac{x^2}{L^2}\right)dx = \dfrac{7khL}{12}$

$M_x = k \displaystyle\int_0^L \int_0^{(h/2)[2-(x/L)-(x^2/L^2)]} y\,dy\,dx$

$= \dfrac{kh^2}{8}\displaystyle\int_0^L \left(2 - \dfrac{x}{L} - \dfrac{x^2}{L^2}\right)^2 dx$

$= \dfrac{kh^2}{8}\displaystyle\int_0^L \left[4 - \dfrac{4x}{L} - \dfrac{3x^2}{L^2} + \dfrac{2x^3}{L^3} + \dfrac{x^4}{L^4}\right]dx$

$= \dfrac{kh^2}{8}\left[4x - \dfrac{2x^2}{L} - \dfrac{x^3}{L^2} + \dfrac{x^4}{2L^3} + \dfrac{x^5}{5L^4}\right]_0^L$

$= \dfrac{kh^2}{8} \cdot \dfrac{17L}{10} = \dfrac{17kh^2 L}{80}$

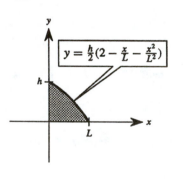

$M_y = k \displaystyle\int_0^L \int_0^{(h/2)[2-(x/L)-(x^2/L^2)]} x\,dy\,dx$

$= \dfrac{kh}{2}\displaystyle\int_0^L \left(2x - \dfrac{x^2}{L} - \dfrac{x^3}{L^2}\right)dx = \dfrac{kh}{2}\left[x^2 - \dfrac{x^3}{3L} - \dfrac{x^4}{4L^2}\right]_0^L = \dfrac{kh}{2} \cdot \dfrac{5L^2}{12} = \dfrac{5khL^2}{24}$

$\bar{x} = \dfrac{M_y}{m} = \dfrac{5khL^2}{24} \cdot \dfrac{12}{7khL} = \dfrac{5L}{14}$

$\bar{y} = \dfrac{M_x}{m} = \dfrac{17kh^2 L}{80} \cdot \dfrac{12}{7khL} = \dfrac{51h}{140}$

35. $S = \displaystyle\iint_R \sqrt{1 + (f_x)^2 + (f_y)^2}\, dA$

$= 4\displaystyle\int_0^4 \int_0^{\sqrt{16-x^2}} \sqrt{1 + 4x^2 + 4y^2}\, dy\, dx = 4\int_0^{\pi/2} \int_0^4 \sqrt{1 + 4r^2}\, r\, dr\, d\theta = \left[\frac{1}{3}(65^{3/2} - 1)\theta\right]_0^{\pi/2} = \frac{\pi}{6}(65\sqrt{65} - 1)$

36. $f(x,\, y) = 16 - x - y^2$

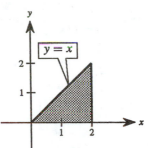

$R = \{(x,\, y):\ 0 \le x \le 2,\ \ 0 \le y \le x\}$

$f_x = -1,\ \ f_y = -2y$

$\sqrt{1 + (f_x)^2 + (f_y)^2} = \sqrt{2 + 4y^2}$

$S = \displaystyle\int_0^2 \int_y^2 \sqrt{2 + 4y^2}\, dx\, dy = \int_0^2 \left[2\sqrt{2 + 4y^2} - y\sqrt{2 + 4y^2}\right] dy$

$= \left[\frac{1}{2}\left(2y\sqrt{2 + 4y^2} + 2\ln\left|2y + \sqrt{2 + 4y^2}\right|\right) - \frac{1}{12}(2 + 4y^2)^{3/2}\right]_0^2$

$= \frac{1}{2}\left[4\sqrt{18} + 2\ln\left|4 + \sqrt{18}\right| - \frac{1}{12}(18\sqrt{18})\right] - \left[\ln\sqrt{2} - \frac{2\sqrt{2}}{12}\right]$

$= 6\sqrt{2} + \ln\left|4 + 3\sqrt{2}\right| - \frac{9\sqrt{2}}{2} - \ln\sqrt{2} + \frac{\sqrt{2}}{6} = \frac{5\sqrt{2}}{3} + \ln\left|2\sqrt{2} + 3\right|$

37. $m = 4k\displaystyle\int_{\pi/4}^{\pi/2} \int_0^{\pi/2} \int_0^{\cos\phi} \rho^2 \sin\phi\, d\rho\, d\theta\, d\phi$

$= \frac{4}{3}k\displaystyle\int_{\pi/4}^{\pi/2} \int_0^{\pi/2} \cos^3\phi \sin\phi\, d\theta\, d\phi = \frac{2}{3}k\pi\int_{\pi/4}^{\pi/2} \cos^3\phi \sin\phi\, d\phi = \left[-\frac{2}{3}k\pi\left(\frac{1}{4}\cos^4\phi\right)\right]_{\pi/4}^{\pi/2} = \frac{k\pi}{24}$

$M_{xy} = 4k\displaystyle\int_{\pi/4}^{\pi/2} \int_0^{\pi/2} \int_0^{\cos\phi} \rho^3 \cos\phi \sin\phi\, d\rho\, d\theta\, d\phi$

$= k\displaystyle\int_{\pi/4}^{\pi/2} \int_0^{\pi/2} \cos^5\phi \sin\phi\, d\theta\, d\phi = \frac{1}{2}k\pi\int_{\pi/4}^{\pi/2} \cos^5\phi \sin\phi\, d\phi = \left[-\frac{1}{12}k\pi \cos^6\phi\right]_{\pi/4}^{\pi/2} = \frac{k\pi}{96}$

$\bar{z} = \dfrac{M_{xy}}{m} = \dfrac{k\pi/96}{k\pi/24} = \dfrac{1}{4}$

$\bar{x} = \bar{y} = 0 \quad \text{by symmetry}$

38. $m = 2k\displaystyle\int_0^{\pi/2} \int_0^a \int_0^{cr\sin\theta} r\, dz\, dr\, d\theta = 2kc\int_0^{\pi/2} \int_0^a r^2 \sin\theta\, dr\, d\theta = \frac{2}{3}kca^3 \int_0^{\pi/2} \sin\theta\, d\theta = \frac{2}{3}kca^3$

$M_{xz} = 2k\displaystyle\int_0^{\pi/2} \int_0^a \int_0^{cr\sin\theta} r^2 \sin\theta\, dz\, dr\, d\theta = 2kc\int_0^{\pi/2} \int_0^a r^3 \sin^2\theta\, dr\, d\theta = \frac{1}{2}kca^4 \int_0^{\pi/2} \sin^2\theta\, d\theta = \frac{1}{8}\pi kca^4$

$M_{xy} = 2k\displaystyle\int_0^{\pi/2} \int_0^a \int_0^{cr\sin\theta} rz\, dz\, dr\, d\theta = kc^2\int_0^{\pi/2} \int_0^a r^3 \sin^2\theta\, dr\, d\theta = \frac{1}{4}kc^2a^4 \int_0^{\pi/2} \sin^2\theta\, d\theta = \frac{1}{16}\pi kc^2a^4$

$\bar{x} = 0$

$\bar{y} = \dfrac{M_{xz}}{m} = \dfrac{\pi kca^4/8}{2kca^3/3} = \dfrac{3\pi a}{16}$

$\bar{z} = \dfrac{M_{xy}}{m} = \dfrac{\pi kc^2a^4/16}{2kca^3/3} = \dfrac{3\pi ca}{32}$

39. $m = k \int_0^{\pi/2} \int_0^{\pi/2} \int_0^a \rho^2 \sin\phi \, d\rho \, d\theta \, d\phi = \dfrac{k\pi a^3}{6}$

$M_{xy} = k \int_0^{\pi/2} \int_0^{\pi/2} \int_0^a (\rho\cos\phi)\rho^2 \sin\phi \, d\rho \, d\theta \, d\phi = \dfrac{k\pi a^4}{16}$

$\bar{x} = \bar{y} = \bar{z} = \dfrac{M_{xy}}{m} = \dfrac{k\pi a^4}{16}\left(\dfrac{6}{k\pi a^3}\right) = \dfrac{3a}{8}$

40. $m = \dfrac{500\pi}{3} - \int_0^3 \int_0^{2\pi} \int_4^{\sqrt{25-r^2}} r \, dz \, d\theta \, dr = \dfrac{500\pi}{3} - \int_0^3 \int_0^{2\pi} \left(r\sqrt{25-r^2} - 4r\right) d\theta \, dr$

$= \dfrac{500\pi}{3} - 2\pi\left[-\dfrac{1}{3}(25-r^2)^{3/2} - 2r^2\right]_0^3 = \dfrac{500\pi}{3} - 2\pi\left[-\dfrac{64}{3} - 18 + \dfrac{125}{3}\right] = \dfrac{500\pi}{3} - \dfrac{14\pi}{3} = 162\pi$

$\bar{x} = \bar{y} = 0$ by symmetry

$M_{xy} = \int_0^{2\pi} \int_0^3 \int_{-\sqrt{25-r^2}}^4 zr \, dz \, dr \, d\theta + \int_0^{2\pi} \int_3^5 \int_{-\sqrt{25-r^2}}^{\sqrt{25-r^2}} zr \, dz \, dr \, d\theta = \int_0^{2\pi} \int_0^3 \left[8 - \dfrac{1}{2}(25-r^2)\right]r \, dr \, d\theta + 0$

$= \int_0^{2\pi} \int_0^3 \left[\dfrac{1}{2}r^3 - \dfrac{9}{2}r\right] dr \, d\theta = \int_0^{2\pi} \left[\dfrac{1}{8}r^4 - \dfrac{9}{4}r^2\right]_0^3 d\theta = \left[-\dfrac{81}{8}\theta\right]_0^{2\pi} = -\dfrac{81}{4}\pi$

$\bar{z} = \dfrac{M_{xy}}{m} = -\dfrac{81\pi}{4}\dfrac{1}{162\pi} = -\dfrac{1}{8}$

41. $I_z = 4k \int_0^{\pi/2} \int_3^4 \int_0^{16-r^2} r^3 \, dz \, dr \, d\theta = 4k \int_0^{\pi/2} \int_3^4 (16r^3 - r^5) \, dr \, d\theta = \dfrac{833\pi k}{3}$

42. $I_z = k \int_0^{\pi} \int_0^{2\pi} \int_0^a \rho^2 \sin^2\phi(\rho)\rho^2 \sin\phi \, d\rho \, d\theta \, d\phi = \dfrac{4k\pi a^6}{9}$

43. $\int_0^2 \int_{-\pi/2}^{\pi/2} \int_0^{6\sin\phi} \rho^2 \sin\phi \, d\rho \, d\phi \, d\theta$

Since $\rho = 6\sin\phi$ represents (in the *yz*-plane) a circle of radius 3 centered at $(0, 3, 0)$, the integral represents the volume of the torus formed by revolving $(0 < \theta < 2\pi)$ this circle about the *z*-axis.

44. $\int_0^{\pi} \int_0^2 \int_0^{1+r^2} r \, dz \, dr \, d\theta$

Since $z = 1 + r^2$ represents a paraboloid with vertex $(0, 0, 1)$, this integral represents the volume of the solid below the paraboloid and above the semi-circle $y = \sqrt{4 - x^2}$ in the *xy*-plane.

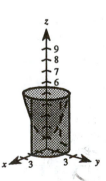

45. True

46. False,

$\int_0^1 \int_0^1 x \, dy \, dx \neq \int_1^2 \int_1^2 x \, dy \, dx$

47. True

48. True,

$\int_0^1 \int_0^1 \dfrac{1}{1+x^2+y^2} \, dx \, dy < \int_0^1 \int_0^1 \dfrac{1}{1+x^2} \, dx \, dy = \dfrac{\pi}{4}$

CHAPTER 15
Vector Analysis

Section 15.1 Vector Fields

1. $\mathbf{F}(x, \ y) = \mathbf{i} + \mathbf{j}$
$\|\mathbf{F}\| = \sqrt{2}$

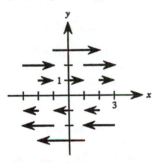

2. $\mathbf{F}(x, \ y) = 2\mathbf{i}$
$\|\mathbf{F}\| = 2$

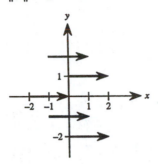

3. $\mathbf{F}(x, \ y) = x\,\mathbf{j}$
$\|\mathbf{F}\| = |x| = c$

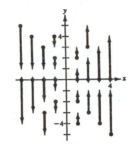

4. $\mathbf{F}(x, \ y) = y\mathbf{i}$
$\|\mathbf{F}\| = |y| = c$

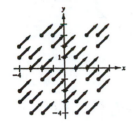

5. $\mathbf{F}(x, \ y) = x\mathbf{i} + y\,\mathbf{j}$
$\|\mathbf{F}\| = \sqrt{x^2 + y^2} = c$
$x^2 + y^2 = c^2$

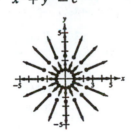

6. $\mathbf{F}(x, \ y) = -x\mathbf{i} + y\,\mathbf{j}$
$\|\mathbf{F}\| = \sqrt{x^2 + y^2} = c$
$x^2 + y^2 = c^2$

7. $\mathbf{F}(x, \ y) = x\mathbf{i} + 3y\,\mathbf{j}$
$\|\mathbf{F}\| = \sqrt{x^2 + 9y^2} = c$
$\dfrac{x^2}{c^2} + \dfrac{y^2}{c^2/9} = 1$

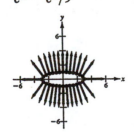

8. $\mathbf{F}(x, \ y) = y\mathbf{i} - x\,\mathbf{j}$
$\|\mathbf{F}\| = \sqrt{y^2 + x^2} = c$
$x^2 + y^2 = c^2$

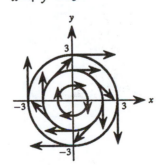

9. $\mathbf{F}(x, \ y, \ z) = 3y\,\mathbf{j}$
$\|\mathbf{F}\| = 3|y| = c$

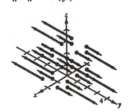

10. $\mathbf{F}(x, y) = x\mathbf{i}$
  $\|\mathbf{F}\| = |x| = c$

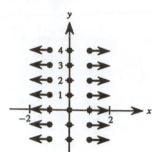

11. $\mathbf{F}(x, y) = 4x\mathbf{i} + y\mathbf{j}$
  $\|\mathbf{F}\| = \sqrt{16x^2 + y^2} = c$
  $$\frac{x^2}{c^2/16} + \frac{y^2}{c^2} = 1$$

12. $\mathbf{F}(x, y) = \mathbf{i} + (x^2 + y^2)\mathbf{j}$
  $\|\mathbf{F}\| = \sqrt{1 + (x^2 + y^2)^2} = c$
  $(x^2 + y^2)^2 = c^2 - 1$
  $$x^2 + y^2 = \sqrt{c^2 - 1}$$

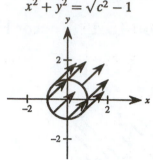

13. $\mathbf{F}(x, y, z) = \mathbf{i} + \mathbf{j} + \mathbf{k}$
  $\|\mathbf{F}\| = \sqrt{3}$

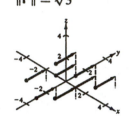

14. $\mathbf{F}(x, y, z) = x\mathbf{i} + y\mathbf{j} + z\mathbf{k}$
  $\|\mathbf{F}\| = \sqrt{x^2 + y^2 + z^2} = c$
  $x^2 + y^2 + z^2 = c^2$

15. $f(x, y) = 5x^2 + 3xy + 10y^2$

  $f_x(x, y) = 10x + 3y$

  $f_y(x, y) = 3x + 20y$

  $\mathbf{F}(x, y) = (10x + 3y)\mathbf{i} + (3x + 20y)\mathbf{j}$

16. $f(x, y) = \sin 3x \cos 4y$

  $f_x(x, y) = 3 \cos 3x \cos 4y$

  $f_y(x, y) = -4 \sin 3x \sin 4y$

  $\mathbf{F}(x, y) = 3 \cos 3x \cos 4y\mathbf{i} - 4 \sin 3x \sin 4y\mathbf{j}$

17. $f(x, y, z) = z - ye^{x^2}$

  $f_x(x, y, z) = -2xye^{x^2}$

  $f_y(x, y, z) = -e^{x^2}$

  $f_z = 1$

  $\mathbf{F}(x, y, z) = -2xye^{x^2}\mathbf{i} - e^{x^2}\mathbf{j} + \mathbf{k}$

18. $f(x, y, z) = \dfrac{y}{z} + \dfrac{z}{x} - \dfrac{xz}{y}$

  $f_x(x, y, z) = -\dfrac{z}{x^2} - \dfrac{z}{y}$

  $f_y(x, y, z) = \dfrac{1}{z} + \dfrac{xz}{y^2}$

  $f_z(x, y, z) = -\dfrac{y}{z^2} + \dfrac{1}{x} - \dfrac{x}{y}$

  $\mathbf{F}(x, y, z) = \left(-\dfrac{z}{x^2} - \dfrac{z}{y}\right)\mathbf{i} + \left(\dfrac{1}{z} + \dfrac{xz}{y^2}\right)\mathbf{j} + \left(-\dfrac{y}{z^2} + \dfrac{1}{x} - \dfrac{x}{y}\right)\mathbf{k}$

19. $g(x,\ y,\ z) = xy\ln(x+y)$

$$g_x(x,\ y,\ z) = y\ln(x+y) + \frac{xy}{x+y}$$

$$g_y(x,\ y,\ z) = x\ln(x+y) + \frac{xy}{x+y}$$

$$g_z(x,\ y,\ z) = 0$$

$$\mathbf{G}(x,\ y,\ z) = \left[\frac{xy}{x+y} + y\ln(x+y)\right]\mathbf{i} + \left[\frac{xy}{x+y} + x\ln(x+y)\right]\mathbf{j}$$

20. $g(x,\ y,\ z) = x\arcsin yz$

$$g_x(x,\ y,\ z) = \arcsin yz$$

$$g_y(x,\ y,\ z) = \frac{xz}{\sqrt{1-y^2z^2}}$$

$$g_z(x,\ y,\ z) = \frac{xy}{\sqrt{1-y^2z^2}}$$

$$\mathbf{G}(x,\ y,\ z) = (\arcsin yz)\mathbf{i} + \frac{xz}{\sqrt{1-y^2z^2}}\mathbf{j} + \frac{xy}{\sqrt{1-y^2z^2}}\mathbf{k}$$

21. $\mathbf{F}(x,\ y) = 2xy\mathbf{i} + x^2\mathbf{j}$

$$\frac{\partial}{\partial y}[2xy] = 2x$$

$$\frac{\partial}{\partial x}[x^2] = 2x$$

Conservative

$$f_x(x,\ y) = 2xy$$

$$f_y(x,\ y) = 2x$$

$$f(x,\ y) = x^2y + K$$

22. $\mathbf{F}(x,\ y) = \frac{1}{y^2}(y\mathbf{i} - 2x\mathbf{j}) = \frac{1}{y}\mathbf{i} - \frac{2x}{y^2}\mathbf{j}$

$$\frac{\partial}{\partial y}\left[\frac{1}{y}\right] = -\frac{1}{y^2}$$

$$\frac{\partial}{\partial x}\left[-\frac{2x}{y^2}\right] = -\frac{2}{y^2}$$

Not conservative

23. $\mathbf{F}(x,\ y) = xe^{x^2y}(2y\mathbf{i} + x\mathbf{j})$

$$\frac{\partial}{\partial y}[2xye^{x^2y}] = 2xe^{x^2y} + 2x^3ye^{x^2y}$$

$$\frac{\partial}{\partial x}[x^2e^{x^2y}] = 2xe^{x^2y} + 2x^3ye^{x^2y}$$

Conservative

$$f_x(x,\ y) = 2xye^{x^2y}$$

$$f_y(x,\ y) = x^2e^{x^2y}$$

$$f(x,\ y) = e^{x^2y} + K$$

24. $\mathbf{F}(x,\ y) = 2xy^3\mathbf{i} + 3y^2x^2\mathbf{j}$

$$\frac{\partial}{\partial y}[2xy^3] = 6xy^2$$

$$\frac{\partial}{\partial x}[3y^2x^2] = 6y^2x$$

Conservative

$$f_x(x,\ y) = 2xy^3$$

$$f_y(x,\ y) = 3y^2x^2$$

$$f(x,\ y) = x^2y^3 + K$$

25. $\mathbf{F}(x, y) = \dfrac{x}{x^2 + y^2}\mathbf{i} + \dfrac{y}{x^2 + y^2}\mathbf{j}$

$\dfrac{\partial}{\partial y}\left[\dfrac{x}{x^2 + y^2}\right] = -\dfrac{2xy}{(x^2 + y^2)^2}$

$\dfrac{\partial}{\partial x}\left[\dfrac{y}{x^2 + y^2}\right] = -\dfrac{2xy}{(x^2 + y^2)^2}$

Conservative

$f_x(x, y) = \dfrac{x}{x^2 + y^2}$

$f_y(x, y) = \dfrac{y}{x^2 + y^2}$

$f(x, y) = \dfrac{1}{2}\ln(x^2 + y^2) + K$

26. $\mathbf{F}(x, y) = \dfrac{2y}{x}\mathbf{i} - \dfrac{x^2}{y^2}\mathbf{j}$

$\dfrac{\partial}{\partial y}\left[\dfrac{2y}{x}\right] = \dfrac{2}{x}$

$\dfrac{\partial}{\partial x}\left[-\dfrac{x^2}{y^2}\right] = -\dfrac{2x}{y^2}$

Not conservative

27. $\mathbf{F}(x, y) = e^x(\cos y\,\mathbf{i} + \sin y\,\mathbf{j})$

$\dfrac{\partial}{\partial y}[e^x \cos y] = -e^x \sin y$

$\dfrac{\partial}{\partial x}[e^x \sin y] = e^x \sin y$

Not conservative

28. $\mathbf{F}(x, y) = \dfrac{2x}{(x^2 + y^2)^2}\mathbf{i} + \dfrac{2y}{(x^2 + y^2)^2}\mathbf{j}$

$\dfrac{\partial}{\partial y}\left[\dfrac{2x}{(x^2 + y^2)^2}\right] = -\dfrac{8xy}{(x^2 + y^2)^3}$

$\dfrac{\partial}{\partial x}\left[\dfrac{2y}{(x^2 + y^2)^2}\right] = -\dfrac{8xy}{(x^2 + y^2)^3}$

Conservative

$f_x(x, y) = \dfrac{2x}{(x^2 + y^2)^2}$

$f_y(x, y) = \dfrac{2y}{(x^2 + y^2)^2}$

$f(x, y) = -\dfrac{1}{x^2 + y^2} + K$

29. $\mathbf{F}(x, y, z) = xyz\,\mathbf{i} + y\,\mathbf{j} + z\,\mathbf{k}, \quad (1, 2, 1)$

$\text{curl } \mathbf{F} = \begin{vmatrix} \mathbf{i} & \mathbf{j} & \mathbf{k} \\ \dfrac{\partial}{\partial x} & \dfrac{\partial}{\partial y} & \dfrac{\partial}{\partial z} \\ xyz & y & z \end{vmatrix} = xy\,\mathbf{j} - xz\,\mathbf{k}$

$\text{curl } \mathbf{F}\,(1, 2, 1) = 2\mathbf{j} - \mathbf{k}$

30. $\mathbf{F}(x, y, z) = x^2 z\,\mathbf{i} - 2xz\,\mathbf{j} + yz\,\mathbf{k}, \quad (2, -1, 3)$

$\text{curl } \mathbf{F} = \begin{vmatrix} \mathbf{i} & \mathbf{j} & \mathbf{k} \\ \dfrac{\partial}{\partial x} & \dfrac{\partial}{\partial y} & \dfrac{\partial}{\partial z} \\ x^2 z & -2xz & yz \end{vmatrix}$

$= (z + 2x)\mathbf{i} - (0 - x^2)\mathbf{j} + (-2z - 0)\mathbf{k}$

$= (z + 2x)\mathbf{i} + x^2\,\mathbf{j} - 2z\,\mathbf{k}$

$\text{curl } \mathbf{F}\,(2, -1, 3) = 7\mathbf{i} + 4\mathbf{j} - 6\mathbf{k}$

31. $\mathbf{F}(x, y, z) = e^x \sin y\,\mathbf{i} - e^x \cos y\,\mathbf{j}, \quad (0, 0, 3)$

$\text{curl } \mathbf{F} = \begin{vmatrix} \mathbf{i} & \mathbf{j} & \mathbf{k} \\ \dfrac{\partial}{\partial x} & \dfrac{\partial}{\partial y} & \dfrac{\partial}{\partial z} \\ e^x \sin y & -e^x \cos y & 0 \end{vmatrix} = -2e^x \cos y\,\mathbf{k}$

$\text{curl } \mathbf{F}\,(0, 0, 3) = -2\mathbf{k}$

32. $\mathbf{F}(x,\ y,\ z) = e^{-xyz}(\mathbf{i}+\mathbf{j}+\mathbf{k}),\quad (3,\ 2,\ 0)$

$$\operatorname{curl}\mathbf{F} = \begin{vmatrix} \mathbf{i} & \mathbf{j} & \mathbf{k} \\ \dfrac{\partial}{\partial x} & \dfrac{\partial}{\partial y} & \dfrac{\partial}{\partial z} \\ e^{-xyz} & e^{-xyz} & e^{-xyz} \end{vmatrix} = (-xz+xy)e^{-xyz}\mathbf{i} - (-yz+xy)e^{-xyz}\mathbf{j} + (-yz+xz)e^{-xyz}\mathbf{k}$$

$\operatorname{curl}\mathbf{F}\ (3,\ 2,\ 0) = 6\mathbf{i}-6\mathbf{j}$

33. $\mathbf{F}(x,\ y,\ z) = \arctan\left(\dfrac{x}{y}\right)\mathbf{i} + \ln\sqrt{x^2+y^2}\,\mathbf{j} + \mathbf{k}$

$$\operatorname{curl}\mathbf{F} = \begin{vmatrix} \mathbf{i} & \mathbf{j} & \mathbf{k} \\ \dfrac{\partial}{\partial x} & \dfrac{\partial}{\partial y} & \dfrac{\partial}{\partial z} \\ \arctan\left(\dfrac{x}{y}\right) & \dfrac{1}{2}\ln(x^2+y^2) & 1 \end{vmatrix} = \left[\dfrac{x}{x^2+y^2} - \dfrac{(-x/y^2)}{1+(x/y)^2}\right]\mathbf{k} = \dfrac{2x}{x^2+y^2}\mathbf{k}$$

34. $\mathbf{F}(x,\ y,\ z) = \dfrac{yz}{y-z}\mathbf{i} + \dfrac{xz}{x-z}\mathbf{j} + \dfrac{xy}{x-y}\mathbf{k}$

$$\operatorname{curl}\mathbf{F} = \begin{vmatrix} \mathbf{i} & \mathbf{j} & \mathbf{k} \\ \dfrac{\partial}{\partial x} & \dfrac{\partial}{\partial y} & \dfrac{\partial}{\partial z} \\ \dfrac{yz}{y-z} & \dfrac{xz}{x-z} & \dfrac{xy}{x-y} \end{vmatrix}$$

$$= \left[\dfrac{x^2}{(x-y)^2} - \dfrac{x^2}{(x-z)^2}\right]\mathbf{i} - \left[\dfrac{-y^2}{(x-y)^2} - \dfrac{y^2}{(y-z)^2}\right]\mathbf{j} + \left[\dfrac{-z^2}{(x-z)^2} - \dfrac{-z^2}{(y-z)^2}\right]\mathbf{k}$$

$$= x^2\left[\dfrac{1}{(x-y)^2} - \dfrac{1}{(x-z)^2}\right]\mathbf{i} + y^2\left[\dfrac{1}{(x-y)^2} + \dfrac{1}{(y-z)^2}\right]\mathbf{j} + z^2\left[\dfrac{1}{(y-z)^2} - \dfrac{1}{(x-z)^2}\right]\mathbf{k}$$

35. $\mathbf{F}(x,\ y,\ z) = \sin(x-y)\mathbf{i} + \sin(y-z)\mathbf{j} + \sin(z-x)\mathbf{k}$

$$\operatorname{curl}\mathbf{F} = \begin{vmatrix} \mathbf{i} & \mathbf{j} & \mathbf{k} \\ \dfrac{\partial}{\partial x} & \dfrac{\partial}{\partial y} & \dfrac{\partial}{\partial z} \\ \sin(x-y) & \sin(y-z) & \sin(z-x) \end{vmatrix} = \cos(y-z)\mathbf{i} + \cos(z-x)\mathbf{j} + \cos(x-y)\mathbf{k}$$

36. $\mathbf{F}(x,\ y,\ z) = \sqrt{x^2+y^2+z^2}(\mathbf{i}+\mathbf{j}+\mathbf{k})$

$$\operatorname{curl}\mathbf{F} = \begin{vmatrix} \mathbf{i} & \mathbf{j} & \mathbf{k} \\ \dfrac{\partial}{\partial x} & \dfrac{\partial}{\partial y} & \dfrac{\partial}{\partial z} \\ \sqrt{x^2+y^2+z^2} & \sqrt{x^2+y^2+z^2} & \sqrt{x^2+y^2+z^2} \end{vmatrix} = \dfrac{(y-z)\mathbf{i} + (z-x)\mathbf{j} + (x-y)\mathbf{k}}{\sqrt{x^2+y^2+z^2}}$$

37. $\mathbf{F}(x,\ y,\ z) = \sin y\,\mathbf{i} - x\cos y\,\mathbf{j} + \mathbf{k}$

$$\operatorname{curl}\mathbf{F} = \begin{vmatrix} \mathbf{i} & \mathbf{j} & \mathbf{k} \\ \dfrac{\partial}{\partial x} & \dfrac{\partial}{\partial y} & \dfrac{\partial}{\partial z} \\ \sin y & -x\cos y & 1 \end{vmatrix} = -2\cos y\,\mathbf{k} \neq 0$$

Not conservative

38. $\mathbf{F}(x,\ y,\ z) = e^z(y\mathbf{i} + x\mathbf{j} + \mathbf{k})$

$$\operatorname{curl}\mathbf{F} = \begin{vmatrix} \mathbf{i} & \mathbf{j} & \mathbf{k} \\ \dfrac{\partial}{\partial x} & \dfrac{\partial}{\partial y} & \dfrac{\partial}{\partial z} \\ ye^z & xe^z & e^z \end{vmatrix} = -xe^z\mathbf{i} + ye^z\mathbf{j} \neq 0$$

Not conservative

39. $F(x, y, z) = e^z(y\mathbf{i} + x\mathbf{j} + xy\mathbf{k})$

$$\text{curl } F = \begin{vmatrix} \mathbf{i} & \mathbf{j} & \mathbf{k} \\ \dfrac{\partial}{\partial x} & \dfrac{\partial}{\partial y} & \dfrac{\partial}{\partial z} \\ ye^z & xe^z & xye^z \end{vmatrix} = 0$$

Conservative

$f_x(x, y, z) = ye^z$

$f_y(x, y, z) = xe^z$

$f_z(x, y, z) = xye^z$

$f(x, y, z) = xye^z + K$

40. $F(x, y, z) = 3x^2y^2z\mathbf{i} + 2x^3yz\mathbf{j} + x^3y^2\mathbf{k}$

$$\text{curl } F = \begin{vmatrix} \mathbf{i} & \mathbf{j} & \mathbf{k} \\ \dfrac{\partial}{\partial x} & \dfrac{\partial}{\partial y} & \dfrac{\partial}{\partial z} \\ 3x^2y^2z & 2x^3yz & x^3y^2 \end{vmatrix} = 0$$

Conservative

$f_x(x, y, z) = 3x^2y^2z$

$f_y(x, y, z) = 2x^3yz$

$f_z(x, y, z) = x^3y^2$

$f(x, y, z) = x^3y^2z + K$

41. $F(x, y, z) = \dfrac{1}{y}\mathbf{i} - \dfrac{x}{y^2}\mathbf{j} + (2z - 1)\mathbf{k}$

$$\text{curl } F = \begin{vmatrix} \mathbf{i} & \mathbf{j} & \mathbf{k} \\ \dfrac{\partial}{\partial x} & \dfrac{\partial}{\partial y} & \dfrac{\partial}{\partial z} \\ \dfrac{1}{y} & -\dfrac{x}{y^2} & 2z - 1 \end{vmatrix} = 0$$

Conservative

$f_x(x, y, z) = \dfrac{1}{y}$

$f_y(x, y, z) = -\dfrac{x}{y^2}$

$f_z(x, y, z) = 2z - 1$

$f(x, y, z) = \displaystyle\int \dfrac{1}{y}\,dx = \dfrac{x}{y} + g(y, z) + K_1$

$f(x, y, z) = \displaystyle\int -\dfrac{x}{y^2}\,dy = \dfrac{x}{y} + h(x, z) + K_2$

$f(x, y, z) = \displaystyle\int (2z - 1)\,dz$

$\qquad\qquad = z^2 - z + p(x, y) + K_3$

$f(x, y, z) = \dfrac{x}{y} + z^2 - z + K$

42. $F(x, y, z) = \dfrac{x}{x^2 + y^2}\mathbf{i} + \dfrac{y}{x^2 + y^2}\mathbf{j} + \mathbf{k}$

$$\text{curl } F = \begin{vmatrix} \mathbf{i} & \mathbf{j} & \mathbf{k} \\ \dfrac{\partial}{\partial x} & \dfrac{\partial}{\partial y} & \dfrac{\partial}{\partial z} \\ \dfrac{x}{x^2 + y^2} & \dfrac{y}{x^2 + y^2} & 1 \end{vmatrix} = 0$$

Conservative

$f_x(x, y, z) = \dfrac{x}{x^2 + y^2}$

$f_y(x, y, z) = \dfrac{y}{x^2 + y^2}$

$f_z(x, y, z) = 1$

$f(x, y, z) = \displaystyle\int \dfrac{x}{x^2 + y^2}\,dx$

$\qquad\qquad = \dfrac{1}{2}\ln(x^2 + y^2) + g(y, z) + K_1$

$f(x, y, z) = \displaystyle\int \dfrac{y}{x^2 + y^2}\,dy$

$\qquad\qquad = \dfrac{1}{2}\ln(x^2 + y^2) + h(x, z) + K_2$

$f(x, y, z) = \displaystyle\int dz = z + p(x, y) + K_3$

$f(x, y, z) = \dfrac{1}{2}\ln(x^2 + y^2) + z + K$

43. $F(x, y, z) = \mathbf{i} + 2x\mathbf{j} + 3y\mathbf{k}$

$G(x, y, z) = x\mathbf{i} - y\mathbf{j} + z\mathbf{k}$

$$F \times G = \begin{vmatrix} \mathbf{i} & \mathbf{j} & \mathbf{k} \\ 1 & 2x & 3y \\ x & -y & z \end{vmatrix} = (2xz + 3y^2)\mathbf{i} - (z - 3xy)\mathbf{j} + (-y - 2x^2)\mathbf{k}$$

$$\text{curl}(F \times G) = \begin{vmatrix} \mathbf{i} & \mathbf{j} & \mathbf{k} \\ \dfrac{\partial}{\partial x} & \dfrac{\partial}{\partial y} & \dfrac{\partial}{\partial z} \\ 2xz + 3y^2 & 3xy - z & -y - 2x^2 \end{vmatrix} = (-1 + 1)\mathbf{i} - (-4x - 2x)\mathbf{j} + (3y - 6y)\mathbf{k} = 6x\mathbf{j} - 3y\mathbf{k}$$

44. $\mathbf{F}(x,\ y,\ z) = x\mathbf{i} - z\mathbf{k}$

$\mathbf{G}(x,\ y,\ z) = x^2\mathbf{i} + y\mathbf{j} + z^2\mathbf{k}$

$$\mathbf{F} \times \mathbf{G} = \begin{vmatrix} \mathbf{i} & \mathbf{j} & \mathbf{k} \\ x & 0 & -z \\ x^2 & y & z^2 \end{vmatrix} = yz\mathbf{i} - (xz^2 + x^2z)\mathbf{j} + xy\mathbf{k}$$

$$\mathbf{curl}(\mathbf{F} \times \mathbf{G}) = \begin{vmatrix} \mathbf{i} & \mathbf{j} & \mathbf{k} \\ \dfrac{\partial}{\partial x} & \dfrac{\partial}{\partial y} & \dfrac{\partial}{\partial z} \\ yz & -xz^2 - x^2z & xy \end{vmatrix}$$

$$= (x + 2xz + x^2)\mathbf{i} - (y - y)\mathbf{j} + (-z^2 - 2xz - z)\mathbf{k}$$

$$= x(x + 2z + 1)\mathbf{i} - z(z + 2x + 1)\mathbf{k}$$

45. $\mathbf{F}(x,\ y,\ z) = xyz\mathbf{i} + y\mathbf{j} + z\mathbf{k}$

$$\mathbf{curl\ F} = \begin{vmatrix} \mathbf{i} & \mathbf{j} & \mathbf{k} \\ \dfrac{\partial}{\partial x} & \dfrac{\partial}{\partial y} & \dfrac{\partial}{\partial z} \\ xyz & y & z \end{vmatrix} = xy\mathbf{j} - xz\mathbf{k}$$

$$\mathbf{curl}(\mathbf{curl\ F}) = \begin{vmatrix} \mathbf{i} & \mathbf{j} & \mathbf{k} \\ \dfrac{\partial}{\partial x} & \dfrac{\partial}{\partial y} & \dfrac{\partial}{\partial z} \\ 0 & xy & -xz \end{vmatrix} = z\mathbf{j} + y\mathbf{k}$$

46. $\mathbf{F}(x,\ y,\ z) = x^2z\mathbf{i} - 2xz\mathbf{j} + yz\mathbf{k}$

$$\mathbf{curl\ F} = \begin{vmatrix} \mathbf{i} & \mathbf{j} & \mathbf{k} \\ \dfrac{\partial}{\partial x} & \dfrac{\partial}{\partial y} & \dfrac{\partial}{\partial z} \\ x^2z & -2xz & yz \end{vmatrix} = (z + 2x)\mathbf{i} + x^2\mathbf{j} - 2z\mathbf{k}$$

$$\mathbf{curl}(\mathbf{curl\ F}) = \begin{vmatrix} \mathbf{i} & \mathbf{j} & \mathbf{k} \\ \dfrac{\partial}{\partial x} & \dfrac{\partial}{\partial y} & \dfrac{\partial}{\partial z} \\ z + 2x & x^2 & -2z \end{vmatrix} = \mathbf{j} + 2x\mathbf{k}$$

47. $\mathbf{F}(x,\ y) = 6x^2\mathbf{i} - xy^2\mathbf{j}$

$$\mathrm{div}\ \mathbf{F}(x,\ y) = \frac{\partial}{\partial x}[6x^2] + \frac{\partial}{\partial y}[-xy^2]$$

$$= 12x - 2xy$$

48. $\mathbf{F}(x,\ y) = xe^x\mathbf{i} + ye^y\mathbf{j}$

$$\mathrm{div}\ \mathbf{F}(x,\ y) = \frac{\partial}{\partial x}[xe^x] + \frac{\partial}{\partial y}[ye^y]$$

$$= xe^x + e^x + ye^y + e^y$$

$$= e^x(x + 1) + e^y(y + 1)$$

49. $\mathbf{F}(x,\ y,\ z) = \sin x\mathbf{i} + \cos y\mathbf{j} + z^2\mathbf{k}$

$$\mathrm{div}\ \mathbf{F}(x,\ y,\ z) = \frac{\partial}{\partial x}[\sin x] + \frac{\partial}{\partial y}[\cos y] + \frac{\partial}{\partial z}[z^2] = \cos x - \sin y + 2z$$

50. $\mathbf{F}(x,\ y,\ z) = \ln(x^2 + y^2)\mathbf{i} + xy\mathbf{j} + \ln(y^2 + z^2)\mathbf{k}$

$$\mathrm{div}\ \mathbf{F}(x,\ y,\ z) = \frac{\partial}{\partial x}[\ln(x^2 + y^2)] + \frac{\partial}{\partial y}[xy] + \frac{\partial}{\partial z}[\ln(y^2 + z^2)] = \frac{2x}{x^2 + y^2} + x + \frac{2z}{y^2 + z^2}$$

51. $\mathbf{F}(x,\ y,\ z) = xyz\mathbf{i} + y\mathbf{j} + z\mathbf{k}$

$\mathrm{div}\ \mathbf{F}(x,\ y,\ z) = yz + 1 + 1 = yz + 2$

$\mathrm{div}\ \mathbf{F}(1,\ 2,\ 1) = 4$

52. $\mathbf{F}(x,\ y,\ z) = x^2z\mathbf{i} - 2xz\mathbf{j} + yz\mathbf{k}$

$\mathrm{div}\ \mathbf{F}(x,\ y,\ z) = 2xz + y$

$\mathrm{div}\ \mathbf{F}(2,\ -1,\ 3) = 11$

53. $\mathbf{F}(x,\ y,\ z) = e^x \sin y\mathbf{i} - e^x \cos y\mathbf{j}$

$\mathrm{div}\ \mathbf{F}(x,\ y,\ z) = e^x \sin y + e^x \sin y$

$\mathrm{div}\ \mathbf{F}(0,\ 0,\ 3) = 0$

54. $\mathbf{F}(x,\ y,\ z) = e^{-xyz}(\mathbf{i} + \mathbf{j} + \mathbf{k})$

$\mathrm{div}\ \mathbf{F}(x,\ y,\ z) = e^{-xyz}(-yz - xz - xy)$

$\mathrm{div}\ \mathbf{F}(3,\ 2,\ 0) = -6$

55. $\mathbf{F}(x,\ y,\ z) = \mathbf{i} + 2x\mathbf{j} + 3y\mathbf{k}$

$\mathbf{G}(x,\ y,\ z) = x\mathbf{i} - y\mathbf{j} + z\mathbf{k}$

$$\mathbf{F} \times \mathbf{G} = \begin{vmatrix} \mathbf{i} & \mathbf{j} & \mathbf{k} \\ 1 & 2x & 3y \\ x & -y & z \end{vmatrix}$$

$$= (2xz + 3y^2)\mathbf{i} - (z - 3xy)\mathbf{j} + (-y - 2x^2)\mathbf{k}$$

$\operatorname{div}(\mathbf{F} \times \mathbf{G}) = 2z + 3x$

56. $\mathbf{F}(x,\ y,\ z) = x\mathbf{i} - z\mathbf{k}$

$\mathbf{G}(x,\ y,\ z) = x^2\mathbf{i} + y\mathbf{j} + z^2\mathbf{k}$

$$\mathbf{F} \times \mathbf{G} = \begin{vmatrix} \mathbf{i} & \mathbf{j} & \mathbf{k} \\ x & 0 & -z \\ x^2 & y & z^2 \end{vmatrix} = yz\mathbf{i} - (xz^2 + x^2z)\mathbf{j} + xy\mathbf{k}$$

$\operatorname{div}(\mathbf{F} \times \mathbf{G}) = 0$

57. $\mathbf{F}(x,\ y,\ z) = xyz\mathbf{i} + y\mathbf{j} + z\mathbf{k}$

$$\operatorname{curl} \mathbf{F} = \begin{vmatrix} \mathbf{i} & \mathbf{j} & \mathbf{k} \\ \dfrac{\partial}{\partial x} & \dfrac{\partial}{\partial y} & \dfrac{\partial}{\partial z} \\ xyz & y & z \end{vmatrix} = xy\mathbf{j} - xz\mathbf{k}$$

$\operatorname{div}(\operatorname{curl} \mathbf{F}) = x - x = 0$

58. $\mathbf{F}(x,\ y,\ z) = x^2z\mathbf{i} - 2xz\mathbf{j} + yz\mathbf{k}$

$$\operatorname{curl} \mathbf{F} = \begin{vmatrix} \mathbf{i} & \mathbf{j} & \mathbf{k} \\ \dfrac{\partial}{\partial x} & \dfrac{\partial}{\partial y} & \dfrac{\partial}{\partial z} \\ x^2z & -2xz & yz \end{vmatrix} = (z + 2x)\mathbf{i} + x^2\mathbf{j} - 2z\mathbf{k}$$

$\operatorname{div}(\operatorname{curl} \mathbf{F}) = 2 - 2 = 0$

59. Let $\mathbf{F} = M\mathbf{i} + N\mathbf{j} + P\mathbf{k}$ and $\mathbf{G} = Q\mathbf{i} + R\mathbf{j} + S\mathbf{k}$ where M, N, P, Q, R, and S have continuous partial derivatives.

$\mathbf{F} + \mathbf{G} = (M + Q)\mathbf{i} + (N + R)\mathbf{j} + (P + S)\mathbf{k}$

$$\operatorname{curl}(\mathbf{F} + \mathbf{G}) = \begin{vmatrix} \mathbf{i} & \mathbf{j} & \mathbf{k} \\ \dfrac{\partial}{\partial x} & \dfrac{\partial}{\partial y} & \dfrac{\partial}{\partial z} \\ M + Q & N + R & P + S \end{vmatrix}$$

$$= \left[\frac{\partial}{\partial y}(P + S) - \frac{\partial}{\partial z}(N + R)\right]\mathbf{i} - \left[\frac{\partial}{\partial x}(P + S) - \frac{\partial}{\partial z}(M + Q)\right]\mathbf{j} + \left[\frac{\partial}{\partial x}(N + R) + \frac{\partial}{\partial y}(M + Q)\right]\mathbf{k}$$

$$= \left(\frac{\partial P}{\partial y} - \frac{\partial N}{\partial z}\right)\mathbf{i} - \left(\frac{\partial P}{\partial x} - \frac{\partial M}{\partial z}\right)\mathbf{j} + \left(\frac{\partial N}{\partial x} - \frac{\partial M}{\partial y}\right)\mathbf{k}$$

$$+ \left(\frac{\partial S}{\partial y} - \frac{\partial R}{\partial z}\right)\mathbf{i} - \left(\frac{\partial S}{\partial x} - \frac{\partial Q}{\partial z}\right)\mathbf{j} + \left(\frac{\partial R}{\partial x} - \frac{\partial Q}{\partial y}\right)\mathbf{k}$$

$$= \operatorname{curl} \mathbf{F} + \operatorname{curl} \mathbf{G}$$

60. Let $f(x,\ y,\ z)$ be a scalar function whose second partial derivatives are continuous.

$$\nabla f = \frac{\partial f}{\partial x}\mathbf{i} + \frac{\partial f}{\partial y}\mathbf{j} + \frac{\partial f}{\partial z}\mathbf{k}$$

$$\operatorname{curl}(\nabla f) = \begin{vmatrix} \mathbf{i} & \mathbf{j} & \mathbf{k} \\ \dfrac{\partial}{\partial x} & \dfrac{\partial}{\partial y} & \dfrac{\partial}{\partial z} \\ \dfrac{\partial f}{\partial x} & \dfrac{\partial f}{\partial y} & \dfrac{\partial f}{\partial z} \end{vmatrix} = \left(\frac{\partial^2 f}{\partial y \partial z} - \frac{\partial^2 f}{\partial z \partial y}\right)\mathbf{i} - \left(\frac{\partial^2 f}{\partial x \partial z} - \frac{\partial^2 f}{\partial z \partial x}\right)\mathbf{j} + \left(\frac{\partial^2 f}{\partial x \partial y} - \frac{\partial^2 f}{\partial y \partial x}\right)\mathbf{k} = \mathbf{0}$$

61. Let $\mathbf{F} = M\mathbf{i} + N\mathbf{j} + P\mathbf{k}$ and $\mathbf{G} = R\mathbf{i} + S\mathbf{j} + T\mathbf{k}$.

$$\operatorname{div}(\mathbf{F} + \mathbf{G}) = \frac{\partial}{\partial x}(M + R) + \frac{\partial}{\partial y}(N + S) + \frac{\partial}{\partial z}(P + T) = \frac{\partial M}{\partial x} + \frac{\partial R}{\partial x} + \frac{\partial N}{\partial y} + \frac{\partial S}{\partial y} + \frac{\partial P}{\partial z} + \frac{\partial T}{\partial z}$$

$$= \left[\frac{\partial M}{\partial x} + \frac{\partial N}{\partial y} + \frac{\partial P}{\partial z}\right] + \left[\frac{\partial R}{\partial x} + \frac{\partial S}{\partial y} + \frac{\partial T}{\partial z}\right]$$

$$= \operatorname{div} \mathbf{F} + \operatorname{div} \mathbf{G}$$

62. Let $\mathbf{F} = M\mathbf{i} + N\mathbf{j} + P\mathbf{k}$ and $\mathbf{G} = R\mathbf{i} + S\mathbf{j} + T\mathbf{k}$.

$$\mathbf{F} \times \mathbf{G} = \begin{vmatrix} \mathbf{i} & \mathbf{j} & \mathbf{k} \\ M & N & P \\ R & S & T \end{vmatrix} = (NT - PS)\mathbf{i} - (MT - PR)\mathbf{j} + (MS - NR)\mathbf{k}$$

$$\operatorname{div}(\mathbf{F} \times \mathbf{G}) = \frac{\partial}{\partial x}(NT - PS) + \frac{\partial}{\partial y}(PR - MT) + \frac{\partial}{\partial z}(MS - NR)$$

$$= N\frac{\partial T}{\partial x} + T\frac{\partial N}{\partial x} - P\frac{\partial S}{\partial x} - S\frac{\partial P}{\partial x} + P\frac{\partial R}{\partial y} + R\frac{\partial P}{\partial y} - M\frac{\partial T}{\partial y}$$

$$- T\frac{\partial M}{\partial y} + M\frac{\partial S}{\partial z} + S\frac{\partial M}{\partial z} - N\frac{\partial R}{\partial z} - R\frac{\partial N}{\partial z}$$

$$= \left[\left(\frac{\partial P}{\partial y} - \frac{\partial N}{\partial z} \right)R + \left(\frac{\partial M}{\partial z} - \frac{\partial P}{\partial x} \right)S + \left(\frac{\partial N}{\partial x} - \frac{\partial M}{\partial y} \right)T \right]$$

$$- \left[M\left(\frac{\partial T}{\partial y} - \frac{\partial S}{\partial z} \right) + N\left(\frac{\partial R}{\partial z} - \frac{\partial T}{\partial x} \right) + P\left(\frac{\partial S}{\partial x} - \frac{\partial R}{\partial y} \right) \right]$$

$$= (\operatorname{curl} \mathbf{F}) \cdot \mathbf{G} - \mathbf{F} \cdot (\operatorname{curl} \mathbf{G})$$

63. $\mathbf{F} = M\mathbf{i} + N\mathbf{j} + P\mathbf{k}$

$$\nabla \times [\nabla f + (\nabla \times \mathbf{F})] = \operatorname{curl}(\nabla f + (\nabla \times \mathbf{F}))$$

$$= \operatorname{curl}(\nabla f) + \operatorname{curl}(\nabla \times \mathbf{F}) \quad \text{(Exercise 59)}$$

$$= \operatorname{curl}(\nabla \times \mathbf{F}) \quad \text{(Exercise 60)}$$

$$= \nabla \times (\nabla \times \mathbf{F})$$

64. Let $\mathbf{F} = M\mathbf{i} + N\mathbf{j} + P\mathbf{k}$.

$$\nabla \times (f\mathbf{F}) = \begin{vmatrix} \mathbf{i} & \mathbf{j} & \mathbf{k} \\ \frac{\partial}{\partial x} & \frac{\partial}{\partial y} & \frac{\partial}{\partial z} \\ fM & fN & fP \end{vmatrix}$$

$$= \left(\frac{\partial f}{\partial y}P + f\frac{\partial P}{\partial y} - \frac{\partial f}{\partial z}N - f\frac{\partial N}{\partial z} \right)\mathbf{i} - \left(\frac{\partial f}{\partial x}P + f\frac{\partial P}{\partial x} - \frac{\partial f}{\partial z}M - f\frac{\partial M}{\partial z} \right)\mathbf{j}$$

$$+ \left(\frac{\partial f}{\partial x}N + f\frac{\partial N}{\partial x} - \frac{\partial f}{\partial y}M - f\frac{\partial M}{\partial y} \right)\mathbf{k}$$

$$= f\left[\left(\frac{\partial P}{\partial y} - \frac{\partial N}{\partial z} \right)\mathbf{i} - \left(\frac{\partial P}{\partial x} - \frac{\partial M}{\partial z} \right)\mathbf{j} + \left(\frac{\partial N}{\partial x} - \frac{\partial M}{\partial y} \right)\mathbf{k} \right] + \begin{vmatrix} \mathbf{i} & \mathbf{j} & \mathbf{k} \\ \frac{\partial f}{\partial x} & \frac{\partial f}{\partial y} & \frac{\partial f}{\partial z} \\ M & N & P \end{vmatrix} = f[\nabla \times \mathbf{F}] + (\nabla f) \times \mathbf{F}$$

65. Let $\mathbf{F} = M\mathbf{i} + N\mathbf{j} + P\mathbf{k}$, then $f\mathbf{F} = fM\mathbf{i} + fN\mathbf{j} + fP\mathbf{k}$.

$$\operatorname{div}(f\mathbf{F}) = \frac{\partial}{\partial x}(fM) + \frac{\partial}{\partial y}(fN) + \frac{\partial}{\partial z}(fP) = f\frac{\partial M}{\partial x} + M\frac{\partial f}{\partial x} + f\frac{\partial N}{\partial y} + N\frac{\partial f}{\partial y} + f\frac{\partial P}{\partial z} + P\frac{\partial f}{\partial z}$$

$$= f\left(\frac{\partial M}{\partial x} + \frac{\partial N}{\partial y} + \frac{\partial N}{\partial z} \right) + \left(\frac{\partial f}{\partial x}M + \frac{\partial f}{\partial y}N + \frac{\partial f}{\partial z}P \right)$$

$$= f\operatorname{div}\mathbf{F} + \nabla f \cdot \mathbf{F}$$

66. Let $\mathbf{F} = M\mathbf{i} + N\mathbf{j} + P\mathbf{k}$.

$$\mathbf{curl\ F} = \left(\frac{\partial P}{\partial y} - \frac{\partial N}{\partial z}\right)\mathbf{i} - \left(\frac{\partial P}{\partial x} - \frac{\partial M}{\partial z}\right)\mathbf{j} + \left(\frac{\partial N}{\partial x} - \frac{\partial M}{\partial y}\right)\mathbf{k}$$

$$\text{div}\ (\mathbf{curl\ F}) = \frac{\partial}{\partial x}\left[\frac{\partial P}{\partial y} - \frac{\partial N}{\partial z}\right] - \frac{\partial}{\partial y}\left[\frac{\partial P}{\partial x} - \frac{\partial M}{\partial z}\right] + \frac{\partial}{\partial z}\left[\frac{\partial N}{\partial x} - \frac{\partial M}{\partial y}\right]$$

$$= \frac{\partial^2 P}{\partial x \partial y} - \frac{\partial^2 N}{\partial x \partial z} - \frac{\partial^2 P}{\partial y \partial x} + \frac{\partial^2 M}{\partial y \partial z} + \frac{\partial^2 N}{\partial z \partial x} - \frac{\partial^2 M}{\partial z \partial y} = 0 \text{ (since the mixed partials are equal)}$$

In Exercises 67–70, $\mathbf{F}(x,\ y,\ z) = x\mathbf{i} + y\mathbf{j} + z\mathbf{k}$ and $f(x,\ y,\ z) = \|\mathbf{F}(x,\ y,\ z)\| = \sqrt{x^2 + y^2 + z^2}$.

67. $\ln f = \dfrac{1}{2}\ln(x^2 + y^2 + z^2)$

$$\nabla(\ln f) = \frac{x}{x^2 + y^2 + z^2}\mathbf{i} + \frac{y}{x^2 + y^2 + z^2}\mathbf{j} + \frac{z}{x^2 + y^2 + z^2}\mathbf{k} = \frac{x\mathbf{i} + y\mathbf{j} + z\mathbf{k}}{x^2 + y^2 + z^2} = \frac{\mathbf{F}}{f^2}$$

68. $\dfrac{1}{f} = \dfrac{1}{\sqrt{x^2 + y^2 + z^2}}$

$$\nabla\left(\frac{1}{f}\right) = \frac{-x}{(x^2 + y^2 + z^2)^{3/2}}\mathbf{i} + \frac{-y}{(x^2 + y^2 + z^2)^{3/2}}\mathbf{j} + \frac{-z}{(x^2 + y^2 + z^2)^{3/2}}\mathbf{k} = \frac{-(x\mathbf{i} + y\mathbf{j} + z\mathbf{k})}{(\sqrt{x^2 + y^2 + z^2})^3} = -\frac{\mathbf{F}}{f^3}$$

69. $f^n = \left(\sqrt{x^2 + y^2 + z^2}\right)^n$

$$\nabla f^n = n\left(\sqrt{x^2 + y^2 + z^2}\right)^{n-1}\frac{x}{\sqrt{x^2 + y^2 + z^2}}\mathbf{i} + n\left(\sqrt{x^2 + y^2 + z^2}\right)^{n-1}\frac{y}{\sqrt{x^2 + y^2 + z^2}}\mathbf{j}$$

$$+ n\left(\sqrt{x^2 + y^2 + z^2}\right)^{n-1}\frac{z}{\sqrt{x^2 + y^2 + z^2}}\mathbf{k}$$

$$= n\left(\sqrt{x^2 + y^2 + z^2}\right)^{n-2}(x\mathbf{i} + y\mathbf{j} + z\mathbf{k}) = nf^{n-2}\mathbf{F}$$

70. $w = \dfrac{1}{f} = \dfrac{1}{\sqrt{x^2 + y^2 + z^2}}$

$$\frac{\partial w}{\partial x} = -\frac{x}{(x^2 + y^2 + z^2)^{3/2}}$$

$$\frac{\partial w}{\partial y} = -\frac{y}{(x^2 + y^2 + z^2)^{3/2}}$$

$$\frac{\partial w}{\partial z} = -\frac{z}{(x^2 + y^2 + z^2)^{3/2}}$$

$$\frac{\partial^2 w}{\partial x^2} = \frac{2x^2 - y^2 - z^2}{(x^2 + y^2 + z^2)^{5/2}}$$

$$\frac{\partial^2 w}{\partial y^2} = \frac{2y^2 - x^2 - z^2}{(x^2 + y^2 + z^2)^{5/2}}$$

$$\frac{\partial^2 w}{\partial z^2} = \frac{2z^2 - x^2 - y^2}{(x^2 + y^2 + z^2)^{5/2}}$$

$$\nabla^2 w = \frac{\partial^2 w}{\partial x^2} + \frac{\partial^2 w}{\partial y^2} + \frac{\partial^2 w}{\partial z^2} = 0$$

Therefore $w = 1/f$ is harmonic.

71. $F(x, \ y) = M(x, \ y)\mathbf{i} + N(x, \ y)\mathbf{j} = \dfrac{m}{(x^2 + y^2)^{5/2}}[3xy\mathbf{i} + (2y^2 - x^2)\mathbf{j}]$

$M = \dfrac{3mxy}{(x^2 + y^2)^{5/2}} = 3mxy(x^2 + y^2)^{-5/2}$

$\dfrac{\partial M}{\partial y} = 3mxy\left[-\dfrac{5}{2}(x^2 + y^2)^{-7/2}(2y)\right] + (x^2 + y^2)^{-5/2}(3mx)$

$\qquad = 3mx(x^2 + y^2)^{-7/2}[-5y^2 + (x^2 + y^2)] = \dfrac{3mx(x^2 - 4y^2)}{(x^2 + y^2)^{7/2}}$

$N = \dfrac{m(2y^2 - x^2)}{(x^2 + y^2)^{5/2}} = m(2y^2 - x^2)(x^2 + y^2)^{-5/2}$

$\dfrac{\partial N}{\partial x} = m(2y^2 - x^2)\left[-\dfrac{5}{2}(x^2 + y^2)^{-7/2}(2x)\right] + (x^2 + y^2)^{-5/2}(-2mx)$

$\qquad = mx(x^2 + y^2)^{-7/2}[(2y^2 - x^2)(-5) + (x^2 + y^2)(-2)]$

$\qquad = mx(x^2 + y^2)^{-7/2}(3x^2 - 12y^2) = \dfrac{3mx(x^2 - 4y^2)}{(x^2 + y^2)^{7/2}}$

Therefore, $\dfrac{\partial N}{\partial x} = \dfrac{\partial M}{\partial y}$ and F is conservative.

Section 15.2   Line Integrals

1. $x^2 + y^2 = 9$

$\dfrac{x^2}{9} + \dfrac{y^2}{9} = 1$

$\cos^2 t + \sin^2 t = 1$

$\cos^2 t = \dfrac{x^2}{9}$

$\sin^2 t = \dfrac{y^2}{9}$

$x = 3\cos t$

$y = 3\sin t$

$\mathbf{r}(t) = 3\cos t\mathbf{i} + 3\sin t\mathbf{j}$

$0 \le t \le 2\pi$

2. $\dfrac{x^2}{16} + \dfrac{y^2}{9} = 1$

$\cos^2 t + \sin^2 t = 1$

$\cos^2 t = \dfrac{x^2}{16}$

$\sin^2 t = \dfrac{y^2}{9}$

$x = 4\cos t$

$y = 3\sin t$

$\mathbf{r}(t) = 4\cos t\mathbf{i} + 3\sin t\mathbf{j}$

$0 \le t \le 2\pi$

3. $\mathbf{r}(t) = \begin{cases} t\mathbf{i}, & 0 \le t \le 3 \\ 3\mathbf{i} + (t - 3)\mathbf{j}, & 3 \le t \le 6 \\ (9 - t)\mathbf{i} + 3\mathbf{j}, & 6 \le t \le 9 \\ (12 - t)\mathbf{j}, & 9 \le t \le 12 \end{cases}$

4. $\mathbf{r}(t) = \begin{cases} t\mathbf{i} + \frac{3}{4}t\mathbf{j}, & 0 \le t \le 4 \\ 4\mathbf{i} + (7 - t)\mathbf{j}, & 4 \le t \le 7 \\ (11 - t)\mathbf{i}, & 7 \le t \le 11 \end{cases}$

5. $\mathbf{r}(t) = \begin{cases} t\mathbf{i} + \sqrt{t}\,\mathbf{j}, & 0 \le t \le 1 \\ (2 - t)\mathbf{i} + (2 - t)\mathbf{j}, & 1 \le t \le 2 \end{cases}$

6. $\mathbf{r}(t) = \begin{cases} t\mathbf{i} + t^2\mathbf{j}, & 0 \le t \le 2 \\ (4 - t)\mathbf{i} + 4\mathbf{j}, & 2 \le t \le 4 \\ (8 - t)\mathbf{j}, & 4 \le t \le 8 \end{cases}$

7. $\mathbf{r}(t) = 4t\mathbf{i} + 3t\mathbf{j}, \quad 0 \le t \le 2; \quad \mathbf{r}'(t) = 4\mathbf{i} + 3\mathbf{j}$

$$\int_C (x - y)\,ds = \int_0^2 (4t - 3t)\sqrt{(4)^2 + (3)^2}\,dt = \int_0^2 5t\,dt = \left[\frac{5t^2}{2}\right]_0^2 = 10$$

8. $\mathbf{r}(t) = t\mathbf{i} + (1 - t)\mathbf{j}, \quad 0 \le t \le 1; \quad \mathbf{r}'(t) = \mathbf{i} - \mathbf{j}$

$$\int_C 4xy\,ds = \int_0^1 4t(1 - t)\sqrt{1 + 1}\,dt = 4\sqrt{2}\int_0^1 (t - t^2)\,dt = \left[4\sqrt{2}\left(\frac{t^2}{2} - \frac{t^3}{3}\right)\right]_0^1 = \frac{4\sqrt{2}}{6} = \frac{2\sqrt{2}}{3}$$

9. $\mathbf{r}(t) = \sin t\,\mathbf{i} + \cos t\,\mathbf{j} + 8t\mathbf{k}, \quad 0 \le t \le \frac{\pi}{2}; \quad \mathbf{r}'(t) = \cos t\,\mathbf{i} - \sin t\,\mathbf{j} + 8\mathbf{k}$

$$\int_C (x^2 + y^2 + z^2)\,ds = \int_0^{\pi/2} (\sin^2 t + \cos^2 t + 64t^2)\sqrt{(\cos t)^2 + (-\sin t)^2 + 64}\,dt$$

$$= \int_0^{\pi/2} \sqrt{65}(1 + 64t^2)\,dt = \left[\sqrt{65}\left(t + \frac{64t^3}{3}\right)\right]_0^{\pi/2} = \sqrt{65}\left(\frac{\pi}{2} + \frac{8\pi^3}{3}\right) = \frac{\sqrt{65}\pi}{6}(3 + 16\pi^2)$$

10. $\mathbf{r}(t) = 3\mathbf{i} + 12t\mathbf{j} + 5t\mathbf{k}, \quad 0 \le t \le 2; \quad \mathbf{r}'(t) = 12\mathbf{j} + 5\mathbf{k}$

$$\int_C 8xyz\,ds = \int_0^2 8(3)(12t)(5t)\sqrt{0 + 144 + 25}\,dt = \int_0^2 18{,}720t^2\,dt = \left[6240t^3\right]_0^2 = 49{,}920$$

11. $\mathbf{r}(t) = t\mathbf{i}, \quad 0 \le t \le 3$

$$\int_C (x^2 + y^2)\,ds = \int_0^3 [t^2 + 0^2]\sqrt{1 + 0}\,dt$$

$$= \int_0^3 t^2\,dt$$

$$= \left[\frac{1}{3}t^3\right]_0^3 = 9$$

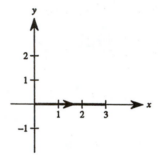

12. $\mathbf{r}(t) = t\mathbf{j}, \quad 1 \le t \le 10$

$$\int_C (x^2 + y^2)\,ds = \int_1^{10} [0 + t^2]\sqrt{0 + 1}\,dt$$

$$= \int_1^{10} t^2\,dt$$

$$= \left[\frac{1}{3}t^3\right]_1^{10} = 333$$

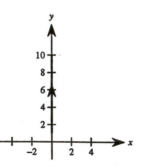

13. $\mathbf{r}(t) = \cos t\,\mathbf{i} + \sin t\,\mathbf{j}, \quad 0 \le t \le \frac{\pi}{2}$

$$\int_C (x^2 + y^2)\,ds = \int_0^{\pi/2} [\cos^2 t + \sin^2 t]\sqrt{(-\sin t)^2 + (\cos t)^2}\,dt$$

$$= \int_0^{\pi/2} dt$$

$$= \frac{\pi}{2}$$

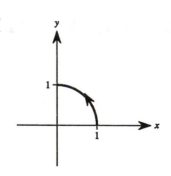

14. $r(t) = 2\cos t\,\mathbf{i} + 2\sin t\,\mathbf{j}, \quad 0 \le t \le \dfrac{\pi}{2}$

$$\int_C (x^2 + y^2)\,ds = \int_0^{\pi/2} [4\cos^2 t + 4\sin^2 t]\sqrt{(-2\sin t)^2 + (2\cos t)^2}\,dt$$

$$= \int_0^{\pi/2} 8\,dt$$

$$= 4\pi$$

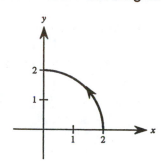

15. $r(t) = t\,\mathbf{i} + t\,\mathbf{j}, \quad 0 \le t \le 1$

$$\int_C (x + 4\sqrt{y})\,ds = \int_0^1 (t + 4\sqrt{t})\sqrt{1+1}\,dt$$

$$= \left[\sqrt{2}\left(\frac{t^2}{2} + \frac{8}{3}t^{3/2}\right)\right]_0^1$$

$$= \frac{19\sqrt{2}}{6}$$

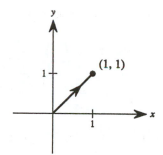

16. $r(t) = t\,\mathbf{i} + 3t\,\mathbf{j}, \quad 0 \le t \le 3$

$$\int_C (x + 4\sqrt{y})\,ds = \int_0^3 (t + 4\sqrt{3t})\sqrt{1+9}\,dt$$

$$= \left[\sqrt{10}\left(\frac{t^2}{2} + \frac{8\sqrt{3}}{3}t^{3/2}\right)\right]_0^3$$

$$= \frac{\sqrt{10}}{6}(27 + 144) = \frac{57\sqrt{10}}{2}$$

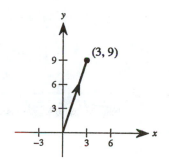

17. $r(t) = \begin{cases} t\,\mathbf{i}, & 0 \le t \le 1 \\ (2-t)\,\mathbf{i} + (t-1)\,\mathbf{j}, & 1 \le t \le 2 \\ (3-t)\,\mathbf{j}, & 2 \le t \le 3 \end{cases}$

$$\int_{C_1} (x + 4\sqrt{y})\,ds = \int_0^1 t\,dt = \frac{1}{2}$$

$$\int_{C_2} (x + 4\sqrt{y})\,ds = \int_1^2 [(2-t) + 4\sqrt{t-1}]\sqrt{1+1}\,dt$$

$$= \sqrt{2}\left[2t - \frac{t^2}{2} + \frac{8}{3}(t-1)^{3/2}\right]_1^2 = \frac{19\sqrt{2}}{6}$$

$$\int_{C_3} (x + 4\sqrt{y})\,ds = \int_2^3 4\sqrt{3-t}\,dt = \left[-\frac{8}{3}(3-t)^{3/2}\right]_2^3 = \frac{8}{3}$$

$$\int_C (x + 4\sqrt{y})\,ds = \frac{1}{2} + \frac{19\sqrt{2}}{6} + \frac{8}{3} = \frac{19 + 19\sqrt{2}}{6} = \frac{19(1 + \sqrt{2})}{6}$$

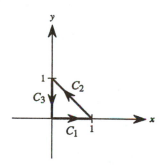

18. $\mathbf{r}(t) = \begin{cases} t\mathbf{i}, & 0 \le t \le 1 \\ \mathbf{i} + (t-1)\mathbf{j}, & 1 \le t \le 2 \\ (3-t)\mathbf{i} + \mathbf{j}, & 2 \le t \le 3 \\ (4-t)\mathbf{j}, & 3 \le t \le 4 \end{cases}$

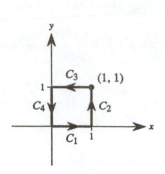

$\displaystyle\int_{C_1} (x + 4\sqrt{y})\,ds = \int_0^1 t\,dt = \frac{1}{2}$

$\displaystyle\int_{C_2} (x + 4\sqrt{y})\,ds = \int_1^2 \left(1 + 4\sqrt{t-1}\right)dt = \left[t + \frac{8}{3}(t-1)^{3/2}\right]_1^2 = \frac{11}{3}$

$\displaystyle\int_{C_3} (x + 4\sqrt{y})\,ds = \int_2^3 [(3-t)+4]\,dt = \left[7t - \frac{t^2}{2}\right]_2^3 = \frac{9}{2}$

$\displaystyle\int_{C_4} (x + 4\sqrt{y})\,ds = \int_3^4 4\sqrt{4-t}\,dt = \left[-\frac{8}{3}(4-t)^{3/2}\right]_3^4 = \frac{8}{3}$

$\displaystyle\int_{C} (x + 4\sqrt{y})\,ds = \frac{1}{2} + \frac{11}{3} + \frac{9}{2} + \frac{8}{3} = \frac{34}{3}$

19. $\rho(x,\ y,\ z) = \dfrac{1}{2}(x^2 + y^2 + z^2)$

$\mathbf{r}(t) = 3\cos t\,\mathbf{i} + 3\sin t\,\mathbf{j} + 2t\,\mathbf{k}, \quad 0 \le t \le 4\pi$

$\mathbf{r}'(t) = -3\sin t\,\mathbf{i} + 3\cos t\,\mathbf{j} + 2\mathbf{k}$

$\|\mathbf{r}'(t)\| = \sqrt{(-3\sin t)^2 + (3\cos t)^2 + (2)^2} = \sqrt{13}$

Mass $= \displaystyle\int_C \rho(x,\ y,\ z)\,ds$

$= \displaystyle\int_0^{4\pi} \frac{1}{2}[(3\cos t)^2 + (3\sin t)^2 + (2t)^2]\sqrt{13}\,dt$

$= \dfrac{\sqrt{13}}{2} \displaystyle\int_0^{4\pi} (9 + 4t^2)\,dt$

$= \left[\dfrac{\sqrt{13}}{2}\left(9t + \dfrac{4t^3}{3}\right)\right]_0^{4\pi}$

$= \dfrac{2\sqrt{13}\,\pi}{3}(27 + 64\pi^2) \approx 4973.8$

20. $\rho(x,\ y,\ z) = 2$

$\mathbf{r}(t) = 3\cos t\,\mathbf{i} + 3\sin t\,\mathbf{j} + 2t\,\mathbf{k}, \quad 0 \le t \le 4\pi$

$\mathbf{r}'(t) = -3\sin t\,\mathbf{i} + 3\cos t\,\mathbf{j} + 2\mathbf{k}$

$\|\mathbf{r}'(t)\| = \sqrt{(-3\sin t)^2 + (3\cos t)^2 + (2)^2} = \sqrt{13}$

Mass $= \displaystyle\int_C \rho(x,\ y,\ z)\,ds = \int_0^{4\pi} 2\sqrt{13}\,dt = 8\sqrt{13}\,\pi$

21. $\mathbf{F}(x,\ y) = xy\mathbf{i} + y\mathbf{j}$

$C: \mathbf{r}(t) = 4t\mathbf{i} + t\mathbf{j}, \quad 0 \le t \le 1$

$\mathbf{F}(t) = 4t^2\mathbf{i} + t\mathbf{j}$

$\mathbf{r}'(t) = 4\mathbf{i} + \mathbf{j}$

$\displaystyle\int_C \mathbf{F} \cdot d\mathbf{r} = \int_0^1 (16t^2 + t)\,dt$

$= \left[\dfrac{16}{3}t^3 + \dfrac{1}{2}t^2\right]_0^1 = \dfrac{35}{6}$

22. $\mathbf{F}(x,\ y) = xy\mathbf{i} + y\mathbf{j}$

$C: \mathbf{r}(t) = 4\cos t\,\mathbf{i} + 4\sin t\,\mathbf{j}, \quad 0 \le t \le \dfrac{\pi}{2}$

$\mathbf{F}(t) = 16\sin t \cos t\,\mathbf{i} + 4\sin t\,\mathbf{j}$

$\mathbf{r}'(t) = -4\sin t\,\mathbf{i} + 4\cos t\,\mathbf{j}$

$\displaystyle\int_C \mathbf{F} \cdot d\mathbf{r} = \int_0^{\pi/2} (-64\sin^2 t \cos t + 16\sin t \cos t)\,dt$

$= \left[-\dfrac{64}{3}\sin^3 t + 8\sin^2 t\right]_0^{\pi/2} = -\dfrac{40}{3}$

23. $\mathbf{F}(x, \ y) = 3x\mathbf{i} + 4y\mathbf{j}$

$C: \mathbf{r}(t) = 2\cos t\mathbf{i} + 2\sin t\mathbf{j}, \quad 0 \le t \le \dfrac{\pi}{2}$

$\mathbf{F}(t) = 6\cos t\mathbf{i} + 8\sin t\mathbf{j}$

$\mathbf{r}'(t) = -2\sin t\mathbf{i} + 2\cos t\mathbf{j}$

$\displaystyle \int_C \mathbf{F} \cdot d\mathbf{r} = \int_0^{\pi/2} (-12\sin t\cos t + 16\sin t\cos t)\, dt$

$\qquad = \Big[2\sin^2 t \Big]_0^{\pi/2} = 2$

24. $\mathbf{F}(x, \ y) = 3x\mathbf{i} + 4y\mathbf{j}$

$C: \mathbf{r}(t) = t\mathbf{i} + \sqrt{4 - t^2}\, \mathbf{j}, \quad -2 \le t \le 2$

$\mathbf{F}(t) = 3t\mathbf{i} + 4\sqrt{4 - t^2}\, \mathbf{j}$

$\mathbf{r}'(t) = \mathbf{i} - \dfrac{t}{\sqrt{4 - t^2}}\, \mathbf{j}$

$\displaystyle \int_C \mathbf{F} \cdot d\mathbf{r} = \int_{-2}^2 (3t - 4t)\, dt = \left[-\dfrac{t^2}{2} \right]_{-2}^2 = 0$

25. $\mathbf{F}(x, \ y, \ z) = x^2 y\mathbf{i} + (x - z)\mathbf{j} + xyz\mathbf{k}$

$C: \mathbf{r}(t) = t\mathbf{i} + t^2\mathbf{j} + 2\mathbf{k}, \quad 0 \le t \le 1$

$\mathbf{F}(t) = t^4\mathbf{i} + (t - 2)\mathbf{j} + 2t^3\mathbf{k}$

$\mathbf{r}'(t) = \mathbf{i} + 2t\mathbf{j}$

$\displaystyle \int_C \mathbf{F} \cdot d\mathbf{r} = \int_0^1 [t^4 + 2t(t - 2)]\, dt$

$\qquad = \left[\dfrac{t^5}{5} + \dfrac{2t^3}{3} - 2t^2 \right]_0^1 = -\dfrac{17}{15}$

26. $\mathbf{F}(x, \ y, \ z) = x^2\mathbf{i} + y^2\mathbf{j} + z^2\mathbf{k}$

$C: \mathbf{r}(t) = \sin t\mathbf{i} + \cos t\mathbf{j} + t^2\mathbf{j}, \quad 0 \le t \le \dfrac{\pi}{2}$

$\mathbf{F}(t) = \sin^2 t\mathbf{i} + \cos^2 t\mathbf{j} + t^4\mathbf{k}$

$\mathbf{r}'(t) = \cos t\mathbf{i} - \sin t\mathbf{j} + 2t\mathbf{k}$

$\displaystyle \int_C \mathbf{F} \cdot d\mathbf{r} = \int_0^{\pi/2} (\sin^2 t\cos t - \cos^2 t\sin t + 2t^5)\, dt$

$\qquad = \left[\dfrac{\sin^3 t}{3} + \dfrac{\cos^3 t}{3} + \dfrac{t^6}{3} \right]_0^{\pi/2} = \dfrac{\pi^6}{192}$

27. $\mathbf{F}(x, \ y, \ z) = x^2 z\mathbf{i} + 6y\mathbf{j} + yz^2\mathbf{k}$

$\mathbf{r}(t) = t\mathbf{i} + t^2\mathbf{j} + \ln t\mathbf{k}, \quad 1 \le t \le 3$

$\mathbf{F}(t) = t^2\ln t\mathbf{i} + 6t^2\mathbf{j} + t^2\ln^2 t\mathbf{k}$

$d\mathbf{r} = \left(\mathbf{i} + 2t\mathbf{j} + \dfrac{1}{t}\mathbf{k} \right) dt$

$\displaystyle \int_C \mathbf{F} \cdot d\mathbf{r} = \int_1^3 [t^2\ln t + 12t^3 + t(\ln t)^2]\, dt$

$\qquad \approx 249.49$

28. $\mathbf{F}(x, \ y, \ z) = \dfrac{x\mathbf{i} + y\mathbf{j} + z\mathbf{k}}{\sqrt{x^2 + y^2 + z^2}}$

$\mathbf{r}(t) = t\mathbf{i} + t\mathbf{j} + e^t\mathbf{k}, \quad 0 \le t \le 2$

$\mathbf{F}(t) = \dfrac{t\mathbf{i} + t\mathbf{j} + e^t\mathbf{k}}{\sqrt{2t^2 + e^{2t}}}$

$d\mathbf{r} = (\mathbf{i} + \mathbf{j} + e^t\mathbf{k})\, dt$

$\displaystyle \int_C \mathbf{F} \cdot d\mathbf{r} = \int_0^2 \dfrac{1}{\sqrt{2t^2 + e^{2t}}}(2t + e^{2t})\, dt \approx 6.91$

29. $\mathbf{F}(x, \ y) = -x\mathbf{i} - 2y\mathbf{j}$

$C: y = x^3$ from $(0, 0)$ to $(2, 8)$

$\mathbf{r}(t) = t\mathbf{i} + t^3\mathbf{j}, \quad 0 \le t \le 2$

$\mathbf{r}'(t) = \mathbf{i} + 3t^2\mathbf{j}$

$\mathbf{F}(t) = -t\mathbf{i} - 2t^3\mathbf{j}$

$\mathbf{F} \cdot \mathbf{r}' = -t - 6t^5$

$\text{Work} = \displaystyle \int_C \mathbf{F} \cdot d\mathbf{r} = \int_0^2 (-t - 6t^5)\, dt = \left[-\dfrac{1}{2}t^2 - t^6 \right]_0^2 = -66$

30. $\mathbf{F}(x,\ y) = x^2\mathbf{i} - xy\mathbf{j}$

$C:\ x = \cos^3 t,\quad y = \sin^3 t$ from $(1, 0)$ to $(0, 1)$

$\mathbf{r}(t) = \cos^3 t\,\mathbf{i} + \sin^3 t\,\mathbf{j},\quad 0 \le t \le \dfrac{\pi}{2}$

$\mathbf{r}'(t) = -3\cos^2 t\,\sin t\,\mathbf{i} + 3\sin^2 t\,\cos t\,\mathbf{j}$

$\mathbf{F}(t) = \cos^6 t\,\mathbf{i} - \cos^3 t\,\sin^3 t\,\mathbf{j}$

$\mathbf{F} \cdot \mathbf{r}' = -3\cos^8 t\,\sin t - 3\cos^4 t\,\sin^5 t$

$\qquad = -3\cos^4 t\,\sin t\,(\cos^4 t + \sin^4 t)$

$\qquad = -3\cos^4 t\,\sin t\,[\cos^4 t + (1 - \cos^2 t)^2]$

$\qquad = -3\cos^4 t\,\sin t\,(2\cos^4 t - 2\cos^2 t + 1)$

$\qquad = -6\cos^8 t\,\sin t + 6\cos^6 t\,\sin t - 3\cos^4 t\,\sin t$

$\text{Work} = \displaystyle\int_C \mathbf{F} \cdot d\mathbf{r} = \int_0^{\pi/2} [-6\cos^8 t\,\sin t + 6\cos^6 t\,\sin t - 3\cos^4 t\,\sin t]\,dt$

$\qquad = \left[\dfrac{2\cos^9 t}{3} - \dfrac{6\cos^7 t}{7} + \dfrac{3\cos^5 t}{5}\right]_0^{\pi/2} = -\dfrac{43}{105}$

31. $\mathbf{F}(x,\ y) = 2x\mathbf{i} + y\mathbf{j}$

$C:$ counterclockwise around the triangle whose vertices are $(0, 0)$, $(1, 0)$, $(1, 1)$

$\mathbf{r}(t) = \begin{cases} t\mathbf{i}, & 0 \le t \le 1 \\ \mathbf{i} + (t - 1)\mathbf{j}, & 1 \le t \le 2 \\ (3 - t)\mathbf{i} + (3 - t)\mathbf{j}, & 2 \le t \le 3 \end{cases}$

On C_1: $\mathbf{F}(t) = 2t\mathbf{i},\quad \mathbf{r}'(t) = \mathbf{i}$

$\qquad \text{Work} = \displaystyle\int_{C_1} \mathbf{F} \cdot d\mathbf{r} = \int_0^1 2t\,dt = 1$

On C_2: $\mathbf{F}(t) = 2\mathbf{i} + (t - 1)\mathbf{j},\quad \mathbf{r}'(t) = \mathbf{j}$

$\qquad \text{Work} = \displaystyle\int_{C_2} \mathbf{F} \cdot d\mathbf{r} = \int_1^2 (t - 1)\,dt = \dfrac{1}{2}$

On C_3: $\mathbf{F}(t) = 2(3 - t)\mathbf{i} + (3 - t)\mathbf{j},\quad \mathbf{r}'(t) = -\mathbf{i} - \mathbf{j}$

$\qquad \text{Work} = \displaystyle\int_{C_3} \mathbf{F} \cdot d\mathbf{r} = \int_2^3 [-2(3 - t) - (3 - t)]\,dt = -\dfrac{3}{2}$

$\text{Total work} = \displaystyle\int_C \mathbf{F} \cdot d\mathbf{r} = 1 + \dfrac{1}{2} - \dfrac{3}{2} = 0$

32. $\mathbf{F}(x,\ y) = -y\mathbf{i} - x\mathbf{j}$

$C:$ counterclockwise along the semicircle $y = \sqrt{4 - x^2}$ from $(2, 0)$ to $(-2,\ 0)$

$\mathbf{r}(t) = 2\cos t\,\mathbf{i} + 2\sin t\,\mathbf{j},\quad 0 \le t \le \pi$

$\mathbf{r}'(t) = -2\sin t\,\mathbf{i} + 2\cos t\,\mathbf{j}$

$\mathbf{F}(t) = -2\sin t\,\mathbf{i} - 2\cos t\,\mathbf{j}$

$\mathbf{F} \cdot \mathbf{r}' = 4\sin^2 t - 4\cos^2 t = -4\cos 2t$

$\text{Work} = \displaystyle\int_C \mathbf{F} \cdot d\mathbf{r} = -4\int_0^\pi \cos 2t\,dt = \left[-2\sin 2t\right]_0^\pi = 0$

33. $\mathbf{F}(x, y, z) = x\mathbf{i} + y\mathbf{j} - 5z\mathbf{k}$

C: $\mathbf{r}(t) = 2\cos t\,\mathbf{i} + 2\sin t\,\mathbf{j} + t\mathbf{k}, \quad 0 \le t \le 2\pi$

$\mathbf{r}'(t) = -2\sin t\,\mathbf{i} + 2\cos t\,\mathbf{j} + \mathbf{k}$

$\mathbf{F}(t) = 2\cos t\,\mathbf{i} + 2\sin t\,\mathbf{j} - 5t\mathbf{k}$

$\mathbf{F} \cdot \mathbf{r}' = -5t$

$\text{Work} = \displaystyle\int_C \mathbf{F} \cdot d\mathbf{r} = \int_0^{2\pi} -5t\,dt = -10\pi^2$

34. $\mathbf{F}(x, y, z) = yz\mathbf{i} + xz\mathbf{j} + xy\mathbf{k}$

C: line from $(0, 0, 0)$ to $(5, 3, 2)$

$\mathbf{r}(t) = 5t\mathbf{i} + 3t\mathbf{j} + 2t\mathbf{k}, \quad 0 \le t \le 1$

$\mathbf{r}'(t) = 5\mathbf{i} + 3\mathbf{j} + 2\mathbf{k}$

$\mathbf{F}(t) = 6t^2\mathbf{i} + 10t^2\mathbf{j} + 15t^2\mathbf{k}$

$\mathbf{F} \cdot \mathbf{r}' = 90t^2$

$\text{Work} = \displaystyle\int_C \mathbf{F} \cdot d\mathbf{r} = \int_0^1 90t^2\,dt = 30$

35. $\mathbf{r}(t) = 3\sin t\,\mathbf{i} + 3\cos t\,\mathbf{j} + \dfrac{10}{2\pi}t\mathbf{k}, \quad 0 \le t \le 2\pi$

$\mathbf{F} = 150\mathbf{k}$

$d\mathbf{r} = \left(3\cos t\,\mathbf{i} - 3\sin t\,\mathbf{j} + \dfrac{10}{2\pi}\mathbf{k}\right)dt$

$\displaystyle\int_C \mathbf{F} \cdot d\mathbf{r} = \int_0^{2\pi} \dfrac{1500}{2\pi}\,dt = \left[\dfrac{1500}{2\pi}t\right]_0^{2\pi} = 1500 \text{ ft} \cdot \text{lb}$

36. $\mathbf{r}(t) = t\mathbf{i} + t^2\mathbf{j}, \quad 0 \le t \le 1$

$\mathbf{r}'(t) = \mathbf{i} + 2t\mathbf{j}$

(x, y)	$(0, 0)$	$\left(\frac{1}{4}, \frac{1}{16}\right)$	$\left(\frac{1}{2}, \frac{1}{4}\right)$	$\left(\frac{3}{4}, \frac{9}{16}\right)$	$(1, 1)$
$\mathbf{F}(x, y)$	$5\mathbf{i}$	$3.5\mathbf{i} + \mathbf{j}$	$2\mathbf{i} + 2\mathbf{j}$	$1.5\mathbf{i} + 3\mathbf{j}$	$\mathbf{i} + 5\mathbf{j}$
$\mathbf{r}'(t)$	$\mathbf{i}$	$\mathbf{i} + 0.5\mathbf{j}$	$\mathbf{i} + \mathbf{j}$	$\mathbf{i} + 1.5\mathbf{j}$	$\mathbf{i} + 2\mathbf{j}$
$\mathbf{F} \cdot \mathbf{r}'$	5	4	4	6	11

$\displaystyle\int_C \mathbf{F} \cdot d\mathbf{r} \approx \dfrac{1-0}{3(4)}[5 + 4(4) + 2(4) + 4(6) + 11]$

$\qquad = \dfrac{16}{3}$

37. $\mathbf{F}(x, y) = y\mathbf{i} - x\mathbf{j}$

C: $\mathbf{r}(t) = t\mathbf{i} - 2t\mathbf{j}$

$\mathbf{r}'(t) = \mathbf{i} - 2\mathbf{j}$

$\mathbf{F}(t) = -2t\mathbf{i} - t\mathbf{j}$

$\mathbf{F} \cdot \mathbf{r}' = -2t + 2t = 0$

Thus, $\displaystyle\int_C \mathbf{F} \cdot d\mathbf{r} = 0.$

38. $\mathbf{F}(x, y) = -3y\mathbf{i} + x\mathbf{j}$

C: $\mathbf{r}(t) = t\mathbf{i} - t^3\mathbf{j}$

$\mathbf{r}'(t) = \mathbf{i} - 3t^2\mathbf{j}$

$\mathbf{F}(t) = 3t^3\mathbf{i} + t\mathbf{j}$

$\mathbf{F} \cdot \mathbf{r}' = 3t^3 - 3t^3 = 0$

Thus, $\displaystyle\int_C \mathbf{F} \cdot d\mathbf{r} = 0.$

39. $\mathbf{F}(x, y) = (x^3 - 2x^2)\mathbf{i} + \left(x - \dfrac{y}{2}\right)\mathbf{j}$

C: $\mathbf{r}(t) = t\mathbf{i} + t^2\mathbf{j}$

$\mathbf{r}'(t) = \mathbf{i} + 2t\mathbf{j}$

$\mathbf{F}(t) = (t^3 - 2t^2)\mathbf{i} + \left(t - \dfrac{t^2}{2}\right)\mathbf{j}$

$\mathbf{F} \cdot \mathbf{r}' = (t^3 - 2t^2) + 2t\left(t - \dfrac{t^2}{2}\right) = 0$

Thus, $\displaystyle\int_C \mathbf{F} \cdot d\mathbf{r} = 0.$

40. $\mathbf{F}(x, y) = x\mathbf{i} + y\mathbf{j}$

C: $\mathbf{r}(t) = 3\sin t\,\mathbf{i} + 3\cos t\,\mathbf{j}$

$\mathbf{r}'(t) = 3\cos t\,\mathbf{i} - 3\sin t\,\mathbf{j}$

$\mathbf{F}(t) = 3\sin t\,\mathbf{i} + 3\cos t\,\mathbf{j}$

$\mathbf{F} \cdot \mathbf{r}' = 9\sin t \cos t - 9\sin t \cos t = 0$

Thus, $\displaystyle\int_C \mathbf{F} \cdot d\mathbf{r} = 0.$

41. $x = 2t$, $y = 10t$, $0 \le t \le 1 \Rightarrow y = 5x$ or $x = \dfrac{y}{5}$, $0 \le y \le 10$

$$\int_C (x + 3y^2)\, dy = \int_0^{10} \left(\frac{y}{5} + 3y^2 \right) dy = \left[\frac{y^2}{10} + y^3 \right]_0^{10} = 1010$$

42. $x = 2t$, $y = 10t$, $0 \le t \le 1 \Rightarrow y = 5x$, $0 \le x \le 2$

$$\int_C (x + 3y^2)\, dx = \int_0^2 (x + 75x^2)\, dx = \left[\frac{x^2}{2} + 25x^3 \right]_0^2 = 202$$

43. $x = 2t$, $y = 10t$, $0 \le t \le 1 \Rightarrow x = \dfrac{y}{5}$, $0 \le y \le 10$, $dx = \dfrac{1}{5}\, dy$

$$\int_C xy\, dx + y\, dy = \int_0^{10} \left(\frac{y^2}{25} + y \right) dy = \left[\frac{y^3}{75} + \frac{y^2}{2} \right]_0^{10} = \frac{190}{3} \quad \text{OR}$$

$y = 5x$, $dy = 5\, dx$, $0 \le x \le 2$

$$\int_C xy\, dx + y\, dy = \int_0^2 (5x^2 + 25x)\, dx = \left[\frac{5x^3}{3} + \frac{25x^2}{2} \right]_0^2 = \frac{190}{3}$$

44. $x = 2t$, $y = 10t$, $0 \le t \le 1 \Rightarrow y = 5x$, $dy = 5\, dx$, $0 \le x \le 2$

$$\int_C (y - 3x)\, dx + x^2\, dy = \int_0^2 (2x + 5x^2)\, dx = \left[x^2 + \frac{5x^3}{3} \right]_0^2 = \frac{52}{3}$$

45. $\mathbf{r}(t) = t\mathbf{i}$, $0 \le t \le 5$

$x(t) = t$, $y(t) = 0$

$dx = dt$, $dy = 0$

$$\int_C (2x - y)\, dx + (x + 3y)\, dy = \int_0^5 2t\, dt = 25$$

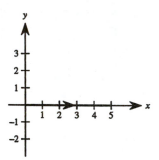

46. $\mathbf{r}(t) = t\mathbf{j}$, $0 \le t \le 2$

$x(t) = 0$, $y(t) = t$

$dx = 0$, $dy = dt$

$$\int_C (2x - y)\, dx + (x + 3y)\, dy = \int_0^2 3t\, dt = \left[\frac{3}{2}t^2 \right]_0^2 = 6$$

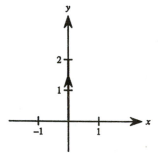

47. $\mathbf{r}(t) = \begin{cases} t\mathbf{i}, & 0 \le t \le 3 \\ 3\mathbf{i} + (t-3)\mathbf{j}, & 3 \le t \le 6 \end{cases}$

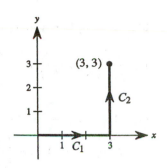

C_1: $\quad x(t) = t, \qquad y(t) = 0,$

$\qquad dx = dt, \qquad dy = 0$

$\displaystyle \int_{C_1} (2x - y)\, dx + (x + 3y)\, dy = \int_0^3 2t\, dt = 9$

C_2: $\quad x(t) = 3, \qquad y = t - 3$

$\qquad dx = 0, \qquad dy = dt$

$\displaystyle \int_{C_2} (2x - y)\, dx + (x + 3y)\, dy = \int_3^6 [3 + 3(t-3)]\, dt = \left[\frac{3t^2}{2} - 6t \right]_3^6 = \frac{45}{2}$

$\displaystyle \int_C (2x - y)\, dx + (x + 3y)\, dy = 9 + \frac{45}{2} = \frac{63}{2}$

48. $\mathbf{r}(t) = \begin{cases} -t\mathbf{j}, & 0 \le t \le 3 \\ (t-3)\mathbf{i} - 3\mathbf{j}, & 3 \le t \le 5 \end{cases}$

C_1: $\quad x(t) = 0, \qquad y(t) = -t$

$\qquad dx = 0, \qquad dy = -dt$

$\displaystyle \int_{C_1} (2x - y)\, dx + (x + 3y)\, dy = \int_0^3 3t\, dt = \frac{27}{2}$

C_2: $\quad x(t) = t - 3, \qquad y(t) = -3$

$\qquad dx = dt, \qquad dy = 0$

$\displaystyle \int_{C_2} (2x - y)\, dx + (x + 3y)\, dy = \int_3^5 [2(t-3) + 3]\, dt = \left[(t-3)^2 + 3t \right]_3^5 = 10$

$\displaystyle \int_C (2x - y)\, dx + (x + 3y)\, dy = \frac{27}{2} + 10 = \frac{47}{2}$

49. $x(t) = t, \qquad y(t) = 2t^2, \qquad 0 \le t \le 2$

$dx = dt, \qquad dy = 4t\, dt$

$\displaystyle \int_C (2x - y)\, dx + (x + 3y)\, dy = \int_0^2 (2t - 2t^2)\, dt + (t + 6t^2)4t\, dt$

$\displaystyle \qquad = \int_0^2 (24t^3 + 2t^2 + 2t)\, dt = \left[6t^4 + \frac{2}{3}t^3 + t^2 \right]_0^2 = \frac{316}{3}$

50. $x(t) = 4\sin t, \qquad y(t) = 3\cos t, \qquad 0 \le t \le \dfrac{\pi}{2}$

$dx = 4\cos t\, dt, \qquad dy = -3\sin t\, dt$

$\displaystyle \int_C (2x - y)\, dx + (x + 3y)\, dy = \int_0^{\pi/2} (8\sin t - 3\cos t)(4\cos t)\, dt + (4\sin t + 9\cos t)(-3\sin t)\, dt$

$\displaystyle \qquad = \int_0^{\pi/2} (5\sin t \cos t - 12\cos^2 t - 12\sin^2 t)\, dt$

$\displaystyle \qquad = \left[\frac{5}{2}\sin^2 t - 12t \right]_0^{\pi/2} = \frac{5}{2} - 6\pi$

51. $f(x, y) = h$

C: line from (0, 0) to (3, 4)

$$\mathbf{r} = 3t\mathbf{i} + 4t\mathbf{j}, \quad 0 \le t \le 1$$

$$\mathbf{r}'(t) = 3\mathbf{i} + 4\mathbf{j}$$

$$\|\mathbf{r}'(t)\| = 5$$

Lateral surface area:

$$\int_C f(x, y)\,ds = \int_0^1 5h\,dt = 5h$$

52. $f(x, y) = y$

C: line from (0, 0) to (4, 4)

$$\mathbf{r}(t) = t\mathbf{i} + t\mathbf{j}, \quad 0 \le t \le 4$$

$$\mathbf{r}'(t) = \mathbf{i} + \mathbf{j}$$

$$\|\mathbf{r}'(t)\| = \sqrt{2}$$

Lateral surface area:

$$\int_C f(x, y)\,ds = \int_0^4 t(\sqrt{2})\,dt = 8\sqrt{2}$$

53. $f(x, y) = xy$

C: $x^2 + y^2 = 1$ from (1, 0) to (0, 1)

$$\mathbf{r}(t) = \cos t\,\mathbf{i} + \sin t\,\mathbf{j}, \quad 0 \le t \le \frac{\pi}{2}$$

$$\mathbf{r}'(t) = -\sin t\,\mathbf{i} + \cos t\,\mathbf{j}$$

$$\|\mathbf{r}'(t)\| = 1$$

Lateral surface area: $\displaystyle\int_C f(x, y)\,ds = \int_0^{\pi/2} \cos t \sin t\,dt = \left[\frac{\sin^2 t}{2}\right]_0^{\pi/2} = \frac{1}{2}$

54. $f(x, y) = x + y$

C: $x^2 + y^2 = 1$ from (1, 0) to (0, 1)

$$\mathbf{r}(t) = \cos t\,\mathbf{i} + \sin t\,\mathbf{j}, \quad 0 \le t \le \frac{\pi}{2}$$

$$\mathbf{r}'(t) = -\sin t\,\mathbf{i} + \cos t\,\mathbf{j}$$

$$\|\mathbf{r}'(t)\| = 1$$

Lateral surface area: $\displaystyle\int_C f(x, y)\,ds = \int_0^{\pi/2} (\cos t + \sin t)\,dt = \left[\sin t - \cos t\right]_0^{\pi/2} = 2$

55. $f(x, y) = h$

C: $y = 1 - x^2$ from (1, 0) to (0, 1)

$$\mathbf{r}(t) = (1 - t)\mathbf{i} + [1 - (1 - t)^2]\mathbf{j}, \quad 0 \le t \le 1$$

$$\mathbf{r}'(t) = -\mathbf{i} + 2(1 - t)\mathbf{j}$$

$$\|\mathbf{r}'(t)\| = \sqrt{1 + 4(1 - t)^2}$$

Lateral surface area:

$$\int_C f(x, y)\,ds = \int_0^1 h\sqrt{1 + 4(1 - t)^2}\,dt$$

$$= -\frac{h}{4}\left[2(1 - t)\sqrt{1 + 4(1 - t)^2} + \ln\left|2(1 - t) + \sqrt{1 + 4(1 - t)^2}\right|\right]_0^1$$

$$= \frac{h}{4}[2\sqrt{5} + \ln(2 + \sqrt{5})] \approx 1.4789h$$

56. $f(x, \ y) = y + 1$

$C: y = 1 - x^2$ from $(1, 0)$ to $(0, 1)$

$\mathbf{r}(t) = (1 - t)\mathbf{i} + [1 - (1 - t)^2]\mathbf{j}, \ \ 0 \le t \le 1$

$\mathbf{r}'(t) = -\mathbf{i} + 2(1 - t)\mathbf{j}$

$\|\mathbf{r}'(t)\| = \sqrt{1 + 4(1 - t)^2}$

Lateral surface area:

$$\int_C f(x, \ y)\, ds = \int_0^1 [2 - (1 - t)^2]\sqrt{1 + 4(1 - t)^2}\, dt$$

$$= 2\int_0^1 \sqrt{1 + 4(1 - t)^2}\, dt - \int_0^1 (1 - t)^2 \sqrt{1 + 4(1 - t)^2}\, dt$$

$$= -\frac{1}{2}\left[2(1 - t)\sqrt{1 + 4(1 - t)^2} + \ln\left| 2(1 - t) + \sqrt{1 + 4(1 - t)^2} \right| \right]_0^1$$

$$+ \frac{1}{64}\left[2(1 - t)[2(4)(1 - t)^2 + 1]\sqrt{1 + 4(1 - t)^2} - \ln\left| 2(1 - t) + \sqrt{1 + 4(1 - t)^2} \right| \right]_0^1$$

$$= \frac{1}{2}[2\sqrt{5} + \ln(2 + \sqrt{5})] - \frac{1}{64}[18\sqrt{5} - \ln(2 + \sqrt{5})]$$

$$= \frac{23}{32}\sqrt{5} + \frac{33}{64}\ln(2 + \sqrt{5}) = \frac{1}{64}[46\sqrt{5} + 33\ln(2 + \sqrt{5})] \approx 2.3515$$

57. $f(x, \ y) = xy$

$C: y = 1 - x^2$ from $(1, 0)$ to $(0, 1)$

You could parameterize the curve C as in Exercises 55 and 56. Alternatively, let $x = \cos t$, then:

$y = 1 - \cos^2 t = \sin^2 t$

$\mathbf{r}(t) = \cos t \, \mathbf{i} + \sin^2 t \, \mathbf{j}, \ \ 0 \le t \le \dfrac{\pi}{2}$

$\mathbf{r}'(t) = -\sin t \, \mathbf{i} + 2\sin t \cos t \, \mathbf{j}$

$\|\mathbf{r}'(t)\| = \sqrt{\sin^2 t + 4\sin^2 t \cos^2 t} = \sin t \sqrt{1 + 4\cos^2 t}$

Lateral surface area:

$$\int_C f(x, \ y)\, ds = \int_0^{\pi/2} \cos t \sin^2 t \left(\sin t \sqrt{1 + 4\cos^2 t} \right) dt = \int_0^{\pi/2} \sin^2 t[(1 + 4\cos^2 t)^{1/2} \sin t \cos t]\, dt$$

Let $u = \sin^2 t$ and $dv = (1 + 4\cos^2 t)^{1/2} \sin t \cos t$, then $du = 2\sin t \cos t \, dt$ and $v = -\frac{1}{12}(1 + 4\cos^2 t)^{3/2}$.

$$\int_C f(x, \ y)\, ds = \left[-\frac{1}{12}\sin^2 t (1 + 4\cos^2 t)^{3/2} \right]_0^{\pi/2} + \frac{1}{6}\int_0^{\pi/2} (1 + 4\cos^2 t)^{3/2} \sin t \cos t \, dt$$

$$= \left[-\frac{1}{12}\sin^2 t (1 + 4\cos^2 t)^{3/2} - \frac{1}{120}(1 + 4\cos^2 t)^{5/2} \right]_0^{\pi/2}$$

$$= \left(-\frac{1}{12} - \frac{1}{120} \right) + \frac{1}{120}(5)^{5/2} = \frac{1}{120}(25\sqrt{5} - 11) \approx 0.3742$$

58. $f(x, y) = x^2 - y^2 + 4$

  $C: x^2 + y^2 = 4$

  $\mathbf{r}(t) = 2\cos t\,\mathbf{i} + 2\sin t\,\mathbf{j}, \quad 0 \le t \le 2\pi$

  $\mathbf{r}'(t) = -2\sin t\,\mathbf{i} + 2\cos t\,\mathbf{j}$

  $\|\mathbf{r}'(t)\| = 2$

Lateral surface area:

$$\int_C f(x, y)\,ds = \int_0^{2\pi} (4\cos^2 t - 4\sin^2 t + 4)(2)\,dt = 8\int_0^{2\pi}(1 + \cos 2t)\,dt = \left[8\left(t + \frac{1}{2}\sin 2t\right)\right]_0^{2\pi} = 16\pi$$

59. a. $f(x, y) = 1 + y^2$

  $\mathbf{r}(t) = 2\cos t\,\mathbf{i} + 2\sin t\,\mathbf{j}, \quad 0 \le t \le 2\pi$

  $\mathbf{r}'(t) = -2\sin t\,\mathbf{i} + 2\cos t\,\mathbf{j}$

  $\|\mathbf{r}'(t)\| = 2$

  $$S = \int_C f(x, y)\,dx = \int_0^{2\pi}(1 + 4\sin^2 t)(2)\,dt$$

  $$= \left[2t + 4(t - \sin t\cos t)\right]_0^{2\pi} = 12\pi \approx 37.70 \text{ cm}^2$$

b. $0.2(12\pi) = \dfrac{12\pi}{5} \approx 7.54 \text{ cm}^3$

c.

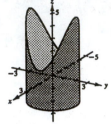

60. $f(x, y) = 20 + \dfrac{1}{4}x$

  $C: y = x^{3/2}, \quad 0 \le x \le 40$

  $\mathbf{r}(t) = t\,\mathbf{i} + t^{3/2}\,\mathbf{j}, \quad 0 \le t \le 40$

  $\mathbf{r}'(t) = \mathbf{i} + \dfrac{3}{2}t^{1/2}\,\mathbf{j}$

  $\|\mathbf{r}'(t)\| = \sqrt{1 + \left(\dfrac{9}{4}\right)t}$

Lateral surface area: $\displaystyle\int_C f(x, y)\,ds = \int_0^{40}\left(20 + \frac{1}{4}t\right)\sqrt{1 + \left(\frac{9}{4}\right)t}\,dt$

Let $u = \sqrt{1 + (\frac{9}{4})t}$, then $t = \frac{4}{9}(u^2 - 1)$ and $dt = \frac{8}{9}u\,du$.

$$\int_0^{40}\left(20 + \frac{1}{4}t\right)\sqrt{1 + \left(\frac{9}{4}\right)t}\,dt = \int_1^{\sqrt{91}}\left[20 + \frac{1}{9}(u^2 - 1)\right](u)\left(\frac{8}{9}u\right)du = \frac{8}{81}\int_1^{\sqrt{91}}(u^4 + 179u^2)\,du$$

$$= \frac{8}{81}\left[\frac{u^5}{5} + \frac{179u^3}{3}\right]_1^{\sqrt{91}} = \frac{850{,}304\sqrt{91} - 7184}{1215} \approx 6670.12$$

61. $f(x, y) = y$
$C: y = x^2$ from $(0, 0)$ to $(2, 4)$
$S \approx 8$
Matches c.

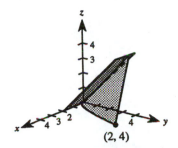

$(2, 4)$

62. $S \approx 75$
Matches d.

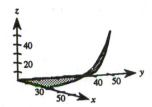

63. False

$$\int_C xy \, ds = \sqrt{2} \int_0^1 t^2 \, dt$$

64. False, the orientation of C does not affect the form

$$\int_C f(x, y) \, ds.$$

65. False, the orientations are different.

66. False. For example, see Example 32.

Section 15.3   Conservative Vector Fields and Independence of Path

1. $\mathbf{F}(x, y) = x^2 \mathbf{i} + xy \mathbf{j}$

a. $\mathbf{r}_1(t) = t \mathbf{i} + t^2 \mathbf{j}, \quad 0 \le t \le 1$

$\mathbf{r}_1'(t) = \mathbf{i} + 2t \mathbf{j}$

$\mathbf{F}(t) = t^2 \mathbf{i} + t^3 \mathbf{j}$

$$\int_C \mathbf{F} \cdot d\mathbf{r} = \int_0^1 (t^2 + 2t^4) \, dt = \frac{11}{15}$$

b. $\mathbf{r}_2(\theta) = \sin\theta \mathbf{i} + \sin^2\theta \mathbf{j}, \quad 0 \le \theta \le \frac{\pi}{2}$

$\mathbf{r}_2'(\theta) = \cos\theta \mathbf{i} + 2\sin\theta\cos\theta \mathbf{j}$

$\mathbf{F}(t) = \sin^2\theta \mathbf{i} + \sin^3\theta \mathbf{j}$

$$\int_C \mathbf{F} \cdot d\mathbf{r} = \int_0^{\pi/2} (\sin^2\theta\cos\theta + 2\sin^4\theta\cos\theta) \, d\theta$$

$$= \left[\frac{\sin^3\theta}{3} + \frac{2\sin^5\theta}{5}\right]_0^{\pi/2} = \frac{11}{15}$$

2. $\mathbf{F}(x, y) = (x^2 + y^2)\mathbf{i} - x\mathbf{j}$

a. $\mathbf{r}_1(t) = t\mathbf{i} + \sqrt{t}\,\mathbf{j}, \quad 0 \le t \le 4$

$\mathbf{r}_1'(t) = \mathbf{i} + \frac{1}{2\sqrt{t}}\mathbf{j}$

$\mathbf{F}(t) = (t^2 + t)\mathbf{i} - t\mathbf{j}$

$$\int_C \mathbf{F} \cdot d\mathbf{r} = \int_0^4 \left(t^2 + t - \frac{1}{2}\sqrt{t}\right) dt$$

$$= \left[\frac{t^3}{3} + \frac{t^2}{2} - \frac{t^{3/2}}{3}\right]_0^4 = \frac{80}{3}$$

b. $\mathbf{r}_2(w) = w^2\mathbf{i} + w\mathbf{j}, \quad 0 \le w \le 2$

$\mathbf{r}_2'(w) = 2w\mathbf{i} + \mathbf{j}$

$\mathbf{F}(w) = (w^4 + w^2)\mathbf{i} - w^2\mathbf{j}$

$$\int_C \mathbf{F} \cdot d\mathbf{r} = \int_0^2 [2w(w^4 + w^2) - w^2] \, dw$$

$$= \left[\frac{w^6}{3} + \frac{w^4}{2} - \frac{w^3}{3}\right]_0^2 = \frac{80}{3}$$

3. $\mathbf{F}(x, \ y) = y\mathbf{i} - x\mathbf{j}$

    a. $\mathbf{r}_1(\theta) = \sec \theta\mathbf{i} + \tan \theta\, \mathbf{j}, \ \ 0 \le \theta \le \dfrac{\pi}{3}$

    $\mathbf{r}_1'(\theta) = \sec \theta \tan \theta\mathbf{i} + \sec^2 \theta\, \mathbf{j}$

    $\mathbf{F}(\theta) = \tan \theta\mathbf{i} - \sec \theta\, \mathbf{j}$

$$\int_C \mathbf{F} \cdot d\mathbf{r} = \int_0^{\pi/3} (\sec \theta \tan^2 \theta - \sec^3 \theta)\, d\theta = \int_0^{\pi/3} [\sec \theta(\sec^2 \theta - 1) - \sec^3 \theta]\, d\theta$$

$$= -\int_0^{\pi/3} \sec \theta\, d\theta = \Big[-\ln|\sec \theta + \tan \theta| \Big]_0^{\pi/3} = -\ln\left(2 + \sqrt{3}\right) \approx -1.317$$

    b. $\mathbf{r}_2(t) = \sqrt{t+1}\,\mathbf{i} + \sqrt{t}\, \mathbf{j}, \ \ 0 \le t \le 3$

    $\mathbf{r}_2'(t) = \dfrac{1}{2\sqrt{t+1}}\mathbf{i} + \dfrac{1}{2\sqrt{t}}\mathbf{j}$

    $\mathbf{F}(t) = \sqrt{t}\,\mathbf{i} - \sqrt{t+1}\, \mathbf{j}$

$$\int_C \mathbf{F} \cdot d\mathbf{r} = \int_0^3 \left[\frac{\sqrt{t}}{2\sqrt{t+1}} - \frac{\sqrt{t+1}}{2\sqrt{t}} \right] dt = -\frac{1}{2}\int_0^3 \frac{1}{\sqrt{t}\sqrt{t+1}}\, dt = -\frac{1}{2}\int_0^3 \frac{1}{\sqrt{t^2 + t + (1/4) - (1/4)}}\, dt$$

$$= -\frac{1}{2}\int_0^3 \frac{1}{\sqrt{[t + (1/2)]^2 - (1/4)}}\, dt = \left[-\frac{1}{2}\ln\left|\left(t + \frac{1}{2}\right) + \sqrt{t^2 + t}\right| \right]_0^3$$

$$= -\frac{1}{2}\left[\ln\left(\frac{7}{2} + 2\sqrt{3} \right) - \ln\left(\frac{1}{2} \right) \right] = -\frac{1}{2}\ln\left(7 + 4\sqrt{3}\right) \approx -1.317$$

4. $\mathbf{F}(x, \ y) = y\mathbf{i} + x^2\mathbf{j}$

    a. $\mathbf{r}_1(t) = (2+t)\mathbf{i} + (3-t)\mathbf{j}, \ \ 0 \le t \le 3$

    $\mathbf{r}_1'(t) = \mathbf{i} - \mathbf{j}$

    $\mathbf{F}(t) = (3 - t)\mathbf{i} + (2+t)^2\mathbf{j}$

$$\int_C \mathbf{F} \cdot d\mathbf{r} = \int_0^3 [(3 - t) - (2+t)^2]\, dt = \left[-\frac{(3 - t)^2}{2} - \frac{(2+t)^3}{3} \right]_0^3 = -\frac{69}{2}$$

    b. $\mathbf{r}_2(w) = (2 + \ln w)\mathbf{i} + (3 - \ln w)\mathbf{j}, \ \ 1 \le w \le e^3$

    $\mathbf{r}_2'(w) = \dfrac{1}{w}\mathbf{i} - \dfrac{1}{w}\mathbf{j}$

    $\mathbf{F}(w) = (3 - \ln w)\mathbf{i} + (2 + \ln w)^2\mathbf{j}$

$$\int_C \mathbf{F} \cdot d\mathbf{r} = \int_1^{e^3} \left[(3 - \ln w)\left(\frac{1}{w} \right) - (2 + \ln w)^2\left(\frac{1}{w} \right) \right] dw = \left[-\frac{(3 - \ln w)^2}{2} - \frac{(2 + \ln w)^3}{3} \right]_1^{e^3} = -\frac{69}{2}$$

5. $\mathbf{F}(x, \ y) = 2xy\mathbf{i} + x^2\mathbf{j}$

    a. $\mathbf{r}_1(t) = t\mathbf{i} + t^2\mathbf{j}, \ \ 0 \le t \le 1$ **b.** $\mathbf{r}_2(t) = t\mathbf{i} + t^3\mathbf{j}, \ \ 0 \le t \le 1$

    $\mathbf{r}_1'(t) = \mathbf{i} + 2t\, \mathbf{j}$ $\mathbf{r}_2'(t) = \mathbf{i} + 3t^2\, \mathbf{j}$

    $\mathbf{F}(t) = 2t^3\mathbf{i} + t^2\mathbf{j}$ $\mathbf{F}(t) = 2t^4\mathbf{i} + t^2\mathbf{j}$

    $\displaystyle\int_C \mathbf{F} \cdot d\mathbf{r} = \int_0^1 4t^3\, dt = 1$ $\displaystyle\int_C \mathbf{F} \cdot d\mathbf{r} = \int_0^1 5t^4\, dt = 1$

6. $F(x, y) = ye^{xy}\mathbf{i} + xe^{xy}\mathbf{j}$

  a. $\mathbf{r}_1(t) = t\mathbf{i} - \dfrac{3}{2}(t - 2)\mathbf{j}, \quad 0 \le t \le 2$

    $\mathbf{r}_1'(t) = \mathbf{i} - \dfrac{3}{2}\mathbf{j}$

    $\mathbf{F}(t) = -\dfrac{3}{2}(t - 2)e^{-3t(t-2)/2}\mathbf{i} + te^{-3t(t-2)/2}\mathbf{j}$

    $\displaystyle\int_C \mathbf{F} \cdot d\mathbf{r} = \int_0^2 \left[-\dfrac{3}{2}(t - 2)e^{-3t(t-2)/2} - \dfrac{3}{2}te^{-3t(t-2)/2} \right] dt$

    $\displaystyle = \int_0^2 e^{-3t(t-2)/2}\left[-\dfrac{3}{2}(t - 2) - \dfrac{3}{2}t \right] dt = \left[e^{-3t(t-2)/2} \right]_0^2 = 1 - 1 = 0$

  b. $F(x, y) = ye^{xy}\mathbf{i} + xe^{xy}\mathbf{j}$ is conservative since

    $\dfrac{\partial M}{\partial y} = \dfrac{\partial N}{\partial x} = xye^{xy} + e^{xy}.$

    The potential function is $f(x, y) = e^{xy} + k.$

    $\displaystyle\int_C \mathbf{F} \cdot d\mathbf{r} = \left[e^{xy} \right]_{(0,3)}^{(2,0)} = 0$

7. $F(x, y) = y\mathbf{i} - x\mathbf{j}$

  a. $\mathbf{r}_1(t) = t\mathbf{i} + t\mathbf{j}, \quad 0 \le t \le 1$

    $\mathbf{r}_1'(t) = \mathbf{i} + \mathbf{j}$

    $\mathbf{F}(t) = t\mathbf{i} - t\mathbf{j}$

    $\displaystyle\int_C \mathbf{F} \cdot d\mathbf{r} = 0$

  b. $\mathbf{r}_2(t) = t\mathbf{i} + t^2\mathbf{j}, \quad 0 \le t \le 1$

    $\mathbf{r}_2'(t) = \mathbf{i} + 2t\mathbf{j}$

    $\mathbf{F}(t) = t^2\mathbf{i} - t\mathbf{j}$

    $\displaystyle\int_C \mathbf{F} \cdot d\mathbf{r} = \int_0^1 -t^2\, dt = -\dfrac{1}{3}$

  c. $\mathbf{r}_3(t) = t\mathbf{i} + t^3\mathbf{j}, \quad 0 \le t \le 1$

    $\mathbf{r}_3'(t) = \mathbf{i} + 3t^2\mathbf{j}$

    $\mathbf{F}(t) = t^3\mathbf{j} - t\mathbf{j}$

    $\displaystyle\int_C \mathbf{F} \cdot d\mathbf{r} = \int_0^1 -2t^3\, dt = -\dfrac{1}{2}$

8. $F(x, y) = xy^2\mathbf{i} + 2x^2y\mathbf{j}$

  a. $\mathbf{r}_1(t) = t\mathbf{i} + \dfrac{1}{t}\mathbf{j}, \quad 1 \le t \le 3$

    $\mathbf{r}_1'(t) = \mathbf{i} - \dfrac{1}{t^2}\mathbf{j}$

    $\mathbf{F}(t) = \dfrac{1}{t}\mathbf{i} + 2t\mathbf{j}$

    $\displaystyle\int_C \mathbf{F} \cdot d\mathbf{r} = \int_1^3 -\dfrac{1}{t}\, dt$

    $\displaystyle = \left[-\ln|t| \right]_1^3 = -\ln 3$

  b. $\mathbf{r}_2(t) = (t + 1)\mathbf{i} - \dfrac{1}{3}(t - 3)\mathbf{j}, \quad 0 \le t \le 2$

    $\mathbf{r}_2'(t) = \mathbf{i} - \dfrac{1}{3}\mathbf{j}$

    $\mathbf{F}(t) = \dfrac{1}{9}(t + 1)(t - 3)^2\mathbf{i} - \dfrac{2}{3}(t + 1)^2(t - 3)\mathbf{j}$

    $\displaystyle\int_C \mathbf{F} \cdot d\mathbf{r} = \int_0^2 \left[\dfrac{1}{9}(t + 1)(t - 3)^2 + \dfrac{2}{9}(t + 1)^2(t - 3) \right] dt$

    $\displaystyle = \dfrac{1}{9}\int_0^2 (3t^3 - 7t^2 - 7t + 3)\, dt$

    $\displaystyle = \dfrac{1}{9}\left[\dfrac{3t^4}{4} - \dfrac{7t^3}{3} - \dfrac{7t^2}{2} + 3t \right]_0^2 = -\dfrac{44}{27}$

9. $\int_C y^2\,dx + 2xy\,dy$

Since $\partial M/\partial y = \partial N/\partial x = 2y$, $\mathbf{F}(x,\ y) = y^2\mathbf{i} + 2xy\mathbf{j}$ is conservative. The potential function is $f(x,\ y) = xy^2 + k$. Therefore, we can use the Fundamental Theorem of Line Integrals.

a. $\displaystyle\int_C y^2\,dx + 2xy\,dy = \left[x^2 y\right]_{(0,0)}^{(4,4)} = 64$

b. $\displaystyle\int_C y^2\,dx + 2xy\,dy = \left[x^2 y\right]_{(-1,0)}^{(1,0)} = 0$

c. and d. Since C is a closed curve, $\displaystyle\int_C y^2\,dx + 2xy\,dy = 0$.

10. $\int_C (2x - 3y + 1)\,dx - (3x + y - 5)\,dy$

Since $\partial M/\partial y = \partial N/\partial x = -3$, $\mathbf{F}(x,\ y) = (2x - 3y + 1)\mathbf{i} - (3x + y - 5)\mathbf{j}$ is conservative. The potential function is $f(x,\ y) = x^2 - 3xy - (y^2/2) + x + 5y + k$.

a. and d. Since C is a closed curve, $\displaystyle\int_C (2x - 3y + 1)\,dx - (3x + y - 5)\,dy = 0$.

b. $\displaystyle\int_C (2x - 3y + 1)\,dx - (3x + y - 5)\,dy = \left[x^2 - 3xy - \frac{y^2}{2} + x + 5y\right]_{(0,-1)}^{(0,1)} = 10$

c. $\displaystyle\int_C (2x - 3y + 1)\,dx - (3x + y - 5)\,dy = \left[x^2 - 3xy - \frac{y^2}{2} + x + 5y\right]_{(0,1)}^{(2,e^2)} = \frac{1}{2}(3 - 2e^2 - e^4)$

11. $\int_C 2xy\,dx + (x^2 + y^2)\,dy$

Since $\partial M/\partial y = \partial N/\partial x = 2x$, $\mathbf{F}(x,\ y) = 2xy\mathbf{i} + (x^2 + y^2)\mathbf{j}$ is conservative. The potential function is $f(x,\ y) = x^2 y + (y^3/3) + k$.

a. $\displaystyle\int_C 2xy\,dx + (x^2 + y^2)\,dy = \left[x^2 y + \frac{y^3}{3}\right]_{(5,0)}^{(0,4)}$

$= \dfrac{64}{3}$

b. $\displaystyle\int_C 2xy\,dx + (x^2 + y^2)\,dy = \left[x^2 y + \frac{y^3}{3}\right]_{(2,0)}^{(0,4)}$

$= \dfrac{64}{3}$

12. $\int_C (x^2 + y^2)\,dx + 2xy\,dy$

Since $\partial M/\partial y = \partial N/\partial x = 2y$, $\mathbf{F}(x,\ y) = (x^2 + y^2)\mathbf{i} + 2xy\mathbf{j}$ is conservative. The potential function is $f(x,\ y) = (x^3/3) + xy^2 + k$.

a. $\displaystyle\int_C (x^2 + y^2)\,dx + 2xy\,dy = \left[\frac{x^3}{3} + xy^2\right]_{(0,0)}^{(8,4)}$

$= \dfrac{896}{3}$

b. $\displaystyle\int_C (x^2 + y^2)\,dx + 2xy\,dy = \left[\frac{x^3}{3} + xy^2\right]_{(2,0)}^{(0,2)}$

$= -\dfrac{8}{3}$

13. $\mathbf{F}(x,\ y,\ z) = yz\mathbf{i} + xz\mathbf{j} + xy\mathbf{k}$

Since **curl** $\mathbf{F} = \mathbf{0}$, $\mathbf{F}(x,\ y,\ z)$ is conservative. The potential function is $f(x,\ y,\ z) = xyz + k$.

a. $\mathbf{r}_1(t) = t\mathbf{i} + 2\mathbf{j} + t\mathbf{k},\quad 0 \le t \le 4$

$\displaystyle\int_C \mathbf{F}\cdot d\mathbf{r} = \left[xyz\right]_{(0,2,0)}^{(4,2,4)} = 32$

b. $\mathbf{r}_2(t) = t^2\mathbf{i} + t\mathbf{j} + t^2\mathbf{k},\quad 0 \le t \le 2$

$\displaystyle\int_C \mathbf{F}\cdot d\mathbf{r} = \left[xyz\right]_{(0,0,0)}^{(4,2,4)} = 32$

14. $\mathbf{F}(x,\ y,\ z) = \mathbf{i} + z\mathbf{j} + y\mathbf{k}$

Since **curl F = 0**, $\mathbf{F}(x,\ y,\ z)$ is conservative. The potential function is $f(x,\ y,\ z) = x + yz + k$.

a. $\mathbf{r}_1(t) = \cos t\,\mathbf{i} + \sin t\,\mathbf{j} + t^2\mathbf{k},\ \ 0 \le t \le \pi$

$$\int_C \mathbf{F} \cdot d\mathbf{r} = \left[x + yz\right]_{(1,0,0)}^{(-1,0,\pi^2)} = -2$$

b. $\mathbf{r}_2(t) = (1 - 2t)\mathbf{i} + \pi^2 t\mathbf{k},\ \ 0 \le t \le 1$

$$\int_C \mathbf{F} \cdot d\mathbf{r} = \left[x + yz\right]_{(1,0,0)}^{(-1,0,\pi^2)} = -2$$

15. $\mathbf{F}(x,\ y,\ z) = (2y + x)\mathbf{i} + (x^2 - z)\mathbf{j} + (2y - 4z)\mathbf{k}$

$\mathbf{F}(x,\ y,\ z)$ is not conservative.

a. $\mathbf{r}_1(t) = t\mathbf{i} + t^2\mathbf{j} + \mathbf{k},\ \ 0 \le t \le 1$

$\mathbf{r}_1'(t) = \mathbf{i} + 2t\,\mathbf{j}$

$\mathbf{F}(t) = (2t^2 + t)\mathbf{i} + (t^2 - 1)\mathbf{j} + (2t^2 - 4)\mathbf{k}$

$$\int_C \mathbf{F} \cdot d\mathbf{r} = \int_0^1 (2t^3 + 2t^2 - t)\,dt = \frac{2}{3}$$

b. $\mathbf{r}_2(t) = t\mathbf{i} + t\mathbf{j} + (2t - 1)^2\mathbf{k},\ \ 0 \le t \le 1$

$\mathbf{r}_2'(t) = \mathbf{i} + \mathbf{j} + 4(2t - 1)\mathbf{k}$

$\mathbf{F}(t) = 3t\mathbf{i} + [t^2 - (2t - 1)^2]\mathbf{j} + [2t - 4(2t - 1)^2]\mathbf{k}$

$$\int_C \mathbf{F} \cdot d\mathbf{r} = \int_0^1 [3t + t^2 - (2t - 1)^2 + 8t(2t - 1) - 16(2t - 1)^3]\,dt$$

$$= \int_0^1 [17t^2 - 5t - (2t - 1)^2 - 16(2t - 1)^3]\,dt = \left[\frac{17t^3}{3} - \frac{5t^2}{2} - \frac{(2t - 1)^3}{6} - 2(2t - 1)^4\right]_0^1 = \frac{17}{6}$$

16. $\mathbf{F}(x,\ y,\ z) = -y\mathbf{i} + x\mathbf{j} + 3xz^2\mathbf{k}$

$\mathbf{F}(x,\ y,\ z)$ is not conservative.

a. $\mathbf{r}_1(t) = \cos t\,\mathbf{i} + \sin t\,\mathbf{j} + t\mathbf{k},\ \ 0 \le t \le \pi$

$\mathbf{r}_1'(t) = -\sin t\,\mathbf{i} + \cos t\,\mathbf{j} + \mathbf{k}$

$\mathbf{F}(t) = -\sin t\,\mathbf{i} + \cos t\,\mathbf{j} + 3t^2 \cos t\,\mathbf{k}$

$$\int_C \mathbf{F} \cdot d\mathbf{r} = \int_0^\pi [\sin^2 t + \cos^2 t + 3t^2 \cos t]\,dt = \int_0^\pi [1 + 3t^2 \cos t]\,dt$$

$$= \left[t\right]_0^\pi + 3\left[t^2 \sin t\right]_0^\pi - 2\int_0^\pi t \sin t\,dt = \left[t + 3t^2 \sin t - 6(\sin t - t \cos t)\right]_0^\pi = -5\pi$$

b. $\mathbf{r}_2(t) = (1 - 2t)\mathbf{i} + \pi t\mathbf{k},\ \ 0 \le t \le 1$

$\mathbf{r}_2'(t) = -2\mathbf{i} + \pi\mathbf{k}$

$\mathbf{F}(t) = (1 - 2t)\mathbf{j} + 3\pi^2 t^2(1 - 2t)\mathbf{k}$

$$\int_C \mathbf{F} \cdot d\mathbf{r} = \int_0^1 3\pi^3 t^2(1 - 2t)\,dt = 3\pi^3 \int_0^1 (t^2 - 2t^3)\,dt = 3\pi^3 \left[\frac{t^3}{3} - \frac{t^4}{2}\right]_0^1 = -\frac{\pi^3}{2}$$

17. $\mathbf{F}(x,\ y,\ z) = e^z(y\mathbf{i} + x\mathbf{j} + xy\mathbf{k})$

$\mathbf{F}(x,\ y,\ z)$ is conservative. The potential function is $f(x,\ y,\ z) = xye^z + k$.

a. $\mathbf{r}_1(t) = 4\cos t\,\mathbf{i} + 4\sin t\,\mathbf{j} + 3\mathbf{k},\ \ 0 \le t \le \pi$

$$\int_C \mathbf{F} \cdot d\mathbf{r} = \left[xye^z\right]_{(4,0,3)}^{(-4,0,3)} = 0$$

b. $\mathbf{r}_2(t) = (4 - 8t)\mathbf{i} + 3\mathbf{k},\ \ 0 \le t \le 1$

$$\int_C \mathbf{F} \cdot d\mathbf{r} = \left[xye^z\right]_{(4,0,3)}^{(-4,0,3)} = 0$$

18. $F(x, \ y, \ z) = y \sin z \mathbf{i} + x \sin z \mathbf{j} + xy \cos z \mathbf{k}$

   a. $\mathbf{r}_1(t) = t^2\mathbf{i} + t^2\mathbf{j}, \ 0 \le t \le 2$

      $\mathbf{r}_1'(t) = 2t\mathbf{i} + 2t\mathbf{j}$

      $F(t) = t^4 \cos t^2 \mathbf{k}$

      $\displaystyle \int_C F \cdot d\mathbf{r} = \int_0^2 0\,dt = 0$

   b. $\mathbf{r}_2(t) = 4t\mathbf{i} + 4t\mathbf{j}, \ 0 \le t \le 1$

      $\mathbf{r}_1(t) = 4\mathbf{i} + 4\mathbf{j}$

      $F(t) = 16t^2 \cos(4t)\mathbf{k}$

      $\displaystyle \int_C F \cdot d\mathbf{r} = \int_0^1 0\,dt = 0$

19. $\displaystyle \int_C (y\mathbf{i} + x\mathbf{j}) \cdot d\mathbf{r} = \Big[xy\Big]_{(0,0)}^{(3,8)} = 24$

20. $\displaystyle \int_C [2(x+y)\mathbf{i} + 2(x+y)\mathbf{j}] \cdot d\mathbf{r} = \Big[(x+y)^2\Big]_{(-1,1)}^{(3,2)}$

$$= 25$$

21. $\displaystyle \int_C \cos x \sin y\,dx + \sin x \cos y\,dy = \Big[\sin x \sin y\Big]_{(0,-\pi)}^{(3\pi/2,\pi/2)} = -1$

22. $\displaystyle \int_C \frac{y\,dx - x\,dy}{x^2 + y^2} = \Big[\arctan\Big(\frac{x}{y}\Big)\Big]_{(1,1)}^{(2\sqrt{3},2)} = \frac{\pi}{3} - \frac{\pi}{4} = \frac{\pi}{12}$

23. $\displaystyle \int_C e^x \sin y\,dx + e^x \cos y\,dy = \Big[e^x \sin y\Big]_{(0,0)}^{(2\pi,0)} = 0$

24. $\displaystyle \int_C \frac{2x}{(x^2+y^2)^2}\,dx + \frac{2y}{(x^2+y^2)^2}\,dy = \Big[-\frac{1}{x^2+y^2}\Big]_{(7,5)}^{(1,5)} = -\frac{1}{26} + \frac{1}{74} = \frac{-12}{481}$

25. $\displaystyle \int_C (z + 2y)\,dx + (2x - z)\,dy + (x - y)\,dz$

Note: Since $F(x, \ y, \ z) = (z + 2y)\mathbf{i} + (2x - z)\mathbf{j} + (x - y)\mathbf{k}$ is conservative and the potential function is $f(x, \ y, \ z) = xz + 2xy - yz + k$, the integral is independent of path as illustrated below.

   a. $\Big[xz + 2xy - yz\Big]_{(0,0,0)}^{(1,1,1)} = 2$

   b. $\Big[xz + 2xy - yz\Big]_{(0,0,0)}^{(0,0,1)} + \Big[xz + 2xy - yz\Big]_{(0,0,1)}^{(1,1,1)} = 0 + 2 = 2$

   c. $\Big[xz + 2xy - yz\Big]_{(0,0,0)}^{(1,0,0)} + \Big[xz + 2xy - yz\Big]_{(1,0,0)}^{(1,1,0)} + \Big[xz + 2xy - yz\Big]_{(1,1,0)}^{(1,1,1)} = 0 + 2 + 0 = 2$

26. $\displaystyle \int_C zy\,dx + xz\,dy + xy\,dz$

Note: Since $F(x, \ y, \ z) = yz\mathbf{i} + xz\mathbf{j} + xy\mathbf{k}$ is conservative and the potential function is $f(x, \ y, \ z) = xyz + k$, the integral is independent of path as illustrated below.

   a. $\Big[xyz\Big]_{(0,0,0)}^{(1,1,1)} = 1$

   b. $\Big[xyz\Big]_{(0,0,0)}^{(0,0,1)} + \Big[xyz\Big]_{(0,0,1)}^{(1,1,1)} = 0 + 1 = 1$

   c. $\Big[xyz\Big]_{(0,0,0)}^{(1,0,0)} + \Big[xyz\Big]_{(1,0,0)}^{(1,1,0)} + \Big[xyz\Big]_{(1,1,0)}^{(1,1,1)} = 0 + 0 + 1 = 1$

27. $\displaystyle\int_C -\sin x\, dx + z\, dy + y\, dz = \Big[\cos x + yz\Big]_{(0,0,0)}^{(\pi/2,3,4)} = 11$

28. $\displaystyle\int_C 6x\, dx - 4z\, dy - (4y - 20z)\, dz = \Big[3x^2 - 4yz + 10z^2\Big]_{(0,0,0)}^{(3,4,0)} = 27$

29. $\mathbf{F}(x,\ y) = 9x^2 y^2 \mathbf{i} + (6x^3 y - 1)\mathbf{j}$ is conservative.

$\text{Work} = \Big[3x^3 y^2 - y\Big]_{(0,0)}^{(5,9)} = 30,366$

30. $\mathbf{F}(x,\ y) = \dfrac{2x}{y}\mathbf{i} - \dfrac{x^2}{y^2}\mathbf{j}$ is conservative.

$\text{Work} = \Big[\dfrac{x^2}{y}\Big]_{(-1,1)}^{(3,2)} = \dfrac{7}{2}$

31. $\mathbf{r}(t) = 2\cos 2\pi t\, \mathbf{i} + 2\sin 2\pi t\, \mathbf{j}$

$\mathbf{r}'(t) = -4\pi \sin 2\pi t\, \mathbf{i} + 4\pi \cos 2\pi t\, \mathbf{j}$

$\mathbf{a}(t) = -8\pi^2 \cos 2\pi t\, \mathbf{i} - 8\pi^2 \sin 2\pi t\, \mathbf{j}$

$\mathbf{F}(t) = m \cdot \mathbf{a}(t) = \dfrac{1}{32}\mathbf{a}(t) = -\dfrac{\pi^2}{4}(\cos 2\pi t\, \mathbf{i} + \sin 2\pi t\, \mathbf{j})$

$W = \displaystyle\int_C \mathbf{F} \cdot d\mathbf{r} = \int_C -\dfrac{\pi^2}{4}(\cos 2\pi t\, \mathbf{i} + \sin 2\pi t\, \mathbf{j}) \cdot 4\pi(-\sin 2\pi t\, \mathbf{i} + \cos 2\pi t\, \mathbf{j})\, dt = -\pi^3 \int_C 0\, dt = 0$

32. $\mathbf{F}(x,\ y,\ z) = a_1 \mathbf{i} + a_2 \mathbf{j} + a_3 \mathbf{k}$

Since $\mathbf{F}(x,\ y,\ z)$ is conservative, the work done in moving a particle along any path from P to Q is

$f(x,\ y,\ z) = \Big[a_1 x + a_2 y + a_3 z\Big]_{P=(p_1,p_2,p_3)}^{Q=(q_1,q_2,q_3)}$

$\qquad = a_1(q_1 - p_1) + a_2(q_2 - p_2) + a_3(q_3 - p_3) = \mathbf{F} \cdot \overrightarrow{PQ}.$

33. Since the sum of the potential and kinetic energies remains constant from point to point, if the kinetic energy is decreasing at a rate of 10 units per minute, then the potential energy is increasing at a rate of 10 units per minute.

34. $\mathbf{F} = -150\mathbf{j}$

a. $\mathbf{r}(t) = t\mathbf{i} + (50 - t)\mathbf{j},\ \ 0 \le t \le 50$

$d\mathbf{r} = (\mathbf{i} - \mathbf{j})\, dt$

$\displaystyle\int_C \mathbf{F} \cdot d\mathbf{r} = \int_0^{50} 150\, dt = 7500\ \text{ft} \cdot \text{lbs}$

b. $\mathbf{r}(t) = t\mathbf{i} + \frac{1}{50}(50 - t)^2 \mathbf{j}$

$d\mathbf{r} = \big(\mathbf{i} - \frac{1}{25}(50 - t)\mathbf{j}\big)\, dt$

$\displaystyle\int_C \mathbf{F} \cdot d\mathbf{r} = 6\int_0^{50} (50 - t)\, dt = 7500\ \text{ft} \cdot \text{lbs}$

35. No. The force field is conservative.

36. $F(x, y) = \dfrac{y}{x^2 + y^2}\mathbf{i} - \dfrac{x}{x^2 + y^2}\mathbf{j}$

a. $M = \dfrac{y}{x^2 + y^2}, \quad \dfrac{\partial M}{\partial y} = \dfrac{(x^2 + y^2)(1) - y(2y)}{(x^2 + y^2)^2} = \dfrac{x^2 - y^2}{(x^2 + y^2)^2}$

$N = -\dfrac{x}{x^2 + y^2}, \quad \dfrac{\partial N}{\partial x} = \dfrac{(x^2 + y^2)(-1) + x(2x)}{(x^2 + y^2)^2} = \dfrac{x^2 - y^2}{(x^2 + y^2)^2}$

Thus, $\dfrac{\partial N}{\partial x} = \dfrac{\partial M}{\partial y}$.

b. $\mathbf{r}(t) = \cos t\,\mathbf{i} + \sin t\,\mathbf{j}, \quad 0 \le t \le \pi$

$\mathbf{F} = \sin t\,\mathbf{i} - \cos t\,\mathbf{j}$

$d\mathbf{r} = (-\sin t\,\mathbf{i} + \cos t\,\mathbf{j})\,dt$

$\displaystyle\int_C \mathbf{F} \cdot d\mathbf{r} = \int_0^\pi (-\sin^2 t - \cos^2 t)\,dt$

$= \Big[-t\Big]_0^\pi = -\pi$

c. $\mathbf{r}(t) = \cos t\,\mathbf{i} - \sin t\,\mathbf{j}, \quad 0 \le t \le \pi$

$\mathbf{F} = -\sin t\,\mathbf{i} - \cos t\,\mathbf{j}$

$d\mathbf{r} = (-\sin t\,\mathbf{i} - \cos t\,\mathbf{j})\,dt$

$\displaystyle\int_C \mathbf{F} \cdot d\mathbf{r} = \int_0^\pi (\sin^2 t + \cos^2 t)\,dt$

$= \Big[t\Big]_0^\pi = \pi$

d. $\mathbf{r}(t) = \cos t\,\mathbf{i} + \sin t\,\mathbf{j}, \quad 0 \le t \le 2\pi$

$\mathbf{F} = \sin t\,\mathbf{i} - \cos t\,\mathbf{j}$

$d\mathbf{r} = (-\sin t\,\mathbf{i} + \cos t\,\mathbf{j})\,dt$

$\displaystyle\int_C \mathbf{F} \cdot d\mathbf{r} = \int_0^{2\pi} (-\sin^2 t - \cos^2 t)\,dt = \Big[-t\Big]_0^{2\pi} = -2\pi$

This does not contradict Theorem 15.7 since $\mathbf{F}$ is not continuous at $(0, 0)$ in R enclosed by curve C.

e. $\nabla\left(\arctan \dfrac{x}{y}\right) = \dfrac{1/y}{1 + (x/y)^2}\mathbf{i} + \dfrac{-x/y^2}{1 + (x/y)^2}\mathbf{j} = \dfrac{y}{x^2 + y^2}\mathbf{i} - \dfrac{x}{x^2 + y^2}\mathbf{j} = \mathbf{F}$

37. False, it would be true if $\mathbf{F}$ were conservative.

38. True

39. True

40. False, the requirement is $\dfrac{\partial M}{\partial y} = \dfrac{\partial N}{\partial x}$.

41. Let

$\mathbf{F} = M\mathbf{i} + N\mathbf{j} = \dfrac{\partial f}{\partial y}\mathbf{i} - \dfrac{\partial f}{\partial x}\mathbf{j}.$

Then

$\dfrac{\partial M}{\partial y} = \dfrac{\partial}{\partial y}\left(\dfrac{\partial f}{\partial y}\right) = \dfrac{\partial^2 f}{\partial y^2} \quad \text{and} \quad \dfrac{\partial N}{\partial x} = \dfrac{\partial}{\partial x}\left(-\dfrac{\partial f}{\partial x}\right) = -\dfrac{\partial^2 f}{\partial x^2}.$

Since

$\dfrac{\partial^2 f}{\partial x^2} + \dfrac{\partial^2 f}{\partial y^2} = 0$ we have $\dfrac{\partial M}{\partial y} = \dfrac{\partial N}{\partial x}$.

Thus, $\mathbf{F}$ is conservative. Therefore, by Theorem 15.7, we have

$\displaystyle\int_C \left(\dfrac{\partial f}{\partial y}\,dx - \dfrac{\partial f}{\partial x}\,dy\right) = \int_C (M\,dx + N\,dy) = \int_C \mathbf{F} \cdot d\mathbf{r} = 0$

for every closed curve in the plane.

Section 15.4 Green's Theorem

1. $r(t) = \begin{cases} t\mathbf{i}, & 0 \le t \le 4 \\ 4\mathbf{i} + (t-4)\mathbf{j}, & 4 \le t \le 8 \\ (12-t)\mathbf{i} + 4\mathbf{j}, & 8 \le t \le 12 \\ (16-t)\mathbf{j}, & 12 \le t \le 16 \end{cases}$

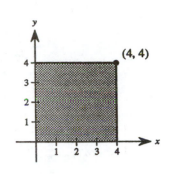

$\displaystyle\int_C y^2\,dx + x^2\,dy = \int_0^4 [0\,dt + t^2(0)] + \int_4^8 [(t-4)^2(0) + 16\,dt]$

$\displaystyle\qquad\qquad + \int_8^{12}[16(-dt) + (12-t)^2(0)] + \int_{12}^{16}[(16-t)^2(0) + 0(-dt)]$

$\displaystyle\qquad = 0 + 64 - 64 + 0 = 0$

By Green's Theorem,

$$\iint_R \left(\frac{\partial N}{\partial x} - \frac{\partial M}{\partial y}\right) dA = \int_0^4\int_0^4 (2x - 2y)\,dy\,dx = \int_0^4 (8x - 16)\,dx = 0.$$

2. $r(t) = \begin{cases} t\mathbf{i}, & 0 \le t \le 4 \\ 4\mathbf{i} + (t-4)\mathbf{j}, & 4 \le t \le 8 \\ (12-t)\mathbf{i} + (12-t)\mathbf{j}, & 8 \le t \le 12 \end{cases}$

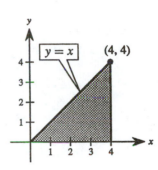

$\displaystyle\int_C y^2\,dx + x^2\,dy = \int_0^4 [0\,dt + t^2(0)] + \int_4^8 [(t-4)^2(0) + 16\,dt]$

$\displaystyle\qquad\qquad + \int_8^{12}[(12-t)^2(-dt) + (12-t)^2(-dt)]$

$\displaystyle\qquad = 0 + 64 - \frac{128}{3} = \frac{64}{3}$

By Green's Theorem,

$$\iint_R \left(\frac{\partial N}{\partial x} - \frac{\partial M}{\partial y}\right) dA = \int_0^4\int_0^x (2x - 2y)\,dy\,dx = \int_0^4 x^2\,dx = \frac{64}{3}.$$

3. $r(t) = \begin{cases} t\mathbf{i} + \dfrac{t^2}{4}\mathbf{j}, & 0 \le t \le 4 \\ (8-t)\mathbf{i} + (8-t)\mathbf{j}, & 4 \le t \le 8 \end{cases}$

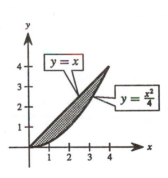

$\displaystyle\int_C y^2\,dx + x^2\,dy$

$\displaystyle\qquad = \int_0^4 \left[\frac{t^4}{16}(dt) + t^2\left(\frac{t}{2}\,dt\right)\right] + \int_4^8 [(8-t)^2(-dt) + (8-t)^2(-dt)]$

$\displaystyle\qquad = \int_0^4 \left[\frac{t^4}{16} + \frac{t^3}{2}\right] dt + \int_4^8 -2(8-t)^2\,dt = \frac{32}{15}$

By Green's Theorem,

$$\iint_R \left(\frac{\partial N}{\partial x} - \frac{\partial M}{\partial y}\right) dA = \int_0^4\int_{x^2/4}^x (2x - 2y)\,dy\,dx = \int_0^4 \left(x^2 - \frac{x^3}{2} + \frac{x^4}{16}\right) dx = \frac{32}{15}.$$

4. $\mathbf{r}(t) = \cos t \, \mathbf{i} + \sin t \, \mathbf{j}, \quad 0 \le t \le 2\pi$

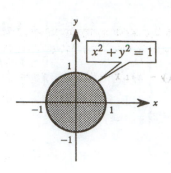

$$\int_C y^2 \, dx + x^2 \, dy = \int_0^{2\pi} [\sin^2 t(-\sin t \, dt) + \cos^2 t(\cos t \, dt)]$$

$$= \int_0^{2\pi} (\cos^3 t - \sin^3 t) \, dt$$

$$= \int_0^{2\pi} [\cos t(1 - \sin^2 t) - \sin t(1 - \cos^2 t)] \, dt$$

$$= \left[\sin t - \frac{\sin^3 t}{3} + \cos t - \frac{\cos^3 t}{3} \right]_0^{2\pi} = 0$$

By Green's Theorem,

$$\iint_R \left(\frac{\partial N}{\partial x} - \frac{\partial M}{\partial y} \right) dA = \int_{-1}^1 \int_{-\sqrt{1-x^2}}^{\sqrt{1-x^2}} (2x - 2y) \, dy \, dx$$

$$= \int_0^{2\pi} \int_0^1 (2r \cos \theta - 2r \sin \theta) r \, dr \, d\theta = \frac{2}{3} \int_0^{2\pi} (\cos \theta - \sin \theta) \, d\theta = \frac{2}{3}(0) = 0.$$

5. $C: x^2 + y^2 = 4$

Let $x = 2 \cos t$ and $y = 2 \sin t, \quad 0 \le t \le 2\pi$.

$$\int_C xe^y \, dx + e^x \, dy = \int_0^{2\pi} [2 \cos t e^{2 \sin t}(-2 \sin t) + e^{2 \cos t}(2 \cos t)] \, dt \approx 19.99$$

$$\iint_R \left(\frac{\partial N}{\partial x} - \frac{\partial M}{\partial y} \right) dA = \int_{-2}^2 \int_{-\sqrt{4-x^2}}^{\sqrt{4-x^2}} (e^x - xe^y) \, dy \, dx$$

$$= \int_{-2}^2 \left[2\sqrt{4-x^2} \, e^x - xe^{\sqrt{4-x^2}} + xe^{-\sqrt{4-x^2}} \right] dx \approx 19.99$$

6. $C:$ boundary of the region lying between the graphs of $y = x$ and $y = x^3$

$$\int_C xe^y \, dx + e^x \, dy = \int_0^1 (xe^{x^3} + 3x^2 e^x) \, dx + \int_1^0 (xe^x + e^x) \, dx \approx 2.936 - 2.718 \approx 0.22$$

$$\iint_R \left(\frac{\partial N}{\partial x} - \frac{\partial M}{\partial y} \right) dA = \int_0^1 \int_{x^3}^x (e^x - xe^y) \, dy \, dx = \int_0^1 (xe^{x^3} - x^3 e^x) \, dx \approx 0.22$$

7.
$$\int_C (y - x) \, dx + (2x - y) \, dy = \int_0^2 \int_{x^2 - x}^x dy \, dx$$

$$= \int_0^2 (2x - x^2) \, dx$$

$$= \frac{4}{3}$$

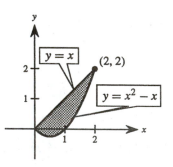

8. Since C is an ellipse with $a = 2$ and $b = 1$, then R is an ellipse of area $\pi ab = 2\pi$. Thus, Green's Theorem yields

$$\int_C (y - x) \, dx + (2x - y) \, dy = \iint_R 1 \, dA = \text{Area of ellipse} = 2\pi.$$

9. From the accompanying figure, we see that R is the shaded region. Thus, Green's Theorem yields

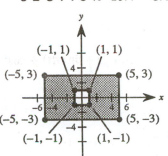

$$\int_C (y-x)\,dx + (2x-y)\,dy = \int\int_R 1\,dA$$

$$= \text{Area of } R$$

$$= 6(10) - 2(2)$$

$$= 56.$$

10. R is the shaded region of the accompanying figure. Thus, Green's Theorem yields

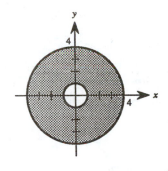

$$\int_C (y-x)\,dx + (2x-y)\,dy = \int\int_R 1\,dA$$

$$= \text{Area between the circles}$$

$$= 16\pi - \pi$$

$$= 15\pi.$$

11. Since the curves $y=0$ and $y=4-x^2$ intersect at $(-2,\ 0)$ and $(2,\ 0)$, Green's Theorem yields

$$\int_C 2xy\,dx + (x+y)\,dy = \int\int_R (1-2x)\,dA = \int_{-2}^{2}\int_{0}^{4-x^2} (1-2x)\,dy\,dx$$

$$= \int_{-2}^{2}\Big[y - 2xy\Big]_{0}^{4-x^2}\,dx$$

$$= \int_{-2}^{2} (4 - 8x - x^2 + 2x^3)\,dx$$

$$= \Big[4x - 4x^2 - \frac{x^3}{3} + \frac{x^4}{2}\Big]_{-2}^{2}$$

$$= -\frac{8}{3} - \frac{8}{3} + 16 = \frac{32}{3}.$$

12. The given curves intersect at $(0,\ 0)$ and $(4,\ 2)$. Thus, Green's Theorem yields

$$\int_C y^2\,dx + xy\,dy = \int\int_R (y - 2y)\,dA$$

$$= \int_0^4\int_0^{\sqrt{x}} -y\,dy\,dx = \int_0^4\Big[\frac{-y^2}{2}\Big]_0^{\sqrt{x}}\,dx = \int_0^4 \frac{-x}{2}\,dx = \Big[\frac{-x^2}{4}\Big]_0^4 = -4.$$

13. Since R is the interior of the circle $x^2 + y^2 = a^2$, Green's Theorem yields

$$\int_C (x^2 - y^2)\,dx + 2xy\,dy = \int\int_R (2y + 2y)\,dA$$

$$= \int_{-a}^{a}\int_{-\sqrt{a^2-x^2}}^{\sqrt{a^2-x^2}} 4y\,dy\,dx = 4\int_{-a}^{a} 0\,dx = 0.$$

14. In this case, let $y = r\sin\theta$, $x = r\cos\theta$. Then $dA = r\,dr\,d\theta$ and Green's Theorem yields

$$\int_C (x^2 - y^2)\,dx + 2xy\,dy = \iint_R 4y\,dA = 4\int_0^{2\pi}\int_0^{1+\cos\theta} r\sin\theta\, r\,dr\,d\theta$$

$$= 4\int_0^{2\pi}\int_0^{1+\cos\theta} r^2\sin\theta\,dr\,d\theta$$

$$= \frac{4}{3}\int_0^{2\pi} \sin\theta(1+\cos\theta)^3\,d\theta$$

$$= \left[-\frac{(1+\cos\theta)^4}{3}\right]_0^{2\pi} = 0.$$

15. Since $\dfrac{\partial M}{\partial y} = \dfrac{2x}{x^2+y^2} = \dfrac{\partial N}{\partial x}$,

we have path independence and

$$\iint_R \left(\frac{\partial N}{\partial x} - \frac{\partial M}{\partial y}\right)dA = 0.$$

16. Since $\dfrac{\partial M}{\partial y} = 2e^x\cos 2y = \dfrac{\partial N}{\partial x}$ we have

$$\iint_R \left(\frac{\partial N}{\partial x} - \frac{\partial M}{\partial y}\right)dA = 0.$$

17. By Green's Theorem,

$$\int_C \sin x\cos y\,dx + (xy + \cos x\sin y)\,dy = \iint_R [(y - \sin x\sin y) - (-\sin x\sin y)]\,dA$$

$$= \int_0^1\int_x^{\sqrt{x}} y\,dy\,dx = \frac{1}{2}\int_0^1 (x - x^2)\,dx = \frac{1}{2}\left[\frac{x^2}{2} - \frac{x^3}{3}\right]_0^1 = \frac{1}{12}.$$

18. By Green's Theorem,

$$\int_C (e^{-x^2/2} - y)\,dx + (e^{-y^2/2} + x)\,dy = \iint_R 2\,dA = 2(\text{Area of }R) = 2[\pi(5)^2 - \pi(2)(1)] = 46\pi.$$

19. By Green's Theorem,

$$\int_C xy\,dx + (x+y)\,dy = \iint_R (1-x)\,dA$$

$$= \int_0^{2\pi}\int_1^3 (1 - r\cos\theta)r\,dr\,d\theta = \int_0^{2\pi}\left(4 - \frac{26}{3}\cos\theta\right)d\theta = 8\pi.$$

20. By Green's Theorem,

$$\int_C 3x^2 e^y\,dx + e^y\,dy = \iint_R -3x^2 e^y\,dA$$

$$= \int_1^2\int_{-2}^2 -3x^2 e^y\,dy\,dx + \int_{-1}^1\int_1^2 -3x^2 e^y\,dy\,dx$$

$$+ \int_{-2}^{-1}\int_{-2}^2 -3x^2 e^y\,dy\,dx + \int_{-1}^1\int_{-2}^{-1} -3x^2 e^y\,dy\,dx$$

$$= -7(e^2 - e^{-2}) - 2(e^2 - e) - 7(e^2 - e^{-2}) - 2(e^{-1} - e^{-2})$$

$$= -16e^2 + 16e^{-2} + 2e - 2e^{-1}.$$

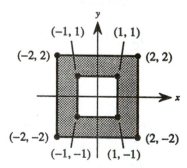

21. $\mathbf{F}(x, \ y) = xy\mathbf{i} + (x + y)\mathbf{j}$

$C: \ x^2 + y^2 = 4$

$$\text{Work} = \int_C xy\,dx + (x + y)\,dy = \iint_R (1 - x)\,dA = \int_0^{2\pi} \int_0^2 (1 - r\cos\theta)r\,dr\,d\theta = \int_0^{2\pi} \left(2 - \frac{8}{3}\cos\theta\right)d\theta = 4\pi$$

22. $\mathbf{F}(x, \ y) = (e^x - 3y)\mathbf{i} + (e^y + 6x)\mathbf{j}$

$C: \ r = 2\cos\theta$

$$\text{Work} = \int_C (e^x - 3y)\,dx + (e^y + 6x)\,dy = \iint_R 9\,dA = 9\pi \text{ since } r = 2\cos\theta \text{ is a circle with a radius of one.}$$

23. $\mathbf{F}(x, \ y) = (x^{3/2} - 3y)\mathbf{i} + \left(6x + 5\sqrt{y}\right)\mathbf{j}$

$C:$ boundary of the triangle with vertices (0, 0), (5, 0), (0, 5)

$$\text{Work} = \int_C (x^{3/2} - 3y)\,dx + \left(6x + 5\sqrt{y}\right)dy = \iint_R 9\,dA = 9(\tfrac{1}{2})(5)(5) = \tfrac{225}{2}$$

24. $\mathbf{F}(x, \ y) = (3x^2 + y)\mathbf{i} + 4xy^2\mathbf{j}$

$C:$ boundary of the region bounded by the graphs of $y = \sqrt{x}, \ \ y = 0, \ \ x = 4$

$$\text{Work} = \int_C (3x^2 + y)\,dx + 4xy^2\,dy = \int_0^4 \int_0^{\sqrt{x}} (4y^2 - 1)\,dy\,dx = \int_0^4 \left(\tfrac{4}{3}x^{3/2} - x^{1/2}\right)dx = \tfrac{176}{15}$$

25. $C:$ let $x = a\cos t, \ \ y = a\sin t, \ \ 0 \le t \le 2\pi$. By Theorem 15.9, we have

$$A = \frac{1}{2}\int_C x\,dy - y\,dx = \frac{1}{2}\int_0^{2\pi} [a\cos t(a\cos t) - a\sin t(-a\sin t)]\,dt = \frac{1}{2}\int_0^{2\pi} a^2\,dt = \left[\frac{a^2}{2}t\right]_0^{2\pi} = \pi a^2.$$

26. From the accompanying figure we see that for

$$C_1: \ y = \frac{2}{3}x, \ \ dy = \frac{2}{3}dx$$

$$C_2: \ y = -\frac{1}{3}x + 3, \ \ dy = -\frac{1}{3}dx$$

$$C_3: \ x = 0, \ \ dx = 0.$$

Thus, by Theorem 15.9, we have

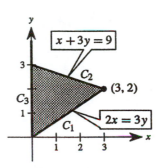

$$A = \frac{1}{2}\int_0^3 \left(\frac{2x}{3} - \frac{2x}{3}\right)dx + \frac{1}{2}\int_3^0 \left[\frac{-x}{3} - \left(\frac{-x}{3} + 3\right)\right]dx + \frac{1}{2}\int_3^0 0\,dy - y(0)$$

$$= \frac{1}{2}(0) + \frac{1}{2}\int_3^0 -3\,dx + \frac{1}{2}(0) = \left[\frac{-3}{2}x\right]_3^0 = \frac{9}{2}.$$

27. From the accompanying figure we see that for

$$C_1: \ y = 2x + 1, \ \ dy = 2\,dx$$

$$C_2: \ y = 4 - x^2, \ \ dy = -2x\,dx.$$

Thus, by Theorem 15.9, we have

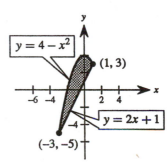

$$A = \frac{1}{2}\int_{-3}^1 [x(2) - (2x + 1)]\,dx + \frac{1}{2}\int_1^{-3} [x(-2x) - (4 - x^2)]\,dx$$

$$= \frac{1}{2}\int_{-3}^1 (-1)\,dx + \frac{1}{2}\int_1^{-3} (-x^2 - 4)\,dx$$

$$= \frac{1}{2}\int_{-3}^1 (-1)\,dx + \frac{1}{2}\int_{-3}^1 (x^2 + 4)\,dx = \frac{1}{2}\int_{-3}^1 (3 + x^2)\,dx = \frac{1}{2}\left[3x + \frac{x^3}{3}\right]_{-3}^1 = \frac{32}{3}.$$

28. Since the loop of the folium is formed on the interval $0 \le t < \infty$,

$$dx = \frac{3(1 - 2t^3)}{(t^3 + 1)^2} \, dt \quad \text{and} \quad dy = \frac{3(2t - t^4)}{(t^3 + 1)^2} \, dt,$$

we have

$$A = \frac{1}{2} \int_0^\infty \left[\left(\frac{3t}{t^3 + 1} \right) \frac{3(2t - t^4)}{(t^3 + 1)^2} - \left(\frac{3t^2}{t^3 + 1} \right) \frac{3(1 - 2t^3)}{(t^3 + 1)^2} \right] dt$$

$$= \frac{9}{2} \int_0^\infty \frac{t^5 + t^2}{(t^3 + 1)^3} \, dt = \frac{9}{2} \int_0^\infty \frac{t^2(t^3 + 1)}{(t^3 + 1)^3} \, dt = \frac{3}{2} \int_0^\infty 3t^2(t^3 + 1)^{-2} \, dt = \left[\frac{-3}{2(t^3 + 1)} \right]_0^\infty = \frac{3}{2}.$$

29. For the moment about the x-axis, $M_x = \iint_R y \, dA$. Let $N = 0$ and $M = -y^2/2$. By Green's Theorem,

$$M_x = \int_C -\frac{y^2}{2} \, dx = -\frac{1}{2} \int_C y^2 \, dx \quad \text{and} \quad \bar{y} = \frac{M_x}{2A} = -\frac{1}{2A} \int_C y^2 \, dx.$$

For the moment about the y-axis, $M_y = \iint_R x \, dA$. Let $N = x^2/2$ and $M = 0$. By Green's Theorem,

$$M_y = \int_C \frac{x^2}{2} \, dy = \frac{1}{2} \int_C x^2 \, dy \quad \text{and} \quad \bar{x} = \frac{M_y}{2A} = \frac{1}{2A} \int_C x^2 \, dy.$$

30. By Theorem 15.9 and the fact that $x = r \cos \theta$, $y = r \sin \theta$, we have

$$A = \frac{1}{2} \int x \, dy - y \, dx = \frac{1}{2} \int (r \cos \theta)(r \cos \theta) \, d\theta - (r \sin \theta)(-r \sin \theta) \, d\theta = \frac{1}{2} \int_C r^2 \, d\theta.$$

31. $A = \int_{-2}^2 (4 - x^2) \, dx = \left[4x - \frac{x^3}{3} \right]_{-2}^2 = \frac{32}{3}$

$$\bar{x} = \frac{1}{2A} \int_{C_1} x^2 \, dy + \frac{1}{2A} \int_{C_2} x^2 \, dy$$

For C_1, $dy = -2x \, dx$ and for C_2, $dy = 0$. Thus,

$$\bar{x} = \frac{1}{2(32/3)} \int_2^{-2} x^2 (-2x \, dx) = \left[\frac{3}{64} \left(-\frac{x^4}{2} \right) \right]_2^{-2} = 0.$$

To calculate $\bar{y}$, note that $y = 0$ along C_2. Thus,

$$\bar{y} = \frac{-1}{2(32/3)} \int_2^{-2} (4 - x^2)^2 \, dx = \frac{3}{64} \int_{-2}^2 (16 - 8x^2 + x^4) \, dx = \frac{3}{64} \left[16x - \frac{8x^3}{3} + \frac{x^5}{5} \right]_{-2}^2 = \frac{8}{5}.$$

32. Since $A = $ area of semicircle $= \frac{\pi a^2}{2}$, we have $\frac{1}{2A} = \frac{1}{\pi a^2}$. Note that $y = 0$ and $dy = 0$ along the boundary $y = 0$.
Let $x = a \cos t$, $y = a \sin t$, $0 \le t \le \pi$, then

$$\bar{x} = \frac{1}{\pi a^2} \int_0^\pi a^2 \cos^2 t \, (a \cos t) \, dt = \frac{a}{\pi} \int_0^\pi \cos^3 t \, dt = \frac{a}{\pi} \int_0^\pi (1 - \sin^2 t) \cos t \, dt = \frac{a}{\pi} \left[\sin t - \frac{\sin^3 t}{3} \right]_0^\pi = 0$$

$$\bar{y} = \frac{-1}{\pi a^2} \int_0^\pi a^2 \sin^2 t \, (-a \sin t \, dt) = \frac{a}{\pi} \int_0^\pi \sin^3 t \, dt = \frac{a}{\pi} \left[-\cos t + \frac{\cos^3 t}{3} \right]_0^\pi = \frac{4a}{3\pi}.$$

33. Since $A = \int_0^1 (x - x^3)\,dx = \left[\frac{x^2}{2} - \frac{x^4}{4}\right]_0^1 = \frac{1}{4}$, we have $\frac{1}{2A} = 2$. On C_1 we have $y = x^3$, $dy = 3x^2\,dx$ and on C_2

we have $y = x$, $dy = dx$. Thus,

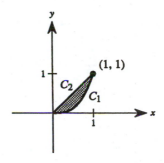

$$\bar{x} = 2\int_C x^2\,dy = 2\int_{C_1} x^2(3x^2\,dx) + 2\int_{C_2} x^2\,dx$$

$$= 6\int_0^1 x^4\,dx + 2\int_1^0 x^2\,dx = \frac{6}{5} - \frac{2}{3} = \frac{8}{15}$$

$$\bar{y} = -2\int_C y^2\,dx$$

$$= -2\int_0^1 x^6\,dx - 2\int_1^0 x^2\,dx = -\frac{2}{7} + \frac{2}{3} = \frac{8}{21}.$$

34. Since $A = \frac{1}{2}(2a)(c) = ac$, we have $\frac{1}{2A} = \frac{1}{2ac}$,

C_1: $y = 0$, $dy = 0$

C_2: $y = \dfrac{c}{b-a}(x-a)$, $dy = \dfrac{c}{b-a}\,dx$

C_3: $y = \dfrac{c}{b+a}(x+a)$, $dy = \dfrac{c}{b+a}\,dx$.

Thus,

$$\bar{x} = \frac{1}{2ac}\int_C x^2\,dy = \frac{1}{2ac}\left[\int_{-a}^a 0 + \int_a^b x^2 \frac{c}{b-a}\,dx + \int_b^{-a} x^2 \frac{c}{b+a}\,dx\right] = \frac{1}{2ac}\left[0 + \frac{2abc}{3}\right] = \frac{b}{3}$$

$$\bar{y} = \frac{-1}{2ac}\int_C y^2\,dx = \frac{-1}{2ac}\left[0 + \int_a^b \left(\frac{c}{b-a}\right)^2 (x-a)^2\,dx + \int_b^{-a} \left(\frac{c}{b+a}\right)^2 (x+a)^2\,dx\right]$$

$$= \frac{-1}{2ac}\left[\frac{c^2(b-a)}{3} - \frac{c^2(b+a)}{3}\right] = \frac{c}{3}.$$

35. $A = \dfrac{1}{2}\displaystyle\int_0^{2\pi} a^2(1 - \cos\theta)^2\,d\theta$

$$= \frac{a^2}{2}\int_0^{2\pi}\left(1 - 2\cos\theta + \frac{1}{2} + \frac{\cos 2\theta}{2}\right)d\theta = \frac{a^2}{2}\left[\frac{3\theta}{2} - 2\sin\theta + \frac{1}{2}\sin 2\theta\right]_0^{2\pi} = \frac{a^2}{2}(3\pi) = \frac{3\pi a^2}{2}$$

36. $A = \dfrac{1}{2}\displaystyle\int_0^{\pi} a^2\cos^2 3\theta\,d\theta = \frac{a^2}{2}\int_0^{\pi}\frac{1 + \cos 6\theta}{2}\,d\theta = \frac{a^2}{4}\left[\theta + \frac{\sin 6\theta}{6}\right]_0^{\pi} = \frac{\pi a^2}{4}$

Note: In this case R is enclosed by $r = a\cos 3\theta$ where $0 \le \theta \le \pi$.

37. In this case the inner loop has domain $\dfrac{2\pi}{3} \le \theta \le \dfrac{4\pi}{3}$. Thus,

$$A = \frac{1}{2}\int_{2\pi/3}^{4\pi/3}(1 + 4\cos\theta + 4\cos^2\theta)\,d\theta$$

$$= \frac{1}{2}\int_{2\pi/3}^{4\pi/3}(3 + 4\cos\theta + 2\cos 2\theta)\,d\theta = \frac{1}{2}\left[3\theta + 4\sin\theta + \sin 2\theta\right]_{2\pi/3}^{4\pi/3} = \pi - \frac{3\sqrt{3}}{2}.$$

38. In this case, $0 \le \theta \le 2\pi$ and we let

$$u = \frac{\sin\theta}{1+\cos\theta}, \quad \cos\theta = \frac{1-u^2}{1+u^2}, \quad d\theta = \frac{2\,du}{1+u^2}.$$

Now $u \Rightarrow \infty$ as $\theta \Rightarrow \pi$ and we have

$$A = 2\left(\frac{1}{2}\right)\int_0^\pi \frac{9}{(2-\cos\theta)^2}\,d\theta = 9\int_0^\infty \frac{\dfrac{2\,du}{1+u^2}}{4 - 4\left(\dfrac{1-u^2}{1+u^2}\right) + \dfrac{(1-u^2)^2}{(1+u^2)^2}} = 18\int_0^\infty \frac{1+u^2}{(1+3u^2)^2}\,du$$

$$= 18\int_0^\infty \frac{1/3}{1+3u^2}\,du + 18\int_0^\infty \frac{2/3}{(1+3u^2)^2}\,du = \left[\frac{6}{\sqrt{3}}\arctan\sqrt{3}\,u\right]_0^\infty + \frac{12}{\sqrt{3}}\left(\frac{1}{2}\right)\left[\frac{u}{1+3u^2} + \int \frac{\sqrt{3}}{1+3u^2}\,du\right]_0^\infty$$

$$= \frac{6}{\sqrt{3}}\left(\frac{\pi}{2}\right) + \frac{6}{\sqrt{3}}\left[\frac{u}{1+3u^2}\right]_0^\infty + \left[\frac{6}{\sqrt{3}}\arctan\sqrt{3}\,u\right]_0^\infty = \frac{3\pi}{\sqrt{3}} + 0 + \frac{3\pi}{\sqrt{3}} = 2\sqrt{3}\,\pi.$$

39. $I = \displaystyle\int_C \frac{y\,dx - x\,dy}{x^2 + y^2}$

a. Let $\mathbf{F} = \dfrac{y}{x^2+y^2}\mathbf{i} - \dfrac{x}{x^2+y^2}\mathbf{j}.$

$\mathbf{F}$ is conservative since $\dfrac{\partial N}{\partial x} = \dfrac{\partial M}{\partial y} = \dfrac{x^2 - y^2}{(x^2+y^2)^2}.$

$\mathbf{F}$ is defined and has continuous first partials everywhere except at the origin. If C is a circle (a closed path) that does not contain the origin, then

$$\int_C \mathbf{F}\cdot d\mathbf{r} = \int_C M\,dx + N\,dy = \iint_R \left(\frac{\partial N}{\partial x} - \frac{\partial M}{\partial y}\right)dA = 0.$$

b. Let $\mathbf{r} = a\cos t\,\mathbf{i} - a\sin t\,\mathbf{j}$, $0 \le t \le 2\pi$ be a circle C_1 oriented clockwise inside C (see figure). Introduce line segments C_2 and C_3 as illustrated in Example 6 of this section in the text. For the region inside C and outside C_1, Green's Theorem applies. Note that since C_2 and C_3 have opposite orientations, the line integrals over them cancel. Thus, $C_4 = C_1 + C_2 + C + C_3$ and

$$\int_{C_4} \mathbf{F}\cdot d\mathbf{r} = \int_{C_1} \mathbf{F}\cdot d\mathbf{r} + \int_C \mathbf{F}\cdot d\mathbf{r} = 0.$$

But,

$$\int_{C_1} \mathbf{F}\cdot d\mathbf{r} = \int_0^{2\pi}\left[\frac{(-a\sin t)(-a\sin t)}{a^2\cos^2 t + a^2\sin^2 t} + \frac{(-a\cos t)(-a\cos t)}{a^2\cos^2 t + a^2\sin^2 t}\right]dt$$

$$= \int_0^{2\pi}(\sin^2 t + \cos^2 t)\,dt = \Big[t\Big]_0^{2\pi} = 2\pi.$$

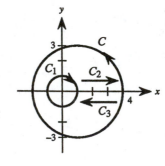

Finally, $\displaystyle\int_C \mathbf{F}\cdot d\mathbf{r} = -\int_{C_1}\mathbf{F}\cdot d\mathbf{r} = -2\pi.$

Note: If C were oriented clockwise, then the answer would have been 2π.

40. a. Let C be the line segment joining (x_1, y_1) and (x_2, y_2).

$$y = \frac{y_2 - y_1}{x_2 - x_1}(x - x_1) + y_1$$

$$dy = \frac{y_2 - y_1}{x_2 - x_1}\,dx$$

$$\int_C -y\,dx + x\,dy = \int_{x_1}^{x_2}\left[-\frac{y_2 - y_1}{x_2 - x_1}(x - x_1) - y_1 + x\left(\frac{y_2 - y_1}{x_2 - x_1}\right)\right]dx = \int_{x_1}^{x_2}\left[x_1\left(\frac{y_2 - y_1}{x_2 - x_1}\right) - y_1\right]dx$$

$$= \left[\left[x_1\left(\frac{y_2 - y_1}{x_2 - x_1}\right) - y_1\right]x\right]_{x_1}^{x_2} = \left[x_1\left(\frac{y_2 - y_1}{x_2 - x_1}\right) - y_1\right](x_2 - x_1)$$

$$= x_1(y_2 - y_1) - y_1(x_2 - x_1) = x_1y_2 - x_2y_1$$

b. Let C be the boundary of the region $A = \dfrac{1}{2}\int_C -y\,dx + x\,dy = \dfrac{1}{2}\iint_R (1 - (-1))\,dA = \iint_R dA.$

Therefore,

$$\iint_R dA = \frac{1}{2}\left[\int_{C_1} -y\,dx + x\,dy + \int_{C_2} -y\,dx + x\,dy + \cdots + \int_{C_n} -y\,dx + x\,dy\right]$$

where C_1 is the line segment joining (x_1, y_1) and (x_2, y_2), C_2 is the line segment joining (x_2, y_2) and $(x_3, y_3), \ldots,$ and C_n is the line segment joining (x_n, y_n) and (x_1, y_1). Thus,

$$\iint_R dA = \frac{1}{2}\left[(x_1y_2 - x_2y_1) + (x_2y_3 - x_3y_2) + \cdots + (x_{n-1}y_n - x_ny_{n-1}) + (x_ny_1 - x_1y_n)\right].$$

41. Pentagon: $(0, 0)$, $(2, 0)$, $(3, 2)$, $(1, 4)$, $(-1, 1)$

$A = \frac{1}{2}[(0 - 0) + (4 - 0) + (12 - 2) + (1 + 4) + (0 - 0)] = \frac{19}{2}$

42. Hexagon: $(0, 0)$, $(2, 0)$, $(3, 2)$, $(2, 4)$, $(0, 3)$, $(-1, 1)$

$A = \frac{1}{2}[(0 - 0) + (4 - 0) + (12 - 4) + (6 - 0) + (0 + 3) + (0 - 0)] = \frac{21}{2}$

43. $\displaystyle\int_C f(x)\,dx + g(y)\,dy = \iint_R\left[\frac{\partial}{\partial x}g(y) - \frac{\partial}{\partial y}f(x)\right]dA = \iint_R (0 - 0)\,dA = 0$

44. $\mathbf{F} = M\mathbf{i} + N\mathbf{j}$

$\text{curl }\mathbf{F} = \left(\dfrac{\partial N}{\partial x} - \dfrac{\partial M}{\partial y}\right)\mathbf{k} = 0 \Rightarrow \dfrac{\partial N}{\partial x} = \dfrac{\partial M}{\partial y}$

$\displaystyle\int_C \mathbf{F}\cdot d\mathbf{r} = \int_C M\,dx + N\,dy = \iint_R\left(\frac{\partial N}{\partial x} - \frac{\partial N}{\partial y}\right)dA = \iint_R (0)\,dA = 0$

45. Since $\displaystyle\int_C \mathbf{F}\cdot\mathbf{N}\,ds = \iint_R \text{div }\mathbf{F}\,dA$, then

$\displaystyle\int_C f\,D_N g\,ds = \int_C f\nabla g\cdot\mathbf{N}\,ds$

$\displaystyle = \iint_R \text{div }(f\nabla g)\,dA = \iint_R (f\,\text{div }(\nabla g) + \nabla f\cdot\nabla g)\,dA = \iint_R (f\nabla^2 g + \nabla f\cdot\nabla g)\,dA.$

46. $\displaystyle\int_C (f\,D_N g - g\,D_N f)\,ds = \int_C f\,D_N g\,ds - \int_C g\,D_N f\,ds$

$\displaystyle = \iint_R (f\nabla^2 g + \nabla f\cdot\nabla g)\,dA - \iint_R (g\nabla^2 f + \nabla g\cdot\nabla f)\,dA = \iint_R (f\nabla^2 g - g\nabla^2 f)\,dA$

Section 15.5   Parametric Surfaces

1. $\mathbf{r}(u, v) = u\mathbf{i} + v\mathbf{j} + uv\mathbf{k}$
$z = xy$
Matches c.

2. $\mathbf{r}(u, v) = u\cos v\mathbf{i} + u\sin v\mathbf{j} + u\mathbf{k}$
$x^2 + y^2 = z^2$
Matches d.

3. $\mathbf{r}(u, v) = 2\cos v\cos u\mathbf{i} + 2\cos v\sin u\mathbf{j} + 2\sin v\mathbf{k}$
$x^2 + y^2 + z^2 = 4$
Matches b.

4. $\mathbf{r}(u, v) = 4\cos u\mathbf{i} + 4\sin u\mathbf{j} + v\mathbf{k}$
$x^2 + y^2 = 16$
Matches a.

5. $\mathbf{r}(u, v) = u\mathbf{i} + v\mathbf{j} + \dfrac{v}{2}\mathbf{k}$
$y - 2z = 0$
Plane

6. $\mathbf{r}(u, v) = u\cos v\mathbf{i} + u\sin v\mathbf{j} + u^2\mathbf{k}$
$z = x^2 + y^2$
Paraboloid

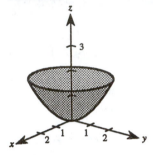

7. $\mathbf{r}(u, v) = 2\cos u\mathbf{i} + v\mathbf{j} + 2\sin u\mathbf{k}$
$x^2 + z^2 = 4$
Cylinder

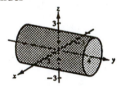

8. $\mathbf{r}(u, v) = 5\cos v\cos u\mathbf{i} + 5\cos v\sin u\mathbf{j} + 5\sin v\mathbf{k}$
$x^2 + y^2 + z^2 = 25$
Sphere

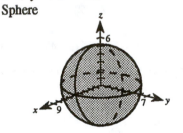

For Exercises 9–12,
$\mathbf{r}(u, v) = u\cos v\mathbf{i} + u\sin v\mathbf{j} + u^2\mathbf{k}$,
$0 \le u \le 2, \ 0 \le v \le 2\pi$. Eliminating
the parameter yields $z = x^2 + y^2$,
$0 \le z \le 4$.

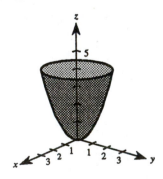

9. $\mathbf{s}(u, v) = u\cos v\mathbf{i} + u\sin v\mathbf{j} - u^2\mathbf{k}, \ 0 \le u \le 2, \ 0 \le v \le 2\pi$
$z = -(x^2 + y^2)$
The paraboloid is reflected (inverted) through the xy-plane.

10. $\mathbf{s}(u, v) = u\cos v\mathbf{i} + u^2\mathbf{j} + u\sin v\mathbf{k}, \ 0 \le u \le 2, \ 0 \le v \le 2\pi$
$y = x^2 + z^2$
The paraboloid opens along the y-axis instead of the z-axis.

11. $\mathbf{s}(u, v) = u\cos v\mathbf{i} + u\sin v\mathbf{j} + u^2\mathbf{k}, \ 0 \le u \le 3, \ 0 \le v \le 2\pi$
The height of the paraboloid is increased from 4 to 9.

12. $\mathbf{s}(u, v) = 4u\cos v\mathbf{i} + 4u\sin v\mathbf{j} + u^2\mathbf{k}, \ 0 \le u \le 2, \ 0 \le v \le 2\pi$
$z = \dfrac{x^2 + y^2}{16}$
The paraboloid is "wider." The top is now the circle
$x^2 + y^2 = 64$. It was $x^2 + y^2 = 4$.

13. $\mathbf{r}(u, v) = 2u \cos v \mathbf{i} + 2u \sin v \mathbf{j} + u^4 \mathbf{k}$,
$0 \le u \le 1, \ 0 \le v \le 2\pi$

$$z = \frac{(x^2 + y^2)^2}{16}$$

14. $\mathbf{r}(u, v) = 2 \cos v \cos u \mathbf{i} + 4 \cos v \sin u \mathbf{j} + \sin v \mathbf{k}$,
$0 \le u \le 2\pi, \ 0 \le v \le 2\pi$

$$\frac{x^2}{4} + \frac{y^2}{16} + \frac{z^2}{1} = 1$$

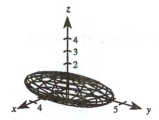

15. $\mathbf{r}(u, v) = 2 \sinh u \cos v \mathbf{i} + \sinh u \sin v \mathbf{j} + \cosh u \mathbf{k}$,
$0 \le u \le 2, \ 0 \le v \le 2\pi$

$$\frac{z^2}{1} - \frac{x^2}{4} - \frac{y^2}{1} = 1$$

16. $\mathbf{r}(u, v) = 2u \cos v \mathbf{i} + 2u \sin v \mathbf{j} + v \mathbf{k}$,
$0 \le u \le 1, \ 0 \le v \le 3\pi$

$$z = \arctan \left(\frac{y}{x} \right)$$

17. $\mathbf{r}(u, v) = (u - \sin u) \cos v \mathbf{i} + (1 - \cos u) \sin v \mathbf{j} + u \mathbf{k}$,
$0 \le u \le \pi, \ 0 \le v \le 2\pi$

18. $\mathbf{r}(u, v) = \cos^3 u \cos v \mathbf{i} + \sin^3 u \sin v \mathbf{j} + u \mathbf{k}$,
$0 \le u \le \frac{\pi}{2}, \ 0 \le v \le 2\pi$

19. **a.** $\mathbf{r}(u, v) = (4 + \cos v) \cos u \mathbf{i} + (4 + \cos v) \sin u \mathbf{j} + \sin v \mathbf{k}, \ 0 \leq u \leq 2\pi, \ 0 \leq v \leq 2\pi$

b. $\mathbf{r}(u, v) = (4 + 2 \cos v) \cos u \mathbf{i} + (4 + 2 \cos v) \sin u \mathbf{j} + 2 \sin v \mathbf{k}, \ 0 \leq u \leq 2\pi,$
$0 \leq v \leq 2\pi$

c. $\mathbf{r}(u, v) = (8 + \cos v) \cos u \mathbf{i} + (8 + \cos v) \sin u \mathbf{j} + \sin v \mathbf{k}, \ 0 \leq u \leq 2\pi, \ 0 \leq v \leq 2\pi$

d. $\mathbf{r}(u, v) = (8 + 3 \cos v) \cos u \mathbf{i} + (8 + 3 \cos v) \sin u \mathbf{j} + 3 \sin v \mathbf{k}, \ 0 \leq u \leq 2\pi,$
$0 \leq v \leq 2\pi$

The radius of the generating circle that is revolved about the z-axis is b, and its center is a units from the axis of revolution.

20. $\mathbf{r}(u, v) = 2u \cos v \mathbf{i} + 2u \sin v \mathbf{j} + v \mathbf{k}, \ 0 \leq u \leq 1, \ 0 \leq v \leq 3\pi$

a. If $u = 1$:

$\mathbf{r}(1, v) = 2 \cos v \mathbf{i} + 2 \sin v \mathbf{j} + v \mathbf{k}$

$x^2 + y^2 = 4$

$0 \leq z \leq 3\pi$

Helix

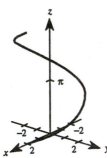

b. If $v = \dfrac{2\pi}{3}$:

$\mathbf{r}\left(u, \dfrac{2\pi}{3}\right) = -u \mathbf{i} + \sqrt{3} u \mathbf{j} + \dfrac{2\pi}{3} \mathbf{k}$

$y = -\sqrt{3} x$

$z = \dfrac{2\pi}{3}$

Line

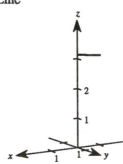

c. If one parameter is held constant, the result is a **curve** in 3-space.

21. $z = y$
$\mathbf{r}(u, v) = u \mathbf{i} + v \mathbf{j} + v \mathbf{k}$

22. $z = 6 - x - y$
$\mathbf{r}(u, v) = u \mathbf{i} + v \mathbf{j} + (6 - u - v) \mathbf{k}$

23. $x^2 + y^2 = 16$
$\mathbf{r}(u, v) = 4\cos u\,\mathbf{i} + 4\sin u\,\mathbf{j} + v\mathbf{k}$

24. $x^2 + 4y^2 = 16$
$\mathbf{r}(u, v) = 4\cos u\,\mathbf{i} + 2\sin u\,\mathbf{j} + v\mathbf{k}$

25. $z = x^2$
$\mathbf{r}(u, v) = u\mathbf{i} + v\mathbf{j} + u^2\mathbf{k}$

26. $\dfrac{x^2}{9} + \dfrac{y^2}{4} + \dfrac{z^2}{1} = 1$
$\mathbf{r}(u, v) = 3\cos v \cos u\,\mathbf{i} + 2\cos v \sin u\,\mathbf{j} + \sin v\,\mathbf{k}$

27. $z = 4$ inside $x^2 + y^2 = 9$.
$\mathbf{r}(u, v) = v\cos u\,\mathbf{i} + v\sin u\,\mathbf{j} + 4\mathbf{k}, \quad 0 \le v \le 3$

28. $z = x^2 + y^2$ inside $x^2 + y^2 = 9$.
$\mathbf{r}(u, v) = v\cos u\,\mathbf{i} + v\sin u\,\mathbf{j} + v^2\mathbf{k}, \quad 0 \le v \le 3$

29. Function: $y = \dfrac{x}{2}, \ 0 \le x \le 6$
Axis of revolution: x-axis
$x = u, \quad y = \dfrac{u}{2}\cos v, \quad z = \dfrac{u}{2}\sin v$
$0 \le u \le 6, \ 0 \le v \le 2\pi$

30. Function: $y = \sqrt{x}, \ 0 \le x \le 4$
Axis of revolution: x-axis
$x = u, \quad y = \sqrt{u}\cos v, \quad z = \sqrt{u}\sin v$
$0 \le u \le 4, \ 0 \le v \le 2\pi$

31. Function: $x = \sin z, \ 0 \le z \le \pi$
Axis of revolution: z-axis
$x = \sin u \cos v, \quad y = \sin u \sin v, \quad z = u$
$0 \le u \le \pi, \ 0 \le v \le 2\pi$

32. Function: $z = 4 - y^2, \ 0 \le y \le 2$
Axis of revolution: y-axis
$x = (4 - u^2)\cos v, \quad y = u, \quad z = (4 - u^2)\sin v$
$0 \le u \le 2, \ 0 \le v \le 2\pi$

33. $\mathbf{r}(u, v) = (u + v)\mathbf{i} + (u - v)\mathbf{j} + v\mathbf{k}, \quad (1, -1, 1)$
$\mathbf{r}_u(u, v) = \mathbf{i} + \mathbf{j}, \quad \mathbf{r}_v(u, v) = \mathbf{i} - \mathbf{j} + \mathbf{k}$
At $(1, -1, 1)$, $u = 0$ and $v = 1$.
$\mathbf{r}_u(0, 1) = \mathbf{i} + \mathbf{j}, \quad \mathbf{r}_v(0, 1) = \mathbf{i} - \mathbf{j} + \mathbf{k}$

$\mathbf{N} = \mathbf{r}_u(0, 1) \times \mathbf{r}_v(0, 1) = \begin{vmatrix} \mathbf{i} & \mathbf{j} & \mathbf{k} \\ 1 & 1 & 0 \\ 1 & -1 & 1 \end{vmatrix} = \mathbf{i} - \mathbf{j} - 2\mathbf{k}$

Tangent plane: $(x - 1) - (y + 1) - 2(z - 1) = 0$
$$x - y - 2z = 0$$

34. $\mathbf{r}(u, v) = u\mathbf{i} + v\mathbf{j} + \sqrt{uv}\,\mathbf{k}, \quad (1, 1, 1)$
$\mathbf{r}_u(u, v) = \mathbf{i} + \dfrac{v}{2\sqrt{uv}}\mathbf{k}, \quad \mathbf{r}_v(u, v) = \mathbf{j} + \dfrac{u}{2\sqrt{uv}}\mathbf{k}$
At $(1, 1, 1)$, $u = 1$ and $v = 1$.
$\mathbf{r}_u(1, 1) = \mathbf{i} + \dfrac{1}{2}\mathbf{k}, \quad \mathbf{r}_v(1, 1) = \mathbf{j} + \dfrac{1}{2}\mathbf{k}$

$\mathbf{N} = \mathbf{r}_u(1, 1) \times \mathbf{r}_v(1, 1) = \begin{vmatrix} \mathbf{i} & \mathbf{j} & \mathbf{k} \\ 1 & 0 & \frac{1}{2} \\ 0 & 1 & \frac{1}{2} \end{vmatrix} = -\dfrac{1}{2}\mathbf{i} - \dfrac{1}{2}\mathbf{j} + \mathbf{k}$

Direction numbers: $1, \ 1, \ -2$
Tangent plane: $(x - 1) + (y - 1) - 2(z - 1) = 0$
$$x + y - 2z = 0$$

35. $\mathbf{r}(u, v) = 2u\cos v\,\mathbf{i} + 3u\sin v\,\mathbf{j} + u^2\mathbf{k}, \quad (0, 6, 4)$
$\mathbf{r}_u(u, v) = 2\cos v\,\mathbf{i} + 3\sin v\,\mathbf{j} + 2u\mathbf{k}$
$\mathbf{r}_v(u, v) = -2u\sin v\,\mathbf{i} + 3u\cos v\,\mathbf{j}$
At $(0, 6, 4)$, $u = 2$ and $v = \pi/2$.
$\mathbf{r}_u\left(2, \dfrac{\pi}{2}\right) = 3\mathbf{j} + 4\mathbf{k}, \quad \mathbf{r}_v\left(2, \dfrac{\pi}{2}\right) = -4\mathbf{i}$

$\mathbf{N} = \mathbf{r}_u\left(2, \dfrac{\pi}{2}\right) \times \mathbf{r}_v\left(2, \dfrac{\pi}{2}\right)$

$= \begin{vmatrix} \mathbf{i} & \mathbf{j} & \mathbf{k} \\ 0 & 3 & 4 \\ -4 & 0 & 0 \end{vmatrix} = -16\mathbf{j} + 12\mathbf{k}$

Direction numbers: $0, \ 4, \ -3$
Tangent plane: $4(y - 6) - 3(z - 4) = 0$
$$4y - 3z = 12$$

36. $\mathbf{r}(u, v) = u\cosh v\,\mathbf{i} + u\sinh v\,\mathbf{j} + u^2\mathbf{k}, \quad (-2, 0, 4)$
$\mathbf{r}_u(u, v) = \cosh v\,\mathbf{i} + \sinh v\,\mathbf{j} + 2u\mathbf{k}$
$\mathbf{r}_v(u, v) = u\sinh v\,\mathbf{i} + u\cosh v\,\mathbf{j}$
At $(-2, 0, 4)$, $u = -2$ and $v = 0$.
$\mathbf{r}_u(-2, 0) = \mathbf{i} - 4\mathbf{k}, \quad \mathbf{r}_v(-2, 0) = -2\mathbf{j}$

$\mathbf{N} = \mathbf{r}_u(-2, 0) \times \mathbf{r}_v(-2, 0)$

$= \begin{vmatrix} \mathbf{i} & \mathbf{j} & \mathbf{k} \\ 1 & 0 & -4 \\ 0 & -2 & 0 \end{vmatrix} = -8\mathbf{i} - 2\mathbf{k}$

Direction numbers: $4, \ 0, \ 1$
Tangent plane: $4(x + 2) + (z - 4) = 0$
$$4x + z = -4$$

37. $\mathbf{r}(u, v) = 2u\mathbf{i} - \dfrac{v}{2}\mathbf{j} + \dfrac{v}{2}\mathbf{k}, \ 0 \le u \le 2, \ 0 \le v \le 1$

$\mathbf{r}_u(u, v) = 2\mathbf{i}, \ \mathbf{r}_v(u, v) = -\dfrac{1}{2}\mathbf{j} + \dfrac{1}{2}\mathbf{k}$

$\mathbf{r}_u \times \mathbf{r}_v = \begin{vmatrix} \mathbf{i} & \mathbf{j} & \mathbf{k} \\ 2 & 0 & 0 \\ 0 & -\frac{1}{2} & \frac{1}{2} \end{vmatrix} = -\mathbf{j} - \mathbf{k}$

$\|\mathbf{r}_u \times \mathbf{r}_v\| = \sqrt{2}$

$A = \displaystyle\int_0^1 \int_0^2 \sqrt{2}\, du\, dv = 2\sqrt{2}$

38. $\mathbf{r}(u, v) = 2u \cos v\mathbf{i} + 2u \sin v\mathbf{j} + u^2\mathbf{k}, \ 0 \le u \le 2, \ 0 \le v \le 2\pi$

$\mathbf{r}_u(u, v) = 2 \cos v\mathbf{i} + 2 \sin v\mathbf{j} + 2u\mathbf{k}$

$\mathbf{r}_v(u, v) = -2u \sin v\mathbf{i} + 2u \cos v\mathbf{j}$

$\mathbf{r}_u \times \mathbf{r}_v = \begin{vmatrix} \mathbf{i} & \mathbf{j} & \mathbf{k} \\ 2 \cos v & 2 \sin v & 2u \\ -2u \sin v & 2u \cos v & 0 \end{vmatrix} = -4u^2 \cos v\mathbf{i} - 4u^2 \sin v\mathbf{j} + 4u\mathbf{k}$

$\|\mathbf{r}_u \times \mathbf{r}_v\| = \sqrt{16u^4 \cos^2 v + 16u^4 \sin^2 v + 16u^2} = 4u\sqrt{u^2 + 1}$

$A = \displaystyle\int_0^{2\pi} \int_0^2 4u\sqrt{u^2 + 1}\, du\, dv = \int_0^{2\pi} \frac{4}{3}(5\sqrt{5} - 1)\, dv = \frac{8\pi}{3}(5\sqrt{5} - 1)$

39. $\mathbf{r}(u, v) = a \cos u\mathbf{i} + a \sin u\mathbf{j} + v\mathbf{k}, \ 0 \le u \le 2\pi, \ 0 \le v \le b$

$\mathbf{r}_u(u, v) = -a \sin u\mathbf{i} + a \cos u\mathbf{j}$

$\mathbf{r}_v(u, v) = \mathbf{k}$

$\mathbf{r}_u \times \mathbf{r}_v = \begin{vmatrix} \mathbf{i} & \mathbf{j} & \mathbf{k} \\ -a \sin u & a \cos u & 0 \\ 0 & 0 & 1 \end{vmatrix} = a \cos u\mathbf{i} + a \sin u\mathbf{j}$

$\|\mathbf{r}_u \times \mathbf{r}_v\| = a$

$A = \displaystyle\int_0^b \int_0^{2\pi} a\, du\, dv = 2\pi ab$

40. $\mathbf{r}(u, v) = a \sin u \cos v\mathbf{i} + a \sin u \sin v\mathbf{j} + a \cos u\mathbf{k}, \ 0 \le u \le \pi, \ 0 \le v \le 2\pi$

$\mathbf{r}_u(u, v) = a \cos u \cos v\mathbf{i} + a \cos u \sin v\mathbf{j} - a \sin u\mathbf{k}$

$\mathbf{r}_v(u, v) = -a \sin u \sin v\mathbf{i} + a \sin u \cos v\mathbf{j}$

$\mathbf{r}_u \times \mathbf{r}_v = \begin{vmatrix} \mathbf{i} & \mathbf{j} & \mathbf{k} \\ a \cos u \cos v & a \cos u \sin v & -a \sin u \\ -a \sin u \sin v & a \sin u \cos v & 0 \end{vmatrix} = a^2 \sin^2 u \cos v\mathbf{i} + a^2 \sin^2 u \sin v\mathbf{j} + a^2 \sin u \cos u\mathbf{k}$

$\|\mathbf{r}_u \times \mathbf{r}_v\| = a^2 \sin u$

$A = \displaystyle\int_0^{2\pi} \int_0^\pi a^2 \sin u\, du\, dv = 4\pi a^2$

41. $\mathbf{r}(u, v) = au\cos v\mathbf{i} + au\sin v\mathbf{j} + u\mathbf{k}$, $0 \le u \le b$, $0 \le v \le 2\pi$

$\mathbf{r}_u(u, v) = a\cos v\mathbf{i} + a\sin v\mathbf{j} + \mathbf{k}$

$\mathbf{r}_v(u, v) = -au\sin v\mathbf{i} + au\cos v\mathbf{j}$

$$\mathbf{r}_u \times \mathbf{r}_v = \begin{vmatrix} \mathbf{i} & \mathbf{j} & \mathbf{k} \\ a\cos v & a\sin v & 1 \\ -au\sin v & au\cos v & 0 \end{vmatrix} = -au\cos v\mathbf{i} - au\sin v\mathbf{j} + a^2u\mathbf{k}$$

$\|\mathbf{r}_u \times \mathbf{r}_v\| = au\sqrt{1+a^2}$

$$A = \int_0^{2\pi} \int_0^b a\sqrt{1+a^2}\, u\, du\, dv = \pi ab^2\sqrt{1+a^2}$$

42. $\mathbf{r}(u, v) = (a + b\cos v)\cos u\mathbf{i} + (a + b\cos v)\sin u\mathbf{j} + b\sin v\mathbf{k}$, $a > b$, $0 \le u \le 2\pi$, $0 \le v \le 2\pi$

$\mathbf{r}_u(u, v) = -(a + b\cos v)\sin u\mathbf{i} + (a + b\cos v)\cos u\mathbf{j}$

$\mathbf{r}_v(u, v) = -b\sin v\cos u\mathbf{i} - b\sin v\sin u\mathbf{j} + b\cos v\mathbf{k}$

$$\mathbf{r}_u \times \mathbf{r}_v = \begin{vmatrix} \mathbf{i} & \mathbf{j} & \mathbf{k} \\ -(a + b\cos v)\sin u & (a + b\cos v)\cos u & 0 \\ -b\sin v\cos u & -b\sin v\sin u & b\cos v \end{vmatrix}$$

$$= b\cos u\cos v(a + b\cos v)\mathbf{i} + b\sin u\cos v(a + b\cos v)\mathbf{j} + b\sin v(a + b\cos v)\mathbf{k}$$

$\|\mathbf{r}_u \times \mathbf{r}_v\| = b(a + b\cos v)$

$$A = \int_0^{2\pi} \int_0^{2\pi} b(a + b\cos v)\, du\, dv = 4\pi^2 ab$$

43. $\mathbf{r}(u, v) = \sqrt{u}\cos v\mathbf{i} + \sqrt{u}\sin v\mathbf{j} + u\mathbf{k}$, $0 \le u \le 4$, $0 \le v \le 2\pi$

$\mathbf{r}_u(u, v) = \dfrac{\cos v}{2\sqrt{u}}\mathbf{i} + \dfrac{\sin v}{2\sqrt{u}}\mathbf{j} + \mathbf{k}$

$\mathbf{r}_v(u, v) = -\sqrt{u}\sin v\mathbf{i} + \sqrt{u}\cos v\mathbf{j}$

$$\mathbf{r}_u \times \mathbf{r}_v = \begin{vmatrix} \mathbf{i} & \mathbf{j} & \mathbf{k} \\ \dfrac{\cos v}{2\sqrt{u}} & \dfrac{\sin v}{2\sqrt{u}} & 1 \\ -\sqrt{u}\sin v & \sqrt{u}\cos v & 0 \end{vmatrix} = -\sqrt{u}\cos v\mathbf{i} - \sqrt{u}\sin v\mathbf{j} + \frac{1}{2}\mathbf{k}$$

$\|\mathbf{r}_u \times \mathbf{r}_v\| = \sqrt{u + \dfrac{1}{4}}$

$$A = \int_0^{2\pi} \int_0^4 \sqrt{u + \frac{1}{4}}\, du\, dv = \frac{\pi}{6}(17\sqrt{17} - 1) \approx 36.177$$

44. $\mathbf{r}(u, v) = u\mathbf{i} + \sin u\cos v\mathbf{j} + \sin u\sin v\mathbf{k}$, $0 \le u \le \pi$, $0 \le v \le 2\pi$

$\mathbf{r}_u(u, v) = \mathbf{i} + \cos u\cos v\mathbf{j} + \cos u\sin v\mathbf{k}$

$\mathbf{r}_v(u, v) = -\sin u\sin v\mathbf{j} + \sin u\cos v\mathbf{k}$

$$\mathbf{r}_u \times \mathbf{r}_v = \begin{vmatrix} \mathbf{i} & \mathbf{j} & \mathbf{k} \\ 1 & \cos u\cos v & \cos u\sin v \\ 0 & -\sin u\sin v & \sin u\cos v \end{vmatrix} = \sin u\cos u\mathbf{i} - \sin u\cos v\mathbf{j} - \sin u\sin v\mathbf{k}$$

$\|\mathbf{r}_u \times \mathbf{r}_v\| = \sin u\sqrt{1 + \cos^2 u}$

$$A = \int_0^{2\pi} \int_0^{\pi} \sin u\sqrt{1 + \cos^2 u}\, du\, dv = \pi \left[2\sqrt{2} + \ln\left| \frac{\sqrt{2} + 1}{\sqrt{2} - 1} \right| \right]$$

45. $\mathbf{r}(u, v) = u\cos v\mathbf{i} + u\sin v\mathbf{j} + 2v\mathbf{k}, \ 0 \le u \le 3, \ 0 \le v \le 2\pi$

$\mathbf{r}_u(u, v) = \cos v\mathbf{i} + \sin v\mathbf{j}$

$\mathbf{r}_v(u, v) = -u\sin v\mathbf{i} + u\cos v\mathbf{j} + 2\mathbf{k}$

$$\mathbf{r}_u \times \mathbf{r}_v = \begin{vmatrix} \mathbf{i} & \mathbf{j} & \mathbf{k} \\ \cos v & \sin v & 0 \\ -u\sin v & u\cos v & 2 \end{vmatrix} = 2\sin v\mathbf{i} - 2\cos v\mathbf{j} + u\mathbf{k}$$

$\|\mathbf{r}_u \times \mathbf{r}_v\| = \sqrt{4 + u^2}$

$$A = \int_0^{2\pi} \int_0^3 \sqrt{4 + u^2}\, du\, dv = \pi\left[3\sqrt{13} + 4\ln\left(\frac{3 + \sqrt{13}}{2}\right)\right]$$

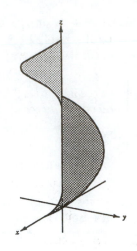

46. $x^2 + y^2 - z^2 = 1$

Let $x = u\cos v, \ y = u\sin v,$ and $z = \sqrt{u^2 - 1}$. Then,

$$\mathbf{r}_u(u, v) = \cos v\mathbf{i} + \sin v\mathbf{j} + \frac{u}{\sqrt{u^2 - 1}}\mathbf{k}$$

$$\mathbf{r}_v(u, v) = -u\sin v\mathbf{i} + u\cos v\mathbf{j}.$$

At $(1, 0, 0)$, $u = 1$ and $v = 0$. $\mathbf{r}_u(1, 0)$ is undefined and $\mathbf{r}_v(1, 0) = \mathbf{j}$. The tangent plane at $(1, 0, 0)$ is $x = 1$.

47. $\mathbf{r}(u, v) = u\mathbf{i} + f(u)\cos v\mathbf{j} + f(u)\sin v\mathbf{k}, \ a \le u \le b, \ 0 \le v \le 2\pi$

$\mathbf{r}_u(u, v) = \mathbf{i} + f'(u)\cos v\mathbf{j} + f'(u)\sin v\mathbf{k}$

$\mathbf{r}_v(u, v) = -f(u)\sin v\mathbf{j} + f(u)\cos v\mathbf{k}$

$$\mathbf{r}_u \times \mathbf{r}_v = \begin{vmatrix} \mathbf{i} & \mathbf{j} & \mathbf{k} \\ 1 & f'(u)\cos v & f'(u)\sin v \\ 0 & -f(u)\sin v & f(u)\cos v \end{vmatrix} = f(u)f'(u)\mathbf{i} - f(u)\cos v\mathbf{j} - f(u)\sin v\mathbf{k}$$

$\|\mathbf{r}_u \times \mathbf{r}_v\| = f(u)\sqrt{1 + [f'(u)]^2}$

$$A = \int_0^{2\pi} \int_a^b f(u)\sqrt{1 + [f'(u)]^2}\, du\, dv = 2\pi \int_a^b f(x)\sqrt{1 + [f'(x)]^2}\, dx \quad \text{(since } u = x\text{)}$$

48. $\mathbf{s}(u, v) = u\cos v\mathbf{i} + u\sin v\mathbf{j} + u\mathbf{k}, \ 0 \le v \le 2\pi, \ 0 \le u, \ x^2 + y^2 = z^2$

Since $z = u \ge 0$, we have $z = \sqrt{x^2 + y^2}$ which is the equation of the cone in Example 3.

Section 15.6   Surface Integrals

1. S: $z = 4 - x, \ 0 \le x \le 4, \ 0 \le y \le 4, \ \dfrac{\partial z}{\partial x} = -1, \ \dfrac{\partial z}{\partial y} = 0$

$$\iint_S (x - 2y + z)\, dS = \int_0^4 \int_0^4 (x - 2y + 4 - x)\sqrt{1 + (-1)^2 + (0)^2}\, dy\, dx = \sqrt{2} \int_0^4 \int_0^4 (4 - 2y)\, dy\, dx = 0$$

2. S: $z = 10 - 2x + 2y, \ 0 \le x \le 2, \ 0 \le y \le 4, \ \dfrac{\partial z}{\partial x} = -2, \ \dfrac{\partial z}{\partial y} = 2$

$$\iint_S (x - 2y + z)\, dS = \int_0^2 \int_0^4 (x - 2y + 10 - 2x + 2y)\sqrt{1 + (-2)^2 + (2)^2}\, dy\, dx = 3 \int_0^2 \int_0^4 (10 - x)\, dy\, dx = 216$$

3. $S: z = 10$, $x^2 + y^2 \le 1$, $\dfrac{\partial z}{\partial x} = \dfrac{\partial z}{\partial y} = 0$

$$\iint_S (x - 2y + z)\, dS = \int_{-1}^{1} \int_{-\sqrt{1-x^2}}^{\sqrt{1-x^2}} (x - 2y + 10)\sqrt{1 + (0)^2 + (0)^2}\, dy\, dx$$

$$= \int_0^{2\pi} \int_0^1 (r\cos\theta - 2r\sin\theta + 10) r\, dr\, d\theta$$

$$= \int_0^{2\pi} \left(\frac{1}{3}\cos\theta - \frac{2}{3}\sin\theta + 5 \right) d\theta = \left[\frac{1}{3}\sin\theta + \frac{2}{3}\cos\theta + 5\theta \right]_0^{2\pi} = 10\pi$$

4. $S: z = \dfrac{2}{3}x^{3/2}$, $0 \le x \le 1$, $0 \le y \le x$, $\dfrac{\partial z}{\partial x} = x^{1/2}$, $\dfrac{\partial z}{\partial y} = 0$

$$\iint_S (x - 2y + z)\, dS = \int_0^1 \int_0^x \left(x - 2y + \frac{2}{3}x^{3/2} \right) \sqrt{1 + (x^{1/2})^2 + (0)^2}\, dy\, dx$$

$$= \int_0^1 \int_0^x \left(x - 2y + \frac{2}{3}x^{3/2} \right) \sqrt{1 + x}\, dy\, dx$$

$$= \frac{2}{3} \int_0^1 x^{5/2}\sqrt{x + 1}\, dx$$

$$= \frac{2}{3} \left[\frac{1}{4} x^{5/2}(1 + x)^{3/2} \right]_0^1 - \frac{5}{3} \int_0^1 x^{3/2}\sqrt{1 + x}\, dx$$

$$= \left[\frac{1}{6} x^{5/2}(1 + x)^{3/2} \right]_0^1 - \frac{5}{12}\left(\frac{1}{3} \right) \left[x^{3/2}(1 + x)^{3/2} \right]_0^1 + \frac{5}{24} \int_0^1 x^{1/2}\sqrt{1 + x}\, dx$$

$$= \frac{\sqrt{2}}{3} - \frac{5\sqrt{2}}{18} + \frac{5}{24} \int_0^1 \sqrt{x + x^2}\, dx$$

$$= \frac{\sqrt{2}}{18} + \frac{5}{24} \int_0^1 \sqrt{\left(x + \frac{1}{2} \right)^2 - \frac{1}{4}}\, dx$$

$$= \frac{\sqrt{2}}{18} + \frac{5}{24}\left(\frac{1}{2} \right) \left[\left(x + \frac{1}{2} \right) \sqrt{x^2 + x} - \frac{1}{4}\ln \left| \left(x + \frac{1}{2} \right) + \sqrt{x^2 + x} \right| \right]_0^1$$

$$= \frac{\sqrt{2}}{18} + \frac{5}{48} \left[\frac{3}{2}\sqrt{2} - \frac{1}{4}\ln \left| \frac{3}{2} + \sqrt{2} \right| + \frac{1}{4}\ln \left| \frac{1}{2} \right| \right]$$

$$= \frac{\sqrt{2}}{18} + \frac{15\sqrt{2}}{96} + \frac{5}{192}\ln \left| \frac{1}{3 + 2\sqrt{2}} \right| = \frac{61\sqrt{2}}{288} - \frac{5}{192}\ln|3 + 2\sqrt{2}| \approx 0.2536$$

5. $S: z = 6 - x - 2y$, (first octant) $\dfrac{\partial z}{\partial x} = -1$, $\dfrac{\partial z}{\partial y} = -2$

$$\iint_S xy\, dS = \int_0^6 \int_0^{3-(x/2)} xy\sqrt{1 + (-1)^2 + (-2)^2}\, dy\, dx$$

$$= \sqrt{6} \int_0^6 \left[\frac{xy^2}{2} \right]_0^{3-(x/2)} dx$$

$$= \frac{\sqrt{6}}{2} \int_0^6 x\left(9 - 3x + \frac{1}{4}x^2 \right) dx$$

$$= \frac{\sqrt{6}}{2} \left[\frac{9x^2}{2} - x^3 + \frac{x^4}{16} \right]_0^6 = \frac{27\sqrt{6}}{2}$$

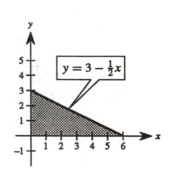

6. $S: z = h, \quad 0 \le x \le 2, \quad 0 \le y \le \sqrt{4-x^2}, \quad \dfrac{\partial z}{\partial x} = \dfrac{\partial z}{\partial y} = 0$

$$\iint_S dx\, dS = \int_0^2 \int_0^{\sqrt{4-x^2}} xy\, dy\, dx = \frac{1}{2}\int_0^2 x(4-x^2)\, dx = \frac{1}{2}\left[2x^2 - \frac{x^4}{4}\right]_0^2 = 2$$

7. $S: z = 9 - x^2, \quad 0 \le x \le 2, \quad 0 \le y \le x, \quad \dfrac{\partial z}{\partial x} = -2x, \quad \dfrac{\partial z}{\partial y} = 0$

$$\iint_S xy\, dS = \int_0^2 \int_y^2 xy\sqrt{1+4x^2}\, dx\, dy = \frac{391\sqrt{17}+1}{240}$$

8. $S: z = xy, \quad 0 \le x \le 2, \quad 0 \le y \le 2, \quad \dfrac{\partial z}{\partial x} = y, \quad \dfrac{\partial z}{\partial y} = x$

$$\iint_S xy\, dS = \int_0^2 \int_0^2 xy\sqrt{1+x^2+y^2}\, dy\, dx = \frac{244 - 50\sqrt{5}}{15}$$

9. $S: \quad z = 10 - x^2 - y^2, \quad 0 \le x \le 2, \quad 0 \le y \le 2$

$$\iint_S (x^2 - 2xy)\, dS = \int_0^2 \int_0^2 (x^2 - 2xy)\sqrt{1+4x^2+4y^2}\, dy\, dx \approx -11.47$$

10. $S: z = \cos x, \quad 0 \le x \le \dfrac{\pi}{2}, \quad 0 \le y \le \dfrac{x}{2}$

$$\iint_S (x^2 - 2xy)\, dS = \int_0^{\pi/2} \int_0^{x/2} (x^2 - 2xy)\sqrt{1+\sin^2 x}\, dy\, dx = \int_0^{\pi/2} \frac{x^3}{4}\sqrt{1+\sin^2 x}\, dx \approx 0.52$$

11. $S: \mathbf{r}(u, v) = u\mathbf{i} + v\mathbf{j} + \dfrac{v}{2}\mathbf{k}, \quad 0 \le u \le 1, \quad 0 \le v \le 2$

$$\|\mathbf{r}_u \times \mathbf{r}_v\| = \left\|-\frac{1}{2}\mathbf{j} + \mathbf{k}\right\| = \frac{\sqrt{5}}{2}$$

$$\iint_S (y + 5)\, dS = \int_0^2 \int_0^1 (v + 5)\frac{\sqrt{5}}{2}\, du\, dv = 6\sqrt{5}$$

12. $S: \mathbf{r}(u, v) = 2\cos u\mathbf{i} + 2\sin u\mathbf{j} + v\mathbf{k}, \quad 0 \le u \le \dfrac{\pi}{2}, \quad 0 \le v \le 2$

$$\|\mathbf{r}_u \times \mathbf{r}_v\| = \|2\cos u\mathbf{i} + 2\sin u\mathbf{j}\| = 2$$

$$\iint_S (x + y)\, dS = \int_0^2 \int_0^{\pi/2} (2\cos u + 2\sin u)2\, du\, dv = 16$$

13. $S: \mathbf{r}(u, v) = 2\cos u\mathbf{i} + 2\sin u\mathbf{j} + v\mathbf{k}, \quad 0 \le u \le \dfrac{\pi}{2}, \quad 0 \le v \le 2$

$$\|\mathbf{r}_u \times \mathbf{r}_v\| = \|2\cos u\mathbf{i} + 2\sin u\mathbf{j}\| = 2$$

$$\iint_S xy\, dS = \int_0^2 \int_0^{\pi/2} 8\cos u \sin u\, du\, dv = 8$$

14. $S: \mathbf{r}(u, v) = 2u\cos v\mathbf{i} + 2u\sin v\mathbf{j} + u\mathbf{k}, \quad 0 \le u \le 4, \quad 0 \le v \le \pi$

$$\|\mathbf{r}_u \times \mathbf{r}_v\| = \|-2u\cos v\mathbf{i} - 2u\sin v\mathbf{j} + 4u\mathbf{k}\| = 2\sqrt{5}u$$

$$\iint_S (x + y)\, dS = \int_0^\pi \int_0^4 (2u\cos v + 2u\sin v)2\sqrt{5}u\, du\, dv = \frac{512\sqrt{5}}{3}$$

15. $f(x, y, z) = x^2 + y^2 + z^2$
$S: z = x + 2, \quad x^2 + y^2 \leq 1$

$$\iint_S f(x, y, z)\, dS = \int_{-1}^{1} \int_{-\sqrt{1-x^2}}^{\sqrt{1-x^2}} [x^2 + y^2 + (x+2)^2]\sqrt{1 + (1)^2 + (0)^2}\, dy\, dx$$

$$= \sqrt{2} \int_0^{2\pi} \int_0^1 [r^2 + (r\cos\theta + 2)^2]\, r\, dr\, d\theta$$

$$= \sqrt{2} \int_0^{2\pi} \int_0^1 [r^2 + r^2 \cos^2\theta + 4r\cos\theta + 4]\, r\, dr\, d\theta$$

$$= \sqrt{2} \int_0^{2\pi} \left[\frac{r^4}{4} + \frac{r^4}{4}\cos^2\theta + \frac{4r^3}{3}\cos\theta + 2r^2\right]_0^1 d\theta$$

$$= \sqrt{2} \int_0^{2\pi} \left[\frac{9}{4} + \left(\frac{1}{4}\right)\frac{1 + \cos 2\theta}{2} + \frac{4}{3}\cos\theta\right] d\theta$$

$$= \sqrt{2}\left[\frac{9}{4}\theta + \frac{1}{8}\left(\theta + \frac{1}{2}\sin 2\theta\right) + \frac{4}{3}\sin\theta\right]_0^{2\pi} = \sqrt{2}\left[\frac{18\pi}{4} + \frac{\pi}{4}\right] = \frac{19\sqrt{2}\,\pi}{4}$$

16. $f(x, y, z) = \dfrac{xy}{z}$
$S: z = x^2 + y^2, \quad 4 \leq x^2 + y^2 \leq 16$

$$\iint_S f(x, y, z)\, dS = \iint_S \frac{xy}{x^2 + y^2}\sqrt{1 + 4x^2 + 4y^2}\, dy\, dx = \int_0^{2\pi} \int_2^4 \frac{r^2 \sin\theta \cos\theta}{r^2}\sqrt{1 + 4r^2}\, r\, dr\, d\theta$$

$$= \int_0^{2\pi} \int_2^4 r\sqrt{1 + 4r^2}\, \sin\theta \cos\theta\, dr\, d\theta = \int_0^{2\pi} \left[\frac{1}{12}(1 + 4r^2)^{3/2}\right]_2^4 \sin\theta \cos\theta\, d\theta$$

$$= \left[\frac{65\sqrt{65} - 17\sqrt{17}}{12}\left(\frac{\sin^2\theta}{2}\right)\right]_0^{2\pi} = 0$$

17. $f(x, y, z) = \sqrt{x^2 + y^2 + z^2}$
$S: z = \sqrt{x^2 + y^2}, \quad x^2 + y^2 \leq 4$

$$\iint_S f(x, y, z)\, dS = \int_{-2}^{2} \int_{-\sqrt{4-x^2}}^{\sqrt{4-x^2}} \sqrt{x^2 + y^2 + \left(\sqrt{x^2 + y^2}\right)^2}\sqrt{1 + \left(\frac{x}{\sqrt{x^2 + y^2}}\right)^2 + \left(\frac{y}{\sqrt{x^2 + y^2}}\right)^2}\, dy\, dx$$

$$= \sqrt{2} \int_{-2}^{2} \int_{-\sqrt{4-x^2}}^{\sqrt{4-x^2}} \sqrt{x^2 + y^2}\sqrt{\frac{x^2 + y^2 + x^2 + y^2}{x^2 + y^2}}\, dy\, dx$$

$$= 2 \int_{-2}^{2} \int_{-\sqrt{4-x^2}}^{\sqrt{4-x^2}} \sqrt{x^2 + y^2}\, dy\, dx = 2 \int_0^{2\pi} \int_0^2 r^2\, dr\, d\theta = 2 \int_0^{2\pi} \left[\frac{r^3}{3}\right]_0^2 d\theta = \left[\frac{16}{3}\theta\right]_0^{2\pi} = \frac{32\pi}{3}$$

18. $f(x, y, z) = \sqrt{x^2 + y^2 + z^2}$
$S: z = \sqrt{x^2 + y^2}, \quad (x - 1)^2 + y^2 \leq 1$

$$\iint_S f(x, y, z)\, dS = \iint_S \sqrt{x^2 + y^2 + \left(\sqrt{x^2 + y^2}\right)^2}\sqrt{1 + \left(\frac{x}{\sqrt{x^2 + y^2}}\right)^2 + \left(\frac{y}{\sqrt{x^2 + y^2}}\right)^2}\, dy\, dx$$

$$= \iint_S \sqrt{2(x^2 + y^2)}\sqrt{\frac{2(x^2 + y^2)}{x^2 + y^2}}\, dy\, dx = 2 \iint_S \sqrt{x^2 + y^2}\, dy\, dx = 2 \int_0^{\pi} \int_0^{2\cos\theta} r^2\, dr\, d\theta$$

$$= \frac{16}{3} \int_0^{\pi} \cos^3\theta\, d\theta = \frac{16}{3} \int_0^{\pi} (1 - \sin^2\theta)\cos\theta\, d\theta = \left[\frac{16}{3}\left(\sin\theta - \frac{\sin^3\theta}{3}\right)\right]_0^{\pi} = 0$$

19. $f(x, y, z) = x^2 + y^2 + z^2$

$S: x^2 + y^2 = 9, \quad 0 \le x \le 3, \quad 0 \le y \le 3, \quad 0 \le z \le 9$

Project the solid onto the yz-plane; $x = \sqrt{9 - y^2}, \quad 0 \le y \le 3, \quad 0 \le z \le 9.$

$$\iint_S f(x, y, z)\, dS = \int_0^3 \int_0^9 [(9 - y^2) + y^2 + z^2]\sqrt{1 + \left(\frac{-y}{\sqrt{9 - y^2}}\right)^2 + (0)^2}\, dz\, dy$$

$$= \int_0^3 \int_0^9 (9 + z^2)\frac{3}{\sqrt{9 - y^2}}\, dz\, dy = \int_0^3 \left[\frac{3}{\sqrt{9 - y^2}}\left(9z + \frac{z^3}{3}\right)\right]_0^9 dy$$

$$= 324 \int_0^3 \frac{3}{\sqrt{9 - y^2}}\, dy = \left[972 \arcsin\left(\frac{y}{3}\right)\right]_0^3 = 972\left(\frac{\pi}{2} - 0\right) = 486\pi$$

20. $f(x, y, z) = x^2 + y^2 + z^2$

$S: x^2 + y^2 = 9, \quad 0 \le x \le 3, \quad 0 \le z \le x$

Project the solid onto the xz-plane; $y = \sqrt{9 - x^2}.$

$$\iint_S f(x, y, z)\, dS = \int_0^3 \int_0^x [x^2 + (9 - x^2) + z^2]\sqrt{1 + \left(\frac{-x}{\sqrt{9 - x^2}}\right)^2 + (0)^2}\, dz\, dx$$

$$= \int_0^3 \int_0^x (9 + z^2)\frac{3}{\sqrt{9 - x^2}}\, dz\, dx = \int_0^3 \left[\frac{3}{\sqrt{9 - x^2}}\left(9z + \frac{z^3}{3}\right)\right]_0^x dx$$

$$= \int_0^3 \frac{3}{\sqrt{9 - x^2}}\left(9x - \frac{x^3}{3}\right) dx = \int_0^3 27x(9 - x^2)^{-1/2}\, dx - \int_0^3 x^3(9 - x^2)^{-1/2}\, dx$$

Let $u = x^2, \quad dv = x(9 - x^2)^{-1/2}\, dx$, then $du = 2x\, dx, \quad v = -\sqrt{9 - x^2}.$

$$= \left[-27\sqrt{9 - x^2}\right]_0^3 - \left[\left[-x^2\sqrt{9 - x^2}\right]_0^3 + \int_0^3 2x\sqrt{9 - x^2}\, dx\right]$$

$$= \left[81 + \frac{2}{3}(9 - x^2)^{3/2}\right]_0^3 = 81 - 18 = 63$$

21. $\mathbf{F}(x, y, z) = 3z\mathbf{i} - 4\mathbf{j} + y\mathbf{k}$

$S: x + y + z = 1$ (first octant)

$G(x, y, z) = x + y + z - 1$

$\nabla G(x, y, z) = \mathbf{i} + \mathbf{j} + \mathbf{k}$

$$\iint_S \mathbf{F} \cdot \mathbf{N}\, dS = \iint_R \mathbf{F} \cdot \nabla G\, dA = \int_0^1 \int_0^{1-x} (3z - 4 + y)\, dy\, dx$$

$$= \int_0^1 \int_0^{1-x} [3(1 - x - y) - 4 + y]\, dy\, dx$$

$$= \int_0^1 \int_0^{1-x} (-1 - 3x - 2y)\, dy\, dx$$

$$= \int_0^1 \left[-y - 3xy - y^2\right]_0^{1-x} dx$$

$$= -\int_0^1 [(1 - x) + 3x(1 - x) + (1 - x)^2]\, dx$$

$$= -\int_0^1 (2 - 2x^2)\, dx = -\frac{4}{3}$$

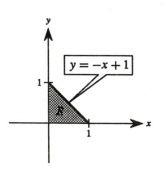

22. $F(x, y, z) = x\mathbf{i} + y\mathbf{j}$

$S: 2x + 3y + z = 6$ (first octant)

$G(x, y, z) = 2x + 3y + z - 6$

$\nabla G(x, y, z) = 2\mathbf{i} + 3\mathbf{j} + \mathbf{k}$

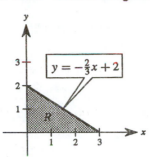

$$\iint_S F \cdot N\, dS = \iint_R F \cdot \nabla G\, dA = \int_0^3 \int_0^{-(2x/3)+2} (2x + 3y)\, dy\, dx$$

$$= \int_0^3 \left[-\frac{4}{3}x^2 + 4x + \frac{3}{2}\left(-\frac{2}{3}x + 2\right)^2 \right] dx$$

$$= \left[-\frac{4}{9}x^3 + 2x^2 - \frac{3}{4}\left(-\frac{2}{3}x + 2\right)^3 \right]_0^3 = 12$$

23. $F(x, y, z) = x\mathbf{i} + y\mathbf{j} + z\mathbf{k}$

$S: z = 9 - x^2 - y^2, \quad 0 \le z$

$G(x, y, z) = x^2 + y^2 + z - 9$

$\nabla G(x, y, z) = 2x\mathbf{i} + 2y\mathbf{j} + \mathbf{k}$

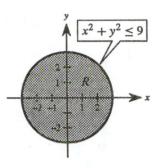

$$\iint_S F \cdot N\, dS = \iint_R F \cdot \nabla G\, dA = \iint_R (2x^2 + 2y^2 + z)\, dA$$

$$= \iint_R [2x^2 + 2y^2 + (9 - x^2 - y^2)]\, dA$$

$$= \iint_R (x^2 + y^2 + 9)\, dA$$

$$= \int_0^{2\pi} \int_0^3 (r^2 + 9) r\, dr\, d\theta$$

$$= \int_0^{2\pi} \left[\frac{r^4}{4} + \frac{9r^2}{2} \right]_0^3 d\theta = \frac{243\pi}{2}$$

24. $F(x, y, z) = x\mathbf{i} + y\mathbf{j} + z\mathbf{k}$

$S: x^2 + y^2 + z^2 = 16$ (first octant)

$z = \sqrt{16 - x^2 - y^2}$

$G(x, y, z) = z - \sqrt{16 - x^2 - y^2}$

$\nabla G(x, y, z) = \dfrac{x}{\sqrt{16 - x^2 - y^2}}\mathbf{i} + \dfrac{y}{\sqrt{16 - x^2 - y^2}}\mathbf{j} + \mathbf{k}$

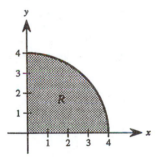

$F \cdot \nabla G = \dfrac{x^2}{\sqrt{16 - x^2 - y^2}} + \dfrac{y^2}{\sqrt{16 - x^2 - y^2}} + z = \dfrac{16}{\sqrt{16 - x^2 - y^2}}$

$$\iint_S F \cdot N\, dS = \iint_R F \cdot \nabla G\, dA = \iint_R \frac{16}{\sqrt{16 - x^2 - y^2}}\, dA$$

$$= \int_0^{\pi/2} \int_0^4 \frac{16}{\sqrt{16 - r^2}} r\, dr\, d\theta \quad \text{(improper integral)}$$

$$= \int_0^{\pi/2} \left[-16\sqrt{16 - r^2} \right]_0^4 d\theta = 32\pi$$

25. $\mathbf{F}(x, y, z) = 4\mathbf{i} - 3\mathbf{j} + 5\mathbf{k}$

$S:\ z = x^2 + y^2,\ \ x^2 + y^2 \leq 4$

$G(x, y, z) = -x^2 - y^2 + z$

$\nabla G(x, y, z) = -2x\mathbf{i} - 2y\mathbf{j} + \mathbf{k}$

$$\iint_S \mathbf{F} \cdot \mathbf{N}\, dS = \iint_R \mathbf{F} \cdot \nabla G\, dA$$

$$= \iint_R (-8x + 6y + 5)\, dA$$

$$= \int_0^{2\pi} \int_0^2 [-8r \cos\theta + 6r \sin\theta + 5] r\, dr\, d\theta$$

$$= \int_0^{2\pi} \left[-\frac{8}{3}r^3 \cos\theta + 2r^3 \sin\theta + \frac{5}{2}r^2 \right]_0^2 d\theta$$

$$= \int_0^{2\pi} \left[-\frac{64}{3} \cos\theta + 16 \sin\theta + 10 \right] d\theta$$

$$= \left[-\frac{64}{3} \sin\theta - 16 \cos\theta + 10\theta \right]_0^{2\pi} = 20\pi$$

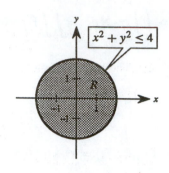

26. $\mathbf{F}(x, y, z) = x\mathbf{i} + y\mathbf{j} - 2z\mathbf{k}$

$S:\ z = \sqrt{a^2 - x^2 - y^2}$

$G(x, y, z) = z - \sqrt{a^2 - x^2 - y^2}$

$\nabla G(x, y, z) = \dfrac{x}{\sqrt{a^2 - x^2 - y^2}}\mathbf{i} + \dfrac{y}{\sqrt{a^2 - x^2 - y^2}}\mathbf{j} + \mathbf{k}$

$\mathbf{F} \cdot \nabla G = \dfrac{x^2}{\sqrt{a^2 - x^2 - y^2}} + \dfrac{y^2}{\sqrt{a^2 - x^2 - y^2}} - 2\sqrt{a^2 - x^2 - y^2}$

$$= \frac{3x^2 + 3y^2 - 2a^2}{\sqrt{a^2 - x^2 - y^2}}$$

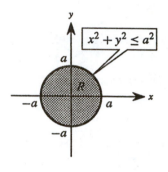

$$\iint_S \mathbf{F} \cdot \mathbf{N}\, dS = \iint_R \mathbf{F} \cdot \nabla G\, dA$$

$$= \iint_R \frac{3x^2 + 3y^2 - 2a^2}{\sqrt{a^2 - x^2 - y^2}}\, dA$$

$$= \int_0^{2\pi} \int_0^a \frac{3r^2 - 2a^2}{\sqrt{a^2 - r^2}} r\, dr\, d\theta$$

$$= 3\int_0^{2\pi} \int_0^a \frac{r^3}{\sqrt{a^2 - r^2}}\, dr\, d\theta - 2a^2 \int_0^{2\pi} \int_0^a \frac{r}{\sqrt{a^2 - r^2}}\, dr\, d\theta$$

$$= 3\left[\int_0^{2\pi} \left[-r^2\sqrt{a^2 - r^2} - \frac{2}{3}(a^2 - r^2)^{3/2} \right]_0^a d\theta \right] - 2a^2 \int_0^{2\pi} \left[-\sqrt{a^2 - r^2} \right]_0^a d\theta$$

$$= 3\int_0^{2\pi} \frac{2}{3}a^3\, d\theta - 2a^2 \int_0^{2\pi} a\, d\theta = 0$$

27. $\mathbf{F}(x,\ y,\ z) = 4xy\mathbf{i} + z^2\mathbf{j} + yz\mathbf{k}$

S: unit cube bounded by $x = 0,\quad x = 1,\quad y = 0,\quad y = 1,\quad z = 0,\quad z = 1$

S_1: The top of the cube

$\mathbf{N} = \mathbf{k},\quad z = 1$

$$\int_{S_1}\int \mathbf{F}\cdot\mathbf{N}\,dS = \int_0^1\int_0^1 y(1)\,dy\,dx = \frac{1}{2}$$

S_2: The bottom of the cube

$\mathbf{N} = -\mathbf{k},\quad z = 0$

$$\int_{S_2}\int \mathbf{F}\cdot\mathbf{N}\,dS = \int_0^1\int_0^1 -y(0)\,dy\,dx = 0$$

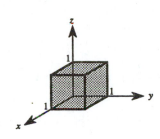

S_3: The front of the cube

$\mathbf{N} = \mathbf{i},\quad x = 1$

$$\int_{S_3}\int \mathbf{F}\cdot\mathbf{N}\,dS = \int_0^1\int_0^1 4(1)y\,dy\,dz = 2$$

S_4: The back of the cube

$\mathbf{N} = -\mathbf{i},\quad x = 0$

$$\int_{S_4}\int \mathbf{F}\cdot\mathbf{N}\,dS = \int_0^1\int_0^1 -4(0)y\,dy\,dz = 0$$

S_5: The right side of the cube

$\mathbf{N} = \mathbf{j},\quad y = 1$

$$\int_{S_5}\int \mathbf{F}\cdot\mathbf{N}\,dS = \int_0^1\int_0^1 z^2\,dz\,dx = \frac{1}{3}$$

S_6: The left side of the cube

$\mathbf{N} = -\mathbf{j},\quad y = 0$

$$\int_{S_6}\int \mathbf{F}\cdot\mathbf{N}\,dS = \int_0^1\int_0^1 -z^2\,dz\,dx = -\frac{1}{3}$$

$$\int_S\int \mathbf{F}\cdot\mathbf{N}\,dS = \frac{1}{2} + 0 + 2 + 0 + \frac{1}{3} - \frac{1}{3} = \frac{5}{2}$$

28. $\mathbf{F}(x,\ y,\ z) = (x + y)\mathbf{i} + y\mathbf{j} + z\mathbf{k}$

S: $z = 1 - x^2 - y^2,\quad z = 0$

$G(x,\ y,\ z) = z + x^2 + y^2 - 1$

$\nabla G(x,\ y,\ z) = 2x\mathbf{i} + 2y\mathbf{j} + \mathbf{k}$

$\mathbf{F}\cdot\nabla G = 2x(x + y) + 2y(y) + (1 - x^2 - y^2) = x^2 + 2xy + y^2 + 1$

$$\int_S\int \mathbf{F}\cdot\mathbf{N}\,dS = \int_R\int \mathbf{F}\cdot\nabla G\,dA$$

$$= \int_R\int (x^2 + 2xy + y^2 + 1)\,dA$$

$$= \int_0^{2\pi}\int_0^1 (r^2 + 2r^2\cos\theta\sin\theta + 1)r\,dr\,d\theta$$

$$= \int_0^{2\pi}\left(\frac{3}{4} + \frac{1}{2}\sin\theta\cos\theta\right)d\theta = \left[\frac{3}{4}\theta + \frac{\sin^2\theta}{4}\right]_0^{2\pi} = \frac{3\pi}{2}$$

The flux across the bottom $z = 0$ is zero.

29. $S: 2x + 3y + 6z = 12$ (first octant) $\Rightarrow z = 2 - \dfrac{1}{3}x - \dfrac{1}{2}y$

$\rho(x, y, z) = x^2 + y^2$

$$m = \iint_R (x^2 + y^2)\sqrt{1 + \left(-\frac{1}{3}\right)^2 + \left(-\frac{1}{2}\right)^2}\, dA$$

$$= \frac{7}{6}\int_0^6 \int_0^{4-(2x/3)} (x^2 + y^2)\, dy\, dx$$

$$= \frac{7}{6}\int_0^6 \left[x^2\left(4 - \frac{2}{3}x\right) + \frac{1}{3}\left(4 - \frac{2}{3}x\right)^3\right] dx = \frac{7}{6}\left[\frac{4}{3}x^3 - \frac{1}{6}x^4 - \frac{1}{8}\left(4 - \frac{2}{3}x\right)^4\right]_0^6 = \frac{364}{3}$$

(graph at top right showing region R with line $y = 4 - \frac{2}{3}x$)

30. $S: z = \sqrt{a^2 - x^2 - y^2}$

$\rho(x, y, z) = kz$

$$m = \iint_S kz\, dS = \iint_R k\sqrt{a^2 - x^2 - y^2}\sqrt{1 + \left(\frac{-x}{\sqrt{a^2 - x^2 - y^2}}\right)^2 + \left(\frac{-y}{\sqrt{a^2 - x^2 - y^2}}\right)^2}\, dA$$

$$= \iint_R k\sqrt{a^2 - x^2 - y^2}\left(\frac{a}{\sqrt{a^2 - x^2 - y^2}}\right) dA$$

$$= \iint_R ka\, dA = ka\iint_R dA = ka(\pi a^2) = ka^3\pi$$

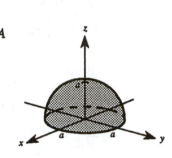

31. $z = \sqrt{x^2 + y^2}, \quad 0 \le z \le a$

$$m = \iint_S k\, dS = k\iint_R \sqrt{1 + \left(\frac{x}{\sqrt{x^2 + y^2}}\right)^2 + \left(\frac{y}{\sqrt{x^2 + y^2}}\right)^2}\, dA = k\iint_R \sqrt{2}\, dA = \sqrt{2}k\pi a^2$$

$$I_z = \iint_S k(x^2 + y^2)\, dS$$

$$= \iint_R k(x^2 + y^2)\sqrt{2}\, dA = \sqrt{2}k\int_0^{2\pi}\int_0^a r^3\, dr\, d\theta = \frac{\sqrt{2}ka^4}{4}(2\pi) = \frac{\sqrt{2}k\pi a^4}{2} = \frac{a^2}{2}\left(\sqrt{2}k\pi a^2\right) = \frac{a^2 m}{2}$$

32. $x^2 + y^2 + z^2 = a^2$

$z = \pm\sqrt{a^2 - x^2 - y^2}$

$$m = 2\iint_S k\, dS = 2k\iint_R \sqrt{1 + \left(\frac{-x}{\sqrt{a^2 - x^2 - y^2}}\right)^2 + \left(\frac{-y}{\sqrt{a^2 - x^2 - y^2}}\right)^2}\, dA$$

$$= 2k\iint_R \frac{a}{\sqrt{a^2 - x^2 - y^2}}\, dA = 2ka\int_0^{2\pi}\int_0^a \frac{r}{\sqrt{a^2 - r^2}}\, dr\, d\theta = 2ka\left[-\sqrt{a^2 - r^2}\right]_0^a (2\pi) = 4\pi ka^2$$

$$I_z = 2\iint_S k(x^2 + y^2)\, dS$$

$$= 2k\iint_R (x^2 + y^2)\frac{a}{\sqrt{a^2 - x^2 - y^2}}\, dA = 2ka\int_0^{2\pi}\int_0^a \frac{r^3}{\sqrt{a^2 - r^2}}\, dr\, d\theta \quad \text{(use integration by parts)}$$

$$= 2ka\left[-r^2\sqrt{a^2 - r^2} - \frac{2}{3}(a^2 - r^2)^{3/2}\right]_0^a (2\pi) = 2ka\left(\frac{2}{3}a^3\right)(2\pi) = \frac{2}{3}a^2(4\pi ka^2) = \frac{2}{3}a^2 m$$

Let $u = r^2, \quad dv = r(a^2 - r^2)^{-1/2}\, dr, \quad du = 2r\, dr, \quad v = -\sqrt{a^2 - r^2}.$

33. $x^2 + y^2 = a^2, \quad 0 \le z \le h$

$\rho(x, y, z) = 1$

$y = \pm\sqrt{a^2 - x^2}$

Project the solid onto the xz-plane.

$$I_z = 4\iint_S (x^2 + y^2)(1)\,dS$$

$$= 4\int_0^h \int_0^a [x^2 + (a^2 - x^2)]\sqrt{1 + \left(\frac{-x}{\sqrt{a^2-x^2}}\right)^2 + (0)^2}\,dx\,dz$$

$$= 4a^3 \int_0^h \int_0^a \frac{1}{\sqrt{a^2 - x^2}}\,dx\,dz$$

$$= 4a^3 \int_0^h \left[\arcsin\frac{x}{a}\right]_0^a dz$$

$$= 4a^3 \left(\frac{\pi}{2}\right)(h) = 2\pi a^3 h$$

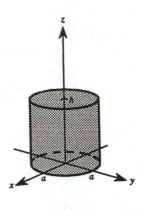

34. $z = x^2 + y^2, \quad 0 \le z \le h$

Project the solid onto the xy-plane.

$$I_z = \iint_S (x^2 + y^2)(1)\,dS$$

$$= \int_{-\sqrt{h}}^{\sqrt{h}} \int_{-\sqrt{h-x^2}}^{\sqrt{h-x^2}} (x^2 + y^2)\sqrt{1 + 4x^2 + 4y^2}\,dy\,dx$$

$$= \int_0^{2\pi} \int_0^{\sqrt{h}} r^2\sqrt{1 + 4r^2}\,r\,dr\,d\theta$$

$$= 2\pi\left[\frac{h}{12}(1 + 4h)^{3/2} - \frac{1}{120}(1 + 4h)^{5/2}\right] + \frac{2\pi}{120}$$

$$= \frac{(1 + 4h)^{3/2}\pi}{60}[10h - (1 + 4h)] + \frac{\pi}{60}$$

$$= \frac{\pi}{60}[(1 + 4h)^{3/2}(6h - 1) + 1]$$

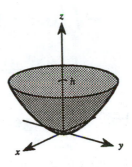

35. $S: z = 16 - x^2 - y^2, \quad z \ge 0$

$\mathbf{F}(x, y, z) = 0.5z\mathbf{k}$

$$\iint_S \rho\mathbf{F}\cdot\mathbf{N}\,dS = \iint_R \rho\mathbf{F}\cdot(-g_x(x, y)\mathbf{i} - g_y(x, y)\mathbf{j} + \mathbf{k})\,dA = \iint_R 0.5\rho z\mathbf{k}\cdot(2x\mathbf{i} + 2y\mathbf{j} + \mathbf{k})\,dA$$

$$= \iint_R 0.5\rho z\,dA = \iint_R 0.5\rho(16 - x^2 - y^2)\,dA$$

$$= 0.5\rho \int_0^{2\pi} \int_0^4 (16 - r^2)r\,dr\,d\theta = 0.5\rho \int_0^{2\pi} 64\,d\theta = 64\pi\rho$$

36. $S: z = \sqrt{16 - x^2 - y^2}$

$\mathbf{F}(x, \ y, \ z) = 0.5z\mathbf{k}$

$$\iint_S \rho\mathbf{F} \cdot \mathbf{N} \, dS = \iint_R \rho\mathbf{F} \cdot (-g_x(x, \ y)\mathbf{i} - g_y(x, \ y)\mathbf{j} + \mathbf{k}) \, dA$$

$$= \iint_R 0.5\rho z\mathbf{k} \cdot \left[\frac{x}{\sqrt{16 - x^2 - y^2}}\mathbf{i} + \frac{y}{\sqrt{16 - x^2 - y^2}}\mathbf{j} + \mathbf{k}\right] dA = \iint_R 0.5\rho z \, dA$$

$$= \iint_R 0.5\rho\sqrt{16 - x^2 - y^2} \, dA = 0.5\rho\int_0^{2\pi}\int_0^4 \sqrt{16 - r^2} \, r \, dr \, d\theta = 0.5\rho\int_0^{2\pi} \frac{64}{3} \, d\theta = \frac{64\pi\rho}{3}$$

37. This surface is not orientable since a normal vector at a point P on the surface will point in the opposite direction if it is moved around the mobius strip one time.

38. Orientable

39. $\mathbf{E} = yz\mathbf{i} + xz\,\mathbf{j} + xy\mathbf{k}$

$S: z = \sqrt{1 - x^2 - y^2}$

$$\iint_S \mathbf{E} \cdot \mathbf{N} \, dS = \iint_R \mathbf{E} \cdot (-g_x(x, \ y)\mathbf{i} - g_y(x, \ y)\,\mathbf{j} + \mathbf{k}) \, dA$$

$$= \iint_R (yz\mathbf{i} + xz\,\mathbf{j} + xy\mathbf{k}) \cdot \left(\frac{x}{\sqrt{1 - x^2 - y^2}}\mathbf{i} + \frac{y}{\sqrt{1 - x^2 - y^2}}\mathbf{j} + \mathbf{k}\right) dA$$

$$= \iint_R \left(\frac{2xyz}{\sqrt{1 - x^2 - y^2}} + xy\right) dA = \iint_R 3xy \, dA = \int_{-1}^1\int_{-\sqrt{1-x^2}}^{\sqrt{1-x^2}} 3xy \, dy \, dx = 0$$

Section 15.7   Divergence Theorem

1. Surface Integral: There are six surfaces to the cube, each with $dS = \sqrt{1} \, dA$.

$z = 0, \quad \mathbf{N} = -\mathbf{k}, \quad \mathbf{F} \cdot \mathbf{N} = -z^2, \quad \int_{S_1}\int 0 \, dA = 0$

$z = a, \quad \mathbf{N} = \mathbf{k}, \quad \mathbf{F} \cdot \mathbf{N} = z^2, \quad \int_{S_2}\int a^2 \, dA = \int_0^a\int_0^a a^2 \, dx \, dy = a^4$

$x = 0, \quad \mathbf{N} = -\mathbf{i}, \quad \mathbf{F} \cdot \mathbf{N} = -2x, \quad \int_{S_3}\int 0 \, dA = 0$

$x = a, \quad \mathbf{N} = \mathbf{i}, \quad \mathbf{F} \cdot \mathbf{N} = 2x, \quad \int_{S_4}\int 2a \, dy \, dz = \int_0^a\int_0^a 2a \, dy \, dz = 2a^3$

$y = 0, \quad \mathbf{N} = -\mathbf{j}, \quad \mathbf{F} \cdot \mathbf{N} = 2y, \quad \int_{S_5}\int 0 \, dA = 0$

$y = a, \quad \mathbf{N} = \mathbf{j}, \quad \mathbf{F} \cdot \mathbf{N} = -2y, \quad \int_{S_6}\int -2a \, dA = \int_0^a\int_0^a -2a \, dz \, dx = -2a^3$

Therefore, $\iint_S \mathbf{F} \cdot \mathbf{N} \, dS = a^4 + 2a^3 - 2a^3 = a^4.$

Divergence Theorem: Since div $\mathbf{F} = 2z$, the Divergence Theorem yields

$$\iiint_Q \text{div } \mathbf{F} \, dV = \int_0^a\int_0^a\int_0^a 2z \, dz \, dy \, dx = \int_0^a\int_0^a a^2 \, dy \, dx = a^4.$$

2. Surface Integral: There are three surfaces to the cylinder.

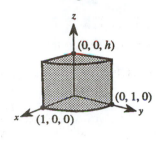

$z = 0, \quad \mathbf{N} = -\mathbf{k}, \quad \mathbf{F} \cdot \mathbf{N} = -z^2$

$$\int_{S_1}\int 0 \, dS = 0$$

$z = h, \quad \mathbf{N} = \mathbf{k}, \quad \mathbf{F} \cdot \mathbf{N} = z^2$

$$\int_{S_2}\int h^2 \, dS = h^2(\text{Area of circle}) = \pi h^2$$

$x^2 + y^2 = 1, \quad \mathbf{N} = \dfrac{2x\mathbf{i} + 2y\mathbf{j}}{2\sqrt{x^2+y^2}} = x\mathbf{i} + y\mathbf{j}, \quad \mathbf{F} \cdot \mathbf{N} = 2x^2 - 2y^2, \quad dS = 2 \, dA$

$$\int_{S_3}\int \mathbf{F} \cdot \mathbf{N} \, dS = \int_{-1}^{1}\int_{-\sqrt{1-x^2}}^{\sqrt{1-x^2}}(4x^2 - 4y^2) \, dy \, dx$$

$$= \int_{0}^{2\pi}\int_{0}^{1}(4\cos^2\theta - 4\sin^2\theta)r \, dr \, d\theta$$

$$= 4\int_{0}^{2\pi}\int_{0}^{1}(\cos 2\theta)r \, dr \, d\theta = 2\int_{0}^{2\pi}\cos 2\theta, \ d\theta = \left[\sin 2\theta\right]_{0}^{2\pi} = 0$$

Therefore, $\displaystyle\iint_{S}\mathbf{F} \cdot \mathbf{N} \, dS = 0 + \pi h^2 + 0 = \pi h^2.$

Divergence Theorem: Since div $\mathbf{F} = 2z$, we have

$$\iiint_{Q} 2z \, dV = \int_{-1}^{1}\int_{-\sqrt{1-x^2}}^{\sqrt{1-x^2}}\int_{0}^{h} 2z \, dz \, dy \, dx = h^2\int_{-1}^{1}\int_{-\sqrt{1-x^2}}^{\sqrt{1-x^2}}dy \, dx = h^2(\text{Area of circle}) = \pi h^2.$$

3. Surface Integral: There are four surfaces to this solid.

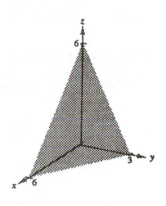

$z = 0, \quad \mathbf{N} = -\mathbf{k}, \quad \mathbf{F} \cdot \mathbf{N} = -z$

$$\int_{S_1}\int 0 \, dS = 0$$

$y = 0, \quad \mathbf{N} = -\mathbf{j}, \quad \mathbf{F} \cdot \mathbf{N} = 2y - z, \quad dS = dA = dx \, dz$

$$\int_{S_2}\int -z \, dS = \int_{0}^{6}\int_{0}^{6-z} -z \, dx \, dz = \int_{0}^{6}(z^2 - 6z) \, dz = -36$$

$x = 0, \quad \mathbf{N} = -\mathbf{i}, \quad \mathbf{F} \cdot \mathbf{N} = y - 2x, \quad dS = dA = dz \, dy$

$$\int_{S_3}\int y \, dS = \int_{0}^{3}\int_{0}^{6-2y} y \, dz \, dy = \int_{0}^{3}(6y - 2y^2) \, dy = 9$$

$x + 2y + z = 6, \quad \mathbf{N} = \dfrac{\mathbf{i} + 2\mathbf{j} + \mathbf{k}}{\sqrt{6}}, \quad \mathbf{F} \cdot \mathbf{N} = \dfrac{2x - 5y + 3z}{\sqrt{6}}, \quad dS = \sqrt{6} \, dA$

$$\int_{S_4}\int (2x - 5y + 3z) \, dz \, dy = \int_{0}^{3}\int_{0}^{6-2y}(18 - x - 11y) \, dx \, dy = \int_{0}^{3}(90 - 90y + 20y^2) \, dy = 45$$

Therefore, $\displaystyle\iint_{S}\mathbf{F} \cdot \mathbf{N} \, dS = 0 - 36 + 9 + 45 = 18.$

Divergence Theorem: Since div $\mathbf{F} = 1$, we have

$$\iiint_{Q} dV = (\text{Volume of solid}) = \frac{1}{3}(\text{Area of base}) \times (\text{Height}) = \frac{1}{3}(9)(6) = 18.$$

4. $\mathbf{F}(x, \; y, \; z) = xy\mathbf{i} + z\mathbf{j} + (x + y)\mathbf{k}$

   S: surface bounded by the planes $y = 4, \;\; z = 4 - x$ and the coordinate planes

   Surface Integral: There are five surfaces to this solid.

$z = 0, \;\; \mathbf{N} = -\mathbf{k}, \;\; \mathbf{F} \cdot \mathbf{N} = -(x + y)$

$$\int_{S_1}\int -(x + y)\, dS = \int_0^4 \int_0^4 -(x + y)\, dy\, dx = -\int_0^4 (4x + 8)\, dx = -64$$

$y = 0, \;\; \mathbf{N} = -\mathbf{j}, \;\; \mathbf{F} \cdot \mathbf{N} = -z$

$$\int_{S_2}\int -z\, dS = \int_0^4 \int_0^{4-x} -z\, dz\, dx = -\int_0^4 \frac{(4 - x)^2}{2}\, dx = -\frac{32}{3}$$

$y = 4, \;\; \mathbf{N} = \mathbf{j}, \;\; \mathbf{F} \cdot \mathbf{N} = z$

$$\int_{S_3}\int z\, dS = \int_0^4 \int_0^{4-x} z\, dz\, dx = \int_0^4 \frac{(4 - x)^2}{2}\, dx = \frac{32}{3}$$

$x = 0, \;\; \mathbf{N} = -\mathbf{i}, \;\; \mathbf{F} \cdot \mathbf{N} = -xy$

$$\int_{S_4}\int -xy\, dS = \int_0^4 \int_0^4 0\, dS = 0$$

$x + z = 4, \;\; \mathbf{N} = \dfrac{\mathbf{i} + \mathbf{k}}{\sqrt{2}}, \;\; \mathbf{F} \cdot \mathbf{N} = \dfrac{1}{\sqrt{2}}[xy + x + y], \;\; dS = \sqrt{2}\, dA$

$$\int_{S_5}\int \frac{1}{\sqrt{2}}[xy + x + y]\sqrt{2}\, dA = \int_0^4 \int_0^4 (xy + x + y)\, dy\, dx = 128$$

Therefore, $\displaystyle\int_S\int \mathbf{F} \cdot \mathbf{N}\, dS = -64 - \frac{32}{3} + \frac{32}{3} + 0 + 128 = 64.$

Divergence Theorem: Since div $\mathbf{F} = y$, we have $\displaystyle\int\int\int_Q \text{div }\mathbf{F}\, dV = \int_0^4 \int_0^4 \int_0^{4-x} y\, dz\, dy\, dx = 64.$

5. Since div $\mathbf{F} = 2x + 2y + 2z$, we have

$$\int\int\int_Q \text{div }\mathbf{F}\, dV = \int_0^a \int_0^a \int_0^a (2x + 2y + 2z)\, dz\, dy\, dx$$

$$= \int_0^a \int_0^a (2ax + 2ay + a^2)\, dy\, dx = \int_0^a (2a^2 x + 2a^3)\, dx = \left[a^2 x^2 + 2a^3 x\right]_0^a = 3a^4.$$

6. Since div $\mathbf{F} = 2xz - 1 + xy$

$$\int\int\int_Q \text{div }\mathbf{F}\, dV = \int_0^a \int_0^a \int_0^a (2xz - 1 + xy)\, dz\, dy\, dx$$

$$= \int_0^a \int_0^a (a^2 x - a + axy)\, dy\, dx$$

$$= \int_0^a \left(a^3 x - a^2 + \frac{a^2}{2}x\right) dx = \frac{a^5}{2} - a^3 + \frac{a^4}{4} = \frac{a^3}{4}(2a^2 + a - 4).$$

7. Since div $\mathbf{F} = 2x - 2x + 2xyz = 2xyz$

$$\iiint_Q \operatorname{div} \mathbf{F}\, dV = \iiint_Q 2xyz\, dV = \int_0^a \int_0^{2\pi} \int_0^{\pi/2} 2(\rho \sin \phi \cos \theta)(\rho \sin \phi \sin \theta)(\rho \cos \phi)\rho^2 \sin \phi\, d\phi\, d\theta\, d\rho$$

$$= \int_0^a \int_0^{2\pi} \int_0^{\pi/2} 2\rho^5 (\sin \theta \cos \theta)(\sin^3 \phi \cos \phi)\, d\phi\, d\theta\, d\rho$$

$$= \int_0^a \int_0^{2\pi} \frac{1}{2}\rho^5 \sin \theta \cos \theta\, d\theta\, d\rho = \int_0^a \left[\left(\frac{\rho^5}{2}\right) \frac{\sin^2 \theta}{2} \right]_0^{2\pi} d\rho = 0.$$

8. Since div $\mathbf{F} = y + z - y = z$, we have

$$\iiint_Q \operatorname{div} \mathbf{F}\, dV = \int_{-a}^a \int_{-\sqrt{a^2-x^2}}^{\sqrt{a^2-x^2}} \int_0^{\sqrt{a^2-x^2-y^2}} z\, dz\, dy\, dx = \int_0^{2\pi} \int_0^a \int_0^{\sqrt{a^2-r^2}} zr\, dz\, dr\, d\theta$$

$$= \int_0^{2\pi} \int_0^a \left[\frac{a^2 r}{2} - \frac{r^3}{2} \right] dr\, d\theta = \int_0^{2\pi} \left[\frac{a^2 r^2}{4} - \frac{r^4}{8} \right]_0^a d\theta = \int_0^{2\pi} \frac{a^4}{8}\, d\theta = \frac{\pi a^4}{4}.$$

9. Since div $\mathbf{F} = 3$, we have $\iiint_Q 3\, dV = 3(\text{Volume of sphere}) = 3\left[\frac{4}{3}\pi (2)^3 \right] = 32\pi.$

10. Since div $\mathbf{F} = xz$, we have

$$\iiint_Q xz\, dV = \int_0^4 \int_{-3}^3 \int_{-\sqrt{9-y^2}}^{\sqrt{9-y^2}} xz\, dx\, dy\, dz = \int_0^4 \int_{-3}^3 \frac{z}{2}(0)\, dy\, dz = 0.$$

11. Since div $\mathbf{F} = 1 + 2y - 1 = 2y$, we have

$$\iiint_Q 2y\, dV = \int_0^4 \int_{-3}^3 \int_{-\sqrt{9-y^2}}^{\sqrt{9-y^2}} 2y\, dx\, dy\, dz = \int_0^4 \int_{-3}^3 4y\sqrt{9-y^2}\, dy\, dz = \int_0^4 \left[-\frac{4}{3}(9-y^2)^{3/2} \right]_{-3}^3 dz = 0.$$

12. Since div $\mathbf{F} = y^2 + x^2 + e^z$, we have

$$\iiint_Q (x^2 + y^2 + e^z)\, dV = \int_0^2 \int_{-\sqrt{4-x^2}}^{\sqrt{4-x^2}} \int_{\sqrt{x^2+y^2}}^4 (x^2 + y^2 + e^z)\, dz\, dy\, dx$$

$$= \int_0^{2\pi} \int_0^2 \int_r^4 (r^2 + e^z)r\, dz\, dr\, d\theta = \int_0^{2\pi} \int_0^2 (4r^3 + re^4 - r^4 - re^r)\, dr\, d\theta$$

$$= \int_0^{2\pi} \left[r^4 + \frac{r^2}{2}e^4 - \frac{r^5}{5} - re^r + e^r \right]_0^2 d\theta = \left[16 + 2e^4 - \frac{32}{5} - 2e^2 + e^2 - 1 \right](2\pi)$$

$$= \left[\frac{43}{5} + 2e^4 - e^2 \right](2\pi) = \frac{2\pi}{5}(10e^4 - 5e^2 + 43).$$

13. Since div $\mathbf{F} = 3x^2 + x^2 + 0 = 4x^2$, we have

$$\iiint_Q 4x^2\, dV = \int_0^6 \int_0^4 \int_0^{4-y} 4x^2\, dz\, dy\, dx = \int_0^6 \int_0^4 4x^2(4-y)\, dy\, dx = \int_0^6 32x^2\, dx = 2304.$$

14. Since div $\mathbf{F} = e^z + e^z + e^z = 3e^z$, we have

$$\iiint_Q 3e^z\, dV = \int_0^6 \int_0^4 \int_0^{4-y} 3e^z\, dz\, dy\, dx = \int_0^6 \int_0^4 3[e^{4-y} - 1]\, dy\, dx = \int_0^6 3(e^4 - 5)\, dx = 18(e^4 - 5).$$

15. Using the triple integral to find volume, we need $\mathbf{F}$ so that

$$\text{div } \mathbf{F} = \frac{\partial M}{\partial x} + \frac{\partial N}{\partial y} + \frac{\partial P}{\partial z} = 1.$$

Hence, we could have $\mathbf{F} = x\mathbf{i}$, $\mathbf{F} = y\mathbf{j}$, or $\mathbf{F} = z\mathbf{k}$.

For $dA = dy\,dz$ consider $\mathbf{F} = x\mathbf{i}$, $x = f(y, z)$, then $\mathbf{N} = \dfrac{\mathbf{i} + f_y\mathbf{j} + f_z\mathbf{k}}{\sqrt{1 + f_y{}^2 + f_z{}^2}}$ and $dS = \sqrt{1 + f_y{}^2 + f_z{}^2}\,dy\,dz$.

For $dA = dz\,dx$ consider $\mathbf{F} = y\mathbf{j}$, $y = f(x, z)$, then $\mathbf{N} = \dfrac{f_x\mathbf{i} + \mathbf{j} + f_z\mathbf{k}}{\sqrt{1 + f_x{}^2 + f_z{}^2}}$ and $dS = \sqrt{1 + f_x{}^2 + f_z{}^2}\,dz\,dx$.

For $dA = dx\,dy$ consider $\mathbf{F} = z\mathbf{k}$, $z = f(x, y)$, then $\mathbf{N} = \dfrac{f_x\mathbf{i} + f_y\mathbf{j} + \mathbf{k}}{\sqrt{1 + f_x{}^2 + f_y{}^2}}$ and $dS = \sqrt{1 + f_x{}^2 + f_y{}^2}\,dx\,dy$.

Correspondingly, we then have $V = \displaystyle\iint_S \mathbf{F} \cdot \mathbf{N}\,dS = \iint_S x\,dy\,dz = \iint_S y\,dz\,dx = \iint_S z\,dx\,dy.$

16. $\displaystyle\int_0^a \int_0^a x\,dy\,dz = \int_0^a \int_0^a a\,dy\,dz = \int_0^a a^2\,dz = a^3$

Similarly, $\displaystyle\int_0^a \int_0^a y\,dz\,dx = \int_0^a \int_0^a z\,dx\,dy = a^3.$

17. Using the Divergence Theorem, we have

$$\iint_S \text{curl } \mathbf{F} \cdot \mathbf{N}\,dS = \iiint_Q \text{div } (\text{curl } \mathbf{F})\,dV$$

$$\text{curl } \mathbf{F}(x, y, z) = \begin{vmatrix} \mathbf{i} & \mathbf{j} & \mathbf{k} \\ \dfrac{\partial}{\partial x} & \dfrac{\partial}{\partial y} & \dfrac{\partial}{\partial z} \\ 4xy + z^2 & 2x^2 + 6yz & 2xz \end{vmatrix} = -6y\mathbf{i} - (2z - 2z)\mathbf{j} + (4x - 4x)\mathbf{k} = -6y\mathbf{i}$$

$$\text{div } (\text{curl } \mathbf{F}) = 0.$$

Therefore, $\displaystyle\iiint_Q \text{div } (\text{curl } \mathbf{F})\,dV = 0.$

18. Using the Divergence Theorem, we have

$$\iint_S \text{curl } \mathbf{F} \cdot \mathbf{N}\,dS = \iiint_Q \text{div } (\text{curl } \mathbf{F})\,dV$$

$$\text{curl } \mathbf{F}(x, y, z) = \begin{vmatrix} \mathbf{i} & \mathbf{j} & \mathbf{k} \\ \dfrac{\partial}{\partial x} & \dfrac{\partial}{\partial y} & \dfrac{\partial}{\partial z} \\ xy\cos z & yz\sin x & xyz \end{vmatrix} = (xz - y\sin x)\mathbf{i} - (yz + xy\sin z)\mathbf{j} + (yz\cos x - x\cos z)\mathbf{k}.$$

Now, div curl $\mathbf{F}(x, y, z) = (z - y\cos x) - (z + x\sin z) + (y\cos x + x\sin z) = 0$. Therefore,

$$\iint_S \text{curl } \mathbf{F} \cdot \mathbf{N}\,dS = \iiint_Q \text{div } (\text{curl } \mathbf{F})\,dV = 0.$$

19. Using the Divergence Theorem, we have $\iint_S \text{curl } \mathbf{F} \cdot \mathbf{N} \, dS = \iiint_Q \text{div (curl F)} \, dV$. Let

$$\mathbf{F}(x, y, z) = M\mathbf{i} + N\mathbf{j} + P\mathbf{k}$$

$$\text{curl } \mathbf{F} = \left(\frac{\partial P}{\partial y} - \frac{\partial N}{\partial z}\right)\mathbf{i} - \left(\frac{\partial P}{\partial x} - \frac{\partial M}{\partial z}\right)\mathbf{j} + \left(\frac{\partial N}{\partial x} - \frac{\partial M}{\partial y}\right)\mathbf{k}$$

$$\text{div (curl F)} = \frac{\partial^2 P}{\partial x \partial y} - \frac{\partial^2 N}{\partial x \partial z} - \frac{\partial^2 P}{\partial y \partial x} + \frac{\partial^2 M}{\partial y \partial z} + \frac{\partial^2 N}{\partial z \partial x} - \frac{\partial^2 M}{\partial z \partial y} = 0.$$

Therefore, $\iint_S \text{curl } \mathbf{F} \cdot \mathbf{N} \, dS = \iiint_Q 0 \, dV = 0.$

20. If $\mathbf{F}(x, y, z) = a_1\mathbf{i} + a_2\mathbf{j} + a_3\mathbf{k}$, then div $\mathbf{F} = 0$.

Therefore, $\iint_S \mathbf{F} \cdot \mathbf{N} \, dS = \iiint_Q \text{div } \mathbf{F} \, dV = \iiint_Q 0 \, dV = 0.$

21. If $\mathbf{F}(x, y, z) = x\mathbf{i} + y\mathbf{j} + z\mathbf{k}$, then div $\mathbf{F} = 3$.

$$\iint_S \mathbf{F} \cdot \mathbf{N} \, dS = \iiint_Q \text{div } \mathbf{F} \, dV = \iiint_Q 3 \, dV = 3V.$$

22. If $\mathbf{F}(x, y, z) = x\mathbf{i} + y\mathbf{j} + z\mathbf{k}$, then div $\mathbf{F} = 3$.

$$\frac{1}{\|\mathbf{F}\|} \iint_S \mathbf{F} \cdot \mathbf{N} \, dS = \frac{1}{\|\mathbf{F}\|} \iiint_Q \text{div } \mathbf{F} \, dV = \frac{1}{\|\mathbf{F}\|} \iiint_Q 3 \, dV = \frac{3}{\|\mathbf{F}\|} \iiint_Q dV$$

23. $\iint_S f D_\mathbf{N} g \, dS = \iint_S f \nabla g \cdot \mathbf{N} \, dS$

$$= \iiint_Q \text{div } (f \nabla g) \, dV$$

$$= \iiint_Q (f \text{div } \nabla g + \nabla f \cdot \nabla g) \, dV$$

$$= \iiint_Q (f \nabla^2 g + \nabla f \cdot \nabla g) \, dV$$

24. $\iint_S (f D_\mathbf{N} g - g D_\mathbf{N} f) \, dS = \iint_S f D_\mathbf{N} g \, dS - \iint_S g D_\mathbf{N} f \, dS$

$$= \iiint_Q (f \nabla^2 g + \nabla f \cdot \nabla g) \, dV - \iiint_Q (g \nabla^2 f + \nabla g \cdot \nabla f) \, dV$$

$$= \iiint_Q (f \nabla^2 g - g \nabla^2 f) \, dV$$

Section 15.8   Stoke's Theorem

1. $F(x, y, z) = (2y - z)\mathbf{i} + xyz\mathbf{j} + e^z\mathbf{k}$

$$\mathbf{curl\ F} = \begin{vmatrix} \mathbf{i} & \mathbf{j} & \mathbf{k} \\ \dfrac{\partial}{\partial x} & \dfrac{\partial}{\partial y} & \dfrac{\partial}{\partial z} \\ 2y - z & xyz & e^z \end{vmatrix}$$

$$= -xy\mathbf{i} - \mathbf{j} + (yz - 2)\mathbf{k}$$

2. $F(x, y, z) = x^2\mathbf{i} + y^2\mathbf{j} + x^2\mathbf{k}$

$$\mathbf{curl\ F} = \begin{vmatrix} \mathbf{i} & \mathbf{j} & \mathbf{k} \\ \dfrac{\partial}{\partial x} & \dfrac{\partial}{\partial y} & \dfrac{\partial}{\partial z} \\ x^2 & y^2 & x^2 \end{vmatrix} = -2x\mathbf{j}$$

3. $F(x, y, z) = 2z\mathbf{i} - 4x^2\mathbf{j} + \arctan x\mathbf{k}$

$$\mathbf{curl\ F} = \begin{vmatrix} \mathbf{i} & \mathbf{j} & \mathbf{k} \\ \dfrac{\partial}{\partial x} & \dfrac{\partial}{\partial y} & \dfrac{\partial}{\partial z} \\ 2z & -4x^2 & \arctan x \end{vmatrix}$$

$$= \left(2 - \dfrac{1}{1 + x^2}\right)\mathbf{j} - 8x\mathbf{k}$$

4. $F(x, y, z) = x \sin y\mathbf{i} - y \cos x\mathbf{j} + yz^2\mathbf{k}$

$$\mathbf{curl\ F} = \begin{vmatrix} \mathbf{i} & \mathbf{j} & \mathbf{k} \\ \dfrac{\partial}{\partial x} & \dfrac{\partial}{\partial y} & \dfrac{\partial}{\partial z} \\ x \sin y & -y \cos x & yz^2 \end{vmatrix}$$

$$= z^2\mathbf{i} + (y \sin x - x \cos y)\mathbf{k}$$

5. $F(x, y, z) = e^{x^2 + y^2}\mathbf{i} + e^{y^2 + z^2}\mathbf{j} + xyz\mathbf{k}$

$$\mathbf{curl\ F} = \begin{vmatrix} \mathbf{i} & \mathbf{j} & \mathbf{k} \\ \dfrac{\partial}{\partial x} & \dfrac{\partial}{\partial y} & \dfrac{\partial}{\partial z} \\ e^{x^2 + y^2} & e^{y^2 + z^2} & xyz \end{vmatrix} = (xz - 2ze^{y^2 + z^2})\mathbf{i} - yz\mathbf{j} - 2ye^{x^2 + y^2}\mathbf{k} = z(x - 2e^{y^2 + z^2})\mathbf{i} - yz\mathbf{j} - 2ye^{x^2 + y^2}\mathbf{k}$$

6. $F(x, y, z) = \arcsin y\mathbf{i} + \sqrt{1 - x^2}\,\mathbf{j} + y^2\mathbf{k}$

$$\mathbf{curl\ F} = \begin{vmatrix} \mathbf{i} & \mathbf{j} & \mathbf{k} \\ \dfrac{\partial}{\partial x} & \dfrac{\partial}{\partial y} & \dfrac{\partial}{\partial z} \\ \arcsin y & \sqrt{1 - x^2} & y^2 \end{vmatrix} = 2y\mathbf{i} + \left[\dfrac{-x}{\sqrt{1 - x^2}} - \dfrac{1}{\sqrt{1 - y^2}}\right]\mathbf{k} = 2y\mathbf{i} - \left[\dfrac{x}{\sqrt{1 - x^2}} + \dfrac{1}{\sqrt{1 - y^2}}\right]\mathbf{k}$$

7. In this case, $M = -y + z$, $N = x - z$, $P = x - y$ and C is the circle $x^2 + y^2 = 1$, $z = 0$, $dz = 0$.

Line Integral: $\displaystyle\int_C \mathbf{F} \cdot d\mathbf{r} = \int_C -y\,dx + x\,dy$

Letting $x = \cos t$, $y = \sin t$, we have $dx = -\sin t\,dt$, $dy = \cos t\,dt$ and

$$\int_C -y\,dx + x\,dy = \int_0^{2\pi}(\sin^2 t + \cos^2 t)\,dt = 2\pi.$$

Double Integral: Consider $F(x, y, z) = x^2 + y^2 + z^2 - 1$. Then

$$\mathbf{N} = \dfrac{\nabla F}{\|\nabla F\|} = \dfrac{2x\mathbf{i} + 2y\mathbf{j} + 2z\mathbf{k}}{2\sqrt{x^2 + y^2 + z^2}} = x\mathbf{i} + y\mathbf{j} + z\mathbf{k}.$$

Since

$$z^2 = 1 - x^2 - y^2, \quad z_x = \dfrac{-2x}{2z} = \dfrac{-x}{z}, \text{ and } z_y = \dfrac{-y}{z}, \quad dS = \sqrt{1 + \dfrac{x^2}{z^2} + \dfrac{y^2}{z^2}}\,dA = \dfrac{1}{z}\,dA.$$

Now, since $\mathbf{curl\ F} = 2\mathbf{k}$, we have

$$\iint_S (\mathbf{curl\ F}) \cdot \mathbf{N}\,dS = \iint_R 2z\left(\dfrac{1}{z}\right)\,dA = \iint_R 2\,dA = 2(\text{Area of circle of radius 1}) = 2\pi.$$

8. In this case C is the circle $x^2 + y^2 = 4$, $z = 0$, $dz = 0$.

Line Integral: $\displaystyle\int_C \mathbf{F} \cdot d\mathbf{r} = \int_C -y\,dx + x\,dy$

Let $x = 2\cos t$, $y = 2\sin t$, then $dx = -2\sin t\,dt$, $dy = 2\cos t\,dt$, and $\displaystyle\int_C -y\,dx + x\,dy = \int_0^{2\pi} 4\,dt = 8\pi$.

Double Integral: $F(x,\ y,\ z) = z + x^2 + y^2 - 4$, $\quad \mathbf{N} = \dfrac{\nabla F}{\|\nabla F\|} = \dfrac{2x\mathbf{i} + 2y\mathbf{j} + \mathbf{k}}{\sqrt{1 + 4x^2 + 4y^2}}$, $\quad dS = \sqrt{1 + 4x^2 + 4y^2}\,dA$

curl F $= 2\mathbf{k}$, therefore

$$\iint_R (\textbf{curl F}) \cdot \mathbf{N}\,dS = \iint_R 2\,dA = \int_{-2}^{2} \int_{-\sqrt{4-x^2}}^{\sqrt{4-x^2}} 2\,dy\,dx = 2\int_{-2}^{2} 2\sqrt{4 - x^2}\,dx$$

$$= 4\int_{-2}^{2} \sqrt{4 - x^2}\,dx = 2\left[x\sqrt{4 - x^2} + 4\arcsin\frac{x}{2} \right]_{-2}^{2} = 8\pi.$$

9. Line Integral: From the accompanying figure we see that for

C_1: $z = 0$, $dz = 0$

C_2: $x = 0$, $dx = 0$

C_3: $y = 0$, $dy = 0$.

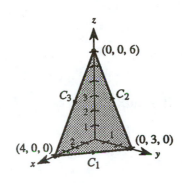

Hence, $\displaystyle\int_C \mathbf{F} \cdot d\mathbf{r} = \int_C xyz\,dx + y\,dy + z\,dz$

$$= \int_{C_1} y\,dy + \int_{C_2} y\,dy + z\,dz + \int_{C_3} z\,dz$$

$$= \int_0^3 y\,dy + \int_3^0 y\,dy + \int_0^6 z\,dz + \int_6^0 z\,dz = 0.$$

Double Integral: curl F $= xy\mathbf{j} - xz\mathbf{k}$

Considering $F(x,\ y,\ z) = 3x + 4y + 2z - 12$, then

$$\mathbf{N} = \frac{\nabla F}{\|\nabla F\|} = \frac{3\mathbf{i} + 4\mathbf{j} + 2\mathbf{k}}{\sqrt{29}} \quad \text{and} \quad dS = \sqrt{29}\,dA.$$

Thus,

$$\iint_S (\textbf{curl F}) \cdot \mathbf{N}\,dS = \iint_R (4xy - 2xz)\,dy\,dx$$

$$= \int_0^4 \int_0^{(-3x+12)/4} \left[4xy - 2x\left(6 - 2y - \frac{3}{2}x \right) \right] dy\,dx$$

$$= \int_0^4 \int_0^{(12-3x)/4} (8xy + 3x^2 - 12x)\,dy\,dx$$

$$= \int_0^4 0\,dx = 0.$$

10. Line Integral: From the accompanying figure we see that for

C_1: $x = 0$, $z = 0$, $dx = dz = 0$

C_2: $z = x^2$, $y = a$, $dy = 0$, $dz = 2x\,dx$

C_3: $x = a$, $z = a$, $dx = dz = 0$

C_4: $z = x^2$, $y = 0$, $dy = 0$, $dz = 2x\,dx$.

Hence,

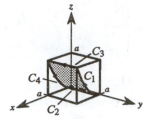

$$\int_C \mathbf{F} \cdot d\mathbf{r} = \int_C z^2\,dx + x^2\,dy + y^2\,dz$$

$$= \int_{C_1} 0 + \int_{C_2} x^4\,dx + a^2(2x\,dx) + \int_{C_3} a^2\,dy + \int_{C_4} x^4\,dx$$

$$= \int_0^a x^4\,dx + \int_0^a 2a^2 x\,dx + \int_a^0 a^2\,dy + \int_a^0 x^4\,dx = \Big[a^2 x^2\Big]_0^a + \Big[a^2 y\Big]_a^0 = a^4 - a^3 = a^3(a-1).$$

Double Integral: Since $F(x,\ y,\ z) = x^2 - z$, we have

$$\mathbf{N} = \frac{2x\mathbf{i} - \mathbf{k}}{\sqrt{1+4x^2}} \text{ and } dS = \sqrt{1+4x^2}\,dA.$$

Furthermore, **curl F** $= 2y\mathbf{i} + 2z\mathbf{j} + 2x\mathbf{k}$. Therefore,

$$\iint_S (\mathbf{curl\ F}) \cdot \mathbf{N}\,dS = \int_0^a \int_0^a (4xy - 2x)\,dy\,dx = \int_0^a (2a^2 x - 2ax)\,dx = \Big[a^2 x^2 - ax^2\Big]_0^a = a^3(a-1).$$

11. Let $A = (0,\ 0,\ 0)$, $B = (1,\ 1,\ 1)$ and $C = (0,\ 2,\ 0)$. Then $\mathbf{U} = \overrightarrow{AB} = \mathbf{i} + \mathbf{j} + \mathbf{k}$ and $\mathbf{V} = \overrightarrow{AC} = 2\mathbf{j}$. Thus,

$$\mathbf{N} = \frac{\mathbf{U} \times \mathbf{V}}{\|\mathbf{U} \times \mathbf{V}\|} = \frac{-2\mathbf{i} + 2\mathbf{k}}{2\sqrt{2}} = \frac{-\mathbf{i} + \mathbf{k}}{\sqrt{2}}.$$

Surface S has direction numbers -1, 0, 1, with equation $z - x = 0$ and $dS = \sqrt{2}\,dA$. Since **curl F** $= -3\mathbf{i} + \mathbf{j} - 2\mathbf{k}$, we have

$$\iint_S (\mathbf{curl\ F}) \cdot \mathbf{N}\,dS = \iint_R \frac{1}{\sqrt{2}}(\sqrt{2})\,dA$$

$$= \iint_R dA = (\text{Area of triangle with } a = 1,\ b = 2) = 1.$$

12. Let $A = (0,\ 0,\ 0)$, $B = (1,\ 1,\ 1)$, and $C = (0,\ 0,\ 2)$. Then $\mathbf{U} = \overrightarrow{AB} = \mathbf{i} + \mathbf{j} + \mathbf{k}$, $\mathbf{V} = \overrightarrow{AC} = 2\mathbf{k}$, and

$$\mathbf{N} = \frac{\mathbf{U} \times \mathbf{V}}{\|\mathbf{U} \times \mathbf{V}\|} = \frac{2\mathbf{i} - 2\mathbf{j}}{2\sqrt{2}} = \frac{\mathbf{i} - \mathbf{j}}{\sqrt{2}}.$$

Hence, $F(x,\ y,\ z) = x - y$ and $dS = \sqrt{2}\,dA$. Since

$$\mathbf{curl\ F} = \frac{2x}{x^2 + y^2}\mathbf{k},$$

we have

$$\iint_S (\mathbf{curl\ F}) \cdot \mathbf{N}\,dS = \iint_R 0\,dS = 0.$$

13. $F(x, y, z) = z^2 i + x^2 j + y^2 k, \quad S: z = 4 - x^2 - y^2, \quad 0 \leq z$

$$\text{curl } F = \begin{vmatrix} i & j & k \\ \dfrac{\partial}{\partial x} & \dfrac{\partial}{\partial y} & \dfrac{\partial}{\partial z} \\ z^2 & x^2 & y^2 \end{vmatrix} = 2y i + 2z j + 2x k$$

$G(x, y, z) = x^2 + y^2 + z - 4$

$\nabla G(x, y, z) = 2x i + 2y j + k$

$$\iint_S (\text{curl } F) \cdot N \, dS = \iint_R (4xy + 4yz + 2x) \, dA = \int_{-2}^{2} \int_{-\sqrt{4-x^2}}^{\sqrt{4-x^2}} [4xy + 4y(4 - x^2 - y^2) + 2x] \, dy \, dx$$

$$= \int_{-2}^{2} \int_{-\sqrt{4-x^2}}^{\sqrt{4-x^2}} [4xy + 16y - 4x^2 y - 4y^3 + 2x] \, dy \, dx$$

$$= \int_{-2}^{2} 4x\sqrt{4 - x^2} \, dx = 0$$

14. $F(x, y, z) = 4xz i + y j + 4xy k, \quad S: z = 4 - x^2 - y^2, \quad 0 \leq z$

$$\text{curl } F = \begin{vmatrix} i & j & k \\ \dfrac{\partial}{\partial x} & \dfrac{\partial}{\partial y} & \dfrac{\partial}{\partial z} \\ 4xz & y & 4xy \end{vmatrix} = 4x i + (4x - 4y) j$$

$G(x, y, z) = x^2 + y^2 + z - 4$

$\nabla G(x, y, z) = 2x i + 2y j + k$

$$\iint_S (\text{curl } F) \cdot N \, dS = \iint_R [8x^2 + 2y(4x - 4y)] \, dA$$

$$= \int_{-2}^{2} \int_{-\sqrt{4-x^2}}^{\sqrt{4-x^2}} [8x^2 + 8xy - 8y^2] \, dy \, dx$$

$$= \int_{-2}^{2} \left[8x^2 y + 4xy^2 - \frac{8}{3} y^3 \right]_{-\sqrt{4-x^2}}^{\sqrt{4-x^2}} dx$$

$$= \int_{-2}^{2} \left[16x^2 \sqrt{4 - x^2} - \frac{16}{3}(4 - x^2)\sqrt{4 - x^2} \right] dx$$

$$= \int_{-2}^{2} \left[\frac{64}{3} x^2 \sqrt{4 - x^2} - \frac{64}{3}\sqrt{4 - x^2} \right] dx$$

$$= \frac{64}{3} \int_{-2}^{2} \left[x^2 \sqrt{4 - x^2} - \sqrt{4 - x^2} \right] dx$$

$$= \frac{64}{3} \left[\frac{1}{8}\left(x(2x^2 - 4)\sqrt{4 - x^2} + 16 \arcsin \frac{x}{2} \right) - \frac{1}{2}\left(x\sqrt{4 - x^2} + 4 \arcsin \frac{x}{2} \right) \right]_{-2}^{2}$$

$$= \frac{64}{3} \left[\frac{1}{8}(8\pi) - \frac{1}{2}(2\pi) - \frac{1}{8}(-8\pi) + \frac{1}{2}(-2\pi) \right] = 0$$

15. $F(x,\ y,\ z) = z^2\mathbf{i} + y\mathbf{j} + xz\mathbf{k},\quad S: z = \sqrt{4 - x^2 - y^2}$

$$\text{curl } \mathbf{F} = \begin{vmatrix} \mathbf{i} & \mathbf{j} & \mathbf{k} \\ \dfrac{\partial}{\partial x} & \dfrac{\partial}{\partial y} & \dfrac{\partial}{\partial z} \\ z^2 & y & xz \end{vmatrix} = z\mathbf{j}$$

$$G(x,\ y,\ z) = z - \sqrt{4 - x^2 - y^2}$$

$$\nabla G(x,\ y,\ z) = \frac{x}{\sqrt{4 - x^2 - y^2}}\mathbf{i} + \frac{y}{\sqrt{4 - x^2 - y^2}}\mathbf{j} + \mathbf{k}$$

$$\iint_S (\text{curl } \mathbf{F}) \cdot \mathbf{F}\,dS = \iint_R \frac{yz}{\sqrt{4 - x^2 - y^2}}\,dA = \iint_R \frac{y\sqrt{4 - x^2 - y^2}}{\sqrt{4 - x^2 - y^2}}\,dA = \int_{-2}^{2}\int_{-\sqrt{4-x^2}}^{\sqrt{4-x^2}} y\,dy\,dx = 0$$

16. $F(x,\ y,\ z) = x^2\mathbf{i} + z^2\mathbf{j} - xyz\mathbf{k},\quad S: z = \sqrt{4 - x^2 - y^2}$

$$\text{curl } \mathbf{F} = \begin{vmatrix} \mathbf{i} & \mathbf{j} & \mathbf{k} \\ \dfrac{\partial}{\partial x} & \dfrac{\partial}{\partial y} & \dfrac{\partial}{\partial z} \\ x^2 & z^2 & -xyz \end{vmatrix} = (-xz - 2z)\mathbf{i} + yz\mathbf{j}$$

$$G(x,\ y,\ z) = z - \sqrt{4 - x^2 - y^2}$$

$$\nabla G(x,\ y,\ z) = \frac{x}{\sqrt{4 - x^2 - y^2}}\mathbf{i} + \frac{y}{\sqrt{4 - x^2 - y^2}}\mathbf{j} + \mathbf{k}$$

$$\iint_S (\text{curl } \mathbf{F}) \cdot \mathbf{N}\,dS = \iint_R \left[\frac{-z(x+2)x}{\sqrt{4 - x^2 - y^2}} + \frac{y^2 z}{\sqrt{4 - x^2 - y^2}} \right] dA$$

$$= \iint_R [-x(x+2) + y^2]\,dA = \int_{-2}^{2}\int_{-\sqrt{4-x^2}}^{\sqrt{4-x^2}} (-x^2 - 2x + y^2)\,dy\,dx$$

$$= \int_{-2}^{2}\left[-x^2 y - 2xy + \frac{y^3}{3} \right]_{-\sqrt{4-x^2}}^{\sqrt{4-x^2}} dx$$

$$= \int_{-2}^{2}\left[-2x^2\sqrt{4 - x^2} - 4x\sqrt{4 - x^2} + \frac{2}{3}(4 - x^2)\sqrt{4 - x^2} \right] dx$$

$$= \int_{-2}^{2}\left[-\frac{8}{3}x^2\sqrt{4 - x^2} - 4x\sqrt{4 - x^2} + \frac{8}{3}\sqrt{4 - x^2} \right] dx$$

$$= \left[-\frac{8}{3}\left(\frac{1}{8}\right)\left[x(2x^2 - 4)\sqrt{4 - x^2} + 16\arcsin\frac{x}{2} \right] \right.$$

$$\left. + \frac{4}{3}(4 - x^2)^{3/2} + \frac{8}{3}\left(\frac{1}{2}\right)\left[x\sqrt{4 - x^2} + 4\arcsin\frac{x}{2} \right] \right]_{-2}^{2}$$

$$= \left[\left(-\frac{1}{3}\right)(8\pi) + \frac{4}{3}(2\pi) + \frac{1}{3}(-8\pi) - \frac{4}{3}(-2\pi) \right] = 0$$

17. $F(x, y, z) = -\ln\sqrt{x^2 + y^2}\,\mathbf{i} + \arctan\dfrac{x}{y}\,\mathbf{j} + \mathbf{k}$

$$\text{curl } F = \begin{vmatrix} \mathbf{i} & \mathbf{j} & \mathbf{k} \\[4pt] \dfrac{\partial}{\partial x} & \dfrac{\partial}{\partial y} & \dfrac{\partial}{\partial z} \\[6pt] -\dfrac{1}{2}\ln(x^2 + y^2) & \arctan\dfrac{x}{y} & 1 \end{vmatrix} = \left[\dfrac{(1/y)}{1 + (x^2/y^2)} + \dfrac{y}{x^2 + y^2}\right]\mathbf{k} = \left[\dfrac{2y}{x^2 + y^2}\right]\mathbf{k}$$

S: $z = 9 - 2x - 3y$ over one petal of $r = 2\sin 2\theta$ in the first octant.

$G(x, y, z) = 2x + 3y + z - 9$

$\nabla G(x, y, z) = 2\mathbf{i} + 3\mathbf{j} + \mathbf{k}$

$$\iint_S (\text{curl } F) \cdot N\,dS = \iint_R \frac{2y}{x^2 + y^2}\,dA$$

$$= \int_0^{\pi/2} \int_0^{2\sin 2\theta} \frac{2r\sin\theta}{r^2}\, r\,dr\,d\theta$$

$$= \int_0^{\pi/2} \int_0^{4\sin\theta\cos\theta} 2\sin\theta\,dr\,d\theta$$

$$= \int_0^{\pi/2} 8\sin^2\theta\cos\theta\,d\theta = \left[\frac{8\sin^3\theta}{3}\right]_0^{\pi/2} = \frac{8}{3}$$

18. $F(x, y, z) = yz\,\mathbf{i} + (2 - 3y)\,\mathbf{j} + (x^2 + y^2)\,\mathbf{k}$

$$\text{curl } F = \begin{vmatrix} \mathbf{i} & \mathbf{j} & \mathbf{k} \\[4pt] \dfrac{\partial}{\partial x} & \dfrac{\partial}{\partial y} & \dfrac{\partial}{\partial z} \\[6pt] yz & 2 - 3y & x^2 + y^2 \end{vmatrix} = 2y\,\mathbf{i} + (y - 2x)\,\mathbf{j} - z\,\mathbf{k}$$

S: the first octant portion of $x^2 + z^2 = 16$ over $x^2 + y^2 = 16$

$G(x, y, z) = z - \sqrt{16 - x^2}$

$\nabla G(x, y, z) = \dfrac{x}{\sqrt{16 - x^2}}\,\mathbf{i} + \mathbf{k}$

$$\iint_S (\text{curl } F) \cdot N\,dS = \iint_R \left[\frac{2xy}{\sqrt{16 - x^2}} - z\right]dA$$

$$= \iint_R \left[\frac{2xy}{\sqrt{16 - x^2}} - \sqrt{16 - x^2}\right]dA$$

$$= \int_0^4 \int_0^{\sqrt{16 - x^2}} \left[\frac{2xy}{\sqrt{16 - x^2}} - \sqrt{16 - x^2}\right]dy\,dx$$

$$= \int_0^4 \left[\frac{x}{\sqrt{16 - x^2}}y^2 - \sqrt{16 - x^2}\,y\right]_0^{\sqrt{16 - x^2}} dx$$

$$= \int_0^4 \left[x\sqrt{16 - x^2} - (16 - x^2)\right]dx$$

$$= \left[-\frac{1}{3}(16 - x^2)^{3/2} - 16x + \frac{x^3}{3}\right]_0^4$$

$$= \left(-64 + \frac{64}{3}\right) - \left(-\frac{64}{3}\right) = -\frac{64}{3}$$

19. From Exercise 10, we have $\mathbf{N} = \dfrac{2x\mathbf{i} - \mathbf{k}}{\sqrt{1 + 4x^2}}$ and $dS = \sqrt{1 + 4x^2}\,dA$. Since **curl F** $= xy\mathbf{j} - xz\mathbf{k}$, we have

$$\iint_S (\text{curl } \mathbf{F}) \cdot \mathbf{N}\,dS = \iint_R xz\,dA = \int_0^a \int_0^a x^3\,dy\,dx = \int_0^a ax^3\,dx = \left[\frac{ax^4}{4}\right]_0^a = \frac{a^5}{4}.$$

20. $\mathbf{F}(x,\ y,\ z) = xyz\mathbf{i} + y\mathbf{j} + z\mathbf{k}$

$$\text{curl } \mathbf{F} = \begin{vmatrix} \mathbf{i} & \mathbf{j} & \mathbf{k} \\ \dfrac{\partial}{\partial x} & \dfrac{\partial}{\partial y} & \dfrac{\partial}{\partial z} \\ xyz & y & z \end{vmatrix} = xy\mathbf{j} - xz\mathbf{k}$$

S: the first octant portion of $z = x^2$ over $x^2 + y^2 = a^2$. From Exercise 10, we have

$$\mathbf{N} = \frac{2x\mathbf{i} - \mathbf{k}}{\sqrt{1 + 4x^2}} \text{ and } dS = \sqrt{1 + 4x^2}\,dA.$$

$$\begin{aligned}
\iint_S (\text{curl } \mathbf{F}) \cdot \mathbf{N}\,dS &= \iint_R xz\,dA = \iint_R x^3\,dA \\
&= \int_0^a \int_0^{\sqrt{a^2 - x^2}} x^3\,dy\,dx \\
&= \int_0^a x^3\sqrt{a^2 - x^2}\,dx \\
&= \left[-\frac{1}{3}x^2(a^2 - x^2)^{3/2} - \frac{2}{15}(a^2 - x^2)^{5/2}\right]_0^a \\
&= \frac{2}{15}a^5
\end{aligned}$$

21. $\mathbf{F}(x,\ y,\ z) = \mathbf{i} + \mathbf{j} - 2\mathbf{k}$

$$\text{curl } \mathbf{F} = \begin{vmatrix} \mathbf{i} & \mathbf{j} & \mathbf{k} \\ \dfrac{\partial}{\partial x} & \dfrac{\partial}{\partial y} & \dfrac{\partial}{\partial z} \\ 1 & 1 & -2 \end{vmatrix} = \mathbf{0}$$

Letting $\mathbf{N} = \mathbf{k}$, we have $\displaystyle\iint_S (\text{curl } \mathbf{F}) \cdot \mathbf{N}\,dS = 0.$

22. $\mathbf{F}(x,\ y,\ z) = -y\mathbf{i} + x\mathbf{j}$

S: $x^2 + y^2 = 1$

$$\text{curl } \mathbf{F} = \begin{vmatrix} \mathbf{i} & \mathbf{j} & \mathbf{k} \\ \dfrac{\partial}{\partial x} & \dfrac{\partial}{\partial y} & \dfrac{\partial}{\partial z} \\ -y & x & 0 \end{vmatrix} = 2\mathbf{k}$$

Letting $\mathbf{N} = \mathbf{k}$, we have $\displaystyle\iint_S (\text{curl } \mathbf{F}) \cdot \mathbf{N}\,dS = \iint_R 2\,dA = \int_0^{2\pi} \int_0^1 2r\,dr\,d\theta = 2\pi.$

23. **a.** $\displaystyle\int_C f\nabla g \cdot d\mathbf{r} = \iint_S \mathbf{curl}\,[f\nabla g] \cdot \mathbf{N}\,dS$ (Stoke's Theorem)

$$f\nabla g = f\frac{\partial g}{\partial x}\mathbf{i} + f\frac{\partial g}{\partial y}\mathbf{j} + f\frac{\partial g}{\partial z}\mathbf{k}$$

$$\mathbf{curl}\,(f\nabla g) = \begin{vmatrix} \mathbf{i} & \mathbf{j} & \mathbf{k} \\ \dfrac{\partial}{\partial x} & \dfrac{\partial}{\partial y} & \dfrac{\partial}{\partial z} \\ f\left(\dfrac{\partial g}{\partial x}\right) & f\left(\dfrac{\partial g}{\partial y}\right) & f\left(\dfrac{\partial g}{\partial z}\right) \end{vmatrix}$$

$$= \left[\left[f\left(\frac{\partial^2 g}{\partial y \partial z}\right) + \left(\frac{\partial f}{\partial y}\right)\left(\frac{\partial g}{\partial z}\right)\right] - \left[f\left(\frac{\partial^2 g}{\partial z \partial y}\right) + \left(\frac{\partial f}{\partial z}\right)\left(\frac{\partial g}{\partial y}\right)\right]\right]\mathbf{i}$$

$$- \left[\left[f\left(\frac{\partial^2 g}{\partial x \partial z}\right) + \left(\frac{\partial f}{\partial x}\right)\left(\frac{\partial g}{\partial z}\right)\right] - \left[f\left(\frac{\partial^2 g}{\partial z \partial x}\right) + \left(\frac{\partial f}{\partial z}\right)\left(\frac{\partial g}{\partial x}\right)\right]\right]\mathbf{j}$$

$$+ \left[\left[f\left(\frac{\partial^2 g}{\partial x \partial y}\right) + \left(\frac{\partial f}{\partial x}\right)\left(\frac{\partial g}{\partial y}\right)\right] - \left[f\left(\frac{\partial^2 g}{\partial y \partial x}\right) + \left(\frac{\partial f}{\partial y}\right)\left(\frac{\partial g}{\partial x}\right)\right]\right]\mathbf{k}$$

$$= \left[\left(\frac{\partial f}{\partial y}\right)\left(\frac{\partial g}{\partial z}\right) - \left(\frac{\partial f}{\partial z}\right)\left(\frac{\partial g}{\partial y}\right)\right]\mathbf{i} - \left[\left(\frac{\partial f}{\partial x}\right)\left(\frac{\partial g}{\partial z}\right) - \left(\frac{\partial f}{\partial z}\right)\left(\frac{\partial g}{\partial x}\right)\right]\mathbf{j}$$

$$+ \left[\left(\frac{\partial f}{\partial x}\right)\left(\frac{\partial g}{\partial y}\right) - \left(\frac{\partial f}{\partial y}\right)\left(\frac{\partial g}{\partial x}\right)\right]\mathbf{k}$$

$$= \begin{vmatrix} \mathbf{i} & \mathbf{j} & \mathbf{k} \\ \dfrac{\partial f}{\partial x} & \dfrac{\partial f}{\partial y} & \dfrac{\partial f}{\partial z} \\ \dfrac{\partial g}{\partial x} & \dfrac{\partial g}{\partial y} & \dfrac{\partial g}{\partial z} \end{vmatrix} = \nabla f \times \nabla g$$

Therefore, $\displaystyle\int_C f\nabla g \cdot d\mathbf{r} = \iint_S \mathbf{curl}\,[f\nabla g] \cdot \mathbf{N}\,dS = \iint_S [\nabla f \times \nabla g] \cdot \mathbf{N}\,dS.$

b. $\displaystyle\int_C (f\nabla f) \cdot d\mathbf{r} = \iint_S (\nabla f \times \nabla f) \cdot \mathbf{N}\,dS$ (using part a.)

$$= 0 \text{ since } \nabla f \times \nabla f = 0.$$

c. $\displaystyle\int_C (f\nabla g + g\nabla f) \cdot d\mathbf{r} = \int_C (f\nabla g) \cdot d\mathbf{r} + \int_C (g\nabla f) \cdot d\mathbf{r}$

$$= \iint_S (\nabla f \times \nabla g) \cdot \mathbf{N}\,dS + \iint_S (\nabla g \times \nabla f) \cdot \mathbf{N}\,dS \quad \text{(using part a.)}$$

$$= \iint_S (\nabla f \times \nabla g) \cdot \mathbf{N}\,dS + \iint_S -(\nabla f \times \nabla g) \cdot \mathbf{N}\,dS = 0$$

24. $f(x, y, z) = xyz$, $g(x, y, z) = z$, $S: z = \sqrt{4 - x^2 - y^2}$

    a. $\nabla g(x, y, z) = \mathbf{k}$

        $f(x, y, z)\nabla g(x, y, z) = xyz\mathbf{k}$

        $\mathbf{r}(t) = 2\cos t\,\mathbf{i} + 2\sin t\,\mathbf{j} + 0\mathbf{k}, \quad 0 \le t \le 2\pi$

        $\int_C [f(x, y, z)\nabla g(x, y, z)] \cdot d\mathbf{r} = 0$

    b. $\nabla f(x, y, z) = yz\mathbf{i} + xz\mathbf{j} + xy\mathbf{k}$

        $\nabla g(x, y, z) = \mathbf{k}$

$$\nabla f \times \nabla g = \begin{vmatrix} \mathbf{i} & \mathbf{j} & \mathbf{k} \\ yz & xz & xy \\ 0 & 0 & 1 \end{vmatrix} = xz\mathbf{i} - yz\mathbf{j}$$

$$\mathbf{N} = \frac{x}{\sqrt{4 - x^2 - y^2}}\mathbf{i} + \frac{y}{\sqrt{4 - x^2 - y^2}}\mathbf{j} + \mathbf{k}$$

$$dS = \sqrt{1 + \left(\frac{-x}{\sqrt{4 - x^2 - y^2}}\right)^2 + \left(\frac{-y}{\sqrt{4 - x^2 - y^2}}\right)^2}\, dA = \frac{2}{\sqrt{4 - x^2 - y^2}}\, dA$$

$$\iint_S [\nabla f(x, y, z) \times \nabla g(x, y, z)] \cdot \mathbf{N}\, dS = \iint_S \left[\frac{x^2 z}{\sqrt{4 - x^2 - y^2}} - \frac{y^2 z}{\sqrt{4 - x^2 - y^2}}\right] \frac{2}{\sqrt{4 - x^2 - y^2}}\, dA$$

$$= \iint_S \frac{2(x^2 - y^2)}{\sqrt{4 - x^2 - y^2}}\, dA$$

$$= \int_0^2 \int_0^{2\pi} \frac{2r^2(\cos^2\theta - \sin^2\theta)}{\sqrt{4 - r^2}}\, r\, d\theta\, dr$$

$$= \int_0^2 \left[\frac{2r^3}{\sqrt{4 - r^2}}\left(\frac{1}{2}\sin 2\theta\right)\right]_0^{2\pi}\, dr = 0$$

25. Let $\mathbf{C} = a\mathbf{i} + b\mathbf{j} + c\mathbf{k}$, then

$$\frac{1}{2}\int_C (\mathbf{C} \times \mathbf{r}) \cdot d\mathbf{r} = \frac{1}{2}\iint_S \mathbf{curl}\,(\mathbf{C} \times \mathbf{r}) \cdot \mathbf{N}\, dS = \frac{1}{2}\iint_S 2\mathbf{C} \cdot \mathbf{N}\, dS = \iint_S \mathbf{C} \cdot \mathbf{N}\, dS$$

since

$$\mathbf{C} \times \mathbf{r} = \begin{vmatrix} \mathbf{i} & \mathbf{j} & \mathbf{k} \\ a & b & c \\ x & y & z \end{vmatrix} = (bz - cy)\mathbf{i} - (az - cx)\mathbf{j} + (ay - bx)\mathbf{k}$$

and

$$\mathbf{curl}(\mathbf{C} \times \mathbf{r}) = \begin{vmatrix} \mathbf{i} & \mathbf{j} & \mathbf{k} \\ \dfrac{\partial}{\partial x} & \dfrac{\partial}{\partial y} & \dfrac{\partial}{\partial z} \\ bz - cy & cx - az & ay - bx \end{vmatrix} = 2(a\mathbf{i} + b\mathbf{j} + c\mathbf{k}) = 2\mathbf{C}.$$

Chapter 15   Review Exercises

1. $\mathbf{F}(x,\ y,\ z) = x\mathbf{i} + \mathbf{j} + 2\mathbf{k}$

2. $\mathbf{F}(x,\ y) = \mathbf{i} - 2y\mathbf{j}$

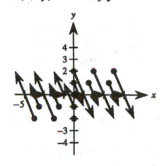

3. $f(x,\ y,\ z) = 8x^2 + xy + z^2$

$\mathbf{F}(x,\ y,\ z) = (16x + y)\mathbf{i} + x\mathbf{j} + 2z\mathbf{k}$

4. $f(x,\ y,\ z) = x^2 e^{yz}$

$\mathbf{F}(x,\ y,\ z) = 2xe^{yz}\mathbf{i} + x^2 z e^{yz}\mathbf{j} + x^2 y e^{yz}\mathbf{k}$

$\qquad\qquad = xe^{yz}(2\mathbf{i} + xz\mathbf{j} + xy\mathbf{k})$

5. Since $\partial M/\partial y = -1/y^2 \ne \partial N/\partial x,$ **F** is not conservative.

6. Since $\partial M/\partial y = -1/x^2 = \partial N/\partial x,$ **F** is conservative. From $M = \partial U/\partial x = -y/x^2$ and $N = \partial U/\partial y = 1/x,$ partial integration yields $U = (y/x) + h(y)$ and $U = (y/x) + g(x)$ which suggests that $U(x,\ y) = (y/x) + C.$

7. Since $\partial M/\partial y = 12xy = \partial N/\partial x,$ **F** is conservative. From $M = \partial U/\partial x = 6xy^2 - 3x^2$ and $N = \partial U/\partial y = 6x^2 y + 3y^2 - 7,$ partial integration yields $U = 3x^2 y^2 - x^3 + h(y)$ and $U = 3x^2 y^2 + y^3 - 7y + g(x)$ which suggests $h(y) = y^3 - 7y,$ $g(x) = -x^3,$ and $U(x,\ y) = 3x^2 y^2 - x^3 + y^3 - 7y + C.$

8. Since $\partial M/\partial y = -6y^2 \sin 2x = \partial N/\partial x,$ **F** is conservative. From $M = \partial U/\partial x = -2y^3 \sin 2x$ and $N = \partial U/\partial y = 3y^2(1 + \cos 2x),$ we obtain $U = y^3 \cos 2x + h(y)$ and $U = y^3(1 + \cos 2x) + g(x)$ which suggests that $h(y) = y^3,$ $g(x) = C,$ and $U(x,\ y) = y^3(1 + \cos 2x) + C.$

9. Since

$$\frac{\partial M}{\partial y} = 4x = \frac{\partial N}{\partial x},$$

$$\frac{\partial M}{\partial z} = 1 \ne \frac{\partial P}{\partial x}.$$

F is not conservative.

10. Since

$$\frac{\partial M}{\partial y} = 4x = \frac{\partial N}{\partial x},$$

$$\frac{\partial M}{\partial z} = 2z = \frac{\partial P}{\partial x},$$

$$\frac{\partial N}{\partial z} = 6y \ne \frac{\partial P}{\partial y},$$

F is not conservative.

11. Since

$$\frac{\partial M}{\partial y} = \frac{-1}{y^2 z} = \frac{\partial N}{\partial x}, \quad \frac{\partial M}{\partial z} = \frac{-1}{yz^2} = \frac{\partial P}{\partial x}, \quad \frac{\partial N}{\partial z} = \frac{x}{y^2 z^2} = \frac{\partial P}{\partial y},$$

F is conservative. From

$$M = \frac{\partial U}{\partial x} = \frac{1}{yz}, \quad N = \frac{\partial U}{\partial y} = \frac{-x}{y^2 z}, \quad P = \frac{\partial U}{\partial z} = \frac{-x}{yz^2},$$

we obtain

$$U = \frac{x}{yz} + f(y, z), \quad U = \frac{x}{yz} + g(x, z), \quad U = \frac{x}{yz} + h(x, y)$$

which suggests that $f(y, z) = C_1$, $g(x, z) = C_2$, $h(x, y) = C_3$. Thus, $U(x, y, z) = (x/yz) + C$.

12. Since

$$\frac{\partial M}{\partial y} = \sin z = \frac{\partial N}{\partial x}, \quad \frac{\partial M}{\partial z} = y \cos z \neq \frac{\partial P}{\partial x},$$

F is not conservative.

13. Since $\mathbf{F} = x^2 \mathbf{i} + y^2 \mathbf{j} + z^2 \mathbf{k}$:

   a. div $\mathbf{F} = 2x + 2y + 2z$

   b. **curl F** $= \left(\frac{\partial P}{\partial y} - \frac{\partial N}{\partial z}\right)\mathbf{i} - \left(\frac{\partial P}{\partial x} - \frac{\partial M}{\partial z}\right)\mathbf{j} + \left(\frac{\partial N}{\partial x} - \frac{\partial M}{\partial y}\right)\mathbf{k} = 0\mathbf{i} - 0\mathbf{j} + 0\mathbf{k} = \mathbf{0}$

14. Since $\mathbf{F} = xy^2 \mathbf{j} - zx^2 \mathbf{k}$:

   a. div $\mathbf{F} = 2xy - x^2$

   b. **curl F** $= 2xz\mathbf{j} + y^2\mathbf{k}$

15. Since $\mathbf{F} = (\cos y + y \cos x)\mathbf{i} + (\sin x - x \sin y)\mathbf{j} + xyz\mathbf{k}$:

   a. div $\mathbf{F} = -y \sin x - x \cos y + xy$

   b. **curl F** $= xz\mathbf{i} - yz\mathbf{j} + (\cos x - \sin y + \sin y - \cos x)\mathbf{k} = xz\mathbf{i} - yz\mathbf{j}$

16. Since $\mathbf{F} = (3x - y)\mathbf{i} + (y - 2z)\mathbf{j} + (z - 3x)\mathbf{k}$:

   a. div $\mathbf{F} = 3 + 1 + 1 = 5$

   b. **curl F** $= 2\mathbf{i} + 3\mathbf{j} + \mathbf{k}$

17. Since $\mathbf{F} = \arcsin x\mathbf{i} + xy^2\mathbf{j} + yz^2\mathbf{k}$:

   a. div $\mathbf{F} = \dfrac{1}{\sqrt{1 - x^2}} + 2xy + 2yz$

   b. **curl F** $= z^2\mathbf{i} + y^2\mathbf{k}$

18. Since $\mathbf{F} = (x^2 - y)\mathbf{i} - (x + \sin^2 y)\mathbf{j}$:

   a. div $\mathbf{F} = 2x - 2 \sin y \cos y$

   b. **curl F** $= \mathbf{0}$

19. Since $\mathbf{F} = \ln(x^2 + y^2)\mathbf{i} + \ln(x^2 + y^2)\mathbf{j} + z\mathbf{k}$:

   a. div $\mathbf{F} = \dfrac{2x}{x^2 + y^2} + \dfrac{2y}{x^2 + y^2} + 1$

         $= \dfrac{2x + 2y}{x^2 + y^2} + 1$

   b. **curl F** $= \dfrac{2x - 2y}{x^2 + y^2}\mathbf{k}$

20. Since $\mathbf{F} = \dfrac{z}{x}\mathbf{i} + \dfrac{z}{y}\mathbf{j} + z^2\mathbf{k}$:

   a. div $\mathbf{F} = -\dfrac{z}{x^2} - \dfrac{z}{y^2} + 2z = z\left(2 - \dfrac{1}{x^2} - \dfrac{1}{y^2}\right)$

   b. **curl F** $= -\dfrac{1}{y}\mathbf{i} + \dfrac{1}{x}\mathbf{j}$

21. a. Let $x = t$, $y = t$, $-1 \le t \le 2$, then $ds = \sqrt{2}\,dt$.

$$\int_C (x^2 + y^2)\,ds = \int_{-1}^{2} 2t^2\sqrt{2}\,dt = \left[2\sqrt{2}\left(\frac{t^3}{3}\right)\right]_{-1}^{2} = 6\sqrt{2}$$

b. Let $x = 4\cos t$, $y = 4\sin t$, $0 \le t \le 2\pi$, then $ds = 4\,dt$.

$$\int_C (x^2 + y^2)\,ds = \int_0^{2\pi} 16(4\,dt) = 128\pi$$

22. a. Let $x = 5t$, $y = 4t$, $0 \le t \le 1$, then $ds = \sqrt{41}\,dt$.

$$\int_C xy\,ds = \int_0^{1} 20t^2\sqrt{41}\,dt = \frac{20\sqrt{41}}{3}$$

b. C_1: $x = t$, $y = 0$, $0 \le t \le 4$, $ds = dt$

C_2: $x = 4 - 4t$, $y = 2t$, $0 \le t \le 1$, $ds = 2\sqrt{5}\,dt$

C_3: $x = 0$, $y = 2 - t$, $0 \le t \le 2$, $ds = dt$

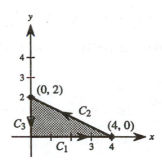

Therefore, $\displaystyle\int_C xy\,ds = \int_0^{4} 0\,dt + \int_0^{1} (8t - 8t^2)2\sqrt{5}\,dt + \int_0^{2} 0\,dt$

$$= 16\sqrt{5}\left[\frac{t^2}{2} - \frac{t^3}{3}\right]_0^{1} = \frac{8\sqrt{5}}{3}.$$

23. $x = \cos t + t\sin t$, $y = \sin t - t\cos t$, $0 \le t \le 2\pi$, $\dfrac{dx}{dt} = t\cos t$, $\dfrac{dy}{dt} = t\sin t$

$$\int_C (x^2 + y^2)\,ds = \int_0^{2\pi} [(\cos t + t\sin t)^2 + (\sin t - t\cos t)^2]\sqrt{t^2\cos^2 t + t^2\sin^2 t}\,dt$$

$$= \int_0^{2\pi} (1 + t^2)t\,dt = \left[\frac{t^2}{2} + \frac{t^4}{4}\right]_0^{2\pi} = 2\pi^2 + 4\pi^4 = 2\pi^2(1 + 2\pi^2)$$

24. $x = t - \sin t$, $y = 1 - \cos t$, $0 \le t \le 2\pi$, $\dfrac{dx}{dt} = 1 - \cos t$, $\dfrac{dy}{dt} = \sin t$

$$\int_C x\,ds = \int_0^{2\pi} (t - \sin t)\sqrt{(1 - \cos t)^2 + (\sin t)^2}\,dt = \int_0^{2\pi} (t - \sin t)\sqrt{2 - 2\cos t}\,dt$$

$$= \sqrt{2}\int_0^{2\pi} [t\sqrt{1 - \cos t} - \sin t\sqrt{1 - \cos t}]\,dt = \sqrt{2}\left[-\frac{2}{3}(1 - \cos t)^{3/2}\right]_0^{2\pi} + \sqrt{2}\int_0^{2\pi} t\sqrt{1 - \cos t}\,dt$$

$$= \sqrt{2}\int_0^{2\pi} t\sqrt{1 - \cos t}\,dt = \sqrt{2}\int_0^{2\pi} \frac{t\sin t}{\sqrt{1 + \cos t}}\,dt = \sqrt{2}\left[-2t\sqrt{1 + \cos t}\right]_0^{2\pi} + 2\int_0^{2\pi} \sqrt{1 + \cos t}\,dt\bigg]$$

$$= \sqrt{2}\left[-4\sqrt{2}\pi + 2\int_0^{2\pi} \frac{\sin t}{\sqrt{1 - \cos t}}\,dt\right] = -8\pi + 2\sqrt{2}\left[2\sqrt{1 - \cos t}\right]_0^{2\pi} = -8\pi$$

25. a. Let $x = 2t$, $y = -3t$, $0 \le t \le 1$

$$\int_C (2x - y)\,dx + (x + 3y)\,dy = \int_0^{1} [7t(2) + (-7t)(-3)]\,dt = \int_0^{1} 35t\,dt = \frac{35}{2}$$

b. $x = 3\cos t$, $y = 3\sin t$, $dx = -3\sin t\,dt$, $dy = 3\cos t\,dt$, $0 \le t \le 2\pi$

$$\int_C (2x - y)\,dx + (x + 3y)\,dy = \int_0^{2\pi} (9 + 9\sin t\cos t)\,dt = 18\pi$$

26. $x = \cos t + t \sin t, \quad y = \sin t - t \sin t, \quad 0 \le t \le \dfrac{\pi}{2}, \quad dx = t \cos t \, dt, \quad dy = (\cos t - t \cos t - \sin t) \, dt$

$$\int_C (2x - y) \, dx + (x + 3y) \, dy = \int_0^{\pi/2} [\sin t \cos t \, (5t^2 - 6t + 2) + \cos^2 t \, (t + 1) + \sin^2 t \, (2t - 3)] \, dt \approx 1.01$$

27. $f(x, \ y) = 5 + \sin(x + y)$
  $C:\ y = 3x$ from $(0, 0)$ to $(2, 6)$
  $\mathbf{r}(t) = t\mathbf{i} + 3t\mathbf{j}, \quad 0 \le t \le 2$
  $\mathbf{r}'(t) = \mathbf{i} + 3\mathbf{j}$
  $\|\mathbf{r}'(t)\| = \sqrt{10}$
  Lateral surface area:

$$\int_{C_2} f(x, \ y) \, ds = \int_0^2 [5 + \sin(t + 3t)]\sqrt{10} \, dt = \sqrt{10} \int_0^2 (5 + \sin 4t) \, dt = \frac{\sqrt{10}}{4}(41 - \cos 8) \approx 32.528$$

28. $f(x, \ y) = 12 - x - y$
  $C:\ y = x^2$ from $(0, 0)$ to $(2, 4)$
  $\mathbf{r}(t) = t\mathbf{i} + t^2\mathbf{j}, \quad 0 \le t \le 2$
  $\mathbf{r}'(t) = \mathbf{i} + 2t\mathbf{j}$
  $\|\mathbf{r}'(t)\| = \sqrt{1 + 4t^2}$
  Lateral surface area:

$$\int_C f(x, \ y) \, ds = \int_0^2 (12 - t - t^2)\sqrt{1 + 4t^2} \, dt \approx 41.532$$

29. $d\mathbf{r} = (2t\mathbf{i} + 3t^2\mathbf{j}) \, dt$

  $\mathbf{F} = t^5\mathbf{i} + t^4\mathbf{j}, \quad 0 \le t \le 1$

$$\int_C \mathbf{F} \cdot d\mathbf{r} = \int_0^1 5t^6 \, dt = \frac{5}{7}$$

30. $d\mathbf{r} = [(-4 \sin t)\mathbf{i} + 3 \cos t \, \mathbf{j}] \, dt$

  $\mathbf{F} = (4 \cos t - 3 \sin t)\mathbf{i} + (4 \cos t + 3 \sin t)\mathbf{j}, \quad 0 \le t \le 2\pi$

$$\int_C \mathbf{F} \cdot d\mathbf{r} = \int_0^{2\pi} (12 - 7 \sin t \cos t) \, dt = \left[12t - \frac{7 \sin^2 t}{2} \right]_0^{2\pi} = 24\pi$$

31. $d\mathbf{r} = [(-2 \sin t)\mathbf{i} + (2 \cos t)\mathbf{j} + \mathbf{k}] \, dt$

  $\mathbf{F} = (2 \cos t)\mathbf{i} + (2 \sin t)\mathbf{j} + t\mathbf{k}, \quad 0 \le t \le 2\pi$

$$\int_C \mathbf{F} \cdot d\mathbf{r} = \int_0^{2\pi} t \, dt = 2\pi^2$$

32. $x = 2 - t, \quad y = 2 - t, \quad z = \sqrt{4t - t^2}, \quad 0 \le t \le 2$

$$d\mathbf{r} = \left[-\mathbf{i} - \mathbf{j} + \frac{2 - t}{\sqrt{4t - t^2}}\mathbf{k} \right] dt$$

$$\mathbf{F} = \left(4 - 2t - \sqrt{4t - t^2} \right)\mathbf{i} + \left(\sqrt{4t - t^2} - 2 + t \right)\mathbf{j} + 0\mathbf{k}$$

$$\int_C \mathbf{F} \cdot d\mathbf{r} = \int_0^2 (t - 2) \, dt = \left[\frac{t^2}{2} - 2t \right]_0^2 = -2$$

33. Let $x = t, \quad y = -t, \quad z = 2t^2, \quad -2 \le t \le 2, \quad d\mathbf{r} = [\mathbf{i} - \mathbf{j} + 4t\mathbf{k}]\,dt.$

$\mathbf{F} = (-t - 2t^2)\mathbf{i} + (2t^2 - t)\mathbf{j} + (2t)\mathbf{k}$

$$\int_C \mathbf{F} \cdot d\mathbf{r} = \int_{-2}^{2} 4t^2\,dt = \left[\frac{4t^3}{3}\right]_{-2}^{2} = \frac{64}{3}$$

34. Let $x = 2\sin t, \quad y = -2\cos t, \quad z = 4\sin^2 t, \quad 0 \le t \le \pi.$

$d\mathbf{r} = [(2\cos t)\mathbf{i} + (2\sin t)\mathbf{j} + (8\sin t \cos t)\mathbf{k}]\,dt$

$\mathbf{F} = 0\mathbf{i} + 4\mathbf{j} + (2\sin t)\mathbf{k}$

$$\int_C \mathbf{F} \cdot d\mathbf{r} = \int_0^{\pi} (8\sin t + 16\sin^2 t \cos t)\,dt = \left[-8\cos t + \frac{16}{3}\sin^3 t\right]_0^{\pi} = 16$$

35. $\mathbf{F} = x\mathbf{i} - \sqrt{y}\,\mathbf{j}$ is conservative.

$$\text{Work} = \left[\frac{1}{2}x^2 - \frac{2}{3}y^{3/2}\right]_{(0,0)}^{(4,8)} = \frac{16}{3}$$

36. $\mathbf{r}(t) = 10\sin t\,\mathbf{i} + 10\cos t\,\mathbf{j} + \dfrac{2000/5280}{\pi/2}t\mathbf{k}, \quad 0 \le t \le \dfrac{\pi}{2}$

$\qquad = 10\sin t\,\mathbf{i} + 10\cos t\,\mathbf{j} + \dfrac{25}{33\pi}t\mathbf{k}$

$\mathbf{F} = 20\mathbf{k}$

$d\mathbf{r} = \left(10\cos t\,\mathbf{i} - 10\sin t\,\mathbf{j} + \dfrac{25}{33\pi}\mathbf{k}\right)$

$$\int_C \mathbf{F} \cdot d\mathbf{r} = \int_0^{\pi/2} \frac{500}{33\pi}\,dt = \frac{250}{33}\ \text{mi} \cdot \text{ton}$$

37. $\displaystyle\int_C 2xyz\,dx + x^2z\,dy + x^2y\,dz = \left[x^2yz\right]_{(0,0,0)}^{(1,4,3)} = 12$

38. $\displaystyle\int_C y\,dx + x\,dy + \frac{1}{z}\,dz = \left[xy + \ln|z|\right]_{(0,0,1)}^{(4,4,4)} = 16 + \ln 4$

39. $\displaystyle\int_C y\,dx + 2x\,dy = \int_0^2 \int_0^2 (2 - 1)\,dy\,dx = \int_0^2 2\,dx = 4$

40. $\displaystyle\int_C xy\,dx + (x^2 + y^2)\,dy = \int_0^2 \int_0^2 (2x - x)\,dy\,dx = \int_0^2 2x\,dx = 4$

41. $\displaystyle\int_C xy^2\,dx + x^2y\,dy = \iint_R (2xy - 2xy)\,dA = 0$

42. $\displaystyle\int_C (x^2 - y^2)\,dx + 2xy\,dy = \int_{-a}^{a} \int_{-\sqrt{a^2-x^2}}^{\sqrt{a^2-x^2}} 4y\,dy\,dx = \int_{-a}^{a} 0\,dx = 0$

43. $\displaystyle\int_C xy\,dx + x^2\,dy = \int_0^1 \int_{x^2}^{x} x\,dy\,dx = \int_0^1 (x^2 - x^3)\,dx = \frac{1}{12}$

44. $\displaystyle\int_C y^2\,dx + x^{2/3}\,dy = \int_{-1}^{1}\int_{-(1-x^{2/3})^{3/2}}^{(1-x^{2/3})^{3/2}}\left(\frac{2}{3}x^{-1/3} - 2y\right)dy\,dx$

$\displaystyle = \int_{-1}^{1}\left[\frac{2y}{3x^{1/3}} - y^2\right]_{-(1-x^{2/3})^{3/2}}^{(1-x^{2/3})^{3/2}}dx$

$\displaystyle = \frac{4}{3}\int_{-1}^{1}\frac{(1-x^{2/3})^{3/2}}{x^{1/3}}\,dx$

$\displaystyle = \left[-\frac{4}{3}\left(-\frac{3}{2}\right)\left(\frac{2}{5}\right)(1-x^{2/3})^{5/2}\right]_{-1}^{1} = \frac{4}{5}[0] = 0$

45. $\mathbf{r}(u, v) = \sec u \cos v\mathbf{i} + (1 + 2\tan u)\sin v\mathbf{j} + 2u\mathbf{k}$

$0 \le u \le \dfrac{\pi}{3},\quad 0 \le v \le 2\pi$

46. $\mathbf{r}(u, v) = e^{-u/4}\cos v\mathbf{i} + e^{-u/4}\sin v\mathbf{j} + \dfrac{u}{6}\mathbf{k}$

$0 \le u \le 4,\quad 0 \le v \le 2\pi$

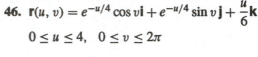

47. *S*: $\mathbf{r}(u, v) = (u + v)\mathbf{i} + (u - v)\mathbf{j} + \sin v\mathbf{k},\quad 0 \le u \le 2,\quad 0 \le v \le \pi$

$\mathbf{r}_u(u, v) = \mathbf{i} + \mathbf{j}$

$\mathbf{r}_v(u, v) = \mathbf{i} - \mathbf{j} + \cos v\mathbf{k}$

$\mathbf{r}_u \times \mathbf{r}_v = \begin{vmatrix} \mathbf{i} & \mathbf{j} & \mathbf{k} \\ 1 & 1 & 0 \\ 1 & -1 & \cos v \end{vmatrix} = \cos v\mathbf{i} - \cos v\mathbf{j} - 2\mathbf{k}$

$\|\mathbf{r}_u \times \mathbf{r}_v\| = \sqrt{2\cos^2 v + 4}$

$\displaystyle\iint_S z\,dS = \int_0^{\pi}\int_0^2 \sin v\sqrt{2\cos^2 v + 4}\,du\,dv = 2\left[\sqrt{6} + \sqrt{2}\ln\left(\frac{\sqrt{6} + \sqrt{2}}{\sqrt{6} - \sqrt{2}}\right)\right]$

48. *S*: $\mathbf{r}(u, v) = u\cos v\mathbf{i} + u\sin v\mathbf{j} + (u - 1)(2 - u)\mathbf{k},\quad 0 \le u \le 2,\quad 0 \le v \le 2\pi$

$\mathbf{r}_u(u, v) = \cos v\mathbf{i} + \sin v\mathbf{j} + (3 - 2u)\mathbf{k}$

$\mathbf{r}_v(u, v) = -u\sin v\mathbf{i} + u\cos v\mathbf{j}$

$\mathbf{r}_u \times \mathbf{r}_v = \begin{vmatrix} \mathbf{i} & \mathbf{j} & \mathbf{k} \\ \cos v & \sin v & 3 - 2u \\ -u\sin v & u\cos v & 0 \end{vmatrix} = (2u - 3)u\cos v\mathbf{i} + (2u - 3)u\sin v\mathbf{j} + u\mathbf{k}$

$\|\mathbf{r}_u \times \mathbf{r}_v\| = u\sqrt{(2u - 3)^2 + 1}$

$\displaystyle\iint_S (x + y)\,dS = \int_0^{2\pi}\int_0^2 (u\cos v + u\sin v)u\sqrt{(2u - 3)^2 + 1}\,du\,dv$

$\displaystyle = \int_0^2\int_0^{2\pi} (\cos v + \sin v)u^2\sqrt{(2u - 3)^2 + 1}\,dv\,du = 0$

49. $\mathbf{F}(x,\ y,\ z) = x^2\mathbf{i} + xy\mathbf{j} + z\mathbf{k}$

Q: solid region bounded by the coordinate planes and the plane $2x + 3y + 4z = 12$

Surface Integral: There are four surfaces for this solid.

$$z = 0, \quad \mathbf{N} = -\mathbf{k}, \quad \mathbf{F} \cdot \mathbf{N} = -z, \quad \int_{S_1}\int 0\,dS = 0$$

$$y = 0, \quad \mathbf{N} = -\mathbf{j}, \quad \mathbf{F} \cdot \mathbf{N} = -xy, \quad \int_{S_2}\int 0\,dS = 0$$

$$x = 0, \quad \mathbf{N} = -\mathbf{i}, \quad \mathbf{F} \cdot \mathbf{N} = -x^2, \quad \int_{S_3}\int 0\,dS = 0$$

$$2x + 3y + 4z = 12, \quad \mathbf{N} = \frac{2\mathbf{i} + 3\mathbf{j} + 4\mathbf{k}}{\sqrt{29}}, \quad dS = \sqrt{1 + \left(\frac{1}{4}\right) + \left(\frac{9}{16}\right)}\,dA = \frac{\sqrt{29}}{4}\,dA$$

$$\int_{S_4}\int \mathbf{F} \cdot \mathbf{N}\,dS = \frac{1}{4}\int_R\int (2x^2 + 3xy + 4z)\,dA$$

$$= \frac{1}{4}\int_0^6 \int_0^{4-(2x/3)} (2x^2 + 3xy + 12 - 2x - 3y)\,dy\,dx$$

$$= \frac{1}{4}\int_0^6 \left[2x^2\left(\frac{12-2x}{3}\right) + \frac{3x}{2}\left(\frac{12-2x}{3}\right)^2 + 12\left(\frac{12-2x}{3}\right) - 2x\left(\frac{12-2x}{3}\right) - \frac{3}{2}\left(\frac{12-2x}{3}\right)^2 \right]dx$$

$$= \frac{1}{6}\int_0^6 (-x^3 + x^2 + 24x + 36)\,dx$$

$$= \frac{1}{6}\left[-\frac{x^4}{4} + \frac{x^3}{3} + 12x^2 + 36x \right]_0^6 = 66$$

Divergence Theorem: Since $\operatorname{div}\mathbf{F} = 2x + x + 1 = 3x + 1$, the Divergence Theorem yields

$$\int\int\int_Q \operatorname{div}\mathbf{F}\,dV = \int_0^6 \int_0^{(12-2x)/3} \int_0^{(12-2x-3y)/4} (3x+1)\,dz\,dy\,dx$$

$$= \int_0^6 \int_0^{(12-2x)/3} (3x+1)\left(\frac{12-2x-3y}{4}\right)dy\,dx$$

$$= \frac{1}{4}\int_0^6 (3x+1)\left[12y - 2xy - \frac{3}{2}y^2 \right]_0^{(12-2x)/3}dx$$

$$= \frac{1}{4}\int_0^6 (3x+1)\left[4(12-2x) - 2x\left(\frac{12-2x}{3}\right) - \frac{3}{2}\left(\frac{12-2x}{3}\right)^2 \right]dx$$

$$= \frac{1}{4}\int_0^6 \frac{2}{3}(3x^3 - 35x^2 + 96x + 36)\,dx$$

$$= \frac{1}{6}\left[\frac{3x^4}{4} - \frac{35x^3}{3} + 48x^2 + 36x \right]_0^6 = 66.$$

50. $\mathbf{F}(x,\ y,\ z) = x\mathbf{i} + y\mathbf{j} + z\mathbf{k}$

Q: solid region bounded by the coordinate planes and the plane $2x + 3y + 4z = 12$

Surface Integral: There are four surfaces for this solid.

$z = 0,\quad \mathbf{N} = -\mathbf{k},\quad \mathbf{F}\cdot\mathbf{N} = -z,\quad \int_{S_1}\!\!\int 0\,dS = 0$

$y = 0,\quad \mathbf{N} = -\mathbf{j},\quad \mathbf{F}\cdot\mathbf{N} = -y,\quad \int_{S_2}\!\!\int 0\,dS = 0$

$x = 0,\quad \mathbf{N} = -\mathbf{i},\quad \mathbf{F}\cdot\mathbf{N} = -x,\quad \int_{S_3}\!\!\int 0\,dS = 0$

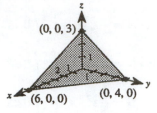

(0, 0, 3)

(6, 0, 0) (0, 4, 0)

$2x + 3y + 4z = 12,\quad \mathbf{N} = \dfrac{2\mathbf{i} + 3\mathbf{j} + 4\mathbf{k}}{\sqrt{29}},\quad dS = \sqrt{1 + \left(\dfrac{1}{4}\right) + \left(\dfrac{9}{16}\right)}\,dA = \dfrac{\sqrt{29}}{4}\,dA$

$$\int_{S_4}\!\!\int \mathbf{N}\cdot\mathbf{F}\,dS = \frac{1}{4}\int_R\!\!\int (2x + 3y + 4z)\,dy\,dx$$

$$= \frac{1}{4}\int_0^6\int_0^{(12-2x)/3} 12\,dy\,dx = 3\int_0^6\left(4 - \frac{2x}{3}\right)dx = 3\left[4x - \frac{x^2}{3}\right]_0^6 = 36$$

Triple Integral: Since div $\mathbf{F} = 3$, the Divergence Theorem yields

$$\iiint_Q \operatorname{div}\mathbf{F}\,dV = \iiint_Q 3\,dV = 3(\text{Volume of solid}) = 3\left[\frac{1}{3}(\text{Area of base})(\text{Height})\right] = \frac{1}{2}(6)(4)(3) = 36.$$

51. $\mathbf{F}(x,\ y,\ z) = (\cos y + y\cos x)\mathbf{i} + (\sin x - x\sin y)\mathbf{j} + xyz\mathbf{k}$

S: portion of $z = y^2$ over the square in the xy-plane with vertices $(0,\ 0)$, $(a,\ 0)$, $(a,\ a)$, $(0,\ a)$

Line Integral: Using the line integral we have:

C_1: $y = 0,\quad dy = 0$

C_2: $x = 0,\quad dx = 0,\quad z = y^2,\quad dz = 2y\,dy$

C_3: $y = a,\quad dy = 0,\quad z = a^2,\quad dz = 0$

C_4: $x = a,\quad dx = 0,\quad z = y^2,\quad dz = 2y\,dy$

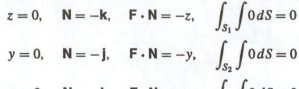

$$\int_C \mathbf{F}\cdot d\mathbf{r} = \int_C (\cos y + y\cos x)\,dx + (\sin x - x\sin y)\,dy + xyz\,dz$$

$$= \int_{C_1} dx + \int_{C_2} 0 + \int_{C_3} (\cos a + a\cos x)\,dx + \int_{C_4} (\sin a - a\sin y)\,dy + ay^3(2y\,dy)$$

$$= \int_a^0 dx + \int_0^a (\cos a + a\cos x)\,dx + \int_a^0 (\sin a - a\sin y)\,dy + \int_a^0 2ay^4\,dy$$

$$= -a + \left[x\cos a + a\sin x\right]_0^a + \left[y\sin a + a\cos y\right]_a^0 + \left[2a\frac{y^5}{5}\right]_a^0$$

$$= -a + a\cos a + a\sin a + a - a\sin a - a\cos a - \frac{2a^6}{5} = -\frac{2a^6}{5}$$

Double Integral: Considering $f(x,\ y,\ z) = y^2 - z$, we have:

$$\mathbf{N} = \frac{\nabla f}{\|\nabla f\|} = \frac{2y\mathbf{j} - \mathbf{k}}{\sqrt{1 + 4y^2}},\quad dS = \sqrt{1 + 4y^2}\,dA,\text{ and } \operatorname{\mathbf{curl}}\mathbf{F} = xz\mathbf{i} - yz\mathbf{j}.$$

Hence,

$$\int_S\!\!\int (\operatorname{\mathbf{curl}}\mathbf{F})\cdot\mathbf{N}\,dS = \int_0^a\int_0^a -2y^2z\,dy\,dx = \int_0^a\int_0^a -2y^4\,dy\,dx = \int_0^a -\frac{2a^5}{5}\,dx = -\frac{2a^6}{5}.$$

52. $F(x, \ y, \ z) = (x - z)\mathbf{i} + (y - z)\mathbf{j} + x^2\mathbf{k}$

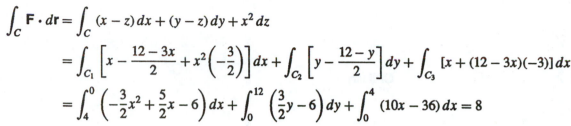

S: first octant portion of the plane $3x + y + 2z = 12$

Line Integral:

C_1: $y = 0$, $dy = 0$, $z = \dfrac{12 - 3x}{2}$, $dz = -\dfrac{3}{2}dx$

C_2: $x = 0$, $dx = 0$, $z = \dfrac{12 - y}{2}$, $dz = -\dfrac{1}{2}dy$

C_3: $z = 0$, $dz = 0$, $y = 12 - 3x$, $dy = -3\,dx$

$$\int_C \mathbf{F} \cdot d\mathbf{r} = \int_C (x - z)\,dx + (y - z)\,dy + x^2\,dz$$

$$= \int_{C_1} \left[x - \frac{12 - 3x}{2} + x^2\left(-\frac{3}{2}\right) \right] dx + \int_{C_2} \left[y - \frac{12 - y}{2} \right] dy + \int_{C_3} [x + (12 - 3x)(-3)]\,dx$$

$$= \int_4^0 \left(-\frac{3}{2}x^2 + \frac{5}{2}x - 6 \right) dx + \int_0^{12} \left(\frac{3}{2}y - 6 \right) dy + \int_0^4 (10x - 36)\,dx = 8$$

Double Integral: $G(x, \ y, \ z) = \dfrac{12 - 3x - y}{2} - z$

$$\nabla G(x, \ y, \ z) = -\frac{3}{2}\mathbf{i} - \frac{1}{2}\mathbf{j} - \mathbf{k}$$

$$\mathbf{curl\ F} = \mathbf{i} - (2x + 1)\mathbf{j}$$

$$\iint_S (\mathbf{curl\ F}) \cdot \mathbf{N}\,dS = \int_0^4 \int_0^{12-3x} (x - 1)\,dy\,dx = \int_0^4 (-3x^2 + 15x - 12)\,dx = 8$$

53. In this case $x = \cos t$, $y = \sin t$, $0 \le t \le 2\pi$, $\mathbf{F} = (-\sin t)\mathbf{i} + (\cos t)\mathbf{j}$, and $d\mathbf{r} = [(-\sin t)\mathbf{i} + (\cos t)\mathbf{j}]\,dt$. Hence,

$$\int_C \mathbf{F} \cdot d\mathbf{r} = \int_0^{2\pi} 1\,dt = 2\pi.$$

Since $M(x, \ y)$ and $N(x, \ y)$ are not defined at the point $(0, 0)$ in region R enclosed by curve C, the condition of a zero value for the integral is not guaranteed.

CHAPTER 16
Differential Equations

Section 16.1   Definitions and Basic Concepts

	Equation	Type	Order
1.	$\dfrac{dy}{dx} + 3xy = x^2$	Ordinary	1
2.	$y'' + 2y' + y = 1$	Ordinary	2
3.	$\dfrac{d^2x}{dt^2} + 2\dfrac{dx}{dt} - 4x = e^t$	Ordinary	2
4.	$\dfrac{d^2u}{dt^2} + \dfrac{du}{dt} = \sec t$	Ordinary	2
5.	$y^{(4)} + 3(y')^2 - 4y = 0$	Ordinary	4
6.	$x^2 y'' + 3xy' = 0$	Ordinary	2
7.	$(y'')^2 + 3y' - 4y = 0$	Ordinary	2
8.	$\dfrac{\partial u}{\partial t} = C^2 \dfrac{\partial^2 u}{\partial x^2}$	Partial	2
9.	$\dfrac{\partial u}{\partial t} + \dfrac{\partial u}{\partial y} = 2u$	Partial	1
10.	$\dfrac{\partial^2 u}{\partial x \partial y} = \dfrac{\partial u}{\partial y}$	Partial	2
11.	$\dfrac{d^2 y}{dx^2} = \sqrt{1 + \left(\dfrac{dy}{dx}\right)^2}$	Ordinary	2
12.	$\sqrt{\dfrac{d^2 y}{dx^2}} = \dfrac{dy}{dx}$	Ordinary	2

13. Differential equation: $y' = 4y$
    Solution: $y = Ce^{4x}$
    Check: $y' = 4Ce^{4x} = 4y$

14. Differential equation: $y' = \dfrac{2xy}{x^2 - y^2}$

    Solution: $x^2 + y^2 = Cy$
    Check: $2x + 2yy' = Cy'$

$$y' = \frac{-2x}{(2y - C)}$$

$$y' = \frac{-2xy}{2y^2 - Cy} = \frac{-2xy}{2y^2 - (x^2 + y^2)} = \frac{-2xy}{y^2 - x^2} = \frac{2xy}{x^2 - y^2}$$

15. Differential equation: $y'' + y = 0$

Solution: $y = C_1 \cos x + C_2 \sin x$

Check: $y' = -C_1 \sin x + C_2 \cos x$

$y'' = -C_1 \cos x - C_2 \sin x$

$y'' + y = -C_1 \cos x - C_2 \sin x + C_1 \cos x + C_2 \sin x = 0$

16. Differential equation: $y'' + 2y' + 2y = 0$

Solution: $y = C_1 e^{-x} \cos x + C_2 e^{-x} \sin x$

Check: $y' = -(C_1 + C_2)e^{-x} \sin x + (-C_1 + C_2)e^{-x} \cos x$

$y'' = 2C_1 e^{-x} \sin x - 2C_2 e^{-x} \cos x$

$y'' + 2y' + 2y = (2C_1 - 2C_1 - 2C_2 + 2C_2)e^{-x} \sin x + (-2C_2 - 2C_1 + 2C_2 + 2C_1)e^{-x} \cos x = 0$

17. Differential equation: $b^2 \dfrac{\partial u}{\partial t} = \dfrac{\partial^2 u}{\partial x^2}$

Solution: $u = e^{-t} \sin bx$

Check: $\dfrac{\partial u}{\partial t} = -e^{-t} \sin bx$

$\dfrac{\partial u}{\partial x} = be^{-t} \cos bx$

$\dfrac{\partial^2 u}{\partial x^2} = -b^2 e^{-t} \sin bx$

$b^2 \dfrac{\partial u}{\partial t} = -b^2 e^{-t} \sin bx = \dfrac{\partial^2 u}{\partial x^2}$

18. Differential equation: $\dfrac{\partial^2 u}{\partial x^2} + \dfrac{\partial^2 u}{\partial y^2} = 0$

Solution: $u = \dfrac{y}{x^2 + y^2}$

Check: $\dfrac{\partial u}{\partial x} = \dfrac{-2xy}{(x^2 + y^2)^2}$

$\dfrac{\partial^2 u}{\partial x^2} = -\dfrac{2y(y^2 - 3x^2)}{(x^2 + y^2)^3}$

$\dfrac{\partial u}{\partial y} = \dfrac{x^2 - y^2}{(x^2 + y^2)^2}$

$\dfrac{\partial^2 u}{\partial y^2} = \dfrac{2y(y^2 - 3x^2)}{(x^2 + y^2)^3}$

$\dfrac{\partial^2 u}{\partial x^2} + \dfrac{\partial^2 u}{\partial y^2} = 0$

In Exercises 19–24, the differential equation is $y^{(4)} - 16y = 0$.

19. $y = 3 \cos x$

$y^{(4)} = 3 \cos x$

$y^{(4)} - 16y = -45 \cos x \neq 0$

20. $y = 3 \cos 2x$

$y^{(4)} = 48 \cos 2x$

$y^{(4)} - 16y = 48 \cos 2x - 48 \cos 2x = 0$

21. $y = e^{-2x}$

$y^{(4)} = 16e^{-2x}$

$y^{(4)} - 16y = 16e^{-2x} - 16e^{-2x} = 0$

22. $y = 5 \ln x$

$y^{(4)} = -\dfrac{30}{x^4}$

$y^{(4)} - 16y = -\dfrac{30}{x^4} - 80 \ln x \neq 0$

23. $y = C_1 e^{2x} + C_2 e^{-2x} + C_3 \sin 2x + C_4 \cos 2x$

$y^{(4)} = 16C_1 e^{2x} + 16C_2 e^{-2x} + 16C_3 \sin 2x + 16C_4 \cos 2x$

$y^{(4)} - 16y = 0$

24. $y = 5e^{-2x} + 3 \cos 2x$

$y^{(4)} = 80e^{-2x} + 48 \cos 2x$

$y^{(4)} - 16y = (80e^{-2x} + 48 \cos 2x) - (80e^{-2x} + 48 \cos 2x) = 0$

In Exercises 25–30, the differential equation is $x\dfrac{\partial u}{\partial x} - y\dfrac{\partial u}{\partial y} = 0$.

25.
$$u = e^{x+y}$$
$$\frac{\partial u}{\partial x} = e^{x+y}$$
$$\frac{\partial u}{\partial y} = e^{x+y}$$
$$x\frac{\partial u}{\partial x} - y\frac{\partial u}{\partial y} \neq 0$$

26.
$$u = 5$$
$$\frac{\partial u}{\partial x} = 0$$
$$\frac{\partial u}{\partial y} = 0$$
$$x\frac{\partial u}{\partial x} - y\frac{\partial u}{\partial y} = 0$$

27.
$$u = x^2 y^2$$
$$\frac{\partial u}{\partial x} = 2xy^2$$
$$\frac{\partial u}{\partial y} = 2x^2 y$$
$$x\frac{\partial u}{\partial x} - y\frac{\partial u}{\partial y} = 2x^2 y^2 - 2x^2 y^2 = 0$$

28.
$$u = \sin xy$$
$$\frac{\partial u}{\partial x} = y\cos xy$$
$$\frac{\partial u}{\partial y} = x\cos xy$$
$$x\frac{\partial u}{\partial x} - y\frac{\partial u}{\partial y} = xy\cos xy - xy\cos xy = 0$$

29.
$$u = (xy)^n$$
$$\frac{\partial u}{\partial x} = ny(xy)^{n-1}$$
$$\frac{\partial u}{\partial y} = nx(xy)^{n-1}$$
$$x\frac{\partial u}{\partial x} - y\frac{\partial u}{\partial y} = nxy(xy)^{n-1} - nxy(xy)^{n-1} = 0$$

30.
$$u = x^2 + y^2$$
$$\frac{\partial u}{\partial x} = 2x$$
$$\frac{\partial u}{\partial y} = 2y$$
$$x\frac{\partial u}{\partial x} - y\frac{\partial u}{\partial y} = 2x^2 - 2y^2 \neq 0$$

31. $y = Ce^{kx}$
$$\frac{dy}{dx} = Cke^{kx}$$
Since $dy/dx = 0.07y$, we have $Cke^{kx} = 0.07Ce^{kx}$.
Thus, $k = 0.07$.

32. $y = A\sin\omega t$
$$\frac{d^2 y}{dt^2} = -A\omega^2 \sin\omega t$$
Since $(d^2 y/dt^2) + 16y = 0$, we have
$-A\omega^2 \sin\omega t + 16A\sin\omega t = 0$. Thus, $\omega^2 = 16$ and
$\omega = \pm 4$.

33. $y^2 = Cx^3$ passes through $(4, 4)$
$16 = C(64) \Rightarrow C = \frac{1}{4}$
Particular solution: $y^2 = \frac{1}{4}x^3$ or $4y^2 = x^3$

34. $2x^2 - y^2 = C$ passes through $(3, 4)$
$2(9) - 16 = C \Rightarrow C = 2$
Particular solution: $2x^2 - y^2 = 2$

35. Differential equation: $4yy' - x = 0$
General solution: $4y^2 - x^2 = C$
Particular solutions: $C = 0$, Two intersecting lines
$\qquad\qquad\qquad C = \pm 1,\quad C = \pm 4$, Hyperbolas

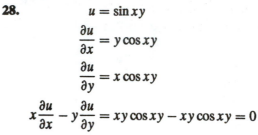

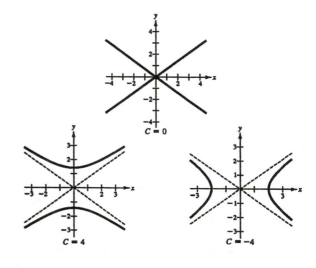
$C = 0$

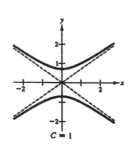
$C = 1$

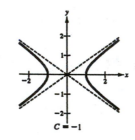
$C = -1$

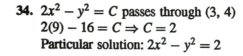
$C = 4$

$C = -4$

36. Differential equation: $yy' + x = 0$
General solution: $x^2 + y^2 = C$
Particular solutions: $C = 0$, Point
$\qquad\qquad\qquad C = 1,\quad C = 4$, Circles

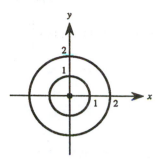

37. Differential equation: $y' + 2y = 0$
General solution: $y = Ce^{-2x}$
Initial condition: $y(0) = 3$, $3 = Ce^0 = C$
Particular solution: $y = 3e^{-2x}$

38. Differential equation: $2x + 3yy' = 0$
General solution: $2x^2 + 3y^2 = C$
Initial condition: $y(1) = 2$, $2(1) + 3(4) = 14 = C$
Particular solution: $2x^2 + 3y^2 = 14$

39. Differential equation: $y'' + 9y = 0$
General solution: $y = C_1 \sin 3x + C_2 \cos 3x$

Initial conditions: $y\left(\dfrac{\pi}{6}\right) = 2$, $\quad y'\left(\dfrac{\pi}{6}\right) = 1$

$$2 = C_1 \sin\left(\frac{\pi}{2}\right) + C_2 \cos\left(\frac{\pi}{2}\right) \Rightarrow C_1 = 2$$

$$y' = 3C_1 \cos 3x - 3C_2 \sin 3x$$

$$1 = 3C_1 \cos\left(\frac{\pi}{2}\right) - 3C_2 \sin\left(\frac{\pi}{2}\right)$$

$$= -3C_2 \Rightarrow C_2 = -\frac{1}{3}$$

Particular solution: $y = 2 \sin 3x - \dfrac{1}{3} \cos 3x$

40. Differential equation: $xy'' + y' = 0$
General solution: $y = C_1 + C_2 \ln x$

Initial conditions: $y(2) = 0$, $\quad y'(2) = \dfrac{1}{2}$

$$0 = C_1 + C_2 \ln 2$$

$$y' = \frac{C_2}{x}$$

$$\frac{1}{2} = \frac{C_2}{2} \Rightarrow C_2 = 1,\quad C_1 = -\ln 2$$

Particular solution: $y = -\ln 2 + \ln x = \ln \dfrac{x}{2}$

41. Differential equation: $x^2 y'' - 3xy' + 3y = 0$
General solution: $y = C_1 x + C_2 x^3$
Initial conditions: $y(2) = 0$, $\quad y'(2) = 4$

$$0 = 2C_1 + 8C_2$$

$$y' = C_1 + 3C_2 x^2$$

$$4 = C_1 + 12C_2$$

$$\left.\begin{array}{l} C_1 + 4C_2 = 0 \\ C_1 + 12C_2 = 4 \end{array}\right\}\quad C_2 = \tfrac{1}{2},\quad C_1 = -2$$

Particular solution: $y = -2x + \tfrac{1}{2}x^3$

42. Differential equation: $9y'' - 12y' + 4y = 0$
General solution: $y = e^{2x/3}(C_1 + C_2 x)$
Initial conditions: $y(0) = 4$, $\quad y(3) = 0$

$$0 = e^2(C_1 + 3C_2)$$

$$4 = (1)(C_1 + 0) \Rightarrow C_1 = 4$$

$$0 = e^2(4 + 3C_2) \Rightarrow C_2 = -\tfrac{4}{3}$$

Particular solution: $y = e^{2x/3}\left(4 - \tfrac{4}{3}x\right)$

43. $\dfrac{dy}{dx} = 3x^2$

$$y = \int 3x^2\, dx = x^3 + C$$

44. $\dfrac{dy}{dx} = \dfrac{1}{1+x^2}$

$$y = \int \frac{1}{1+x^2}\, dx = \arctan x + C$$

45. $\dfrac{dy}{dx} = \dfrac{x-2}{x} = 1 - \dfrac{2}{x}$

$$y = \int \left[1 - \frac{2}{x}\right] dx$$

$$= x - 2\ln|x| + C = x - \ln x^2 + C$$

46. $\dfrac{dy}{dx} = x \cos x$

$$y = \int x \cos x\, dx = x \sin x + \cos x + C$$

47. $\dfrac{dy}{dx} = e^x \sin 2x$

$$y = \int e^x (\sin 2x)\, dx = \frac{e^x}{5}[\sin 2x - 2\cos 2x] + C$$

48. $\dfrac{dy}{dx} = \tan^2 x = \sec^2 x - 1$

$$y = \int (\sec^2 x - 1)\, dx = \tan x - x + C$$

49. $\dfrac{dy}{dx} = x\sqrt{x-3}$

$$y = \int x\sqrt{x-3}\, dx = \int (u^2 + 3)(u)(2u)\, du$$

$$= 2\int (u^4 + 3u^2)\, du = 2\left(\frac{u^5}{5} + u^3\right) + C = \frac{2}{5}(x-3)^{5/2} + 2(x-3)^{3/2} + C$$

Let $u = \sqrt{x-3}$, then $x = u^2 + 3$ and $dx = 2u\, du$.

50. $\dfrac{dy}{dx} = xe^x$

$$y = \int xe^x\, dx = xe^x - e^x + C$$

51. a. $N = L - Ce^{-kt}, \quad L = 750$

When $t = 0, \quad N = 100$.

$$100 = 750 - Ce^{-k(0)}$$

$$-650 = -C(1) \Rightarrow C = 650$$

$$N = 750 - 650e^{-kt}$$

When $t = 2, \quad N = 160$.

$$160 = 750 - 650e^{-k(2)}$$

$$\frac{590}{650} = e^{-2k}$$

$$\ln\left(\frac{59}{65}\right) = -2k$$

$$-\frac{1}{2}\ln\left(\frac{59}{65}\right) = k$$

$$k = \frac{1}{2}\ln\left(\frac{65}{59}\right)$$

$$N = 750 - 650e^{-(1/2)\ln(65/59)t}$$

$$\approx 750 - 650e^{-0.0484t}$$

c.

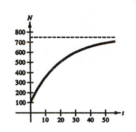

b. $\dfrac{dN}{dt} = -650\left[-\frac{1}{2}\ln\left(\frac{65}{59}\right)\right]e^{-(1/2)\ln(65/59)t} = 325\ln\left(\frac{65}{59}\right)e^{-(1/2)\ln(65/59)t}$

$$k(L - N) = k\left[750 - \left(750 - 650e^{-(1/2)\ln(65/59)t}\right)\right]$$

$$= \frac{1}{2}\ln\left(\frac{65}{59}\right)\left[650e^{-(1/2)\ln(65/59)t}\right] = 325\ln\left(\frac{65}{59}\right)e^{-(1/2)\ln(65/59)t}$$

Therefore, $\dfrac{dN}{dt} = k(L - N)$.

52. a. $A = Ce^{kt}$

$$\frac{dA}{dt} = Cke^{kt} = k(Ce^{kt}) = kA$$

b. $A = Ce^{kt}$

When $t = 0$, $A = \$1000$.

$$1000 = Ce^{k(0)} \Rightarrow C = 1000$$

$$A = 1000e^{kt}$$

When $t = 10$, $A = \$3320.12$.

$$3320.12 = 1000e^{k(10)}$$

$$3.32012 = e^{10k}$$

$$\ln 3.32012 = 10k$$

$$k = \frac{\ln 3.32012}{10} \approx 0.12$$

Particular solution: $A = 1000e^{0.12t}$

53. $\dfrac{dy}{dx} = x$

a.

b. $y = \displaystyle\int x\,dx = \frac{1}{2}x^2 + C$

54. $\dfrac{dy}{dx} = -\dfrac{x}{y}$

a.

b. $x^2 + y^2 = C$

$$2x + 2y\frac{dy}{dx} = 0$$

$$\frac{dy}{dx} = -\frac{2x}{2y} = -\frac{x}{y}$$

55. $\dfrac{dy}{dx} = 4 - y$

a.

b. $\displaystyle\lim_{x\to\infty} y = 4$

56. $\dfrac{dy}{dx} = 0.25y(4 - y)$

a.

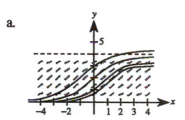

b. $\displaystyle\lim_{x\to\infty} y = 4$

c. $\displaystyle\lim_{x\to-\infty} y = 0$

57. True

58. False. Consider Example 3. $y = x^3$ is a solution to $xy' - 3y = 0$, but $y = x^3 + 1$ is not a solution.

Section 16.2   Separation of Variables in First-Order Equations

1. $\dfrac{dy}{dx} = \dfrac{x}{y}$

$\displaystyle\int y\,dy = \int x\,dx$

$\dfrac{y^2}{2} = \dfrac{x^2}{2} + C_1$

$y^2 - x^2 = C$

2. $\dfrac{dy}{dx} = \dfrac{x^2 + 2}{3y^2}$

$\displaystyle\int 3y^2\,dy = \int (x^2 + 2)\,dx$

$y^3 = \dfrac{x^3}{3} + 2x + C$

3. $\dfrac{dr}{ds} = 0.05r$

$\displaystyle\int \dfrac{dr}{r} = \int 0.05\,ds$

$\ln r = 0.05s + C_1$

$r = e^{0.05s + C_1} = Ce^{0.05s}$

4. $\dfrac{dr}{ds} = 0.05s$

$\displaystyle\int dr = \int 0.05s\,ds$

$r = 0.025s^2 + C$

5. $(2 + x)y' = 3y$

$\displaystyle\int \dfrac{dy}{y} = \int \dfrac{3}{2 + x}\,dx$

$\ln y = 3\ln(2 + x) + \ln C = \ln C(2 + x)^3$

$y = C(x + 2)^3$

6. $xy' = y$

$\displaystyle\int \dfrac{dy}{y} = \int \dfrac{dx}{x}$

$\ln y = \ln x + \ln C = \ln Cx$

$y = Cx$

7. $yy' = \sin x$

$\displaystyle\int y\,dy = \int \sin x\,dx$

$\dfrac{y^2}{2} = -\cos x + C_1$

$y^2 = -2\cos x + C$

8. $\sqrt{1 - 4x^2}\,y' = 1$

$\displaystyle\int dy = \int \dfrac{1}{\sqrt{1 - 4x^2}}\,dx$

$y = \dfrac{1}{2}\arcsin 2x + C$

9. $y\ln x - xy' = 0$

$\displaystyle\int \dfrac{dy}{y} = \int \dfrac{\ln x}{x}\,dx$

$\ln y = \dfrac{1}{2}(\ln x)^2 + C_1$

$y = e^{(1/2)(\ln x)^2 + C_1} = Ce^{(\ln x)^2/2}$

10. $yy' - 2xe^x = 0$

$\displaystyle\int y\,dy = \int 2xe^x\,dx$

$\dfrac{1}{2}y^2 = 2e^x(x - 1) + C_1$

$y^2 = 4e^x(x - 1) + C$

11. $yy' - e^x = 0$

$\displaystyle\int y\,dy = \int e^x\,dx$

$\dfrac{y^2}{2} = e^x + C_1$

$y^2 = 2e^x + C$

Initial condition: $y(0) = 4$, $16 = 2 + C$, $C = 14$

Particular solution: $y^2 = 2e^x + 14$

12. $\sqrt{x} + \sqrt{y}\,y' = 0$

$\displaystyle\int y^{1/2}\,dy = -\int x^{1/2}\,dx$

$\dfrac{2}{3}y^{3/2} = -\dfrac{2}{3}x^{3/2} + C_1$

$y^{3/2} + x^{3/2} = C$

Initial condition:

$y(1) = 4$, $(4)^{3/2} + (1)^{3/2} = 8 + 1 = 9 = C$

Particular solution: $y^{3/2} + x^{3/2} = 9$

13. $y(x+1) + y' = 0$

$$\int \frac{dy}{y} = -\int (x+1)\,dx$$

$$\ln y = -\frac{(x+1)^2}{2} + C_1$$

$$y = Ce^{-(x+1)^2/2}$$

Initial condition: $y(-2) = 1$, $\quad 1 = Ce^{-1/2}$, $\quad C = e^{1/2}$

Particular solution: $y = e^{[1-(x+1)^2]/2} = e^{-(x^2+2x)/2}$

14. $xyy' - \ln x = 0$

$$\int y\,dy = \int \frac{\ln x}{x}\,dx$$

$$\frac{y^2}{2} = \frac{\ln^2 x}{2} + C_1$$

$$y^2 = \ln^2 x + C$$

Initial condition: $y(1) = 0$, $\quad 0 = \ln^2(1) + C = C$

Particular solution: $y^2 = \ln^2 x$

15. $(1+x^2)y' - (1+y^2) = 0$

$$\int \frac{dy}{1+y^2} = \int \frac{dx}{1+x^2}$$

$$\arctan y = \arctan x + C_1$$

$$C_1 = \arctan y - \arctan x$$

$$C = \tan C_1 = \tan(\arctan y - \arctan x)$$

$$= \frac{\tan(\arctan y) - \tan(\arctan x)}{1 + \tan(\arctan y)\tan(\arctan x)}$$

$$C = \frac{y-x}{1+xy}$$

Initial condition: $y(0) = \sqrt{3}$, $\quad C = \frac{\sqrt{3}-0}{1+0} = \sqrt{3}$

Particular solution: $\dfrac{y-x}{1+xy} = \sqrt{3}$, $\quad y = \dfrac{x+\sqrt{3}}{1-\sqrt{3}x}$

16. $\sqrt{1-x^2}\,y' - \sqrt{1-y^2} = 0$

$$\int \frac{dy}{\sqrt{1-y^2}} = \int \frac{dx}{\sqrt{1-x^2}}$$

$$\arcsin y = \arcsin x + C$$

Initial condition: $y(0) = 1$, $\quad \dfrac{\pi}{2} = 0 + C = C$

Particular solution:

$$\arcsin y = \arcsin x + \frac{\pi}{2}$$

$$y = \sin\left(\arcsin x + \frac{\pi}{2}\right)$$

$$y = x \cos\frac{\pi}{2} + [\cos(\arcsin x)]\sin\frac{\pi}{2}$$

$$y = \cos(\arcsin x) = \sqrt{1-x^2}$$

17. $\dfrac{du}{dv} = uv \sin v^2$

$$\int \frac{du}{u} = \int v \sin v^2\,dv$$

$$\ln u = -\frac{1}{2}\cos v^2 + C_1$$

$$u = Ce^{-(\cos v^2)/2}$$

Initial condition: $u(0) = 1$, $\quad C = \dfrac{1}{e^{-1/2}} = e^{1/2}$

Particular solution: $u = e^{(1-\cos v^2)/2}$

18. $\dfrac{dr}{ds} = e^{r+s} = e^r e^s$

$$\int e^{-r}\,dr = \int e^s\,ds$$

$$-e^{-r} = e^s + C_1$$

$$e^s + e^{-r} = C$$

Initial condition: $r(1) = 0$, $\quad C = e+1$

Particular solution: $e^s + e^{-r} = e+1$

$$r = -\ln(e+1-e^s)$$

19. $dP - kP\,dt = 0$

$$\int \frac{dP}{P} = k \int dt$$

$$\ln P = kt + C_1$$

$$P = Ce^{kt}$$

Initial condition: $P(0) = P_0$, $\quad P_0 = Ce^0 = C$

Particular solution: $P = P_0 e^{kt}$

20. $dT + k(T-70)\,dt = 0$

$$\int \frac{dT}{T-70} = -k \int dt$$

$$\ln(T-70) = -kt + C_1$$

$$T - 70 = Ce^{-kt}$$

Initial condition:

$T(0) = 140$, $\quad 140 - 70 = 70 = Ce^0 = C$

Particular solution:

$T - 70 = 70e^{-kt}$, $\quad T = 70(1 + e^{-kt})$

21. $\dfrac{dy}{dx} = \dfrac{-9x}{16y}$

$$\int 16y\,dy = -\int 9x\,dx$$

$$8y^2 = \dfrac{-9}{2}x^2 + C$$

Initial condition:

$$y(1) = 1, \quad 8 = -\dfrac{9}{2} + C, \quad C = \dfrac{25}{2}$$

Particular solution:

$$8y^2 = \dfrac{-9}{2}x^2 + \dfrac{25}{2}, \quad 16y^2 + 9x^2 = 25$$

22. $\dfrac{dy}{dx} = \dfrac{2y}{3x}$

$$\int \dfrac{3}{y}\,dy = \int \dfrac{2}{x}\,dx$$

$$\ln y^3 = \ln x^2 + \ln C$$

$$y^3 = Cx^2$$

Initial condition: $y(8) = 2$, $\quad 2^3 = C(8^2)$, $\quad C = \dfrac{1}{8}$

Particular solution: $8y^3 = x^2$, $\quad y = \dfrac{1}{2}x^{2/3}$

23. $m = \dfrac{dy}{dx} = \dfrac{0 - y}{(x+2) - x} = -\dfrac{y}{2}$

$$\int \dfrac{dy}{y} = \int -\dfrac{1}{2}\,dx$$

$$\ln y = -\dfrac{1}{2}x + C_1$$

$$y = Ce^{-x/2}$$

24. $m = \dfrac{dy}{dx} = \dfrac{y - 0}{x - 0} = \dfrac{y}{x}$

$$\int \dfrac{dy}{y} = \int \dfrac{dx}{x}$$

$$\ln y = \ln x + C_1 = \ln x + \ln C = \ln Cx$$

$$y = Cx$$

25. $\dfrac{dy}{dx} = 0.05y$

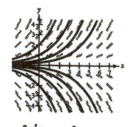

$$\int \dfrac{dy}{y} = \int 0.05\,dx$$

$$\ln y = 0.05x + C_1$$

$$y = Ce^{0.05x}$$

26. $\dfrac{dy}{dx} = 1 + y^2$

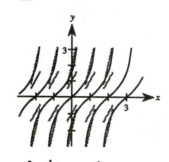

$$\int \dfrac{dy}{1 + y^2} = \int dx$$

$$\arctan y = x + C$$

$$y = \tan(x + C)$$

27. $f(x,\ y) = x^3 - 4xy^2 + y^3$

$$f(tx,\ ty) = t^3x^3 - 4txt^2y^2 + t^3y^3$$

$$= t^3(x^3 - 4xy^2 + y^3)$$

Homogeneous of degree 3

28. $f(x,\ y) = \dfrac{xy}{\sqrt{x^2 + y^2}}$

$$f(tx,\ ty) = \dfrac{txty}{\sqrt{t^2x^2 + t^2y^2}}$$

$$= \dfrac{t^2xy}{t\sqrt{x^2 + y^2}} = t\dfrac{xy}{\sqrt{x^2 + y^2}}$$

Homogeneous of degree 1

29. $f(x,\ y) = 2\ln xy$

$f(tx,\ ty) = 2\ln txty$

$= 2\ln t^2 xy = 2(\ln t^2 + \ln xy)$

Not homogeneous

30. $f(x,\ y) = \tan(x+y)$

$f(tx,\ ty) = \tan(tx+ty) = \tan[t(x+y)]$

Not homogeneous

31. $f(x,\ y) = 2\ln\dfrac{x}{y}$

$f(tx,\ ty) = 2\ln\dfrac{tx}{ty} = 2\ln\dfrac{x}{y}$

Homogeneous of degree 0

32. $f(x,\ y) = \tan\dfrac{y}{x}$

$f(tx,\ ty) = \tan\dfrac{ty}{tx} = \tan\dfrac{y}{x}$

Homogeneous of degree 0

33. $y' = \dfrac{x+y}{2x}$

$v + x\dfrac{dv}{dx} = \dfrac{x+vx}{2x}$

$2\displaystyle\int\dfrac{dv}{1-v} = \int\dfrac{dx}{x}$

$-\ln(1-v)^2 = \ln x + \ln C = \ln Cx$

$\dfrac{1}{(1-v)^2} = Cx$

$\dfrac{1}{[1-(y/x)]^2} = Cx$

$\dfrac{x^2}{(x-y)^2} = Cx$

$x = C(x-y)^2$

34. $y' = \dfrac{2x+y}{y}$

$v + x\dfrac{dv}{dx} = \dfrac{2x+xv}{xv}$

$\displaystyle\int\dfrac{v}{v^2-v-2}\,dv = -\int\dfrac{dx}{x}$

$\dfrac{2}{3}\ln(v-2) + \dfrac{1}{3}\ln(v+1) = -\ln x + \ln C_1 = \ln\dfrac{C_1}{x}$

$(v-2)^2(v+1) = \dfrac{C}{x^3}$

$\left(\dfrac{y}{x}-2\right)^2\left(\dfrac{y}{x}+1\right) = \dfrac{C}{x^3}$

$(y-2x)^2(y+x) = C$

35. $y' = \dfrac{x-y}{x+y}$

$v + x\dfrac{dv}{dx} = \dfrac{x-xv}{x+xv}$

$\displaystyle\int\dfrac{v+1}{v^2+2v-1}\,dv = -\int\dfrac{dx}{x}$

$\dfrac{1}{2}\ln(v^2+2v-1) = -\ln x + \ln C_1 = \ln\dfrac{C_1}{x}$

$v^2 + 2v - 1 = \dfrac{C}{x^2}$

$\left(\dfrac{y^2}{x^2} + 2\dfrac{y}{x} - 1\right) = \dfrac{C}{x^2}$

$y^2 + 2xy - x^2 = C$

36. $y' = \dfrac{x^2+y^2}{2xy}$

$v + x\dfrac{dv}{dx} = \dfrac{x^2+v^2x^2}{2x^2v}$

$\displaystyle\int\dfrac{2v}{v^2-1}\,dv = -\int\dfrac{dx}{x}$

$\ln(v^2-1) = -\ln x + \ln C = \ln\dfrac{C}{x}$

$v^2 - 1 = \dfrac{C}{x}$

$\dfrac{y^2}{x^2} - 1 = \dfrac{C}{x}$

$y^2 - x^2 = Cx$

37.
$$y' = \frac{xy}{x^2 - y^2}$$

$$v + x\frac{dv}{dx} = \frac{x^2 v}{x^2 - x^2 v^2}$$

$$\int \frac{1 - v^2}{v^3}\, dv = \int \frac{dx}{x}$$

$$-\frac{1}{2v^2} - \ln v = \ln x + \ln C_1 = \ln C_1 x$$

$$\frac{-1}{2v^2} = \ln C_1 x v$$

$$\frac{-x^2}{2y^2} = \ln C_1 y$$

$$y = Ce^{-x^2/2y^2}$$

38.
$$y' = \frac{3x + 2y}{x}$$

$$v + x\frac{dv}{dx} = \frac{3x + 2vx}{x} = 3 + 2v$$

$$\int \frac{dv}{v + 3} = \int \frac{dx}{x}$$

$$\ln(v + 3) = \ln x + \ln C$$

$$v + 3 = Cx$$

$$\frac{y}{x} + 3 = Cx$$

$$y = Cx^2 - 3x$$

39.
$$x\, dy - (2xe^{-y/x} + y)\, dx = 0$$

$$x(v\, dx + x\, dv) - (2xe^{-v} + vx)\, dx = 0$$

$$\int e^v\, dv = \int \frac{2}{x}\, dx$$

$$e^v = \ln C_1 x^2$$

$$e^{y/x} = \ln C_1 + \ln x^2$$

$$e^{y/x} = C + \ln x^2$$

Initial condition: $y(1) = 0$, $1 = C$
Particular solution: $e^{y/x} = 1 + \ln x^2$

40.
$$-y^2\, dx + x(x + y)\, dy = 0$$

$$-x^2 v^2\, dx + (x^2 + x^2 v)(v\, dx + x\, dv) = 0$$

$$\int \frac{1 + v}{v}\, dv = -\int \frac{dx}{x}$$

$$v + \ln v = -\ln x + \ln C_1 = \ln \frac{C_1}{x}$$

$$v = \ln \frac{C_1}{xv}$$

$$\frac{C_1}{vx} = e^v$$

$$\frac{C_1}{y} = e^{y/x}$$

$$y = Ce^{-y/x}$$

Initial condition: $y(1) = 1$, $1 = Ce^{-1}$, $C = e$
Particular solution: $y = e^{1-y/x}$

41.
$$\left(x \sec \frac{y}{x} + y\right) dx - x\, dy = 0\;,$$

$$(x \sec v + xv)\, dx - x(v\, dx + x\, dv) = 0$$

$$\int \cos v\, dv = \int \frac{dx}{x}$$

$$\sin v = \ln x + \ln C_1$$

$$x = Ce^{\sin v}$$

$$= Ce^{\sin(y/x)}$$

Initial condition: $y(1) = 0$, $1 = Ce^0 = C$
Particular solution: $x = e^{\sin(y/x)}$

42.
$$\left(y - \sqrt{x^2 - y^2}\right) dx - x\, dy = 0$$

$$\left(xv - \sqrt{x^2 - x^2 v^2}\right) dx - x(v\, dx + x\, dv) = 0$$

$$\int \frac{dv}{\sqrt{1 - v^2}} = -\int \frac{dx}{x}$$

$$\arcsin v = -\ln x + \ln C_1$$

$$\frac{C_1}{x} = e^{\arcsin v} = e^{\arcsin(y/x)}$$

$$x = Ce^{-\arcsin(y/x)}$$

Initial condition: $y(1) = 0$, $1 = Ce^0 = C$
Particular solution: $x = e^{-\arcsin(y/x)}$

43. a. $\dfrac{dA}{dt} = kA$, $\displaystyle\int \dfrac{dA}{A} = \int k\,dt$, $\ln A = kt + C_1$,

$A = Pe^{kt}$

b. $P = 1000$, $R = 0.11$,

$A = 1000e^{0.11(10)} \approx \$3004.17$

c. $2000 = 1000e^{0.11t}$, $0.11t = \ln 2$,

$t = \dfrac{1}{0.11}\ln 2 \approx 6.3$ years

44. $\dfrac{dP}{dt} = kP$, $\displaystyle\int \dfrac{dP}{P} = \int k\,dt$,

$\ln P = kt + C_1$, $P = Ce^{kt}$

Initial conditions: $P(2) = 180$, $P(4) = 300$

$$180 = Ce^{2k}$$

$$300 = Ce^{4k} = C(e^{2k})^2$$

$$300 = C\left(\dfrac{180}{C}\right)^2 = \dfrac{32{,}400}{C}$$

$$C = 108$$

$$\dfrac{180}{108} = e^{2k}$$

$$k = \dfrac{1}{2}\ln \dfrac{180}{108} = \ln\sqrt{\dfrac{5}{3}}$$

Particular solution: $P = 108e^{t\,\ln\sqrt{5/3}} \approx 108e^{0.2554t}$

There were 108 flies in the original population.

45. $\dfrac{dy}{dt} = ky$, $y = Ce^{kt}$

Initial conditions: $y(0) = y_0$

$$y(1600) = \dfrac{y_0}{2}$$

$$C = y_0$$

$$\dfrac{y_0}{2} = y_0 e^{1600k}$$

$$k = \dfrac{\ln(1/2)}{1600}$$

Particular solution: $y = y_0 e^{-t\,(\ln 2)/1600}$

When $t = 25$, $y = y_0 e^{-(\ln 2)/64} \approx 0.989 y_0$,

$y = 98.9\%$ of y_0.

46. $\dfrac{dy}{dt} = ky$, $y = Ce^{kt}$

Initial conditions: $y(0) = 20$, $y(1) = 16$

$$20 = Ce^0 = C$$

$$16 = 20e^k$$

$$k = \ln \dfrac{4}{5}$$

Particular solution: $y = 20e^{t\,\ln(4/5)}$

When 75% has been changed:

$5 = 20e^{t\,\ln(4/5)}$

$$\dfrac{1}{4} = e^{t\,\ln(4/5)}$$

$$t = \dfrac{\ln(1/4)}{\ln(4/5)} \approx 6.2 \text{ hr}$$

47. $\dfrac{dT}{dt} = k(T - 70)$

$\displaystyle\int \dfrac{dT}{T - 70} = \int k\,dt$

$T - 70 = Ce^{kt}$

Initial conditions: $T(0) = 350$, $T(45) = 150$

$350 - 70 = 280 = Ce^0 = C$

$150 - 70 = 80 = 280e^{45k}$

$$k = \dfrac{\ln(2/7)}{45}$$

Particular solution: $T = 70 + 280e^{t[\ln(2/7)]/45}$

When $T = 80$, $80 = 70 + 280e^{t[\ln(2/7)]/45}$

$$\dfrac{t\ln(2/7)}{45} = \ln\left(\dfrac{1}{28}\right)$$

$$t = \dfrac{45\ln(1/28)}{\ln(2/7)} \approx 119.7 \text{ minutes.}$$

48. a. $y = x - x^2$

$$\frac{dy}{dx} = 1 - 2x$$

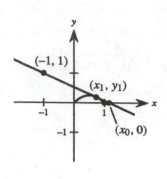

At (x_1, y_1), $m = 1 - 2x_1$. Also, $m = (y_1 - 1)/(x_1 + 1)$.

$$\frac{y_1 - 1}{x_1 + 1} = 1 - 2x_1$$

$$(x_1 - x_1^2) - 1 = (1 - 2x_1)(x_1 + 1)$$

$$-x_1^2 + x_1 - 1 = -2x_1^2 - x_1 + 1$$

$$x_1^2 + 2x_1 - 2 = 0$$

$$x_1 = \frac{-2 \pm \sqrt{2^2 - 4(1)(-2)}}{2(1)} = -1 \pm \sqrt{3}$$

Choosing the positive value for x_1, we have $x_1 = -1 + \sqrt{3}$ and $m = 1 - 2(-1 + \sqrt{3}) = 3 - 2\sqrt{3}$. Also,

$$\frac{0 - 1}{x_0 + 1} = 3 - 2\sqrt{3}$$

$$\frac{-1}{3 - 2\sqrt{3}} = x_0 + 1$$

$$x_0 = \frac{2\sqrt{3}}{3}.$$

The closest the receiver can be to the hill is $\dfrac{2\sqrt{3}}{3} - 1 \approx 0.15$.

b. If the transmitter is located at $(-1, \ h)$, we have:

$$\frac{y_1 - h}{x_1 + 1} = 1 - 2x_1$$

$$(x - x_1^2) - h = (1 - 2x_1)(x_1 + 1)$$

$$-x_1^2 + x_1 - h = -2x_1^2 - x_1 + 1$$

$$x_1^2 + 2x_1 - (h + 1) = 0$$

$$x_1 = \frac{-2 \pm \sqrt{2^2 + 4(1)(h + 1)}}{2(1)}$$

$$= -1 \pm \sqrt{2 + h}$$

Choosing the positive value for x_1, we have $x_1 = -1 + \sqrt{2 + h}$ and $m = 1 - 2(-1 + \sqrt{2 + h}) = 3 - 2\sqrt{2 + h}$. Also,

$$\frac{0 - 1}{x_0 + 1} = 3 - 2\sqrt{2 + h}$$

$$\frac{-1}{3 - 2\sqrt{2 + h}} = x_0 + 1$$

$$x_0 = \frac{-1}{3 - 2\sqrt{2 + h}} - 1.$$

The closest the receiver can be to the hill is

$$x = x_0 - 1 = \frac{-1}{3 - 2\sqrt{2 + h}} - 2.$$

c. $x = \dfrac{-1}{3 - 2\sqrt{2 + h}} - 2$

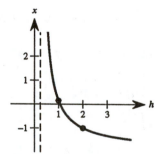

The vertical asymptote is at $h = \frac{1}{4}$. This corresponds to the height of the hill.

49. a.

$$\frac{dS}{dt} = kS(L - S)$$

$$\int \frac{dS}{S(L - S)} = \int k\,dt$$

$$\frac{1}{L}\ln\left(\frac{S}{L - S}\right) = kt + C_1$$

$$\frac{S}{L - S} = C_2 e^{Lk_1 t}$$

$$S = \frac{LC_2 e^{Lk_1 t}}{1 + C_2 e^{Lk_1 t}}$$

$$= \frac{L}{1 + (1/C_2)e^{-Lk_1 t}} = \frac{L}{1 + Ce^{-kt}}$$

Initial conditions:
$L = 100$, $S = 10$ when $t = 0$ and
$S = 20$ when $t = 1$.
$C = 9$, $k = -\ln\left(\frac{4}{9}\right)$

Particular solution:

$$S = \frac{100}{1 + 9e^{\ln(4/9)t}} \approx \frac{100}{1 + 9e^{-0.8109t}}$$

b.

$$\frac{dS}{dt} = \ln\left(\frac{4}{9}\right)S(100 - S)$$

$$\frac{d^2S}{dt^2} = \ln\left(\frac{4}{9}\right)\left[S\left(-\frac{dS}{dt}\right) + (100 - S)\frac{dS}{dt}\right]$$

$$= \ln\left(\frac{4}{9}\right)(100 - 2S)\frac{dS}{dt}$$

$= 0$ when $S = 50$ or $\frac{dS}{dt} = 0$.

Choosing $S = 50$, we have:

$$50 = \frac{100}{1 + 9e^{\ln(4/9)t}}$$

$$2 = 1 + 9e^{\ln(4/9)t}$$

$$\frac{\ln(1/9)}{\ln(4/9)} = t$$

$$t \approx 2.7 \text{ months}$$

c.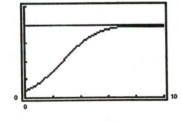

d.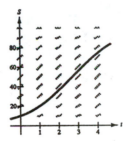

e. Sales will decrease toward the line $S = L$.

50. a.

$$\frac{dP}{dn} = kP(L - P)$$

$$\int \frac{dP}{dn} = \int k\,dn$$

$$P = \frac{L}{1 + Ce^{-kn}} \quad \text{(see Exercise 49a.)}$$

b. Initial conditions:
$L = 1$, $P = 0.25$ when $n = 0$ and
$P = 0.60$ when $n = 10$.
$C = 3$ and $k = -\frac{1}{10}\ln\left(\frac{2}{9}\right)$

Particular solution:

$$P = \frac{1}{1 + 3e^{n\ln(2/9)/10}} \approx \frac{1}{1 + 3e^{-0.1504n}}$$

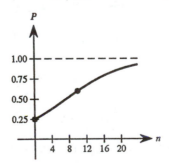

c. The graph has an inflection point when $n \approx 7.3$. After 7.3 trials the rate of growth of correct responses is increasing most rapidly.

51. a.
$$\frac{dv}{dt} = k(W - v)$$

$$\int \frac{dv}{W - v} = \int k\, dt$$

$$-\ln(W - v) = kt + C_1$$

$$v = W - Ce^{-kt}$$

Initial conditions:
$W = 20$, $v = 0$ when $t = 0$,
and $v = 5$ when $t = 1$.
$C = 20$, $k = -\ln\left(\frac{3}{4}\right)$

Particular solution:
$$v = 20(1 - e^{\ln(3/4)t}) \approx 20(1 - e^{-0.2877t})$$

b. $S = \displaystyle\int 20(1 - e^{-0.2877t})\, dt$

$$\approx 20[t + 3.4761e^{-0.2877t}] + C$$

Since $S(0) = 0$, $C \approx -69.5$ and we have
$S \approx 20t + 69.5(e^{-0.2877t} - 1)$.

52.
$$F = mv\frac{dv}{ds} = \frac{w}{32}v\frac{dv}{ds}$$

$$\frac{640}{32}\frac{dv}{dt} = 60 - 3v$$

$$20\int \frac{dv}{20 - v} = 3\int dt$$

$$-20\ln(20 - v) = 3t + C_1$$

$$20 - v = Ce^{-3t/20}$$

$$v = 20 - Ce^{-3t/20}$$

Initial condition: $v(0) = 0$, $0 = 20 - C$, $C = 20$
Particular solution: $v = 20(1 - e^{-3t/20})$
Limiting speed:
$$\lim_{t\to\infty} 20(1 - e^{-3t/20}) = 20(1 - 0) = 20 \text{ ft/sec}$$

53. Given family (circles): $x^2 + y^2 = C$

$$2x + 2yy' = 0$$

$$y' = -\frac{x}{y}$$

Orthogonal trajectory (lines): $y' = \dfrac{y}{x}$

$$\int \frac{dy}{y} = \int \frac{dx}{x}$$

$$\ln y = \ln x + \ln K$$

$$y = Kx$$

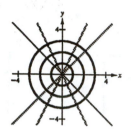

54. Given family (hyperbolas): $2x^2 - y^2 = C$

$$4x - 2yy' = 0$$

$$y' = \frac{2x}{y}$$

Orthogonal trajectory (hyperbolas): $y' = -\dfrac{y}{2x}$

$$2\int \frac{dy}{y} = -\int \frac{dx}{x}$$

$$2\ln y = -\ln x + \ln K$$

$$y^2 = \frac{K}{x}$$

55. Given family (parabolas): $x^2 = Cy$

$$2x = Cy'$$

$$y' = \frac{2x}{C} = \frac{2x}{x^2/y} = \frac{2y}{x}$$

Orthogonal trajectory (ellipses): $\qquad y' = -\dfrac{x}{2y}$

$$2\int y\,dy = -\int x\,dx$$

$$y^2 = -\frac{x^2}{2} + K_1$$

$$x^2 + 2y^2 = K$$

56. Given family (parabolas): $\quad y^2 = 2Cx$

$$2yy' = 2C$$

$$y' = \frac{C}{y} = \frac{y^2}{2x}\left(\frac{1}{y}\right) = \frac{y}{2x}$$

Orthogonal trajectory (ellipses): $\qquad y' = -\dfrac{2x}{y}$

$$\int y\,dy = -\int 2x\,dx$$

$$\frac{y^2}{2} = -x^2 + K_1$$

$$2x^2 + y^2 = K$$

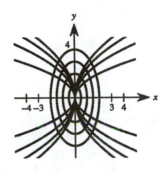

57. Given family: $\quad y^2 = Cx^3$

$$2yy' = 3Cx^2$$

$$y' = \frac{3Cx^2}{2y} = \frac{3x^2}{2y}\left(\frac{y^2}{x^3}\right) = \frac{3y}{2x}$$

Orthogonal trajectory (ellipses): $\qquad y' = -\dfrac{2x}{3y}$

$$3\int y\,dy = -2\int x\,dx$$

$$\frac{3y^2}{2} = -x^2 + K_1$$

$$3y^2 + 2x^2 = K$$

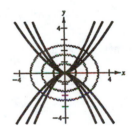

58. Given family (exponential functions): $y = Ce^x$

$$y' = Ce^x = y$$

Orthogonal trajectory (parabolas): $\quad y' = -\dfrac{1}{y}$

$$\int y\,dy = -\int dx$$

$$\frac{y^2}{2} = -x + K_1$$

$$y^2 = -2x + K$$

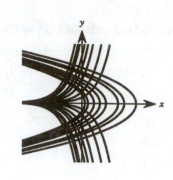

59. True

$$\frac{dy}{dx} = (x-2)(y+1)$$

60. True

$$\frac{dy}{dx} = \sin(x+y) + \sin(x-y)$$

$$= 2\sin x \cos y$$

61. False

$$f(tx, ty) = t^2x^2 + t^2xy + 2$$

$$\neq t^2 f(x, y)$$

62. True

$$x^2 + y^2 = 2Cy \qquad x^2 + y^2 = 2Kx$$

$$\frac{dy}{dx} = \frac{x}{C-y} \qquad \frac{dy}{dx} = \frac{K-x}{y}$$

$$\frac{x}{C-y} \cdot \frac{K-x}{y} = \frac{Kx - x^2}{Cy - y^2} = \frac{2Kx - 2x^2}{2Cy - 2y^2} = \frac{x^2 + y^2 - 2x^2}{x^2 + y^2 - 2y^2} = \frac{y^2 - x^2}{x^2 - y^2} = -1$$

63. a.

$$3500\frac{dw}{dt} = C - 17.5w$$

$$\int \frac{dw}{C - 17.5w} = \int \frac{1}{3500}\,dt$$

$$-\frac{1}{17.5}\ln(C - 17.5w) = \frac{1}{3500}t + K_1$$

$$C - 17.5w = e^{-(t/200 + K_1)}$$

$$w = \frac{2}{35}(C + K_2 e^{-t/200}) = \frac{1}{35}(2C + Ke^{-t/200})$$

b. $C = 2500$ when $t = 0$, $w = 180$. Thus, $K = 1300$. Therefore,

$$w = \frac{1}{35}(5000 + 1300e^{-t/200}) = \frac{20}{7}(50 + 13e^{-t/200}).$$

Solving for t, we have

$$e^{-t/200} = \frac{(7w/20) - 50}{13}$$

$$t = -200\ln\left(\frac{7w - 1000}{260}\right).$$

When $w = 170$ (10 pounds lost): $t \approx 63$ days
When $t = 145$ (35 pounds lost): $t \approx 571$ days

c. $\displaystyle\lim_{t \to \infty} \frac{20}{7}(50 + 13e^{-t/200}) = \frac{1000}{7} \approx 142.9$ pounds

Section 16.3   Exact First-Order Equations

1. $(2x - 3y)\,dx + (2y - 3x)\,dy = 0$

$\dfrac{\partial M}{\partial y} = -3 = \dfrac{\partial N}{\partial x}$

$U(x,\ y) = x^2 - 3xy + f(y)$

$f'(y) = 2y$

$f(y) = y^2 + C_1$

$x^2 - 3xy + y^2 = C$

2. $ye^x\,dx + e^x\,dy = 0$

$\dfrac{\partial M}{\partial y} = e^x = \dfrac{\partial N}{\partial x}$

$U(x,\ y) = ye^x + g(x)$

$g'(x) = 0$

$g(x) = C_1$

$ye^x = C$

3. $(3y^2 + 10xy^2)\,dx + (6xy - 2 + 10x^2y)\,dy = 0$

$\dfrac{\partial M}{\partial y} = 6y + 20xy = \dfrac{\partial N}{\partial x}$

$U(x,\ y) = 3y^2x + 5x^2y^2 + f(y)$

$f'(y) = -2$

$f(y) = -2y + C_1$

$3y^2x + 5x^2y^2 - 2y = C$

4. $2\cos(2x - y)\,dx - \cos(2x - y)\,dy = 0$

$\dfrac{\partial M}{\partial y} = 2\sin(2x - y) = \dfrac{\partial N}{\partial x}$

$U(x,\ y) = \sin(2x - y) + f(y)$

$f'(y) = 0$

$f(y) = C_1$

$\sin(2x - y) = C$

5. $(4x^3 - 6xy^2)\,dx + (4y^3 - 6xy)\,dy = 0$

$\dfrac{\partial M}{\partial y} = -12xy$

$\dfrac{\partial N}{\partial x} = -6y$

Not exact

6. $2y^2e^{xy^2}\,dx + 2xye^{xy^2}\,dy = 0$

$\dfrac{\partial M}{\partial y} = 4(xy^3 + y)e^{xy^2}$

$\dfrac{\partial N}{\partial x} = 2(xy^3 + y)e^{xy^2}$

Not exact

7. $\dfrac{-y}{x^2 + y^2}\,dx + \dfrac{x}{x^2 + y^2}\,dy = 0$

$\dfrac{\partial M}{\partial y} = \dfrac{y^2 - x^2}{(x^2 + y^2)^2} = \dfrac{\partial N}{\partial x}$

$U(x,\ y) = -\arctan\dfrac{x}{y} + f(y)$

$f'(y) = 0$

$f(y) = C_1$

$\arctan\dfrac{x}{y} = C$

8. $xe^{-(x^2 + y^2)}\,dx + ye^{-(x^2 + y^2)}\,dy = 0$

$\dfrac{\partial M}{\partial y} = -2xye^{-(x^2 + y^2)} = \dfrac{\partial N}{\partial x}$

$U(x,\ y) = -\dfrac{1}{2}e^{-(x^2 + y^2)} + f(y)$

$f'(y) = 0$

$f(y) = C_1$

$C = e^{-(x^2 + y^2)}$

9. $\left(\dfrac{y}{x - y}\right)^2 dx + \left(\dfrac{x}{x - y}\right)^2 dy = 0$

$\dfrac{\partial M}{\partial y} = \dfrac{2xy}{(x - y)^3}$

$\dfrac{\partial N}{\partial x} = \dfrac{-2xy}{(x - y)^3}$

Not exact

10. $(ye^y \cos xy)\,dx + e^y(x\cos xy + \sin xy)\,dy = 0$

$\dfrac{\partial M}{\partial y} = -xye^y \sin xy + (y + 1)e^y \cos xy = \dfrac{\partial N}{\partial x}$

$U(x,\ y) = e^y \sin xy + f(y)$

$f'(y) = 0$

$f(y) = C_1$

$e^y \sin xy = C$

11. $\dfrac{y}{x-1}\,dx + [\ln(x-1) + 2y]\,dy = 0$

$\dfrac{\partial M}{\partial y} = \dfrac{1}{x-1} = \dfrac{\partial N}{\partial x}$

$U(x,\ y) = y\ln(x-1) + f(y)$

$f'(y) = 2y$

$f(y) = y^2 + C_1$

$y\ln(x-1) + y^2 = C$

Initial condition:

$\quad y(2) = 4, \quad 4\ln(1) + 16 = C, \quad C = 16$

Particular solution: $y\ln(x-1) + y^2 = 16$

12. $\dfrac{x}{\sqrt{x^2+y^2}}\,dx + \dfrac{y}{\sqrt{x^2+y^2}}\,dy = 0$

$\dfrac{\partial M}{\partial y} = \dfrac{-xy}{\sqrt{(x^2+y^2)^3}} = \dfrac{\partial N}{\partial x}$

$U(x,\ y) = \sqrt{x^2+y^2} + f(y)$

$f'(y) = 0$

$f(y) = C_1$

$\sqrt{x^2+y^2} = C$

Initial condition:

$\quad y(4) = 3, \quad \sqrt{16+9} = \sqrt{25} = 5 = C$

Particular solution: $\sqrt{x^2+y^2} = 5$

13. $\dfrac{x}{x^2+y^2}\,dx + \dfrac{y}{x^2+y^2}\,dy = 0$

$\dfrac{\partial M}{\partial y} = \dfrac{-2xy}{(x^2+y^2)^2} = \dfrac{\partial N}{\partial x}$

$U(x,\ y) = \dfrac{1}{2}\ln(x^2+y^2) + f(y)$

$f'(y) = 0$

$f(y) = C_1$

$\ln(x^2+y^2) = C_2$

$x^2 + y^2 = C$

Initial condition: $y(0) = 4, \quad 16 = C$

Particular solution: $x^2 + y^2 = 16$

14. $(e^{3x}\sin 3y)\,dx + (e^{3x}\cos 3y)\,dy = 0$

$\dfrac{\partial M}{\partial y} = 3e^{3x}\cos 3y = \dfrac{\partial N}{\partial x}$

$U(x,\ y) = \dfrac{1}{3}e^{3x}\sin 3y + f(y)$

$f'(y) = 0$

$f(y) = C_1$

$e^{3x}\sin 3y = C$

Initial condition: $y(0) = \pi, \quad C = 0$

Particular solution: $e^{3x}\sin 3y = 0$

15. $(2x\tan y + 5)\,dx + (x^2\sec^2 y)\,dy = 0$

$\dfrac{\partial M}{\partial y} = 2x\sec^2 y = \dfrac{\partial N}{\partial x}$

$U(x,\ y) = x^2\tan y + 5x + f(y)$

$f'(y) = 0$

$f(y) = C_1$

$x^2\tan y + 5x = C$

Initial condition: $y(0) = 0, \quad C = 0$

Particular solution: $x^2\tan y + 5x = 0$

16. $(x^2 + y^2)\,dx + 2xy\,dy = 0$

$\dfrac{\partial M}{\partial y} = 2y = \dfrac{\partial N}{\partial x}$

$U(x,\ y) = \dfrac{x^3}{3} + xy^2 + f(y)$

$f'(y) = 0$

$f(y) = C_1$

$x^3 + 3xy^2 = C$

Initial condition: $y(3) = 1, \quad C = 36$

Particular solution: $x^3 + 3xy^2 = 36$

17. $y\,dx - (x + 6y^2)\,dy = 0$

$\dfrac{(\partial N/\partial x) - (\partial M/\partial y)}{M} = -\dfrac{2}{y} = k(y)$

Integrating factor: $e^{\int k(y)\,dy} = e^{\ln y^{-2}} = \dfrac{1}{y^2}$

Exact equation: $\dfrac{1}{y}\,dx - \left(\dfrac{x}{y^2} + 6\right)dy = 0$

$U(x,\ y) = \dfrac{x}{y} + f(y)$

$f'(y) = -6$

$f(y) = -6y + C_1$

$\dfrac{x}{y} - 6y = C$

18. $(2x^3 + y)\,dx - x\,dy = 0$

$\dfrac{(\partial M/\partial y) - (\partial N/\partial x)}{N} = -\dfrac{2}{x} = h(x)$

Integrating factor: $e^{\int h(x)\,dx} = e^{\ln x^{-2}} = \dfrac{1}{x^2}$

Exact equation: $\left(2x + \dfrac{y}{x^2}\right)dx - \dfrac{1}{x}\,dy = 0$

$U(x,\ y) = x^2 - \dfrac{y}{x} + f(y)$

$f'(y) = 0$

$f(y) = C_1$

$x^2 - \dfrac{y}{x} = C$

19. $(5x^2 - y)\,dx + x\,dy = 0$

$\dfrac{(\partial M/\partial y) - (\partial N/\partial x)}{N} = \dfrac{-2}{x} = h(x)$

Integrating factor: $e^{\int h(x)\,dx} = e^{\ln x^{-2}} = \dfrac{1}{x^2}$

Exact equation: $\left(5 - \dfrac{y}{x^2}\right)dx + \dfrac{1}{x}\,dy = 0$

$U(x,\ y) = 5x + \dfrac{y}{x} + f(y)$

$f'(y) = 0$

$f(y) = C_1$

$5x + \dfrac{y}{x} = C$

20. $(5x^2 - y^2)\,dx + 2y\,dy = 0$

$\dfrac{(\partial M/\partial y) - (\partial N/\partial x)}{N} = -1 = h(x)$

Integrating factor: $e^{\int h(x)\,dx} = e^{-x}$

Exact equation: $(5x^2 - y^2)e^{-x}\,dx + 2ye^{-x}\,dy = 0$

$U(x,\ y) = -5x^2 e^{-x} - 10xe^{-x} - 10e^{-x} + y^2 e^{-x} + f(y)$

$f'(y) = 0$

$f(y) = C_1$

$y^2 e^{-x} - 5x^2 e^{-x} - 10xe^{-x} - 10e^{-x} = C$

21. $(x + y)\,dx + (\tan x)\,dy = 0$

$\dfrac{(\partial M/\partial y) - (\partial N/\partial x)}{N} = -\tan x = h(x)$

Integrating factor: $e^{\int h(x)\,dx} = e^{\ln \cos x} = \cos x$

Exact equation: $(x + y)\cos x\,dx + \sin x\,dy = 0$

$U(x,\ y) = x \sin x + \cos x + y \sin x + f(y)$

$f'(y) = 0$

$f(y) = C_1$

$x \sin x + \cos x + y \sin x = C$

22. $(2x^2 y - 1)\,dx + x^3\,dy = 0$

$\dfrac{(\partial M/\partial y) - (\partial N/\partial x)}{N} = -\dfrac{1}{x} = h(x)$

Integrating factor: $e^{\int h(x)\,dx} = e^{\ln(1/x)} = \dfrac{1}{x}$

Exact equation: $\left(2xy - \dfrac{1}{x}\right)dx + x^2\,dy = 0$

$U(x,\ y) = x^2 y - \ln|x| + f(y)$

$f'(y) = 0$

$f(y) = C_1$

$x^2 y - \ln|x| = C$

23. $y^2\,dx + (xy - 1)\,dy = 0$

$\dfrac{(\partial N/\partial x) - (\partial M/\partial y)}{M} = -\dfrac{1}{y} = k(y)$

Integrating factor: $e^{\int k(y)\,dy} = e^{\ln(1/y)} = \dfrac{1}{y}$

Exact equation: $y\,dx + \left(x - \dfrac{1}{y}\right)dy = 0$

$U(x,\ y) = xy + f(y)$

$f'(y) = -\dfrac{1}{y}$

$f(y) = -\ln|y| + C_1$

$xy - \ln|y| = C$

24. $(x^2 + 2x + y)\,dx + 2\,dy = 0$

$\dfrac{(\partial M/\partial y) - (\partial N/\partial x)}{N} = \dfrac{1}{2} = h(x)$

Integrating factor: $e^{\int h(x)\,dx} = e^{x/2}$

Exact equation: $(x^2 + 2x + y)e^{x/2}\,dx + 2e^{x/2}\,dy = 0$

$U(x,\ y) = 2(x^2 - 2x + 4 + y)e^{x/2} + f(y)$

$f'(y) = 0$

$f(y) = C_1$

$(x^2 - 2x + 4 + y)e^{x/2} = C$

25. $2y\,dx + \left(x - \sin\sqrt{y}\right)dy = 0$

$$\frac{(\partial N/\partial x) - (\partial M/\partial y)}{M} = \frac{-1}{2y} = k(y)$$

Integrating factor: $e^{\int k(y)\,dy} = e^{\ln(1/\sqrt{y})} = \dfrac{1}{\sqrt{y}}$

Exact equation: $2\sqrt{y}\,dx + \left(\dfrac{x}{\sqrt{y}} - \dfrac{\sin\sqrt{y}}{\sqrt{y}}\right)dy = 0$

$U(x,\ y) = 2\sqrt{y}\,x + f(y)$

$f'(y) = -\dfrac{\sin\sqrt{y}}{\sqrt{y}}$

$f(y) = 2\cos\sqrt{y} + C_1$

$\sqrt{y}\,x + \cos\sqrt{y} = C$

26. $\left(-2y^3 + 1\right)dx + \left(3xy^2 + x^3\right)dy = 0$

$$\frac{(\partial M/\partial y) - (\partial N/\partial x)}{N} = \frac{-3}{x} = h(x)$$

Integrating factor: $e^{\int h(x)\,dx} = e^{\ln(1/x^3)} = \dfrac{1}{x^3}$

Exact equation: $\left(\dfrac{-2y^3}{x^3} + \dfrac{1}{x^3}\right)dx + \left(\dfrac{3y^2}{x^2} + 1\right)dy = 0$

$U(x,\ y) = \dfrac{y^3}{x^2} - \dfrac{1}{2x^2} + f(y)$

$f'(y) = 1$

$f(y) = y + C_1$

$\dfrac{y^3}{x^2} - \dfrac{1}{2x^2} + y = C$

27. $\left(4x^2 y + 2y^2\right)dx + \left(3x^3 + 4xy\right)dy = 0$

Integrating factor: xy^2

Exact equation:

$\quad \left(4x^3 y^3 + 2xy^4\right)dx + \left(3x^4 y^2 + 4x^2 y^3\right)dy = 0$

$U(x,\ y) = x^4 y^3 + x^2 y^4 + f(y)$

$f'(y) = 0$

$f(y) = C_1$

$x^4 y^3 + x^2 y^4 = C$

28. $\left(3y^2 + 5x^2 y\right)dx + \left(3xy + 2x^3\right)dy = 0$

Integrating factor: $x^2 y$

Exact equation:

$\quad \left(3x^2 y^3 + 5x^4 y^2\right)dx + \left(3x^3 y^2 + 2x^5 y\right)dy = 0$

$U(x,\ y) = x^3 y^3 + x^5 y^2 + f(y)$

$f'(y) = 0$

$f(y) = C_1$

$x^3 y^3 + x^5 y^2 = C$

29. $\left(-y^5 + x^2 y\right)dx + \left(2xy^4 - 2x^3\right)dy = 0$

Integrating factor: $x^{-2} y^{-3}$

Exact equation: $\left(-\dfrac{y^2}{x^2} + \dfrac{1}{y^2}\right)dx + \left(2\dfrac{y}{x} - 2\dfrac{x}{y^3}\right)dy = 0$

$U(x,\ y) = \dfrac{y^2}{x} + \dfrac{x}{y^2} + f(y)$

$f'(y) = 0$

$f(y) = C_1$

$\dfrac{y^2}{x} + \dfrac{x}{y^2} = C$

30. $-y^3\,dx + \left(xy^2 - x^2\right)dy = 0$

Integrating factor: $x^{-2} y^{-2}$

Exact equation: $\dfrac{-y}{x^2}\,dx + \left(\dfrac{1}{x} - \dfrac{1}{y^2}\right)dy = 0$

$U(x,\ y) = \dfrac{y}{x} + f(y)$

$f'(y) = \dfrac{-1}{y^2}$

$f(y) = \dfrac{1}{y} + C_1$

$\dfrac{y}{x} + \dfrac{1}{y} = C$

31. $y\,dx - x\,dy = 0$

    a. $\dfrac{1}{x^2}, \quad \dfrac{y}{x^2}\,dx - \dfrac{1}{x}\,dy = 0, \quad \dfrac{\partial M}{\partial y} = \dfrac{1}{x^2} = \dfrac{\partial N}{\partial x}$

    b. $\dfrac{1}{y^2}, \quad \dfrac{1}{y}\,dx - \dfrac{x}{y^2}\,dy = 0, \quad \dfrac{\partial M}{\partial y} = \dfrac{-1}{y^2} = \dfrac{\partial N}{\partial x}$

    c. $\dfrac{1}{xy}, \quad \dfrac{1}{x}\,dx - \dfrac{1}{y}\,dy = 0, \quad \dfrac{\partial M}{\partial y} = 0 = \dfrac{\partial N}{\partial x}$

    d. $\dfrac{1}{x^2 + y^2}, \quad \dfrac{y}{x^2 + y^2}\,dx - \dfrac{x}{x^2 + y^2}\,dy = 0, \quad \dfrac{\partial M}{\partial y} = \dfrac{x^2 - y^2}{(x^2 + y^2)^2} = \dfrac{\partial N}{\partial x}$

32. $(axy^2 + by) \, dx + (bx^2y + ax) \, dy = 0$

Exact equation: $\dfrac{\partial M}{\partial y} = 2axy + b, \quad \dfrac{\partial N}{\partial x} = 2bxy + a, \quad \dfrac{\partial M}{\partial y} = \dfrac{\partial M}{\partial x}$ only if $a = b$.

Integrating factor: $x^m y^n$

$(ax^{m+1}y^{n+2} + bx^m y^{n+1}) \, dx + (bx^{m+2}y^{n+1} + ax^{m+1}y^n) \, dy = 0$

$\left.\begin{array}{l} \dfrac{\partial M}{\partial y} = a(n+2)x^{m+1}y^{n+1} + b(n+1)x^m y^n \\[2mm] \dfrac{\partial N}{\partial x} = b(m+2)x^{m+1}y^{n+1} + a(m+1)x^m y^n \end{array}\right\} \quad \begin{array}{l} a(n+2) = b(m+2) \\[2mm] b(n+1) = a(m+1) \end{array}$

$\left.\begin{array}{l} an - bm = 2(b-a) \\[2mm] bn - am = a - b \end{array}\right\} \quad \begin{array}{l} abn - b^2m = 2b(b-a) \\[2mm] \dfrac{abn - a^2m = a(a-b)}{(a^2 - b^2)m = -(2b+a)(a-b)} \\[2mm] m = -\dfrac{2b+a}{a+b} \end{array}$

$bn - a\left(-\dfrac{2b+a}{a+b}\right) = a - b$

$$bn = \frac{-2ab - a^2 + a^2 - b^2}{a+b} = \frac{-b(2a+b)}{a+b}$$

$$n = -\frac{2a+b}{a+b}$$

33. $\mathbf{F}(x, \ y) = \dfrac{y}{\sqrt{x^2+y^2}}\mathbf{i} - \dfrac{x}{\sqrt{x^2+y^2}}\mathbf{j}$

$\dfrac{dy}{dx} = -\dfrac{x}{y}$

$y \, dy + x \, dx = 0$

$y^2 + x^2 = C$

Family of circles

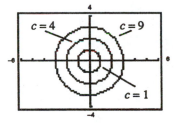

34. $\mathbf{F}(x, \ y) = \dfrac{x}{\sqrt{x^2+y^2}}\mathbf{i} - \dfrac{y}{\sqrt{x^2+y^2}}\mathbf{j}$

$\dfrac{dy}{dx} = -\dfrac{y}{x}$

$x \, dy + y \, dx = 0$

$xy = C$

Family of hyperbolas

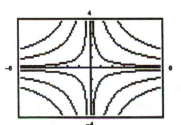

35. $\mathbf{F}(x, \ y) = 4x^2y\mathbf{i} - \left(2xy^2 + \dfrac{x}{y^2}\right)\mathbf{j}$

$\dfrac{dy}{dx} = \dfrac{-y}{2x} - \dfrac{1}{4xy^3}$

$\dfrac{8y^3}{2y^4+1} \, dy = -\dfrac{2}{x} \, dx$

$\ln(2y^4 + 1) = \ln\left(\dfrac{1}{x^2}\right) + \ln C$

$2y^4 + 1 = \dfrac{C}{x^2}$

$2x^2y^4 + x^2 = C$

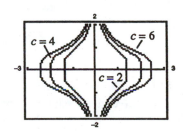

36. $\mathbf{F}(x, \ y) = (1 + x^2)\mathbf{i} - 2xy\mathbf{j}$

$$\frac{dy}{dx} = \frac{-2xy}{1+x^2}$$

$$\frac{1}{y}\,dy = -\frac{2x}{1+x^2}\,dx$$

$$\ln y = \ln\left(\frac{1}{1+x^2}\right) + \ln C$$

$$y = \frac{C}{1+x^2}$$

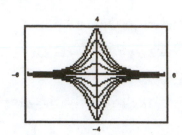

37. $\dfrac{dy}{dx} = \dfrac{y-x}{3y-x}$

$(x - y)\,dx + (3y - x)\,dy = 0$

$\dfrac{\partial M}{\partial y} = -1 = \dfrac{\partial N}{\partial x}$

$U(x, \ y) = \dfrac{x^2}{2} - xy + f(y)$

$f'(y) = 3y$

$f(y) = \dfrac{3y^2}{2} + C_1$

$x^2 - 2xy + 3y^2 = C$

Initial condition:

$y(2) = 1, \ 4 - 4 + 3 = C, \ C = 3$

Particular solution: $x^2 - 2xy + 3y^2 = 3$

38. $\dfrac{dy}{dx} = \dfrac{-2xy}{x^2 + y^2}$

$2xy\,dx + (x^2 + y^2)\,dy = 0$

$\dfrac{\partial M}{\partial y} = 2x = \dfrac{\partial N}{\partial x}$

$U(x, \ y) = x^2 y + f(y)$

$f'(y) = y^2$

$f(y) = \dfrac{y^3}{3} + C_1$

$3x^2 y + y^3 = C$

Initial condition: $y(0) = 2, \ 8 = C$

Particular solution: $3x^2 y + y^3 = 8$

39. $E(x) = \dfrac{20x - y}{2y - 10x} = \dfrac{x}{y}\dfrac{dy}{dx}$

$(20xy - y^2)\,dx + (10x^2 - 2xy)\,dy = 0$

$\dfrac{\partial M}{\partial y} = 20x - 2y = \dfrac{\partial N}{\partial x}$

$U(x, \ y) = 10x^2 y - xy^2 + f(y)$

$f'(y) = 0$

$f(y) = C_1$

$10x^2 y - xy^2 = K$

Initial condition: $C(100) = 500, \ 100 \le x, \ K = 25,000,000$

$$10x^2 y - xy^2 = 25,000,000$$

$xy^2 - 10x^2 y + 25,000,000 = 0$ Quadratic Formula

$$y = \frac{10x^2 + \sqrt{100x^4 - 4x(25,000,000)}}{2x} = \frac{5\left(x^2 + \sqrt{x^4 - 1,000,000x}\right)}{x}$$

40. a. The interval $[a, \ b]$ is divided into n equal subintervals of width Δx. Euler's Method provides approximate solution values to the differential equation at the endpoints of each subinterval. Given $y_0 = y(a)$, compute $y_1 \approx y(x_1)$, $y_2 \approx y(x_2)$, ..., $y_n \approx y(b)$. You now have n solution points and can graph the curve through these points. This curve approximates the solution to the differential equation over this interval.

b. Decreasing the value of Δx will increase the number of iterations and will provide a better estimate to the solution of the differential equation.

41. $y' = x + \sqrt{y}$, $y(0) = 2$ and $\Delta x = 0.50$

$y_1 = 2 + (0 + \sqrt{2})(0.50) \approx 2.7071$

$y_2 = 2.7071 + (0.5 + \sqrt{2.7071})(0.50) \approx 3.7798$

$y(1) \approx 3.7798$

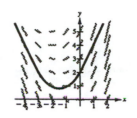

$y' = x + \sqrt{y}$, $y(0) = 2$ and $\Delta x = 0.25$

$y_1 = 2 + (0 + \sqrt{2})(0.25) \approx 2.3536$

$y_2 = 2.3536 + (0.25 + \sqrt{2.3536})(0.25) \approx 2.7996$

$y_3 = 2.7996 + (0.50 + \sqrt{2.7996})(0.25) \approx 3.3429$

$y_4 = 3.3429 + (0.75 + \sqrt{3.3429})(0.25) \approx 3.9875$

$y(1) \approx 3.9875$

$y' = x + \sqrt{y}$, $y(0) = 2$ and $\Delta x = 0.10$

$y_1 = 2 + (0 + \sqrt{2})(0.10) \approx 2.1414$

$y_2 = 2.1414 + (0.10 + \sqrt{2.1414})(0.10) \approx 2.2977$

$y_3 = 2.2977 + (0.20 + \sqrt{2.2977})(0.10) \approx 2.4693$

$y_4 = 2.4693 + (0.30 + \sqrt{2.4693})(0.10) \approx 2.6564$

$y_5 = 2.6564 + (0.40 + \sqrt{2.6564})(0.10) \approx 2.8594$

$y_6 = 2.8594 + (0.50 + \sqrt{2.8594})(0.10) \approx 3.0785$

$y_7 = 3.0785 + (0.60 + \sqrt{3.0785})(0.10) \approx 3.3140$

$y_8 = 3.3140 + (0.70 + \sqrt{3.3140})(0.10) \approx 3.5660$

$y_9 = 3.5660 + (0.80 + \sqrt{3.5660})(0.10) \approx 3.8348$

$y_{10} = 3.8348 + (0.90 + \sqrt{3.8348})(0.10) \approx 4.1206$

$y(1) \approx 4.1206$

Δx	0.50	0.25	0.10
Estimate	3.7798	3.9875	4.1206

42. a. Your program will vary depending on the type of computer that you have. Since $\Delta x = 0.01$, your program will need to calculate 100 iterations to approximate $y(2)$.

b.
$$\frac{dy}{dx} = x\sqrt[3]{y}$$

$$\int y^{-1/3}\, dy = \int x\, dx$$

$$\frac{3}{2}y^{2/3} = \frac{1}{2}x^2 + C_1$$

$$y^{2/3} = \frac{1}{3}x^2 + C$$

$$y = \left[\frac{1}{3}x^2 + C\right]^{3/2}$$

Initial condition: $y(1) = 1$, $1 = \left[\frac{1}{3} + C\right]^{3/2} \Rightarrow C = \frac{2}{3}$

Particular solution: $y = \left[\frac{1}{3}x^2 + \frac{2}{3}\right]^{3/2}$

Thus, $y(2) = 2\sqrt{2} \approx 2.8284$.

43. False

$$\frac{\partial M}{\partial y} = 2x \text{ and } \frac{\partial N}{\partial x} = -2x$$

44. False

$y\,dx + x\,dy = 0$ is exact, but $xy\,dx + x^2\,dy = 0$ is not exact.

45. True

$$\frac{\partial}{\partial y}[f(x) + M] = \frac{\partial M}{\partial y} \text{ and } \frac{\partial}{\partial x}[g(y) + N] = \frac{\partial N}{\partial x}$$

46. True

$$\frac{\partial}{\partial y}[f(x)] = 0 \text{ and } \frac{\partial}{\partial x}[g(y)] = 0$$

Section 16.4   First-Order Linear Differential Equations

1. False

$y' + xy = x^2$ is first-order linear.

2. True

$y' + y(x - e^x) = 0$ is first-order linear.

3. $\dfrac{dy}{dx} + \left(\dfrac{1}{x}\right)y = 3x + 4$

Integrating factor: $e^{\int (1/x)\,dx} = e^{\ln x} = x$

$$xy = \int x(3x + 4)\,dx = x^3 + 2x^2 + C$$

$$y = x^2 + 2x + \frac{C}{x}$$

4. $\dfrac{dy}{dx} + \left(\dfrac{2}{x}\right)y = 3x + 1$

Integrating factor: $e^{\int (2/x)\,dx} = e^{\ln x^2} = x^2$

$$x^2y = \int x^2(3x + 1)\,dx = \frac{3}{4}x^4 + \frac{1}{3}x^3 + C$$

$$y = \frac{3x^2}{4} + \frac{x}{3} + \frac{C}{x^2}$$

5. $\dfrac{dy}{dx} = e^x - y, \quad \dfrac{dy}{dx} + y = e^x$

Integrating factor: $e^{\int dx} = e^x$

$$ye^x = \int e^{2x}\,dx = \frac{1}{2}e^{2x} + C$$

$$y = \frac{1}{2}e^x + Ce^{-x}$$

6. $y' + 2y = \sin x$

Integrating factor: $e^{\int 2\,dx} = e^{2x}$

$$ye^{2x} = \int e^{2x} \sin x\,dx = \frac{1}{5}e^{2x}(2 \sin x - \cos x) + C$$

$$y = \frac{1}{5}(2 \sin x - \cos x) + Ce^{-2x}$$

7. $y' - y = \cos x$

Integrating factor: $e^{\int -1\,dx} = e^{-x}$

$$ye^{-x} = \int e^{-x} \cos x\,dx$$

$$= \frac{1}{2}e^{-x}(-\cos x + \sin x) + C$$

$$y = \frac{1}{2}(\sin x - \cos x) + Ce^x$$

8. $y' + 2xy = 2x$

Integrating factor: $e^{\int 2x\,dx} = e^{x^2}$

$$ye^{x^2} = \int 2xe^{x^2}\,dx = e^{x^2} + C$$

$$y = 1 + Ce^{-x^2}$$

9. $(3y + \sin 2x)\,dx - dy = 0$

$y' - 3y = \sin 2x$

Integrating factor: $e^{\int -3\,dx} = e^{-3x}$

$$ye^{-3x} = \int e^{-3x} \sin 2x\,dx$$

$$= \frac{1}{13}e^{-3x}(-3 \sin 2x - 2 \cos 2x) + C$$

$$y = -\frac{1}{13}(3 \sin 2x + 2 \cos 2x) + Ce^{3x}$$

10. $[(y - 1) \sin x]\,dx - dy = 0$

$y' - (\sin x)y = -\sin x$

Integrating factor: $e^{\int -\sin x\,dx} = e^{\cos x}$

$$ye^{\cos x} = \int -\sin x\, e^{\cos x}\,dx = e^{\cos x} + C$$

$$y = 1 + Ce^{-\cos x}$$

11. $(x-1)y' + y = x^2 - 1$

$$y' + \left(\frac{1}{x-1}\right)y = x + 1$$

Integrating factor: $e^{\int [1/(x-1)]\,dx} = e^{\ln|x-1|} = x - 1$

$$y(x-1) = \int (x^2 - 1)\,dx = \frac{1}{3}x^3 - x + C_1$$

$$y = \frac{x^3 - 3x + C}{3(x-1)}$$

12. $y' + 5y = e^{5x}$

Integrating factor: $e^{\int 5\,dx} = e^{5x}$

$$ye^{5x} = \int e^{10x}\,dx = \frac{1}{10}e^{10x} + C$$

$$y = \frac{1}{10}e^{5x} + Ce^{-5x}$$

13. $y'\cos^2 x + y - 1 = 0$

$$y' + (\sec^2 x)y = \sec^2 x$$

Integrating factor: $e^{\int \sec^2 x\,dx} = e^{\tan x}$

$$ye^{\tan x} = \int \sec^2 x\, e^{\tan x}\,dx = e^{\tan x} + C$$

$$y = 1 + Ce^{-\tan x}$$

Initial condition: $y(0) = 5,\ \ C = 4$

Particular solution: $y = 1 + 4e^{-\tan x}$

14. $x^3 y' + 2y = e^{1/x^2}$

$$y' + \left(\frac{2}{x^3}\right)y = \frac{1}{x^3}e^{1/x^2}$$

Integrating factor: $e^{\int (2/x^3)\,dx} = e^{-(1/x^2)}$

$$ye^{-1/x^2} = \int \frac{1}{x^3}\,dx = -\frac{1}{2x^2} + C_1$$

$$y = e^{1/x^2}\left(\frac{Cx^2 - 1}{2x^2}\right)$$

Initial condition: $y(1) = e,\ \ C = 3$

Particular solution: $y = e^{1/x^2}\left(\frac{3x^2 - 1}{2x^2}\right)$

15. $y' + y\tan x = \sec x + \cos x$

Integrating factor: $e^{\int \tan x\,dx} = e^{\ln|\sec x|} = \sec x$

$$y\sec x = \int \sec x(\sec x + \cos x)\,dx = \tan x + x + C$$

$$y = \sin x + x\cos x + C\cos x$$

Initial condition: $y(0) = 1,\ \ 1 = C$

Particular solution: $y = \sin x + (x + 1)\cos x$

16. $y' + y\sec x = \sec x$

Integrating factor: $e^{\int \sec x\,dx} = e^{\ln|\sec x + \tan x|} = \sec x + \tan x$

$$y(\sec x + \tan x) = \int (\sec x + \tan x)\sec x\,dx = \sec x + \tan x + C$$

$$y = 1 + \frac{C}{\sec x + \tan x}$$

Initial condition: $y(0) = 4,\ \ 4 = 1 + \frac{C}{1+0},\ \ C = 3$

Particular solution: $y = 1 + \frac{3}{\sec x + \tan x} = 1 + \frac{3\cos x}{1 + \sin x}$

17. $y' + \left(\dfrac{1}{x}\right)y = 0$

Integrating factor: $e^{\int (1/x)\,dx} = e^{\ln|x|} = x, \quad xy = C$

Separation of variables:

$$\frac{dy}{dx} = -\frac{y}{x}$$

$$\int \frac{1}{y}\,dy = \int -\frac{1}{x}\,dx$$

$$\ln y = -\ln x + \ln C$$

$$\ln xy = \ln C$$

$$xy = C$$

Initial condition: $y(2) = 2, \quad C = 4$

Particular solution: $xy = 4$

18. $y' + (2x - 1)y = 0$

Integrating factor: $e^{\int (2x-1)\,dx} = e^{x^2 - x}$

$$ye^{x^2-x} = C$$

$$y = Ce^{x-x^2}$$

Separation of variables:

$$\int \frac{1}{y}\,dy = \int (1 - 2x)\,dx$$

$$\ln y + \ln C_1 = x - x^2$$

$$yC_1 = e^{x-x^2}$$

$$y = Ce^{x-x^2}$$

Initial condition: $y(1) = 2, \quad 2 = C$

Particular solution: $y = 2e^{x-x^2}$

19. $y' + 3x^2y = x^2y^3$

$n = 3, \quad Q = x^2, \quad P = 3x^2$

$$y^{-2}e^{\int(-2)3x^2\,dx} = \int (-2)x^2 e^{\int(-2)3x^2\,dx}\,dx$$

$$y^{-2}e^{-2x^3} = -\int 2x^2 e^{-2x^3}\,dx$$

$$y^{-2}e^{-2x^3} = \frac{1}{3}e^{-2x^3} + C$$

$$y^{-2} = \frac{1}{3} + Ce^{2x^3}$$

$$\frac{1}{y^2} = Ce^{2x^3} + \frac{1}{3}$$

20. $y' + 2xy = xy^2$

$n = 2, \quad Q = x, \quad P = 2x, \quad e^{\int -2x\,dx} = e^{-x^2}$

$$y^{-1}e^{-x^2} = \int -xe^{-x^2}\,dx = \frac{1}{2}e^{-x^2} + C_1$$

$$\frac{1}{y} = \frac{1 + Ce^{x^2}}{2}$$

$$y = \frac{2}{1 + Ce^{x^2}}$$

21. $y' + \left(\dfrac{1}{x}\right)y = xy^2$

$n = 2, \quad Q = x, \quad P = x^{-1}$

$$e^{\int -(1/x)\,dx} = e^{-\ln|x|} = x^{-1}$$

$$y^{-1}x^{-1} = \int -x(x^{-1})\,dx = -x + C$$

$$\frac{1}{y} = -x^2 + Cx$$

$$y = \frac{1}{Cx - x^2}$$

22. $y' + \left(\dfrac{1}{x}\right)y = x\sqrt{y}$

$n = \dfrac{1}{2}, \quad Q = x, \quad P = x^{-1}$

$$e^{\int (1/2)(1/x)\,dx} = e^{(1/2)\ln x} = \sqrt{x}$$

$$y^{1/2}x^{1/2} = \int \frac{1}{2}x^{1/2}(x)\,dx$$

$$= \frac{1}{5}x^{5/2} + C_1 = \frac{x^{5/2} + C}{5}$$

$$y = \frac{(x^{5/2} + C)^2}{25x}$$

23. $y' - y = x^3 \sqrt[3]{y}$

$n = \dfrac{1}{3}, \quad Q = x^3, \quad P = -1$

$e^{\int -(2/3)\,dx} = e^{-(2/3)x}$

$y^{2/3}e^{-(2/3)x} = \displaystyle\int \frac{2}{3}x^3 e^{-(2/3)x}\,dx$

$y^{2/3} = -\dfrac{1}{4}(4x^3 + 18x^2 + 54x + 81) + Ce^{2x/3}$

24. $yy' - 2y^2 = e^x$

$y' - 2y = e^x y^{-1}$

$n = -1, \quad Q = e^x, \quad P = -2$

$e^{\int 2(-2)\,dx} = e^{-4x}$

$y^2 e^{-4x} = \displaystyle\int 2e^{-4x}e^x\,dx = -\frac{2}{3}e^{-3x} + C$

$y^2 = -\dfrac{2}{3}e^x + Ce^{4x}$

25. $L\dfrac{dI}{dt} + RI = E_0, \quad I' + \dfrac{R}{L}I = \dfrac{E_0}{L}$

Integrating factor: $e^{\int (R/L)\,dt} = e^{Rt/L}$

$Ie^{Rt/L} = \displaystyle\int \frac{E_0}{L}e^{Rt/L}\,dt = \frac{E_0}{R}e^{Rt/L} + C$

$I = \dfrac{E_0}{R} + Ce^{-Rt/L}$

26. $I(0) = 0, \quad E_0 = 110$ volts

$R = 550$ ohms, $\quad L = 4$ henrys

$0 = \dfrac{110}{550} + Ce^0, \quad C = -\dfrac{1}{5}$

$y = \dfrac{E_0}{R} - \dfrac{1}{5}e^{-Rt/L} = \dfrac{1}{5}(1 - e^{-137.5t})$

$\displaystyle\lim_{t\to\infty} \frac{1}{5}(1 - e^{-137.5t}) = \frac{1}{5}$ amp

$(0.90)\left(\dfrac{1}{5}\right) = 0.18$

$0.18 = \dfrac{1}{5}(1 - e^{-137.5t})$

$0.90 = 1 - e^{-137.5t}$

$-137.5t = \ln(0.10)$

$t = \dfrac{\ln(0.10)}{-137.5} \approx 0.0167$

27. $L\dfrac{dI}{dt} + RI = E_0 \sin \omega t$

$\dfrac{dI}{dt} + \dfrac{R}{L}I = \dfrac{E_0}{L}\sin \omega t$

Integrating factor: $e^{\int (R/L)\,dt} = e^{Rt/L}$

$Ie^{Rt/L} = \displaystyle\int \frac{E_0}{L}e^{Rt/L}\sin \omega t\,dt$

$= \dfrac{E_0}{L}\left[\dfrac{L^2 e^{Rt/L}}{R^2 + L^2\omega^2}\left(\dfrac{R}{L}\sin \omega t - \omega \cos \omega t\right)\right] + C = \dfrac{E_0 e^{Rt/L}}{R^2 + \omega^2 L^2}(R \sin \omega t - \omega L \cos \omega t) + C$

$I = \dfrac{E_0}{R^2 + \omega^2 L^2}(R \sin \omega t - \omega L \cos \omega t) + Ce^{-Rt/L}$

28. $\sin(\omega t - \phi) = \sin \omega t \cos \phi - \cos \omega t \sin \phi$

$= \dfrac{R}{\sqrt{R^2 + \omega^2 L^2}}\sin \omega t - \dfrac{\omega L}{\sqrt{R^2 \omega^2 L^2}}\cos \omega t$

$= \dfrac{1}{\sqrt{R^2 + \omega^2 L^2}}(R \sin \omega t - \omega L \cos \omega t)$

From Exercise 27, we have:

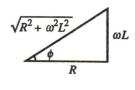

$I = Ce^{-Rt/L} + \dfrac{E_0}{\sqrt{R^2 + \omega^2 L^2}} \cdot \dfrac{1}{\sqrt{R^2 + \omega^2 L^2}}(R \sin \omega t - \omega L \cos \omega t) = Ce^{-Rt/L} + \dfrac{E_0}{\sqrt{R^2 + \omega^2 L^2}}\sin(\omega t - \phi)$

29.

$$\frac{dP}{dt} = kP + N, \quad N \text{ constant}$$

$$\frac{dP}{kP + N} = dt$$

$$\int \frac{1}{kP + N} \, dP = \int dt$$

$$\frac{1}{k} \ln(kP + N) = t + C_1$$

$$\ln(kP + N) = kt + C_2$$

$$kP + N = e^{kt + C_2}$$

$$P = \frac{C_3 e^{kt} - N}{k}$$

$$P = Ce^{kt} - \frac{N}{k}$$

When $t = 0$: $P = P_0$

$$P_0 = C - \frac{N}{k} \Rightarrow C = P_0 + \frac{N}{k}$$

$$P = \left(P_0 + \frac{N}{k} \right) e^{kt} - \frac{N}{k}$$

30.

$$\frac{dA}{dt} = rA + P$$

$$\frac{dA}{rA + P} = dt$$

$$\int \frac{dA}{rA + P} = \int dt$$

$$\frac{1}{r} \ln(rA + P) = t + C_1$$

$$\ln(rA + P) = rt + C_2$$

$$rA + P = e^{rt + C_2}$$

$$A = \frac{C_3 e^{rt} - P}{r}$$

$$A = Ce^{rt} - \frac{P}{r}$$

When $t = 0$: $A = 0$

$$0 = C - \frac{P}{r} \Rightarrow C = \frac{P}{r}$$

$$A = \frac{P}{r}(e^{rt} - 1)$$

31. a. $A = \dfrac{P}{r}(e^{rt} - 1)$

$$A = \frac{100,000}{0.12}(e^{0.12(5)} - 1) \approx \$685,099.00$$

b. $A = \dfrac{250,000}{0.15}(e^{0.15(10)} - 1) \approx \$5,802,815.12$

32.

$$800,000 = \frac{75,000}{0.13}(e^{0.13t} - 1)$$

$$2.386666667 = e^{0.13t}$$

$$t = \frac{\ln 2.386666667}{0.13}$$

$$t \approx 6.69 \text{ years}$$

33. a. $\dfrac{dQ}{dt} = q - kQ, \quad q$ constant

b. $Q' + kQ = q$

Let $P(t) = k, \quad Q(t) = q$, then the integrating factor is $u(t) = e^{kt}$.

$$Q = e^{-kt} \int q e^{kt} \, dt = e^{-kt} \left(\frac{q}{k} e^{kt} + C \right) = \frac{q}{k} + Ce^{-kt}$$

When $t = 0$: $Q = Q_0$

$$Q_0 = \frac{q}{k} + C \Rightarrow C = Q_0 - \frac{q}{k}$$

$$Q = \frac{q}{k} + \left(Q_0 - \frac{q}{k} \right) e^{-kt}$$

c. $\displaystyle\lim_{t \to \infty} Q = \frac{q}{k}$

34. a. $\dfrac{dN}{dt} = k(30 - N)$

b. $N' + kN = 30k$

Let $P(t) = k$, $Q(t) = 30k$, then the integrating factor is $u(t) = e^{kt}$.

$$N = e^{-kt} \int 30k e^{kt} \, dt = e^{-kt}(30e^{kt} + C) = 30 + Ce^{-kt}$$

c. When $t = 1$: $N = 10$

$10 = 30 + Ce^{-k}$ $\qquad$ and $\qquad$ when $t = 20$: $N = 19$ $\qquad$ $19 = 30 + Ce^{-20k}$

$C = -20e^k$ $\hspace{5cm}$ $C = -11e^{20k}$

$$-20e^k = -11e^{20k}$$

$$\frac{20}{11} = e^{19k}$$

$$k = \frac{\ln(20/11)}{19} \approx 0.0315$$

$$C = -20e^{[\ln(20/11)]/19} \approx -20.6393$$

$N = 30 - 20.6393e^{-0.0315t}$

35. Let Q be the number of pounds of concentrate in the solution at any time t. Since the number of gallons of solution in the tank at any time t is $v_0 + (r_1 - r_2)t$ and since the tank loses r_2 gallons of solution per minute, it must lose concentrate at the rate

$$\left[\frac{Q}{v_0 + (r_1 - r_2)t} \right] r_2.$$

The solution gains concentrate at the rate $r_1 q_1$. Therefore, the net rate of change is

$$\frac{dQ}{dt} = q_1 r_1 - \left[\frac{Q}{v_0 + (r_1 - r_2)t} \right] r_2 \quad \text{or} \quad \frac{dQ}{dt} + \frac{r_2 Q}{v_0 + (r_1 - r_2)t} = q_1 r_1.$$

36. a. $Q' + \dfrac{r_2 Q}{v_0 + (r_1 - r_2)t} = q_1 r_1$

$Q(0) = q_0$, $\quad q_0 = 25$, $\quad q_1 = 0$, $\quad v_0 = 200$, $\quad r_1 = 10$, $\quad r_2 = 10$, $\quad Q' + \dfrac{1}{20}Q = 0$

$$\int \frac{1}{Q} \, dQ = \int -\frac{1}{20} \, dt$$

$$\ln Q = -\frac{1}{20}t + \ln C_1$$

$$Q = Ce^{-(1/20)t}$$

Initial condition: $Q(0) = 25$, $\quad C = 25$

Particular solution: $Q = 25e^{-(1/20)t}$

b. $\qquad 15 = 25e^{-(1/20)t}$

$$\ln\left(\frac{3}{5}\right) = -\frac{1}{20}t$$

$$t = -20 \ln\left(\frac{3}{5}\right) \approx 10.2 \text{ min}$$

c. $\displaystyle\lim_{t \to \infty} 25e^{-(1/20)t} = 0$

37. a. $Q' + \dfrac{r_2 Q}{v_0 + (r_1 - r_2)t} = q_1 r_1$

$Q(0) = q_0, \quad q_0 = 25, \quad q_1 = 0.05, \quad v_0 = 200, \quad r_1 = 10, \quad r_2 = 10, \quad Q' + \dfrac{1}{20}Q = 0.5$

Integrating factor: $e^{\int (1/20)\, dt} = e^{(1/20)t}$

$Qe^{(1/20)t} = \displaystyle\int 0.5 e^{(1/20)t}\, dt = 10 e^{(1/20)t} + C$

$Q = 10 + Ce^{-(1/20)t}$

Initial condition: $Q(0) = 25, \quad 25 = 10 + C, \quad C = 15$

Particular solution: $Q = 10 + 15 e^{-(1/20)t}$

b. $15 = 10 + 15 e^{-(1/20)t}$

$\ln\left(\dfrac{1}{3}\right) = -\dfrac{1}{20}t$

$t = -20\ln\left(\dfrac{1}{3}\right) \approx 21.97 \text{ min}$

c. $\displaystyle\lim_{t \to \infty} (10 + 15 e^{-(1/20)t}) = 10 \text{ lbs}$

38. a. The volume of the solution in the tank is given by $v_0 + (r_1 - r_2)t$. Therefore, $100 + (5-3)t = 200$ or $t = 50$ minutes.

b. $Q' + \dfrac{r_2 Q}{v_0 + (r_1 - r_2)t} = q_1 r_1$

$Q(0) = q_0, \quad q_0 = 0, \quad q_1 = 0.5, \quad v_0 = 100, \quad r_1 = 5, \quad r_2 = 3, \quad Q' + \dfrac{3}{100 + 2t}Q = 2.5$

Integrating factor: $e^{\int [3/(100+2t)]\, dt} = (50 + t)^{3/2}$

$Q(50 + t)^{3/2} = \displaystyle\int 2.5(50 + t)^{3/2}\, dt = (50 + t)^{5/2} + C$

$Q = (50 + t) + C(50 + t)^{-3/2}$

Initial condition: $Q(0) = 0, \quad 0 = 50 + C(50^{-3/2}), \quad C = -50^{5/2}$

Particular solution: $Q = (50 + t) - 50^{-5/2}(50 + t)^{-3/2}$

$Q(50) = 100 - 50^{5/2}(100)^{-3/2} = 100 - \dfrac{25}{\sqrt{2}} \approx 82.32 \text{ lbs}$

39. $y' - 2x = 0$

$\displaystyle\int dy = \int 2x\, dx$

$y = x^2 + C$

Matches c.

40. $y' - 2y = 0$

$\displaystyle\int \dfrac{dy}{y} = \int 2\, dx$

$\ln y = 2x + C_1$

$y = Ce^{2x}$

Matches d.

41. $y' - 2xy = 0$

$$\int \frac{dy}{y} = \int 2x \, dx$$

$$\ln y = x^2 + C_1$$

$$y = Ce^{x^2}$$

Matches a.

42. $y' - 2xy = x$

$$\int \frac{dy}{2y + 1} = \int x \, dx$$

$$\frac{1}{2} \ln(2y + 1) = \frac{1}{2}x^2 + C_1$$

$$2y + 1 = C_2 e^{x^2}$$

$$y = -\frac{1}{2} + Ce^{x^2}$$

Matches b.

43. $e^{2x+y} \, dx - e^{x-y} \, dy = 0$

Separation of variables:

$$e^{2x}e^y \, dx = e^x e^{-y} \, dy$$

$$\int e^x \, dx = \int e^{-2y} \, dy$$

$$e^x = -\frac{1}{2}e^{-2y} + C_1$$

$$2e^x + e^{-2y} = C$$

44. $(x + 1) \, dx - (y^2 + 2y) \, dy = 0$

Separation of variables:

$$\int (x + 1) \, dx = \int (y^2 + 2y) \, dy$$

$$\frac{1}{2}x^2 + x = \frac{1}{3}y^3 + y^2 + C_1$$

$$3x^2 + 6x - 2y^3 - 6y^2 = C$$

45. $(1 + y^2) \, dx + (2xy + y + 2) \, dy = 0$

Exact: $\dfrac{\partial M}{\partial y} = 2y = \dfrac{\partial N}{\partial x}$

$$U(x, y) = \int (1 + y^2) \, dx = x + xy^2 + f(y)$$

$$U_y(x, y) = 2xy + f'(y)$$

$$= 2xy + y + 2 \text{ or } f'(y) = y + 2$$

$$f(y) = \frac{1}{2}y^2 + 2y + C_1$$

$$U(x, y) = x + xy^2 + \frac{1}{2}y^2 + 2y + C_1$$

$$x + xy^2 + \frac{1}{2}y^2 + 2y = C$$

46. $(1 + 2e^{2x+y}) \, dx + e^{2x+y} \, dy = 0$

Exact: $\dfrac{\partial M}{\partial y} = 2e^{2x+y} = \dfrac{\partial N}{\partial x}$

$$U(x, y) = \int e^{2x+y} \, dy = e^{2x+y} + g(x)$$

$$U_x(x, y) = 2e^{2x+y} + g'(x) = 1 + 2e^{2x+y}$$

$$g'(x) = 1 \text{ or } g(x) = x + C_1$$

$$U(x, y) = e^{2x+y} + x + C_1 \text{ or } x + e^{2x+y} = C$$

47. $(y \cos x - \cos x) \, dx + dy = 0$

Separation of variables:

$$\int \cos x \, dx = \int \frac{-1}{y - 1} \, dy$$

$$\sin x = -\ln(y - 1) + \ln C$$

$$\ln(y - 1) = -\sin x + \ln C$$

$$y = Ce^{-\sin x} + 1$$

48. $(x + 1) \, dy + (y - e^x) \, dx = 0$

Exact: $\dfrac{\partial M}{\partial y} = 1 = \dfrac{\partial N}{\partial x}$

$$U(x, y) = \int (x + 1) \, dy = xy + y + g(x)$$

$$U_x(x, y) = y + g'(x) = y - e^x \text{ or } g'(x) = -e^x$$

$$g(x) = -e^x + C_1$$

$$U(x, y) = xy + y - e^x + C_1$$

$$xy + y - e^x = C$$

49. $2xy\,dx + (x^2 + \cos y)\,dy = 0$

Exact: $\dfrac{\partial M}{\partial y} = 2x = \dfrac{\partial N}{\partial x}$

$$U(x,\ y) = \int 2xy\,dx = x^2 y + f(y)$$

$$U_y(x,\ y) = x^2 + f'(y) = x^2 + \cos y$$

$$\text{or } f'(y) = \cos y$$

$$f(y) = \sin y + C_1$$

$$U(x,\ y) = x^2 y + \sin y + C_1$$

$$x^2 y + \sin y = C$$

50. $y' = 2x\sqrt{1 - y^2}$

Separation of variables:

$$\int \frac{1}{\sqrt{1 - y^2}}\,dy = \int 2x\,dx$$

$$\arcsin y = x^2 + C$$

$$y = \sin(x^2 + C)$$

51. $(3y^2 + 4xy)\,dx + (2xy + x^2)\,dy = 0$

Homogeneous: $y = vx, \quad dy = v\,dx + x\,dv$

$$(3v^2 x^2 + 4vx^2)\,dx + (2vx^2 + x^2)(v\,dx + x\,dv) = 0$$

$$\int \frac{5}{x}\,dx + \int \left(\frac{2v + 1}{v^2 + v}\right)dv = 0$$

$$\ln x^5 + \ln |v^2 + v| = \ln C$$

$$x^5(v^2 + v) = C$$

$$x^3 y^2 + x^4 y = C$$

52. $(x + y)\,dx - x\,dy = 0$

Linear: $y' - \dfrac{1}{x}y = 1$

Integrating factor: $e^{\int -(1/x)\,dx} = e^{\ln |x^{-1}|} = \dfrac{1}{x}$

$$y\frac{1}{x} = \int \frac{1}{x}\,dx = \ln |x| + C$$

$$y = x(\ln |x| + C)$$

53. $(2y - e^x)\,dx + x\,dy = 0$

Linear: $y' + \left(\dfrac{2}{x}\right)y = \dfrac{1}{x}e^x$

Integrating factor: $e^{\int (2/x)\,dx} = e^{\ln x^2} = x^2$

$$yx^2 = \int x^2 \frac{1}{x}e^x\,dx = e^x(x - 1) + C$$

$$y = \frac{e^x}{x^2}(x - 1) + \frac{C}{x^2}$$

54. $(y^2 + xy)\,dx - x^2\,dy = 0$

Homogeneous: $y = vx, \quad dy = v\,dx + x\,dv$

$$(v^2 x^2 + vx^2)\,dx - x^2(v\,dx + x\,dv) = 0$$

$$v^2\,dx - x\,dv = 0$$

$$\int \frac{1}{x}\,dx = \int \frac{1}{v^2}\,dv$$

$$\ln x = -\frac{1}{v} + C$$

$$y = \frac{x}{C - \ln |x|}$$

55. $(x^2 y^4 - 1)\,dx + x^3 y^3\,dy = 0$

$$y' + \left(\frac{1}{x}\right)y = x^{-3}y^{-3}$$

Bernoulli: $n = -3, \quad Q = x^{-3}, \quad P = x^{-1}, \quad e^{\int (4/x)\,dx} = e^{\ln x^4} = x^4$

$$y^4 x^4 = \int 4(x^{-3})(x^4)\,dx = 2x^2 + C$$

$$x^4 y^4 - 2x^2 = C$$

56. $y\,dx + (3x + 4y)\,dy = 0$

Homogeneous: $x = vy$, $dx = v\,dy + y\,dv$

$y(v\,dy + y\,dv) + (3vy + 4y)\,dy = 0$

$$\int \frac{1}{v+1}\,dv = \int -\frac{4}{y}\,dy$$

$$\ln|v + 1| = -\ln y^4 + \ln C$$

$$y^4(v + 1) = C$$

$$y^3(x + y) = C$$

57. $3y\,dx - (x^2 + 3x + y^2)\,dy = 0$

Multiplying by the integrating factor, $1/(x^2 + y^2)$, and regrouping, we have

$$3\left[\frac{y\,dx - x\,dy}{x^2 + y^2}\right] - dy = 0$$

$$\int 3d\left[\arctan \frac{x}{y}\right] - \int dy = 0$$

$$3\arctan \frac{x}{y} - y = C.$$

58. $x\,dx + (y + e^y)(x^2 + 1)\,dy = 0$

Separation of variables:

$$\int \frac{x}{x^2 + 1}\,dx = \int -(y + e^y)\,dy$$

$$\frac{1}{2}\ln(x^2 + 1) = -\frac{1}{2}y^2 - e^y + C_1$$

$$\ln(x^2 + 1) + y^2 + 2e^y = C$$

Section 16.5   Second-Order Homogeneous Linear Equations

1.
$$y = C_1 e^{-3x} + C_2 x e^{-3x}$$
$$y' = -3C_1 e^{-3x} + C_2 e^{-3x} - 3C_2 x e^{-3x}$$
$$y'' = 9C_1 e^{-3x} - 6C_2 e^{-3x} + 9C_2 x e^{-3x}$$
$$y'' + 6y' + 9y = (9C_1 e^{-3x} - 6C_2 e^{-3x} + 9C_2 x e^{-3x})$$
$$+ (-18C_1 e^{-3x} + 6C_2 e^{-3x} - 18C_2 x e^{-3x}) + (9C_1 e^{-3x} + 9C_2 x e^{-3x}) = 0$$

2.
$$y = C_1 e^{2x} + C_2 e^{-2x}$$
$$y' = 2C_1 e^{2x} - 2C_2 e^{-2x}$$
$$y'' = 4C_1 e^{2x} + 4C_2 e^{-2x} = 4y$$
$$y'' - 4y = 4y - 4y = 0$$

3.
$$y = C_1 \cos 2x + C_2 \sin 2x$$
$$y' = -2C_1 \sin 2x + 2C_2 \cos 2x$$
$$y'' = -4C_1 \cos 2x - 4C_2 \sin 2x = -4y$$
$$y'' + 4y = -4y + 4y = 0$$

4.
$$y = e^{-x} \sin 3x$$
$$y' = 3e^{-x} \cos 3x - e^{-x} \sin 3x$$
$$y'' = -8e^{-x} \sin 3x - 6e^{-x} \cos 3x$$
$$y'' + 2y' + 10y = (-8e^{-x} \sin 3x - 6e^{-x} \cos 3x) + (6e^{-x} \cos 3x - 2e^{-x} \sin 3x) + 10e^{-x} \sin 3x = 0$$

5. $y'' - y' = 0$

Characteristic equation: $m^2 - m = 0$

Roots: $m = 0, \ 1$

$y = C_1 + C_2 e^x$

6. $y'' + 2y' = 0$

Characteristic equation: $m^2 + 2m = 0$

Roots: $m = 0, \ -2$

$y = C_1 + C_2 e^{-2x}$

7. $y'' - y' - 6y = 0$

Characteristic equation: $m^2 - m - 6 = 0$

Roots: $m = 3, \ -2$

$y = C_1 e^{3x} + C_2 e^{-2x}$

8. $y'' + 6y' + 5y = 0$

Characteristic equation: $m^2 + 6m + 5 = 0$

Roots: $m = -1, \ -5$

$y = C_1 e^{-x} + C_2 e^{-5x}$

9. $2y'' + 3y' - 2y = 0$

Characteristic equation: $2m^2 + 3m - 2 = 0$

Roots: $m = \frac{1}{2}, \ -2$

$y = C_1 e^{(1/2)x} + C_2 e^{-2x}$

10. $16y'' - 16y' + 3y = 0$

Characteristic equation: $16m^2 - 16m + 3 = 0$

Roots: $m = \frac{1}{4}, \ \frac{3}{4}$

$y = C_1 e^{(1/4)x} + C_2 e^{(3/4)x}$

11. $y'' + 6y' + 9y = 0$

Characteristic equation: $m^2 + 6m + 9 = 0$

Roots: $m = -3, \ -3$

$y = C_1 e^{-3x} + C_2 x e^{-3x}$

12. $y'' - 10y' + 25y = 0$

Characteristic equation: $m^2 - 10m + 25 = 0$

Roots: $m = 5, \ 5$

$y = C_1 e^{5x} + C_2 x e^{5x}$

13. $16y'' - 8y' + y = 0$

Characteristic equation: $16m^2 - 8m + 1 = 0$

Roots: $m = \frac{1}{4}, \ \frac{1}{4}$

$y = C_1 e^{(1/4)x} + C_2 x e^{(1/4)x}$

14. $9y'' - 12y' + 4y = 0$

Characteristic equation: $9m^2 - 12m + 4 = 0$

Roots: $m = \frac{2}{3}, \ \frac{2}{3}$

$y = C_1 e^{(2/3)x} + C_2 x e^{(2/3)x}$

15. $y'' + y = 0$

Characteristic equation: $m^2 + 1 = 0$

Roots: $m = -i, \ i$

$y = C_1 \cos x + C_2 \sin x$

16. $y'' + 4y = 0$

Characteristic equation: $m^2 + 4 = 0$

Roots: $m = -2i, \ 2i$

$y = C_1 \cos 2x + C_2 \sin 2x$

17. $y'' - 9y = 0$

Characteristic equation: $m^2 - 9 = 0$

Roots: $m = -3, \ 3$

$y = C_1 e^{3x} + C_2 e^{-3x}$

18. $y'' - 2y = 0$

Characteristic equation: $m^2 - 2 = 0$

Roots: $m = -\sqrt{2}, \ \sqrt{2}$

$y = C_1 e^{\sqrt{2}x} + C_2 {}^{-\sqrt{2}x}$

19. $y'' - 2y' + 4y = 0$

Characteristic equation: $m^2 - 2m + 4 = 0$

Roots: $m = 1 - \sqrt{3}i, \ 1 + \sqrt{3}i$

$y = e^x(C_1 \cos \sqrt{3}x + C_2 \sin \sqrt{3}x)$

20. $y'' - 4y' + 21y = 0$

Characteristic equation: $m^2 - 4m + 21 = 0$

Roots: $m = 2 - \sqrt{17}i, \ 2 + \sqrt{17}i$

$y = e^{2x}(C_1 \cos \sqrt{17}x + C_2 \sin \sqrt{17}x)$

21. $y'' - 3y' + y = 0$

Characteristic equation: $m^2 - 3m + 1 = 0$

Roots: $m = \dfrac{3 - \sqrt{5}}{2}, \ \dfrac{3 + \sqrt{5}}{2}$

$y = C_1 e^{[(3+\sqrt{5})/2]x} + C_2 e^{[(3-\sqrt{5})/2]x}$

22. $3y'' + 4y' - y = 0$

Characteristic equation: $3m^2 + 4m - 1 = 0$

Roots: $m = \dfrac{-2 - \sqrt{7}}{3}, \ \dfrac{-2 + \sqrt{7}}{3}$

$y = C_1 e^{[(-2+\sqrt{7})/3]x} + C_2 e^{[(-2-\sqrt{7})/3]x}$

23. $9y'' - 12y' + 11y = 0$
Characteristic equation: $9m^2 - 12m + 11 = 0$

Roots: $m = \dfrac{2 + \sqrt{7}i}{3}, \dfrac{2 - \sqrt{7}i}{3}$

$y = e^{(2/3)x}\left[C_1 \cos\left(\dfrac{\sqrt{7}}{3}x\right) + C_2 \sin\left(\dfrac{\sqrt{7}}{3}x\right)\right]$

24. $2y'' - 6y' + 7y = 0$
Characteristic equation: $2m^2 - 6m + 7 = 0$

Roots: $m = \dfrac{3 + \sqrt{5}i}{2}, \dfrac{3 - \sqrt{5}i}{2}$

$y = e^{(3/2)x}\left[C_1 \cos\left(\dfrac{\sqrt{5}}{2}x\right) + C_2 \sin\left(\dfrac{\sqrt{5}}{2}x\right)\right]$

25. $y^{(4)} - y = 0$
Characteristic equation: $m^4 - 1 = 0$
Roots: $m = -1, 1, -i, i$
$y = C_1 e^x + C_2 e^{-x} + C_3 \cos x + C_4 \sin x$

26. $y^{(4)} - y'' = 0$
Characteristic equation: $m^4 - m^2 = 0$
Roots: $m = 0, 0, -1, 1$
$y = C_1 + C_2 x + C_3 e^x + C_4 e^{-x}$

27. $y''' - 6y'' + 11y' - 6y = 0$
Characteristic equation: $m^3 - 6m^2 + 11m - 6 = 0$
Roots: $m = 1, 2, 3$
$y = C_1 e^x + C_2 e^{2x} + C_3 e^{3x}$

28. $y''' - y'' - y' + y = 0$
Characteristic equation: $m^3 - m^2 - m + 1 = 0$
Roots: $m = -1, 1, 1$
$y = C_1 e^x + C_2 x e^x + C_3 e^{-x}$

29. $y''' - 3y'' + 7y' - 5y = 0$
Characteristic equation: $m^3 - 3m^2 + 7m - 5 = 0$
Roots: $m = 1, 1 - 2i, 1 + 2i$
$y = C_1 e^x + e^x(C_2 \cos 2x + C_3 \sin 2x)$

30. $y''' - 3y'' + 3y' - y = 0$
Characteristic equation: $m^3 - 3m^2 + 3m - 1 = 0$
Roots: $m = 1, 1, 1$
$y = C_1 e^x + C_2 x e^x + C_3 x^2 e^x$

31. $y'' + 100y = 0$
$y = C_1 \cos 10x + C_2 \sin 10x$

$y' = -10C_1 \sin 10x + 10C_2 \cos 10x$

a. $y(0) = 2$: $2 = C_1$

$y'(0) = 0$: $0 = 10C_2 \Rightarrow C_2 = 0$

Particular solution: $y = 2 \cos 10x$

b. $y(0) = 0$: $0 = C_1$

$y'(0) = 2$: $2 = 10C_2 \Rightarrow C_2 = \frac{1}{5}$

Particular solution: $y = \frac{1}{5} \sin 10x$

c. $y(0) = -1$: $-1 = C_1$

$y'(0) = 3$: $3 = 10C_2 \Rightarrow C_2 = \frac{3}{10}$

Particular solution: $y = -\cos 10x + \frac{3}{10} \sin 10x$

32.
$y = C \sin \sqrt{3}\, t$

$y' = \sqrt{3}\, C \cos \sqrt{3}\, t$

$y'' = -3C \sin \sqrt{3}\, t$

$y'' + \omega y = -3C \sin \sqrt{3}\, t + \omega \sin \sqrt{3}\, t$

$= 0 \Rightarrow \omega = 3C$

$y'(0) = -5$: $-5 = \sqrt{3}\, C \Rightarrow C = \dfrac{5\sqrt{3}}{3}$

and $\omega = -5\sqrt{3}$

33. $y'' - y' - 30y = 0$, $y(0) = 1$, $y'(0) = -4$
Characteristic equation: $m^2 - m - 30 = 0$
Roots: $m = 6, -5$
$y = C_1 e^{6x} + C_2 e^{-5x}$, $y' = 6C_1 e^{6x} - 5C_2 e^{-5x}$
Initial conditions: $y(0) = 1$, $y'(0) = -4$, $1 = C_1 + C_2$, $-4 = 6C_1 - 5C_2$
Solving simultaneously: $C_1 = \frac{1}{11}$, $C_2 = \frac{10}{11}$
Particular solution: $y = \frac{1}{11}(e^{6x} + 10e^{-5x})$

34. $y'' + 2y' + 3y = 0$, $y(0) = 2$, $y'(0) = 1$

Characteristic equation: $m^2 + 2m + 3 = 0$

Roots: $m = -1 + \sqrt{2}\,i$, $-1 - \sqrt{2}\,i$

$$y = e^{-x}\left(C_1 \cos \sqrt{2}\,x + C_2 \sin \sqrt{2}\,x\right)$$

$$y' = e^{-x}\left(-\sqrt{2}\,C_1 \sin \sqrt{2}\,x + \sqrt{2}\,C_2 \cos \sqrt{2}\,x\right) - e^{-x}\left(C_1 \cos \sqrt{2}\,x + C_2 \sin \sqrt{2}\,x\right)$$

Initial conditions: $y(0) = 2$, $y'(0) = 1$, $2 = C_1$, $1 = \sqrt{2}\,C_2 - C_1$, $C_2 = \dfrac{3}{\sqrt{2}}$

Particular solution: $y = e^{-x}\left(2 \cos \sqrt{2}\,x + \dfrac{3}{\sqrt{2}} \sin \sqrt{2}\,x\right)$

35. By Hooke's Law, $F = kx$

$$k = \frac{F}{x} = \frac{32}{2/3} = 48.$$

Also, $F = ma$, and

$$m = \frac{F}{a} = \frac{32}{32} = 1.$$

Therefore, $y = \frac{1}{2} \cos \left(4\sqrt{3}\,t\right)$.

36. By Hooke'e Law, $F = kx$

$$k = \frac{F}{x} = \frac{32}{2/3} = 48.$$

Also, $F = ma$, and

$$m = \frac{F}{a} = \frac{32}{32} = 1.$$

Therefore, $y = -\frac{2}{3} \cos \left(4\sqrt{3}\,t\right)$.

37. $y = C_1 \cos \left(\sqrt{k/m}\,t\right) + C_2 \sin \left(\sqrt{k/m}\,t\right)$, $\sqrt{k/m} = \sqrt{48} = 4\sqrt{3}$

Initial conditions: $y(0) = \dfrac{2}{3}$, $y'(0) = -\dfrac{1}{2}$

$$y = C_1 \cos \left(4\sqrt{3}\,t\right) + C_2 \sin \left(4\sqrt{3}\,t\right)$$

$$y(0) = C_1 = \frac{2}{3}$$

$$y'(t) = -4\sqrt{3}\,C_1 \sin \left(4\sqrt{3}\,t\right) + 4\sqrt{3}\,C_2 \cos \left(4\sqrt{3}\,t\right)$$

$$y'(0) = 4\sqrt{3}\,C_2 = -\frac{1}{2} \Rightarrow C_2 = -\frac{1}{8\sqrt{3}} = -\frac{\sqrt{3}}{24}$$

$$y(t) = \frac{2}{3} \cos \left(4\sqrt{3}\,t\right) - \frac{\sqrt{3}}{24} \sin \left(4\sqrt{3}\,t\right)$$

38. $y = C_1 \cos \left(4\sqrt{3}\,t\right) + C_2 \sin \left(4\sqrt{3}\,t\right)$

Initial conditions: $y(0) = -\dfrac{1}{2}$, $y'(0) = \dfrac{1}{2}$

$$y(0) = C_1 = -\frac{1}{2}$$

$$y'(t) = -4\sqrt{3}\,C_1 \sin \left(4\sqrt{3}\,t\right) + 4\sqrt{3}\,C_2 \cos \left(4\sqrt{3}\,t\right)$$

$$y'(0) = 4\sqrt{3}\,C_2 = \frac{1}{2} \Rightarrow C_2 = \frac{1}{8\sqrt{3}}$$

$$y(t) = -\frac{1}{2} \cos \left(4\sqrt{3}\,t\right) + \frac{1}{8\sqrt{3}} \sin \left(4\sqrt{3}\,t\right)$$

39. By Hooke's Law, $32 = k(2/3)$ so that $k = 48$. Moreover, since the weight w is given by mg, it follows that $m = w/g = 32/32 = 1$. Also, the damping force is given by $(-1/8)(dy/dt)$. Thus, the differential equation for the oscillations of the weight is

$$m\left(\frac{d^2y}{dt^2}\right) = -\frac{1}{8}\left(\frac{dy}{dt}\right) - 48y$$

$$m\left(\frac{d^2y}{dt^2}\right) + \frac{1}{8}\left(\frac{dy}{dt}\right) + 48y = 0.$$

In this case the characteristic equation is $8m^2 + m + 384 = 0$ with complex roots $m = (-1/16) \pm \left(\sqrt{12{,}287}/16\right)i$. Therefore, the general solution is

$$y(t) = e^{-t/16}\left(C_1 \cos\frac{\sqrt{12{,}287}\,t}{16} + C_2 \sin\frac{\sqrt{12{,}287}\,t}{16}\right).$$

Using the initial conditions, we have

$$y(0) = C_1 = \frac{1}{2}$$

$$y'(t) = e^{-t/16}\left[\left(-\frac{\sqrt{12{,}287}}{16}C_1 - \frac{C_2}{16}\right)\sin\frac{\sqrt{12{,}287}\,t}{16} + \left(\frac{\sqrt{12{,}287}}{16}C_2 - \frac{C_1}{16}\right)\cos\frac{\sqrt{12{,}287}\,t}{16}\right]$$

$$y'(0) = \frac{\sqrt{12{,}287}}{16}C_2 - \frac{C_1}{16} = 0 \Rightarrow C_2 = \frac{\sqrt{12{,}287}}{24{,}574}$$

and the particular solution is

$$y(t) = \frac{e^{-t/16}}{2}\left(\cos\frac{\sqrt{12{,}287}\,t}{16} + \frac{\sqrt{12{,}287}}{12{,}287}\sin\frac{\sqrt{12{,}287}\,t}{16}\right).$$

40. By Hooke's Law, $32 = k(2/3)$ so $k = 48$. Also, $m = w/g = 32/32 = 1$. The damping force is given by $(-1/4)(dy/dt)$. Thus,

$$m\left(\frac{d^2y}{dt^2}\right) = -\frac{1}{4}\left(\frac{dy}{dt}\right) - 48y$$

$$m\left(\frac{d^2y}{dt^2}\right) + \frac{1}{4}\left(\frac{dy}{dt}\right) + 48y = 0.$$

The characteristic equation is $4m^2 + m + 192 = 0$ with complex roots $m = (-1/8) \pm \left(\sqrt{3071}/8\right)i$. Therefore, the general solution is

$$y(t) = e^{-t/8}\left(C_1 \cos\frac{\sqrt{3071}\,t}{8} + C_2 \sin\frac{\sqrt{3071}\,t}{8}\right).$$

Using the initial conditions $y(0) = C_1 = 1/2$,

$$y'(t) = e^{-t/8}\left[\left(-\frac{\sqrt{3071}}{8}C_1 - \frac{C_2}{8}\right)\sin\frac{\sqrt{3071}\,t}{8} + \left(\frac{\sqrt{3071}\,C_2}{8} - \frac{C_1}{8}\right)\cos\frac{\sqrt{3071}\,t}{8}\right]$$

$$y'(0) = \frac{\sqrt{3071}}{8}C_2 - \frac{C_1}{8} = 0 \Rightarrow C_2 = \frac{\sqrt{3071}}{6142}$$

and the particular solution is

$$y(t) = \frac{e^{-t/8}}{2}\left[\cos\frac{\sqrt{3071}\,t}{8} + \frac{\sqrt{3071}}{3071}\sin\frac{\sqrt{3071}\,t}{8}\right].$$

41. $y'' + 9y = 0$
Undamped vibration

Period: $\dfrac{2\pi}{3}$

Matches (b)

42. $y'' + 25y = 0$
Undamped vibration

Period: $\dfrac{2\pi}{5}$

Matches (d)

43. $y'' + 2y' + 10y = 0$
Damped vibration
Matches (c)

44. $y'' + y' + \frac{37}{4}y = 0$
Damped vibration
Matches (a)

45. Since $m = -a/2$ is a double root of the characteristic equation, we have

$$\left(m + \frac{a}{2}\right)^2 = m^2 + am + \frac{a^2}{4} = 0$$

and the differential equation is $y'' + ay' + (a^2/4)y = 0$. The solution is

$$y = (C_1 + C_2 x)e^{-(a/2)x}$$

$$y' = \left(-\frac{C_1 a}{2} + C_2 - \frac{C_2 a}{2}x\right)e^{-(a/2)x}$$

$$y'' = \left(\frac{C_1 a^2}{4} - aC_2 + \frac{C_2 a^2}{4}x\right)e^{-(a/2)x}$$

$$y'' + ay' + \frac{a^2}{4}y = \left(\frac{C_1 a^2}{4} - C_2 a + \frac{C_2 a^2}{4}x\right) + \left(-\frac{C_1 a^2}{2} + C_2 a - \frac{C_2 a^2}{2}x\right)e^{-(a/2)x}$$

$$+ \left(\frac{C_1 a^2}{4} + \frac{C_2 a^2}{4}x\right)e^{-(a/2)x} = 0.$$

46. Since $m = \alpha \pm \beta i$ are roots to the characteristic equation, we have

$$[m - (\alpha + \beta i)][m - (\alpha - \beta i)] = m^2 - 2\alpha m + (\alpha^2 + \beta^2) = 0$$

and the differential equation is $y'' - 2\alpha y' + (\alpha^2 + \beta^2)y = 0$. (**Note:** $i^2 = -1$.) The solution is

$$y = e^{\alpha x}(C_1 \cos \beta x + C_2 \sin \beta x)$$

$$y' = e^{\alpha x}[(C_1\alpha + C_2\beta)\cos \beta x + (C_2\alpha - C_1\beta)\sin \beta x]$$

$$y'' = e^{\alpha x}[(C_1\alpha^2 - C_1\beta^2 + 2C_2\alpha\beta)\cos \beta x + (C_2\alpha^2 - C_2\beta^2 - 2C_1\alpha\beta)\sin \beta x]$$

$$-2\alpha y' = e^{\alpha x}[(-2C_1\alpha^2 - 2C_2\alpha\beta)\cos \beta x + (-2C_2\alpha^2 + 2C_1\alpha\beta)\sin \beta x]$$

$$(\alpha^2 + \beta^2)y = e^{\alpha x}[(C_1\alpha^2 + C_1\beta^2)\cos \beta x + (C_2\alpha^2 + C_2\beta^2)\sin \beta x]$$

Therefore, $y'' - 2\alpha y' + (\alpha^2 + \beta^2)y = 0$.

47. False. The general solution is
$y = C_1 e^{3x} + C_2 x e^{3x}$.

48. True

49. True

50. False. The solution $y = x^2 e^x$ requires that $m = 1$ is a triple root of the characteristic equation. Since the characteristic equation is quadratic, $m = 1$ can be at most a double root.

51. $y_1 = e^{ax}, \quad y_2 = e^{bx}, \quad a \neq b$

$$W(y_1, y_2) = \begin{vmatrix} e^{ax} & e^{bx} \\ ae^{ax} & be^{bx} \end{vmatrix}$$

$$= (b - a)e^{ax+bx} \neq 0 \text{ for any value of } x.$$

52. $y_1 = e^{ax}, \quad y_2 = xe^{ax}$

$$W(y_1, y_2) = \begin{vmatrix} e^{ax} & xe^{ax} \\ ae^{ax} & e^{ax} + axe^{ax} \end{vmatrix}$$

$$= e^{2ax} \neq 0 \text{ for any value of } x.$$

53. $y_1 = e^{ax} \sin bx$, $y_2 = e^{ax} \cos bx$, $b \neq 0$

$$W(y_1, y_2) = \begin{vmatrix} e^{ax} \sin bx & e^{ax} \cos bx \\ ae^{ax} \sin bx + be^{ax} \cos bx & ae^{ax} \cos bx - be^{ax} \sin bx \end{vmatrix}$$

$$= -be^{2ax} \sin^2 bx - be^{2ax} \cos^2 bx$$

$$= -be^{2ax} \neq 0 \text{ for any value of } x.$$

54. $y_1 = x$, $y_2 = x^2$

$$W(y_1, y_2) = \begin{vmatrix} x & x^2 \\ 1 & 2x \end{vmatrix} = x^2 \neq 0 \text{ for } x \neq 0.$$

55. $x^2 y'' + axy' + by = 0$, $x > 0$

Let $x = e^t$.

a. $\dfrac{dy}{dx} = \dfrac{dy/dt}{dx/dt} = e^{-t} \dfrac{dy}{dt}$

$$\frac{d^2 y}{dx^2} = \frac{d/dt[e^{-t}(dy/dt)]}{e^t} = e^{-t}\left[e^{-t}\frac{d^2 y}{dt^2} - e^{-t}\frac{dy}{dt}\right] = e^{-2t}\left[\frac{d^2 y}{dt^2} - \frac{dy}{dt}\right]$$

$$x^2 y'' + axy' + by = 0$$

$$e^{2t}\left[e^{-2t}\left(\frac{d^2 y}{dt^2} - \frac{dy}{dt}\right)\right] + ae^t\left(e^{-t}\frac{dy}{dt}\right) + by = 0$$

$$\frac{d^2 y}{dt^2} + (a-1)\frac{dy}{dt} + by = 0$$

b. $x^2 y'' + 6xy' + 6y = 0$

Let $x = e^t$. From part a. we have:

$$\frac{d^2 y}{dt^2} + 5\frac{dy}{dt} + 6y = 0$$

$$m^2 + 5m + 6 = 0$$

$$(m+3)(m+2) = 0$$

$$m_1 = -3, \quad m_2 = -2$$

$$y = C_1 e^{-3t} + C_2 e^{-2t} = C_1 e^{-3\ln x} + C_2 e^{-2\ln x} = C_1 e^{\ln(1/x^3)} + C_2 e^{\ln(1/x^2)} = \frac{C_1}{x^3} + \frac{C_2}{x^2}$$

56. $y'' + Ay = 0$

If $A > 0$, then $y = C_1 \cos \sqrt{A}\, x + C_2 \sin \sqrt{A}\, x$.

$y(0) = 0 \Rightarrow C_1 = 0$ and $y = C_2 \sin \sqrt{A}\, x$

$y(\pi) = 0 \Rightarrow 0 = C_2 \sin \sqrt{A}\, \pi \Rightarrow \sqrt{A} = \pm 1, \pm 2, \pm 3, \ldots \Rightarrow A = 1, 4, 9, 16, \ldots$

Thus, A is a perfect square integer.

If $A < 0$, then $y = C_1 e^{\sqrt{-A}\, x} + C_2 e^{-\sqrt{-A}\, x}$.

$y(0) = 0 \Rightarrow C_1 + C_2 = 0$

$y(\pi) = 0 \Rightarrow C_1 e^{\sqrt{-A}\, \pi} + C_2 e^{-\sqrt{-A}\, \pi} = 0$ $\Big\}$ $C_1 = C_2 = 0 \Rightarrow y = 0$

If $A = 0$, then $y'' = 0$ and the initial condition gives $y = 0$.

Section 16.6   Second-Order Nonhomogeneous Linear Equations

1. $y = 2e^{2x} - 2\cos x$

$y' = 4e^{2x} + 2\sin x$

$y'' = 8e^{2x} + 2\cos x$

$y'' + y = (8e^{2x} + 2\cos x) + (2e^{2x} - \cos x) = 10e^{2x}$

2. $y = 2\sin x + \dfrac{1}{2}x\sin x$

$y' = 2\cos x + \dfrac{1}{2}x\cos x + \dfrac{1}{2}\sin x$

$y'' = -2\sin x - \dfrac{1}{2}x\sin x + \cos x$

$y'' + y = \left(-2\sin x - \dfrac{1}{2}x\sin x + \cos x\right) + \left(2\sin x + \dfrac{1}{2}x\sin x\right) = \cos x$

3. $y = 3\sin x - \cos x \ln|\sec x + \tan x|$

$y' = 3\cos x - 1 + \sin x \ln|\sec x + \tan x|$

$y'' = -3\sin x + \tan x + \cos x \ln|\sec x + \tan x|$

$y'' + y = (-3\sin x + \tan x + \cos x \ln|\sec x \tan x|) + (3\sin x - \cos x \ln|\sec x + \tan x|) = \tan x$

4. $y = (5 - \ln|\sin x|)\cos x - x\sin x$

$y' = -(5 - \ln|\sin x|)\sin x - \cos x \cot x - \sin x - x\cos x$

$\quad = -6\sin x + \sin x \ln|\sin x| - \cos x(\cot x + x)$

$y'' = -6\cos x + \cos x + \cos x \ln|\sin x| - \cos x(-\csc^2 x + 1) + \sin x(\cot x + x)$

$\quad = -5\cos x + \cos x \ln|\sin x| + \csc x \cot x + x\sin x$

$y'' + y = \cos x(-5 + \ln|\sin x|) + \csc x \cot x + x\sin x + (5 - \ln|\sin x|)\cos x - x\sin x$

$\quad = \csc x \cot x$

5. $y'' - 3y' + 2y = 2x$

$y'' - 3y' + 2y = 0$

$m^2 - 3m + 2 = 0$ when $m = 1,\ 2.$

$$y_h = C_1 e^x + C_2 e^{2x}$$

$$y_p = A_0 + A_1 x$$

$$y_p{}' = A_1$$

$$y_p{}'' = 0$$

$y_p{}'' - 3y_p{}' + 2y_p = (2A_0 - 3A_1) + 2A_1 x = 2x$

$\left.\begin{array}{l} 2A_0 - 3A_1 = 0 \\ 2A_1 = 2 \end{array}\right\}\ A_1 = 1,\quad A_0 = \tfrac{3}{2}$

$y = C_1 e^x + C_2 e^{2x} + x + \tfrac{3}{2}$

6. $y'' - 2y' - 3y = x^2 - 1$

$y'' - 2y' - 3y = 0$

$m^2 - 2m - 3 = 0$ when $m = -1, 3$.

$$y_h = C_1 e^{-x} + C_2 e^{3x}$$

$$y_p = A_0 + A_1 x + A_2 x^2$$

$$y_p' = A_1 + 2A_2 x$$

$$y_p'' = 2A_2$$

$$y_p'' - 2y_p' - 3y_p = (-3A_2)x^2 + (-3A_1 - 4A_2)x + (-3A_0 - 2A_1 + 2A_2) = x^2 - 1$$

$$\left. \begin{array}{l} -3A_2 = 1 \\ -3A_1 - 4A_2 = 0 \\ -3A_0 - 2A_1 + 2A_2 = -1 \end{array} \right\} A_0 = -\tfrac{5}{27}, \quad A_1 = \tfrac{4}{9}, \quad A_2 = -\tfrac{1}{3}$$

$$y = C_1 e^{-x} + C_2 e^{3x} - \tfrac{1}{3}x^2 + \tfrac{4}{9}x - \tfrac{5}{27}$$

7. $y'' + y = x^3$, $y(0) = 1$, $y'(0) = 0$

$y'' + y = 0$

$m^2 + 1 = 0$ when $m = i, -i$.

$$y_h = C_1 \cos x + C_2 \sin x$$

$$y_p = A_0 + A_1 x + A_2 x^2 + A_3 x^3$$

$$y_p' = A_1 + 2A_2 x + 3A_3 x^2$$

$$y_p'' = 2A_2 + 6A_3 x$$

$$y_p'' + y_p = A_3 x^3 + A_2 x^2 + (A_1 + 6A_3)x + (A_0 + 2A_2) = x^3 \quad \text{or}$$

$$A_3 = 1, \quad A_2 = 0, \quad A_1 = -6, \quad A_0 = 0$$

$$y = C_1 \cos x + C_2 \sin x + x^3 - 6x$$

$$y' = -C_1 \sin x + C_2 \cos x + 3x^2 - 6$$

Initial conditions: $y(0) = 1$, $y'(0) = 0$, $1 = C_1$, $0 = C_2 - 6$, $C_2 = 6$

Particular solution: $y = \cos x + 6\sin x + x^3 - 6x$

8. $y'' + 4y = 4$, $y(0) = 1$, $y'(0) = 6$

$y'' + 4y = 0$

$m^2 + 4 = 0$ when $m = 2i, -2i$.

$$y_h = C_1 \cos 2x + C_2 \sin 2x$$

$$y_p = A_0$$

$$y_p'' = 0$$

$$y_p'' + 4y_p = 4A_0 = 4 \text{ or } A_0 = 1$$

$$y = C_1 \cos 2x + C_2 \sin 2x + 1$$

$$y' = -2C_1 \sin 2x + 2C_2 \cos 2x$$

Initial conditions: $y(0) = 1$, $y'(0) = 6$, $1 = C_1 + 1$, $C_1 = 0$, $6 = 2C_2$, $C_2 = 3$

Particular solution: $y = 3\sin 2x + 1$

9. $y'' + 2y' = 2e^x$
$y'' + 2y' = 0$
$m^2 + 2m = 0$ when $m = 0, -2$.

$$y_h = C_1 + C_2 e^{-2x}$$

$$y_p = Ae^x = y_p{'} = y_p{''}$$

$$y_p{''} + 2y_p{'} = 3Ae^x = 2e^x \text{ or } A = \tfrac{2}{3}$$

$$y = C_1 + C_2 e^{-2x} + \tfrac{2}{3}e^x$$

10. $y'' - 9y = 5e^{3x}$
$y'' - 9y = 0$
$m^2 - 9 = 0$ when $m = -3, 3$.

$$y_h = C_1 e^{-3x} + C_2 e^{3x}$$

$$y_p = Axe^{3x}$$

$$y_p{'} = Ae^{3x}(3x + 1)$$

$$y_p{''} = Ae^{3x}(9x + 6)$$

$$y_p{''} - 9y_p = 6Ae^{3x} = 5e^{3x} \text{ or } A = \tfrac{5}{6}$$

$$y = C_1 e^{-3x} + \left(C_2 + \tfrac{5}{6}x\right)e^{3x}$$

11. $y'' - 10y' + 25y = 5 + 6e^x$
$y'' - 10y' + 25y = 0$
$m^2 - 10m + 25 = 0$ when $m = 5, 5$.

$$y_h = C_1 e^{5x} + C_2 x e^{5x}$$

$$y_p = A_0 + A_1 e^x$$

$$y_p{'} = y_p{''} = A_1 e^x$$

$$y_p{''} - 10y_p{'} + 25y_p = 25A_0 + 16A_1 e^x = 5 + 6e^x \text{ or } A_0 = \tfrac{1}{5}, \quad A_1 = \tfrac{3}{8}$$

$$y = (C_1 + C_2 x)e^{5x} + \tfrac{3}{8}e^x + \tfrac{1}{5}$$

12. $16y'' - 8y' + y = 4(x + e^x)$
$16y'' - 8y' + y = 0$
$16m^2 - 8m + 1 = 0$ when $m = \tfrac{1}{4}, \tfrac{1}{4}$.

$$y_h = (C_1 + C_2 x)e^{(1/4)x}$$

$$y_p = A_0 + A_1 x + A_2 e^x$$

$$y_p{'} = A_1 + A_2 e^x$$

$$y_p{''} = A_2 e^x$$

$$16y_p{''} - 8y_p{'} + y_p = (A_0 - 8A_1) + A_1 x + 9A_2 e^x = 4x + 4e^x \text{ or } A_2 = \tfrac{4}{9}, \quad A_1 = 4, \quad A_0 = 32$$

$$y = (C_1 + C_2 x)e^{(1/4)x} + 32 + 4x + \tfrac{4}{9}e^x$$

13. $y'' + y' = 2\sin x$, $y(0) = 0$, $y'(0) = -3$
$y'' + y' = 0$
$m^2 + m = 0$ when $m = 0, -1$.

$$y_h = C_1 + C_2 e^{-x}$$

$$y_p = A\cos x + B\sin x$$

$$y_p{'} = -A\sin x + B\cos x$$

$$y_p{''} = -A\cos x - B\sin x$$

$$y_p{''} + y_p{'} = (-A + B)\cos x + (-A - B)\sin x = 2\sin x$$

$$\left.\begin{array}{r} -A + B = 0 \\ -A - B = 2 \end{array}\right\} A = -1, \quad B = -1$$

$$y = C_1 + C_2 e^x - (\cos x + \sin x)$$

$$y' = -C_2 e^{-x} - (-\sin x + \cos x)$$

Initial conditions: $y(0) = 0$, $y'(0) = -3$, $0 = C_1 + C_2 - 1$, $-3 = -C_2 - 1$, $C_2 = 2$, $C_1 = -1$
Particular solution: $y = -1 + 2e^{-x} - (\cos x + \sin x)$

14. $y'' + y' - 2y = 3\cos 2x, \quad y(0) = -1, \quad y'(0) = 2$

$y'' + y' - 2y = 0$

$m^2 + m - 2 = 0$ when $m = 1, \; -2.$

$$y_h = C_1 e^x + C_2 e^{-2x}$$

$$y_p = A\cos 2x + B\sin 2x$$

$$y_p{}' = -2A\sin 2x + 2B\cos 2x$$

$$y_p{}'' = -4A\cos 2x - 4B\sin 2x$$

$$y_p{}'' + y_p{}' - 2y_p = (-6A + 2B)\cos 2x + (-2A - 6B)\sin 2x = 3\cos 2x$$

$$\left.\begin{array}{l} -6A + 2B = 3 \\[4pt] -2A - 6B = 0 \end{array}\right\} A = -\tfrac{9}{20}, \quad B = \tfrac{3}{20}$$

$$y = C_1 e^x + C_2 e^{-2x} - \tfrac{9}{20}\cos 2x + \tfrac{3}{20}\sin 2x$$

$$y' = C_1 e^x - 2C_2 e^{-2x} + \tfrac{9}{10}\sin 2x + \tfrac{3}{10}\cos 2x$$

Initial conditions: $y(0) = -1, \quad y'(0) = 2, \quad -1 = C_1 + C_2 - \tfrac{9}{20}, \quad 2 = C_1 - 2C_2 + \tfrac{3}{10}$

$$\left.\begin{array}{l} C_1 + C_2 = \tfrac{-11}{20} \\[4pt] C_1 - 2C_2 = \tfrac{17}{10} \end{array}\right\} C_1 = \tfrac{1}{5}, \quad C_2 = -\tfrac{3}{4}$$

Particular solution: $y = \tfrac{1}{20}(4e^x - 15e^{-2x} - 9\cos 2x + 3\sin 2x)$

15. $y'' + 9y = \sin 3x$

$y'' + 9y = 0$

$m^2 + 9 = 0$ when $m = -3i, \; 3i.$

$$y_h = C_1 \cos 3x + C_2 \sin 3x$$

$$y_p = A_0 \sin 3x + A_1 x \sin 3x + A_2 \cos 3x + A_3 x \cos 3x$$

$$y_p{}'' = (-9A_0 - 6A_3)\sin 3x - 9A_1 x \sin 3x + (6A_1 - 9A_2)\cos 3x - 9A_3 x \cos 3x$$

$$y_p{}'' + 9y_p = -6A_3 \sin 3x + 6A_1 \cos 3x = \sin 3x, \quad A_1 = 0, \quad A_3 = -\tfrac{1}{6}$$

$$y = \left(C_1 - \tfrac{1}{6}x\right)\cos 3x + C_2 \sin 3x$$

16. $y'' + 4y' + 5y = \sin x + \cos x$

$y'' + 4y' + 5y = 0$

$m^2 + 4m + 5 = 0$ when $m = -2 - i, \; -2 + i.$

$$y_h = e^{-2x}(C_1 \cos x + C_2 \sin x)$$

$$y_p = A\cos x + B\sin x$$

$$y_p{}' = -A\sin x + B\cos x$$

$$y_p{}'' = -A\cos x - B\sin x$$

$$y_p{}'' + 4y_p{}' + 5y_p = (4A + 4B)\cos x + (-4A + 4B)\sin x = \sin x + \cos x$$

$$\left.\begin{array}{l} 4A + 4B = 1 \\[4pt] -4A + 4B = 1 \end{array}\right\} A = 0, \quad B = \tfrac{1}{4}$$

$$y = e^{-2x}(C_1 \cos x + C_2 \sin x) + \tfrac{1}{4}\sin x$$

17. $y''' - 3y' + 2y = 2e^{-2x}$

$y''' - 3y' + 2y = 0$

$m^3 - 3m + 2 = 0$ when $m = 1,\ 1,\ -2$.

$$y_h = C_1 e^x + C_2 x e^x + C_3 e^{-2x}$$

$$y_p = A_0 e^{-2x} + A_1 x e^{-2x}$$

$$y_p' = (-2A_0 + A_1)e^{-2x} - 2A_1 x e^{-2x}$$

$$y_p'' = (4A_0 - 4A_1)e^{-2x} + 4A_1 x e^{-2x}$$

$$y_p''' = (-8A_0 + 12A_1)e^{-2x} - 8A_1 x e^{-2x}$$

$$y_p''' - 3y_p' + 2y_p = 9A_1 e^{-2x} = 2e^{-2x} \text{ or } A_1 = \tfrac{2}{9}$$

$$y = C_1 e^x + C_2 x e^x + \left(C_3 + \tfrac{2}{9}x\right)e^{-2x}$$

18. $y''' - y'' = 4x^2,\quad y(0) = 1,\quad y'(0) = 1,\quad y''(0) = 1$

$y''' - y'' = 0$

$m^3 - m^2 = 0$ when $m = 0,\ 0,\ 1$.

$$y_h = C_1 + C_2 x + C_3 e^x$$

$$y_p = A_0 x^2 + A_1 x^3 + A_2 x^4$$

$$y_p' = 2A_0 x + 3A_1 x^2 + 4A_2 x^3$$

$$y_p'' = 2A_0 + 6A_1 x + 12A_2 x^2$$

$$y_p''' = 6A_1 + 24A_2 x$$

$$y_p''' - y_p'' = (-2A_0 + 6A_1) + (-6A_1 + 24A_2)x - 12A_2 x^2 = 4x^2 \text{ or } A_0 = -4,\quad A_1 = -\tfrac{4}{3},\quad A_2 = -\tfrac{1}{3}$$

$$y = C_1 + C_2 x + C_3 e^x - 4x^2 - \tfrac{4}{3}x^3 - \tfrac{1}{3}x^4$$

$$y' = C_2 + C_3 e^x - 8x - 4x^2 - \tfrac{4}{3}x^3$$

$$y'' = C_3 e^x - 8 - 8x - 4x^2$$

Initial conditions:

$y(0) = 1,\quad y'(0) = 1,\quad y''(0) = 1,\quad 1 = C_1 + C_3,\quad 1 = C_2 + C_3,\quad 1 = C_3 - 8,\quad C_1 = -8,\quad C_2 = -8,\quad C_3 = 9$

Particular solution: $y = -8 - 8x - 4x^2 - \tfrac{4}{3}x^3 - \tfrac{1}{3}x^4 + 9e^x$

19. $y' - 4y = xe^x - xe^{4x},\quad y(0) = \tfrac{1}{3}$

$y' - 4y = 0$

$m - 4 = 0$ when $m = 4$.

$$y_h = Ce^{4x}$$

$$y_p = (A_0 + A_1 x)e^x + (A_2 x + A_3 x^2)e^{4x}$$

$$y_p' = (A_0 + A_1 x)e^x + A_1 e^x + 4(A_2 x + A_3 x^2)e^{4x} + (A_2 + 2A_3 x)e^{4x}$$

$$y_p' - 4y_p = (-3A_0 - 3A_1 x)e^x + A_1 e^x + A_2 e^{4x} + 2A_3 x e^{4x} = xe^x - xe^{4x}$$

$$A_0 = -\tfrac{1}{9},\quad A_1 = -\tfrac{1}{3},\quad A_2 = 0,\quad A_3 = -\tfrac{1}{2}$$

$$y = \left(C - \tfrac{1}{2}x^2\right)e^{4x} - \tfrac{1}{9}(1 + 3x)e^x$$

Initial condition: $y(0) = \tfrac{1}{3},\quad \tfrac{1}{3} = C - \tfrac{1}{9},\quad C = \tfrac{4}{9}$

Particular solution: $y = \left(\tfrac{4}{9} - \tfrac{1}{2}x^2\right)e^{4x} - \tfrac{1}{9}(1 + 3x)e^x$

20. $y' + 2y = \sin x,$ $y\left(\dfrac{\pi}{2}\right) = \dfrac{2}{5}$

$y' + 2y = 0$

$m + 2 = 0$ when $m = -2.$

$\qquad y_h = Ce^{-2x}$

$\qquad y_p = A\cos x + B\sin x$

$\qquad y_p' = -A\sin x + B\cos x$

$y_p' + 2y_p = (-A\sin x + B\cos x) + 2(A\cos x + B\sin x) = (2B - A)\sin x + (2A + B)\cos x = \sin x$

$\quad 2B - A = 1,\quad 2A + B = 0 \Rightarrow B = \dfrac{2}{5},\quad A = -\dfrac{1}{5}$

$\qquad y = y_h + y_p = Ce^{-2x} - \dfrac{1}{5}\cos x + \dfrac{2}{5}\sin x$

Initial condition: $y\left(\dfrac{\pi}{2}\right) = \dfrac{2}{5},\quad \dfrac{2}{5} = Ce^{-\pi} + \dfrac{2}{5},\quad C = 0$

Particular solution: $y = \dfrac{2}{5}\sin x - \dfrac{1}{5}\cos x$

21. $y'' + y = \sec x$

$y'' + y = 0$

$m^2 + 1 = 0$ when $m = -i,\ i.$

$y_h = C_1\cos x + C_2\sin x$

$y_p = v_1\cos x + v_2\sin x$

$\quad v_1'\cos x + v_2'\sin x = 0$

$\quad v_1'(-\sin x) + v_2'(\cos x) = \sec x$

$$v_1' = \frac{\begin{vmatrix} 0 & \sin x \\ \sec x & \cos x \end{vmatrix}}{\begin{vmatrix} \cos x & \sin x \\ -\sin x & \cos x \end{vmatrix}} = -\tan x$$

$$v_1 = \int -\tan x\,dx = \ln|\cos x|$$

$$v_2' = \frac{\begin{vmatrix} \cos x & 0 \\ -\sin x & \sec x \end{vmatrix}}{\begin{vmatrix} \cos x & \sin x \\ -\sin x & \cos x \end{vmatrix}} = 1$$

$$v_2 = \int dx = x$$

$y = (C_1 + \ln|\cos x|)\cos x + (C_2 + x)\sin x$

22. $y'' + y = \sec x \tan x$

$y'' + y = 0$

$m^2 + 1 = 0$ when $m = \pm i$

$y_h = C_1\cos x + C_2\sin x$

$y_p = v_1\cos x + v_2\sin x$

$\quad v_1'\cos x + v_2'\sin x = 0$

$\quad v_1'(-\sin x) + v_2'\cos x = \sec x\tan x$

$$v_1' = \frac{\begin{vmatrix} 0 & \sin x \\ \sec x \tan x & \cos x \end{vmatrix}}{\begin{vmatrix} \cos x & \sin x \\ -\sin x & \cos x \end{vmatrix}} = -\tan^2 x$$

$$v_1 = \int -\tan^2 x\,dx = -\int (\sec^2 x - 1)\,dx$$
$$= -\tan x + x$$

$$v_2' = \frac{\begin{vmatrix} \cos x & 0 \\ -\sin x & \sec x \tan x \end{vmatrix}}{\begin{vmatrix} \cos x & \sin x \\ -\sin x & \cos x \end{vmatrix}} = \tan x$$

$$v_2 = \int \tan x\,dx = -\ln|\cos x| = \ln|\sec x|$$

$y = y_h + y_p$

$\quad = C_1\cos x + C_2\sin x + (x - \tan x)\cos x$

$\qquad + \ln|\sec x|\sin x$

$\quad = (C_1 + x - \tan x)\cos x + (C_2 + \ln|\sec x|)\sin x$

23. $y'' + 4y = \csc 2x$

$y'' + 4y = 0$

$m^2 + 4 = 0$ when $m = -2i, \ 2i.$

$y_h = C_1 \cos 2x + C_2 \sin 2x$

$y_p = v_1 \cos 2x + v_2 \sin 2x = 0$

$\qquad v_1{}' \cos 2x + v_2{}' \sin 2x = 0$

$\qquad v_1{}'(-2 \sin 2x) + v_2{}'(2 \cos 2x) = \csc 2x$

$$v_1{}' = \frac{\begin{vmatrix} 0 & \sin 2x \\ \csc 2x & 2\cos 2x \end{vmatrix}}{\begin{vmatrix} \cos 2x & \sin 2x \\ -2\sin 2x & 2\cos 2x \end{vmatrix}} = -\frac{1}{2}$$

$$v_1 = \int -\frac{1}{2}\,dx = -\frac{1}{2}x$$

$$v_2{}' = \frac{\begin{vmatrix} \cos 2x & 0 \\ -2\sin 2x & \csc 2x \end{vmatrix}}{\begin{vmatrix} \cos 2x & \sin 2x \\ -2\sin 2x & 2\cos 2x \end{vmatrix}} = \frac{1}{2}\cot 2x$$

$$v_2 = \int \frac{1}{2}\cot 2x\,dx = \frac{1}{4}\ln|\sin 2x|$$

$$y = \left(C_1 - \frac{1}{2}x\right)\cos 2x + \left(C_2 + \frac{1}{4}\ln|\sin 2x|\right)\sin 2x$$

24. $y'' - 4y' + 4y = x^2 e^{2x}$

$y'' - 4y' + 4y = 0$

$m^2 - 4m + 4 = 0$ when $m = 2, \ 2.$

$y_h = (C_1 + C_2 x)e^{2x}$

$y_p = (v_1 + v_2 x)e^{2x}$

$\qquad v_1{}'e^{2x} + v_2{}'xe^{2x} = 0$

$\qquad v_1{}'(2e^{2x}) + v_2{}'(2x+1)e^{2x} = x^2 e^{2x}$

$$v_1{}' = \frac{\begin{vmatrix} 0 & xe^{2x} \\ x^2 e^{2x} & (2x+1)e^{2x} \end{vmatrix}}{\begin{vmatrix} e^{2x} & xe^{2x} \\ 2e^{2x} & (2x+1)e^{2x} \end{vmatrix}} = \frac{-x^3 e^{4x}}{e^{4x}} = -x^3$$

$$v_1 = \int -x^3\,dx = -\frac{1}{4}x^4$$

$$v_2{}' = \frac{\begin{vmatrix} e^{2x} & 0 \\ 2e^{2x} & x^2 e^{2x} \end{vmatrix}}{e^{4x}} = \frac{x^2 e^{4x}}{e^{4x}} = x^2$$

$$v_2 = \int x^2\,dx = \frac{1}{3}x^3$$

$$y = \left(C_1 + C_2 x + \frac{1}{12}x^4\right)e^{2x}$$

25. $y'' - 2y' + y = e^x \ln x$

$y'' - 2y' + y = 0$

$m^2 - 2m + 1 = 0$ when $m = 1, \ 1.$

$y_h = (C_1 + C_2 x)e^x$

$y_p = (v_1 + v_2 x)e^x$

$\qquad v_1{}'e^x + v_2{}'xe^x = 0$

$\qquad v_1{}'e^x + v_2{}'(x+1)e^x = e^x \ln x$

$\qquad v_1{}' = -x \ln x$

$$v_1 = \int -x \ln x\,dx = -\frac{x^2}{2}\ln x + \frac{x^2}{4}$$

$\qquad v_2{}' = \ln x$

$$v_2 = \int \ln x\,dx = x \ln x - x$$

$$y = (C_1 + C_2 x)e^x + \frac{x^2 e^x}{4}(\ln x^2 - 3)$$

26. $y'' - 4y' + 4y = \dfrac{e^{2x}}{x}$

$y'' - 4y' + 4y = 0$

$m^2 - 4m + 4 = 0$ when $m = 2, \ 2.$

$y_h = (C_1 + C_2 x)e^{2x}$

$y_p = (v_1 + v_2 x)e^{2x}$

$\qquad v_1{}'e^{2x} + v_2{}'xe^{2x} = 0$

$\qquad v_1{}'e^{2x}(2) + v_2{}'(2x+1)e^{2x} = \dfrac{e^{2x}}{x}$

$\qquad v_1{}' = -1$

$$v_1 = \int -1\,dx = -x$$

$\qquad v_2{}' = \dfrac{1}{x}$

$$v_2 = \int \frac{1}{x}\,dx = \ln|x|$$

$$y = (C_1 + C_2 x - x + x\ln|x|)e^{2x}$$

27. $q'' + 10q' + 25q = 6\sin 5t, \quad q(0) = 0, \quad q'(0) = 0$
$m^2 + 10m + 25 = 0$ when $m = -5, -5.$

$$q_h = (C_1 + C_2 t)e^{-5t}$$

$$q_p = A\cos 5t + B\sin 5t$$

$$q_p{}' = -5A\sin 5t + 5B\cos 5t$$

$$q_p{}'' = -25A\cos 5t - 25B\sin 5t$$

$$q_p{}'' + 10q_p{}' + 25q_p = 50B\cos 5t - 50A\sin 5t = 6\sin 5t, \quad A = -\tfrac{3}{25}, \quad B = 0$$

$$q = (C_1 + C_2 t)e^{-5t} - \tfrac{3}{25}\cos 5t$$

Initial conditions: $q(0) = 0, \quad q'(0) = 0, \quad C_1 - \tfrac{3}{25} = 0, \quad -5C_1 + C_2 = 0, \quad C_1 = \tfrac{3}{25}, \quad C_2 = \tfrac{3}{5}$

Particular solution: $q = \tfrac{3}{25}(e^{-5t} + 5te^{-5t} - \cos 5t)$

28. $q'' + 20q' + 50q = 10\sin 5t$
$m^2 + 20m + 50 = 0$ when $m = -10 \pm 5\sqrt{2}.$

$$q_h = C_1 e^{(-10+5\sqrt{2})t} + C_2 e^{(-10-5\sqrt{2})t}$$

$$q_p = A\cos 5t + B\sin 5t$$

$$q_p{}' = 5B\cos 5t - 5A\sin 5t$$

$$q_p{}'' = -25A\cos 5t - 25B\sin 5t$$

$$q_p{}'' + 20q_p{}' + 50q_p = (25A + 100B)\cos 5t + (25B - 100A)\sin 5t = 10\sin 5t$$

$$\left.\begin{array}{l} 25A + 100B = 0 \\ 25B - 100A = 10 \end{array}\right\} B = \frac{2}{85}, \quad A = -\frac{8}{85}$$

$$q = C_1 e^{(-10+5\sqrt{2})t} + C_2 e^{(-10-5\sqrt{2})t} - \frac{8}{85}\cos 5t + \frac{2}{85}\sin 5t$$

Initial conditions: $q(0) = 0, \quad q'(0) = 0, \quad C_1 + C_2 - \dfrac{8}{85} = 0,$

$$(-10 + 5\sqrt{2})C_1 + (-10 - 5\sqrt{2})C_2 + \frac{2}{17} = 0,$$

$$C_1 = \frac{8 + 7\sqrt{2}}{170}, \quad C_2 = \frac{8 - 7\sqrt{2}}{170}$$

Particular solution: $q = \dfrac{8 + 7\sqrt{2}}{170}e^{(-10+5\sqrt{2})t} + \dfrac{8 - 7\sqrt{2}}{170}e^{(-10-5\sqrt{2})t} - \dfrac{8}{85}\cos 5t + \dfrac{2}{85}\sin 5t$

29. $\tfrac{24}{32}y'' + 48y = \tfrac{24}{32}(48\sin 4t), \quad y(0) = \tfrac{1}{4}, \quad y'(0) = 0$

$\tfrac{24}{32}m^2 + 48 = 0$ when $m = \pm 8i.$

$$y_h = C_1\cos 8t + C_2\sin 8t$$

$$y_p = A\sin 4t + B\cos 4t$$

$$y_p{}' = 4A\cos 4t - 4B\sin 4t$$

$$y_p{}'' = -16A\sin 4t - 16B\cos 4t$$

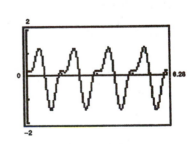

$$\tfrac{24}{32}y_p{}'' + 48y_p = 36A\sin 4t + 36B\cos 4t = \tfrac{24}{32}(48\sin 4t), \quad B = 0, \quad A = 1$$

$$y = y_h + y_p = C_1\cos 8t + C_2\sin 8t + \sin 4t$$

Initial conditions: $y(0) = \tfrac{1}{4}, \quad y'(0) = 0, \quad \tfrac{1}{4} = C_1, \quad 0 = 8C_2 + 4 \Rightarrow C_2 = -\tfrac{1}{2}$

Particular solution: $y = \tfrac{1}{4}\cos 8t - \tfrac{1}{2}\sin 8t + \sin 4t$

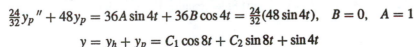

30. $\frac{2}{32}y'' + 4y = \frac{2}{32}(4\sin 8t)$, $y(0) = \frac{1}{4}$, $y'(0) = 0$

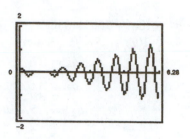

$\frac{2}{32}m^2 + 4 = 0$ when $m = \pm 8i$.

$$y_h = C_1\cos 8t + C_2\sin 8t$$

$$y_p = At\sin 8t + Bt\cos 8t$$

$$y_p'' = (-64At - 16B)\sin 8t + (16A - 64Bt)\cos 8t$$

$$\frac{2}{32}y_p'' + 4y_p = -B\sin 8t + A\cos 8t = \frac{2}{32}(4\sin 8t), \quad A = 0, \quad B = -\frac{1}{4}$$

$$y = C_1\cos 8t + C_2\sin 8t - \frac{1}{4}t\cos 8t$$

Initial conditions: $y(0) = \frac{1}{4}$, $y'(0) = 0$, $\frac{1}{4} = C_1$, $0 = 8C_2 - \frac{1}{4} \Rightarrow C_2 = \frac{1}{32}$

Particular solution: $y = \frac{1}{4}\cos 8t + \frac{1}{32}\sin 8t - \frac{1}{4}t\cos 8t$

31. $\frac{2}{32}y'' + y' + 4y = \frac{2}{32}(4\sin 8t)$, $y(0) = \frac{1}{4}$, $y'(0) = -3$

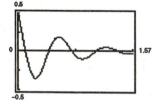

$\frac{1}{16}m^2 + m + 4 = 0$ when $m = -8, -8$.

$$y_h = (C_1 + C_2t)e^{-8t}$$

$$y_p = A\sin 8t + B\cos 8t$$

$$y_p' = 8A\cos 8t - 8B\sin 8t$$

$$y_p'' = -64A\sin 8t - 64B\cos 8t$$

$$\frac{2}{32}y_p'' + y_p' + 4y_p = -8B\sin 8t + 8A\cos 8t = \frac{2}{32}(4\sin 8t)$$

$$-8B = \frac{1}{4} \Rightarrow B = -\frac{1}{32}, \quad 8A = 0 \Rightarrow A = 0$$

$$y = y_h + y_p = (C_1 + C_2t)e^{-8t} - \frac{1}{32}\cos 8t$$

Initial conditions: $y(0) = \frac{1}{4}$, $y'(0) = -3$, $\frac{1}{4} = C_1 - \frac{1}{32} \Rightarrow C_1 = \frac{9}{32}$, $-3 = -8C_1 + C_2 \Rightarrow C_2 = -\frac{3}{4}$

Particular solution: $y = \left(\frac{9}{32} - \frac{3}{4}t\right)e^{-8t} - \frac{1}{32}\cos 8t$

32. $\frac{4}{32}y'' + \frac{1}{2}y' + \frac{25}{2}y = 0$, $y(0) = \frac{1}{2}$, $y'(0) = -4$

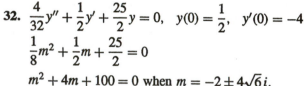

$\frac{1}{8}m^2 + \frac{1}{2}m + \frac{25}{2} = 0$

$m^2 + 4m + 100 = 0$ when $m = -2 \pm 4\sqrt{6}\,i$.

$y = C_1e^{-2t}\cos\left(4\sqrt{6}\,t\right) + C_2e^{-2t}\sin\left(4\sqrt{6}\,t\right)$

Initial conditions:

$$y(0) = \frac{1}{2}, \quad y'(0) = -4, \quad \frac{1}{2} = C_1, \quad -4 = -2C_1 + 4\sqrt{6}\,C_2, \quad C_2 = -\frac{3}{4\sqrt{6}} = -\frac{\sqrt{6}}{8}$$

Particular solution: $y = \frac{1}{2}e^{-2t}\cos\left(4\sqrt{6}\,t\right) - \frac{\sqrt{6}}{8}e^{-2t}\sin\left(4\sqrt{6}\,t\right)$

33. In Exercise 29,

$$y_h = \frac{1}{4}\cos 8t - \frac{1}{2}\sin 8t = \frac{\sqrt{5}}{4}\sin\left[8t + \arctan\left(-\frac{1}{2}\right)\right] = \frac{\sqrt{5}}{4}\sin\left(8t - \arctan\frac{1}{2}\right) \approx \frac{\sqrt{5}}{4}\sin(8t - 0.4636).$$

34. a. $\frac{4}{32}y'' + \frac{25}{2}y = 0$

$y = C_1 \cos 10x + C_2 \sin 10x$

$y(0) = \frac{1}{2}: \ \frac{1}{2} = C_1$

$y'(0) = -4: \ -4 = 10C_2 \Rightarrow C_2 = -\frac{2}{5}$

$y = \frac{1}{2}\cos 10x - \frac{2}{5}\sin 10x$

The motion is undamped.

b. If $b > 0$, the motion is damped.

c. If $b > \frac{5}{2}$, the solution to the differential equation is of the form $y = C_1 e^{m_1 x} + C_2 e^{m_2 x}$. There would be no oscillations in this case.

35. $-5y'' - 8y' = 160$

$-5m^2 - 8m = 0$ when $m = 0, \ -\frac{8}{5}$.

$y_h = C_1 + C_2 e^{-1.6t}$

$y_p = At + B$

$y_p{}' = A$

$y_p{}'' = 0$

$-5y'' - 8y' = -8A = 160 \Rightarrow A = -20$

$y = C_1 + C_2 e^{-1.6t} - 20t$

Initial conditions:

$y(0) = 2000, \quad y'(0) = -100, \quad 2000 = C_1 + C_2, \quad -100 = -1.6C_2 - 20, \quad C_2 = 50 \Rightarrow C_1 = 1950$

Particular solution: $y = 1950 + 50e^{-1.6t} - 20t$

36. $-6y'' - 9y' = 192$

$-6m^2 - 9m = 0$ when $m = 0, \ -1.5$.

$y_h = C_1 + C_2 e^{-1.5t}$

$y_p = At + B$

$-6y'' - 9y' = -9A = 192 \Rightarrow A = -\frac{64}{3}$

$y = C_1 + C_2 e^{-1.5t} - \frac{64}{3}t$

Initial conditions:

$y(0) = 2000, \quad y'(0) = -100, \quad 2000 = C_1 + C_2, \quad -100 = -1.5C_2 - \frac{64}{3}, \quad C_2 = \frac{472}{9} \Rightarrow C_1 = \frac{17{,}528}{9}$

$y = \frac{17{,}528}{9} + \frac{472}{9}e^{-1.5t} - \frac{64}{3}t$

37. $x^2 y'' - xy' + y = 4x \ln x$

$y_1 = x$ and $y_2 = x \ln x$

$u_1'x + u_2'x \ln x = 0 \Rightarrow u_1' = -u_2' \ln x$

$u_1' + u_2'(1 + \ln x) = \frac{4}{x}\ln x \Rightarrow u_2' = \frac{4}{x}\ln x$ and $u_1' = -\frac{4}{x}(\ln x)^2$

$u_1 = -\frac{4}{3}(\ln x)^3$ and $u_2 = 2(\ln x)^2$

$y_p = -\frac{4}{3}x(\ln x)^3 + 2x(\ln x)^3 = \frac{2}{3}x(\ln x)^3$

$y = y_n + y_p = C_1 x + C_2 x \ln x + \frac{2}{3}x(\ln x)^3$

Section 16.7   Series Solutions of Differential Equations

1. $y' - y = 0$. Letting $y = \sum\limits_{n=0}^{\infty} a_n x^n$:

$$y' - y = \sum_{n=1}^{\infty} n a_n x^{n-1} - \sum_{n=0}^{\infty} a_n x^n$$

$$= \sum_{n=0}^{\infty} (n+1) a_{n+1} x^n - \sum_{n=0}^{\infty} a_n x^n = 0$$

$$(n+1) a_{n+1} = a_n$$

$$a_{n+1} = \frac{a_n}{n+1}$$

$$a_1 = a_0, \quad a_2 = \frac{a_1}{2} = \frac{a_0}{2}, \quad a_3 = \frac{a_2}{3} = \frac{a_0}{1 \cdot 2 \cdot 3}, \quad \ldots, \quad a_n = \frac{a_0}{n!}$$

$$y = \sum_{n=0}^{\infty} \frac{a_0}{n!} x^n = a_0 e^x$$

Check: By separation of variables, we have:

$$\int \frac{dy}{y} = \int dx$$

$$\ln y = x + C_1$$

$$y = Ce^x$$

2. $y' - ky = 0$. Letting $y = \sum\limits_{n=0}^{\infty} a_n x^n$:

$$y' - ky = \sum_{n=1}^{\infty} n a_n x^{n-1} - k \sum_{n=0}^{\infty} a_n x^n$$

$$= \sum_{n=0}^{\infty} (n+1) a_{n+1} x^n - \sum_{n=0}^{\infty} k a_n x^n = 0$$

$$(n+1) a_{n+1} = k a_n$$

$$a_{n+1} = \frac{k a_n}{n+1}$$

$$a_1 = k a_0, \quad a_2 = \frac{k a_1}{2} = \frac{k^2 a_0}{2}, \quad a_3 = \frac{k a_2}{3} = \frac{k^3 a_0}{1 \cdot 2 \cdot 3}, \quad \ldots, \quad a_n = \frac{k^n}{n!} a_0$$

$$y = \sum_{n=0}^{\infty} \frac{k^n}{n!} a_0 x^n = a_0 \sum_{n=0}^{\infty} \frac{(kx)^n}{n!} = a_0 e^{kx}$$

Check: By separation of variables, we have:

$$\int \frac{dy}{y} = \int k \, dx$$

$$\ln y = kx + C_1$$

$$y = Ce^{kx}$$

3. $y'' - 9y = 0$. Letting $y = \sum_{n=0}^{\infty} a_n x^n$:

$$y'' - 9y = \sum_{n=2}^{\infty} n(n-1)a_n x^{n-2} - 9\sum_{n=0}^{\infty} a_n x^n = \sum_{n=0}^{\infty} (n+2)(n+1)a_{n+2}x^n - \sum_{n=0}^{\infty} 9a_n x^n = 0$$

$$(n+2)(n+1)a_{n+2} = 9a_n$$

$$a_{n+2} = \frac{9a_n}{(n+2)(n+1)}$$

$$a_0 = a_0 \qquad\qquad\qquad a_1 = a_1$$

$$a_2 = \frac{9a_0}{2} \qquad\qquad\qquad a_3 = \frac{9a_1}{3 \cdot 2}$$

$$a_4 = \frac{9a_2}{4 \cdot 3} = \frac{9^2 a_0}{4 \cdot 3 \cdot 2 \cdot 1} \qquad\qquad a_5 = \frac{9a_3}{5 \cdot 4} = \frac{9^2 a_1}{5 \cdot 4 \cdot 3 \cdot 2 \cdot 1}$$

$$\vdots \qquad\qquad\qquad\qquad \vdots$$

$$a_{2n} = \frac{9^n a_0}{(2n)!} \qquad\qquad\qquad a_{2n+1} = \frac{9^n a_1}{(2n+1)!}$$

$$y = \sum_{n=0}^{\infty} \frac{9^n a_0}{(2n)!}x^{2n} + \sum_{n=0}^{\infty} \frac{9^n a_1}{(2n+1)!}x^{2n+1} = a_0\sum_{n=0}^{\infty} \frac{(3x)^{2n}}{(2n)!} + \frac{a_1}{3}\sum_{n=0}^{\infty} \frac{(3x)^{2n+1}}{(2n+1)!} = C_0\sum_{n=0}^{\infty} \frac{(3x)^n}{n!} + C_1\sum_{n=0}^{\infty} \frac{(-3x)^n}{n!}$$

$$= C_0 e^{3x} + C_1 e^{-3x} \text{ where } C_0 + C_1 = a_0 \text{ and } C_0 - C_1 = \frac{a_1}{3}.$$

Check: $y'' - 9y = 0$ is a second-order homogeneous linear equation.

$$m^2 - 9 = 0 \Rightarrow m_1 = 3 \text{ and } m_2 = -3$$

$$y = C_1 e^{3x} + C_2 e^{-3x}$$

4. $y = C_0 e^{kx} + C_1 e^{-kx}$. Follow the solution to Exercise 3 with 9 replaced by k^2.

5. $y'' + 4y = 0$. Letting $y = \sum_{n=0}^{\infty} a_n x^n$:

$$y'' + 4y = \sum_{n=2}^{\infty} n(n-1)a_n x^{n-2} + 4\sum_{n=0}^{\infty} a_n x^n = \sum_{n=0}^{\infty} (n+2)(n+1)a_{n+2}x^n + \sum_{n=0}^{\infty} 4a_n x^n = 0$$

$$(n+2)(n+1)a_{n+2} = -4a_n$$

$$a_{n+2} = \frac{-4a_n}{(n+2)(n+1)}$$

$$a_0 = a_0 \qquad\qquad\qquad a_1 = a_1$$

$$a_2 = \frac{-4a_0}{2} \qquad\qquad\qquad a_3 = \frac{-4a_1}{3 \cdot 2}$$

$$a_4 = \frac{-4a_2}{4 \cdot 3} = \frac{(-4)^2}{4!}a_0 \qquad\qquad a_5 = \frac{-4a_3}{5 \cdot 4} = \frac{(-4)^2 a_1}{5!}$$

$$\vdots \qquad\qquad\qquad\qquad \vdots$$

$$a_{2n} = \frac{(-1)^n 4^n}{(2n)!}a_0 \qquad\qquad a_{2n+1} = \frac{(-1)^n 4^n}{(2n+1)!}a_1$$

5. —CONTINUED—

$$y = \sum_{n=0}^{\infty} \frac{(-1)^n 4^n a_0}{(2n)!} x^{2n} + \sum_{n=0}^{\infty} \frac{(-1)^n 4^n a_1}{(2n+1)!} x^{2n+1} = a_0 \sum_{n=0}^{\infty} \frac{(-1)^n (2x)^{2n}}{(2n)!} + \frac{a_1}{4} \sum_{n=0}^{\infty} \frac{(-1)^n (2x)^{2n+1}}{(2n+1)!}$$

$$= C_0 \cos 2x + C_1 \sin 2x$$

Check: $y'' + 4y = 0$ is a second-order homogeneous linear equation.

$$m^2 + 4 = 0 \Rightarrow m = \pm 2i$$

$$y = C_1 \cos 2x + C_2 \sin 2x$$

6. $y = C_0 \cos kx + C_1 \sin kx$. Follow the solution to Exercise 5 with 4 replaced by k^2.

7. $y' + 3xy = 0$. Letting $y = \sum_{n=0}^{\infty} a_n x^n$:

$$y' + 3xy = \sum_{n=1}^{\infty} n a_n x^{n-1} + \sum_{n=0}^{\infty} 3 a_n x^{n+1} = 0$$

$$\sum_{n=-1}^{\infty} (n+2) a_{n+2} x^{n+1} = \sum_{n=0}^{\infty} -3 a_n x^{n+1}$$

$$a_{n+2} = \frac{-3 a_n}{n+2}$$

$$a_0 = a_0 \qquad\qquad\qquad a_1 = a_1$$

$$a_2 = -\frac{3 a_0}{2} \qquad\qquad\qquad a_3 = -\frac{3 a_1}{3}$$

$$a_4 = -\frac{3}{4}\left(-\frac{3 a_0}{2}\right) = \frac{3^2}{2^3} a_0 \qquad\qquad a_5 = -\frac{3}{5}\left(-\frac{3 a_1}{3}\right) = \frac{3^2 a_1}{3 \cdot 5}$$

$$a_6 = -\frac{3}{6}\left(\frac{3^2}{2^3} a_0\right) = -\frac{3^3 a_0}{2^3 (3 \cdot 2)} \qquad\qquad a_7 = -\frac{3}{7}\left(\frac{3^2 a_1}{3 \cdot 5}\right) = -\frac{3^3 a_1}{3 \cdot 5 \cdot 7}$$

$$a_8 = -\frac{3}{8}\left(-\frac{3^3 a_0}{2^3 (3 \cdot 2)}\right) = \frac{3^4 a_0}{2^4 (4 \cdot 3 \cdot 2)} \qquad a_9 = -\frac{3}{9}\left(-\frac{3^3 a_1}{3 \cdot 5 \cdot 7}\right) = \frac{3^4 a_1}{3 \cdot 5 \cdot 7 \cdot 9}$$

$$y = a_0 \sum_{n=0}^{\infty} \frac{(-3)^n x^{2n}}{2^n n!} + a_1 \sum_{n=0}^{\infty} \frac{(-3)^n x^{2n+1}}{1 \cdot 3 \cdot 5 \cdot 7 \cdots (2n+1)}$$

$$\lim_{n \to \infty} \left|\frac{u_{n+1}}{u_n}\right| = \lim_{n \to \infty} \left|\frac{(-3)^{n+1} x^{2n+2}}{2^{n+1}(n+1)!} \cdot \frac{2^n n!}{(-3)^n x^{2n}}\right| = \lim_{n \to \infty} \frac{3x^2}{2(n+1)} = 0$$

$$\lim_{n \to \infty} \left|\frac{u_{n+1}}{u_n}\right| = \lim_{n \to \infty} \left|\frac{(-3)^{n+1} x^{2n+3}}{1 \cdot 3 \cdot 5 \cdot 7 \cdots (2n+3)} \cdot \frac{1 \cdot 3 \cdot 5 \cdot 7 \cdots (2n+1)}{(-3)^n x^{2n+1}}\right| = \lim_{n \to \infty} \frac{3x^2}{2n+3} = 0$$

Since the interval of convergence for each series is $(-\infty, \infty)$, the interval of convergence for the solution is $(-\infty, \infty)$.

8. $y' - 2xy = 0$. Letting $y = \sum_{n=0}^{\infty} a_n x^n$:

$$y' - 2xy = \sum_{n=0}^{\infty} n a_n x^{n-1} - \sum_{n=0}^{\infty} 2 a_n x^{n+1} = 0$$

$$\sum_{n=-1}^{\infty} (n+2) a_{n+2} x^{n+1} = \sum_{n=0}^{\infty} 2 a_n x^{n+1}$$

$$a_{n+2} = \frac{2 a_n}{n+2}$$

$a_0 = a_0$

$a_1 = a_1$

$a_2 = \dfrac{2 a_0}{2} = a_0$

$a_3 = \dfrac{2 a_1}{3}$

$a_4 = \dfrac{2}{4}\left(\dfrac{2 a_0}{2}\right) = \dfrac{2^2 a_0}{2^2 \cdot 2} = \dfrac{a_0}{2}$

$a_5 = \dfrac{2}{5}\left(\dfrac{2 a_1}{3}\right) = \dfrac{2^2 a_1}{3 \cdot 5}$

$a_6 = \dfrac{2}{6}\left(\dfrac{2^2 a_0}{2^2 \cdot 2}\right) = \dfrac{2^3 a_0}{2^3 3 \cdot 2} = \dfrac{a_0}{3!}$

$a_7 = \dfrac{2}{7}\left(\dfrac{2^2 a_1}{3 \cdot 5}\right) = \dfrac{2^3 a_1}{3 \cdot 5 \cdot 7}$

$a_8 = \dfrac{2}{8}\left(\dfrac{a_0}{3!}\right) = \dfrac{a_0}{4!}$

$a_9 = \dfrac{2}{9}\left(\dfrac{2^3 a_1}{3 \cdot 5 \cdot 7}\right) = \dfrac{2^4 a_1}{3 \cdot 5 \cdot 7 \cdot 9}$

$$y = a_0 \sum_{n=0}^{\infty} \frac{x^{2n}}{n!} + a_1 \sum_{n=0}^{\infty} \frac{2^n x^{2n+1}}{1 \cdot 3 \cdot 5 \cdot 7 \cdots (2n+1)}$$

$$\lim_{n \to \infty} \left| \frac{u_{n+1}}{u_n} \right| = \lim_{n \to \infty} \left| \frac{x^{2n+2}}{(n+1)!} \cdot \frac{n!}{2^n} \right| = \lim_{n \to \infty} \frac{x^2}{n+1} = 0$$

$$\lim_{n \to \infty} \left| \frac{u_{n+1}}{u_n} \right| = \lim_{n \to \infty} \left| \frac{2^{n+1} x^{2n+3}}{1 \cdot 3 \cdot 5 \cdot 7 \cdots (2n+3)} \cdot \frac{1 \cdot 3 \cdot 5 \cdot 7 \cdots (2n+1)}{2^n x^{2n+1}} \right| = \lim_{n \to \infty} \frac{2 x^2}{2n+3} = 0$$

Since the interval of convergence for each series is $(-\infty, \infty)$, the interval of convergence for the solution is $(-\infty, \infty)$.

9. $y'' - xy' = 0$. Letting $y = \sum_{n=0}^{\infty} a_n x^n$:

$$y'' - xy' = \sum_{n=2}^{\infty} n(n-1) a_n x^{n-2} - x \sum_{n=1}^{\infty} n a_n x^{n-1} = 0$$

$$\sum_{n=2}^{\infty} n(n-1) a_n x^{n-2} = \sum_{n=0}^{\infty} n a_n x^n$$

$$\sum_{n=0}^{\infty} (n+2)(n+1) a_{n+2} x^n = \sum_{n=0}^{\infty} n a_n x^n$$

$$a_{n+2} = \frac{n a_n}{(n+2)(n+1)}$$

$a_0 = a_0$

$a_1 = a_1$

$a_2 = 0$

$a_3 = \dfrac{a_1}{3 \cdot 2}$

There are no even-powered terms.

$a_5 = \dfrac{3 a_3}{5 \cdot 4} = \dfrac{3 a_1}{5!}$

$a_7 = \dfrac{5 a_5}{7 \cdot 6} = \dfrac{5 \cdot 3 a_1}{7!}$

9. —CONTINUED—

$$y = a_1 \sum_{n=0}^{\infty} \frac{1 \cdot 3 \cdot 5 \cdot 7 \cdots (2n-1)x^{2n+1}}{(2n+1)!} = a_1 \sum_{n=0}^{\infty} \frac{(2n)!x^{2n+1}}{2^n n!(2n+1)!} = a_1 \sum_{n=0}^{\infty} \frac{x^{2n+1}}{2^n n!(2n+1)}$$

$$\lim_{n \to \infty} \left| \frac{u_{n+1}}{u_n} \right| = \lim_{n \to \infty} \left| \frac{x^{2n+3}}{2^{n+1}(n+1)!(2n+3)} \cdot \frac{2^n n!(2n+1)}{x^{2n+1}} \right| = \lim_{n \to \infty} \frac{(2n+1)x^2}{2(n+1)(2n+3)} = 0$$

Interval of convergence: $(-\infty, \infty)$

10. $y'' - xy' - y = 0$. Letting $y = \sum_{n=0}^{\infty} a_n x^n$,

$$y'' - xy' - y = \sum_{n=2}^{\infty} n(n-1)a_n x^{n-2} - x \sum_{n=1}^{\infty} n a_n x^{n-1} - \sum_{n=0}^{\infty} a_n x^n = 0$$

$$\sum_{n=0}^{\infty} (n+2)(n+1)a_{n+2} x^n = \sum_{n=0}^{\infty} (n+1)a_n x^n$$

$$a_{n+2} = \frac{a_n}{n+2}$$

$$a_0 = a_0 \qquad\qquad\qquad a_1 = a_1$$

$$a_2 = \frac{a_0}{2} \qquad\qquad\qquad a_3 = \frac{a_1}{3}$$

$$a_4 = \frac{a_2}{4} = \frac{a_0}{8} = \frac{a_0}{2^2 2!} \qquad a_5 = \frac{a_3}{5} = \frac{a_1}{3 \cdot 5}$$

$$a_6 = \frac{a_4}{6} = \frac{a_0}{2^3 3!} \qquad\qquad a_7 = \frac{a_5}{7} = \frac{a_1}{3 \cdot 5 \cdot 7}$$

$$a_8 = \frac{a^6}{8} = \frac{a_0}{2^4 4!} \qquad\qquad a_9 = \frac{a_7}{9} = \frac{a_1}{3 \cdot 5 \cdot 7 \cdot 9}$$

$$y = a_0 \sum_{n=0}^{\infty} \frac{x^{2n}}{2^n n!} + a_1 \sum_{n=0}^{\infty} \frac{x^{2n+1}}{1 \cdot 3 \cdot 5 \cdot 7 \cdots (2n+1)}$$

$$\lim_{n \to \infty} \left| \frac{u_{n+1}}{u_n} \right| = \lim_{n \to \infty} \left| \frac{x^{2n+2}}{2^{n+1}(n+1)!} \cdot \frac{2^n n!}{x^{2n}} \right| = \lim_{n \to \infty} \frac{x^2}{2(n+1)} = 0$$

$$\lim_{n \to \infty} \left| \frac{u_{n+1}}{u_n} \right| = \lim_{n \to \infty} \left| \frac{x^{2n+3}}{1 \cdot 3 \cdot 5 \cdot 7 \cdots (2n+3)} \cdot \frac{1 \cdot 3 \cdot 5 \cdot 7 \cdots (2n+1)}{x^{2n+1}} \right| = \lim_{n \to \infty} \frac{x^2}{2n+3} = 0$$

Since the interval of convergence for each series is $(-\infty, \infty)$, the interval of convergence for the solution is $(-\infty, \infty)$.

11. $(x^2 + 4)y'' + y = 0$. Letting $y = \displaystyle\sum_{n=0}^{\infty} a_n x^n$:

$$(x^2 + 4)y'' + y = \sum_{n=2}^{\infty} n(n-1)a_n x^n + 4\sum_{n=2}^{\infty} n(n-1)a_n x^{n-2} + \sum_{n=0}^{\infty} a_n x^n$$

$$= \sum_{n=0}^{\infty} (n^2 - n + 1)a_n x^n + \sum_{n=0}^{\infty} 4(n+2)(n+1)a_{n+2}x^n = 0$$

$$a_{n+2} = \frac{-(n^2 - n + 1)a_n}{4(n+2)(n+1)}$$

$a_0 = a_0$ 　　　　　　　　　$a_1 = a_1$

$a_2 = \dfrac{-a_0}{4 \cdot 2 \cdot 1}$ 　　　　　$a_3 = \dfrac{-a_1}{4(3)(2)} = \dfrac{-a_1}{24}$

$a_4 = \dfrac{-3a_2}{4(4)(3)} = \dfrac{a_0}{128}$ 　　$a_5 = \dfrac{-7a_3}{4(5)(4)} = \dfrac{7a_1}{1920}$

$$y = a_0\left(1 - \frac{x^2}{8} + \frac{x^4}{128} - \cdots\right) + a_1\left(x - \frac{x^3}{24} + \frac{7x^5}{1920} - \cdots\right)$$

12. $y'' + x^2 y = 0$. Letting $y = \displaystyle\sum_{n=0}^{\infty} a_n x^n$:

$$y'' + x^2 y = \sum_{n=2}^{\infty} n(n-1)a_n x^{n-2} + \sum_{n=0}^{\infty} a_n x^{n+2} = 0$$

$$\sum_{n=-2}^{\infty} (n+4)(n+3)a_{n+4}x^{n+2} = -\sum_{n=0}^{\infty} a_n x^{n+2}$$

$$a_{n+4} = \frac{-a_n}{(n+4)(n+3)}$$

Also: 　　　$y = a_0 + a_1 x + a_2 x^2 + a_3 x^3 + \cdots + a_n x^n + \cdots$

$$y'' = 2a_2 + 3 \cdot 2a_3 x + \cdots + n(n-1)a_n x^{n-2} + \cdots$$

$$y'' + x^2 y = 2a_2 + 3 \cdot 2a_3 x + (a_0 + 4 \cdot 3a_4)x^2 + (a_1 + 5 \cdot 4a_5)x^3 + \cdots = 0$$

$$2a_2 = 0, \quad 6a_3 = 0, \quad 12a_4 + a_0 = 0, \quad 20a_5 + a_1 = 0$$

Thus, $a_2 = 0$ and $a_3 = 0 \Rightarrow a_6 = 0$, $a_7 = 0$, $a_{10} = 0$, and $a_{11} = 0$. Therefore, $a_{4n+2} = 0$ and $a_{4n+3} = 0$.

$a_0 = a_0$ 　　　　　　　　　　　　　　$a_1 = a_1$

$a_4 = -\dfrac{a_0}{4 \cdot 3}$ 　　　　　　　　　　$a_5 = -\dfrac{a_1}{5 \cdot 4}$

$a_8 = -\dfrac{a_4}{8 \cdot 7} = \dfrac{a_0}{8 \cdot 7 \cdot 4 \cdot 3}$ 　　　$a_9 = -\dfrac{a_5}{9 \cdot 8} = \dfrac{a_1}{9 \cdot 8 \cdot 5 \cdot 4}$

$a_{12} = -\dfrac{a_8}{12 \cdot 11} = -\dfrac{a_0}{12 \cdot 11 \cdot 8 \cdot 7 \cdot 4 \cdot 3}$ 　$a_{13} = \dfrac{a_9}{13 \cdot 12} = \dfrac{a_1}{13 \cdot 12 \cdot 9 \cdot 8 \cdot 5 \cdot 4}$

$$y'' + x^2 y = a_0\left(1 - \frac{x^4}{4 \cdot 3} + \frac{x^8}{8 \cdot 7 \cdot 4 \cdot 3} - \frac{x^{12}}{12 \cdot 11 \cdot 8 \cdot 7 \cdot 4 \cdot 3} + \cdots\right)$$

$$+ a_1\left(x - \frac{x^5}{5 \cdot 4} + \frac{x^7}{9 \cdot 8 \cdot 5 \cdot 4} - \frac{x^9}{13 \cdot 12 \cdot 9 \cdot 8 \cdot 5 \cdot 4} + \cdots\right)$$

13. $y' + (2x - 1)y = 0, \quad y(0) = 2$

$$y' = (1 - 2x)y \qquad\qquad y'(0) = 2$$

$$y'' = (1 - 2x)y' - 2y \qquad y''(0) = -2$$

$$y''' = (1 - 2x)y'' - 4y' \qquad y'''(0) = -10$$

$$y^{(4)} = (1 - 2x)y''' - 6y'' \qquad y^{(4)}(0) = 2$$

$$\vdots \qquad\qquad\qquad \vdots$$

$$y(x) = 2 + \frac{2}{1!}x - \frac{2}{2!}x^2 - \frac{10}{3!}x^3 + \frac{2}{4!}x^4 + \cdots$$

Using the first five terms of the series, $y\left(\frac{1}{2}\right) = \frac{163}{64} \approx 2.547.$

Using Euler's Method with $\Delta x = 0.1$ we have $y' = (1 - 2x)y.$

i	x_i	y_i
0	0	2
1	0.1	2.2
2	0.2	2.376
3	0.3	2.51856
4	0.4	2.61930
5	0.5	2.67169

Therefore, $y\left(\frac{1}{2}\right) \approx 2.672.$

14. $y' - 2xy = 0, \quad y(0) = 1$

$$y' = 2xy \qquad\qquad y'(0) = 0$$

$$y'' = 2(xy' + y) \qquad y''(0) = 2$$

$$y''' = 2(xy'' + 2y') \qquad y'''(0) = 0$$

$$y^{(4)} = 2(xy''' + 3y'') \qquad y^{(4)}(0) = 12$$

$$y^{(5)} = 2(xy^{(4)} + 4y''') \qquad y^{(5)}(0) = 0$$

$$y^{(6)} = 2(xy^{(5)} + 5y^{(4)}) \qquad y^{(6)}(0) = 120$$

$$\vdots \qquad\qquad\qquad \vdots$$

$$y(x) = 1 + \frac{2}{2!}x^2 + \frac{12}{4!}x^4 + \frac{120}{6!}x^6 + \cdots = 1 + x^2 + \frac{1}{2}x^4 + \frac{1}{6}x^6 + \cdots$$

Using the first four terms of the series, $y(1) = \frac{8}{3} \approx 2.667.$

Using Euler's Method with $\Delta x = 0.1$ we have $y' = 2xy.$

i	x_i	y_i
0	0	1
1	0.1	1
2	0.2	1.02
3	0.3	1.0608
4	0.4	1.1244
5	0.5	1.2144
6	0.6	1.3358
7	0.7	1.4961
8	0.8	1.7056
9	0.9	1.9785
10	1.0	2.3346

Therefore, $y(1) \approx 2.335.$

15. $y'' - 2xy = 0, \quad y(0) = 1, \quad y'(0) = -3$

$$y'' = 2xy \qquad\qquad y''(0) = 0$$

$$y''' = 2(xy' + y) \qquad y'''(0) = 2$$

$$y^{(4)} = 2(xy'' + 2y') \qquad y^{(4)}(0) = -12$$

$$y^{(5)} = 2(xy''' + 3y'') \qquad y^{(5)}(0) = 0$$

$$y^{(6)} = 2(xy^{(4)} + 4y''') \qquad y^{(6)}(0) = 16$$

$$y^{(7)} = 2(xy^{(5)} + 5y^{(4)}) \qquad y^{(7)}(0) = -120$$

$$\vdots \qquad\qquad\qquad \vdots$$

$$y(x) = 1 - \frac{3}{1!}x + \frac{2}{3!}x^3 - \frac{12}{4!}x^4 + \frac{16}{6!}x^6 - \frac{120}{7!}x^7 + \cdots$$

Using the first six terms of the series, $y\left(\frac{1}{4}\right) \approx 0.253.$

16. $y'' - 2xy' + y = 0,$ $y(0) = 1,$ $y'(0) = 2$

$$y'' = 2xy' - y \qquad\qquad y''(0) = -1$$

$$y''' = 2xy'' + y' \qquad\qquad y'''(0) = 2$$

$$y^{(4)} = 2xy''' + 3y'' \qquad\qquad y^{(4)}(0) = -3$$

$$y^{(5)} = 2xy^{(4)} + 5y''' \qquad\qquad y^{(5)}(0) = 10$$

$$y^{(6)} = 2xy^{(5)} + 7y^{(4)} \qquad\qquad y^{(6)}(0) = -21$$

$$y^{(7)} = 2xy^{(6)} + 9y^{(5)} \qquad\qquad y^{(7)}(0) = 90$$

$$\vdots \qquad\qquad\qquad\qquad \vdots$$

$$y(x) = 1 + \frac{2}{1!}x - \frac{1}{2!}x^2 + \frac{2}{3!}x^3 - \frac{3}{4!}x^4 + \frac{10}{5!}x^5 - \frac{21}{6!}x^6 + \frac{90}{7!}x^7 - \cdots$$

Using the first eight terms of the series, $y(\frac{1}{2}) \approx 1.911$.

17. $f(x) = e^x,$ $f'(x) = e^x,$ $y' - y = 0.$ Assume $y = \sum_{n=0}^{\infty} a_n x^n$, then:

$$y' = \sum_{n=0}^{\infty} n a_n x^{n-1}$$

$$\sum_{n=1}^{\infty} n a_n x^{n-1} = \sum_{n=0}^{\infty} a_n x^n$$

$$\sum_{n=0}^{\infty} (n+1)a_{n+1}x^n = \sum_{n=0}^{\infty} a_n x^n$$

$$a_{n+1} = \frac{a_n}{n+1}, \quad n \geq 0$$

$n = 0, \quad a_1 = a_0$

$n = 1, \quad a_2 = \dfrac{a_1}{2} = \dfrac{a_0}{2}$

$n = 2, \quad a_3 = \dfrac{a_2}{3} = \dfrac{a_0}{2(3)}$

$n = 3, \quad a_4 = \dfrac{a_3}{4} = \dfrac{a_0}{2(3)(4)}$

$n = 4, \quad a_5 = \dfrac{a_4}{5} = \dfrac{a_0}{2(3)(4)(5)}$

$$\vdots$$

$$a_{n+1} = \frac{a_0}{(n+1)!} \Rightarrow a_n = \frac{a_0}{n!}$$

$y = a_0 \sum_{n=0}^{\infty} \dfrac{x^n}{n!}$ which converges on $(-\infty, \infty)$. When $a_0 = 1$, we have the Maclaurin Series for $f(x) = e^x$.

18. $f(x) = \cos x$, $f'(x) = -\sin x$, $f''(x) = -\cos x$, $y'' + y = 0$. Assume $y = \sum_{n=0}^{\infty} a_n x^n$, then:

$$y'' = \sum_{n=2}^{\infty} n(n-1)a_n x^{n-2}$$

$$\sum_{n=2}^{\infty} n(n-1)a_n x^{n-2} + \sum_{n=0}^{\infty} a_n x^n = 0$$

$$\sum_{n=0}^{\infty} (n+2)(n+1)a_{n+2}x^n = -\sum_{n=0}^{\infty} a_n x^n$$

$$a_{n+2} = -\frac{a_n}{(n+1)(n+2)}, \quad n \ge 0$$

$a_0 = a_0$ $\qquad\qquad\qquad$ $a_1 = a_1$

$a_2 = -\dfrac{a_0}{(1)(2)}$ $\qquad\qquad$ $a_3 = -\dfrac{a_1}{(2)(3)}$

$a_4 = -\dfrac{a_2}{(3)(4)} = \dfrac{a_0}{4!}$ $\qquad$ $a_5 = -\dfrac{a_3}{(4)(5)} = \dfrac{a_1}{5!}$

$\qquad \vdots \qquad\qquad\qquad\qquad \vdots$

$a_{2n} = \dfrac{(-1)^n a_0}{(2n)!}$ $\qquad\qquad$ $a_{2n+1} = \dfrac{(-1)^n a_1}{(2n+1)!}$

$$y = a_0 \sum_{n=0}^{\infty} \frac{(-1)^n x^{2n}}{(2n)!} + a_1 \sum_{n=0}^{\infty} \frac{(-1)^n x^{2n+1}}{(2n+1)!} \text{ which converges on } (-\infty, \ \infty)$$

When $a_0 = 1$ and $a_1 = 0$, we have the Maclaurin Series for $f(x) = \cos x$.

19.
$$f(x) = \arctan x$$

$$f'(x) = \frac{1}{1+x^2}$$

$$f''(x) = \frac{-2x}{(1+x^2)^2}$$

$$y'' = \frac{-2x}{1+x^2}y'$$

$$(1+x^2)y'' + 2xy' = 0$$

Assume $y = \sum_{n=0}^{\infty} a_n x^n$, then:

$$y' = \sum_{n=0}^{\infty} na_n x^{n-1}$$

$$y'' = \sum_{n=2}^{\infty} n(n-1)a_n x^{n-2}$$

$$(1+x^2)y'' + 2xy' = \sum_{n=2}^{\infty} n(n-1)a_n x^{n-2} + \sum_{n=0}^{\infty} n(n-1)a_n x^n + \sum_{n=0}^{\infty} 2na_n x^n = 0$$

$$\sum_{n=2}^{\infty} n(n-1)a_n x^{n-2} = -\sum_{n=0}^{\infty} n(n-1)a_n x^n - \sum_{n=0}^{\infty} 2na_n x^n$$

$$\sum_{n=0}^{\infty} (n+2)(n+1)a_{n+2}x^n = -\sum_{n=0}^{\infty} n(n+1)a_n x^n$$

$$(n+2)(n+1)a_{n+2} = -n(n+1)a_n$$

$$a_{n+2} = -\frac{n}{n+2}a_n, \quad n \ge 0$$

19. **—CONTINUED—**

$n = 0 \Rightarrow a_2 = 0 \Rightarrow$ all the even-powered terms have a coefficient of 0.

$n = 1, \quad a_3 = -\frac{1}{3}a_1$

$n = 3, \quad a_5 = -\frac{3}{5}a_3 = \frac{1}{5}a_1$

$n = 5, \quad a_7 = -\frac{5}{7}a_5 = -\frac{1}{7}a_1$

$n = 7, \quad a_9 = -\frac{7}{9}a_7 = \frac{1}{9}a_1$

$$\vdots$$

$$a_{2n+1} = \frac{(-1)^n a_1}{2n+1}$$

$y = a_1 \sum_{n=0}^{\infty} \frac{(-1)^n x^{2n+1}}{2n+1}$ which converges on $(-1, \ 1)$.

When $a_1 = 1$, we have the Maclaurin Series for $f(x) = \arctan x$.

20.
$$f(x) = \arcsin x$$

$$f'(x) = \frac{1}{\sqrt{1-x^2}}$$

$$f''(x) = \frac{x}{(1-x^2)^{3/2}}$$

$$y'' = \frac{1}{\sqrt{1-x^2}} \cdot \frac{x}{1-x^2} = \frac{x}{1-x^2}y'$$

$$(1-x^2)y'' - xy' = 0$$

Assume $y = \sum_{n=0}^{\infty} a_n x^n$, then:

$$\sum_{n=2}^{\infty} a_n n(n-1)x^{n-2} - \sum_{n=0}^{\infty} a_n n(n-1)x^n - \sum_{n=0}^{\infty} a_n n x^n = 0$$

$$\sum_{n=0}^{\infty} (n+2)(n+1)a_{n+2}x^n = \sum_{n=0}^{\infty} n^2 a_n x^n$$

$$a_{n+2} = \frac{n^2}{(n+1)(n+2)}a_n, \quad n \geq 0$$

20. —CONTINUED—

$n = 0 \Rightarrow a_2 = 0 \Rightarrow$ all the even-powered terms have a coefficient of 0.

$$a_1 = a_1$$

$$n = 1, \quad a_3 = \frac{1}{(2)(3)}a_1$$

$$n = 3, \quad a_5 = \frac{9}{(4)(5)}a_3 = \frac{9}{(2)(3)(4)(5)}a_1 = \frac{3}{(2)(4)(5)}a_1$$

$$n = 5, \quad a_7 = \frac{25}{(6)(7)}a_5 = \frac{(9)(25)}{(2)(3)(4)(5)(6)(7)}a_1 = \frac{(3)(5)}{(2)(4)(6)(7)}a_1$$

$$n = 7, \quad a_9 = \frac{49}{(8)(9)}a_7 = \frac{(9)(25)(49)}{(2)(3)(4)(5)(6)(7)(8)(9)}a_1 = \frac{(3)(5)(7)}{(2)(4)(6)(8)(9)}a_1$$

$$n = 9, \quad a_{11} = \frac{81}{(10)(11)}a_9 = \frac{(9)(25)(49)(81)}{(2)(3)(4)(5)(6)(7)(8)(9)(10)(11)}a_1 = \frac{(3)(5)(7)(9)}{(2)(4)(6)(8)(10)(11)}a_1$$

$$\vdots$$

$$a_{2n+1} = \frac{(2n)!}{(2^n n!)^2(2n + 1)}a_1$$

$$y = a_1 \sum_{n=0}^{\infty} \frac{(2n)!}{(2^n n!)^2(2n + 1)}x^{2n+1} \text{ which converges on } (-1, \ 1).$$

When $a_1 = 1$, we have the Maclaurin Series for $f(x) = \arcsin x$.

21. $y'' - xy = 0.$ Let $y = \displaystyle\sum_{n=0}^{\infty} a_n x^n.$

$$y'' - xy = \sum_{n=2}^{\infty} n(n - 1)a_n x^{n-2} - x\sum_{n=0}^{\infty} a_n x^n = \sum_{n=-1}^{\infty} (n + 3)(n + 2)a_{n+3}x^{n+1} - \sum_{n=0}^{\infty} a_n x^{n+1} = 0$$

$$2a_2 + \sum_{n=0}^{\infty}[(n + 3)(n + 2)a_{n+3} - a_n]x^{n+1} = 0$$

Hence, $a_2 = 0$ and

$$a_{n+3} = \frac{a_n}{(n + 3)(n + 2)} \text{ for } n = 0, \ 1, \ 2, \ \ldots.$$

The constants a_0 and a_1 are arbitrary.

$$a_0 = a_0 \qquad\qquad a_1 = a_1$$

$$a_3 = \frac{a_0}{3 \cdot 2} \qquad\qquad a_4 = \frac{a_1}{4 \cdot 3}$$

$$a_6 = \frac{a_3}{6 \cdot 5} = \frac{a_0}{6 \cdot 5 \cdot 3 \cdot 2} \qquad a_7 = \frac{a_4}{7 \cdot 6} = \frac{a_1}{7 \cdot 6 \cdot 4 \cdot 3}$$

Therefore, $y = a_0 + a_1 x + \dfrac{a_0}{6}x^3 + \dfrac{a_1}{12}x^4 + \dfrac{a_0}{180}x^6 + \dfrac{a_1}{504}x^7.$

Chapter 16   Review Exercises

Differential Equation	Type	Order
1. $\dfrac{\partial^2 u}{\partial t^2} = c^2 \dfrac{\partial^2 u}{\partial x^2}$	Partial	2
2. $yy'' = x + 1$	Ordinary	2
3. $y'' + 3y' - 10 = 0$	Ordinary	2
4. $(y'')^2 + 4y' = 0$	Ordinary	2

5. a.

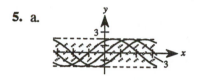

b.
$$\frac{dy}{dx} = \frac{y}{x}$$
$$\int \frac{dy}{y} = \int \frac{dx}{x}$$
$$\ln y = \ln x + \ln C$$
$$y = Cx$$

6. a.

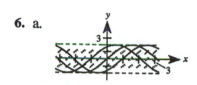

b. The rate of change is greatest when $y = 0$. The rate of change is least when $y = \pm 1$.

c.
$$\frac{dy}{dx} = \sqrt{1 - y^2}$$
$$\int \frac{dy}{\sqrt{1 - y^2}} = \int dx$$
$$\arcsin y = x + C$$
$$y = \sin(x + C)$$

7. $y' - 4 = 0$
$$\int dy = \int 4\, dx$$
$$y = 4x + C$$
Matches b.

8. $y' - 4y = 0$
$$\int \frac{dy}{y} = \int 4\, dx$$
$$\ln y = 4x + C_1$$
$$y = Ce^{4x}$$
Matches d.

9. $y'' - 4y = 0$
$$m^2 - 4 = 0$$
$$m_1 = 2, \quad m_2 = -2$$
$$y = C_1 e^{2x} + C_2 e^{-2x}$$
Matches a.

10. $y'' + 4y = 0$
$$m^2 + 4 = 0$$
$$m_1 = 2i, \quad m_2 = -2i$$
$$y = C_1 \cos 2x + C_2 \sin 2x$$
Matches c.

11. $\dfrac{dy}{dx} - \dfrac{y}{x} = 2 + \sqrt{x}$

Integrating factor: $e^{-\int (1/x)\, dx} = e^{-\ln |x|} = \dfrac{1}{x}$
$$y\left(\frac{1}{x}\right) = \int \frac{1}{x}(2 + \sqrt{x})\, dx = \ln x^2 + 2\sqrt{x} + C$$
$$y = x \ln x^2 + 2x^{3/2} + Cx$$

12. $\dfrac{dy}{dx} + xy = 2y$
$$\frac{dy}{dx} = -(x - 2)y$$
$$\int \frac{1}{y}\, dy = \int -(x - 2)\, dx$$
$$\ln y = -\frac{1}{2}(x - 2)^2 + \ln C$$
$$y = Ce^{-(x-2)^2/2}$$

13. $y' - \dfrac{2y}{x} = \dfrac{1}{x}y'$

$(x-1)\dfrac{dy}{dx} = 2y$

$\displaystyle\int \dfrac{1}{y}\,dy = \int \dfrac{2}{x-1}\,dx$

$\ln|y| = \ln(x-1)^2 + \ln C$

$y = C(x-1)^2$

14. $\dfrac{dy}{dx} - 3x^2 y = e^{x^3}$

Integrating factor: $e^{\int -3x^2\,dx} = e^{-x^3}$

$ye^{-x^3} = \displaystyle\int dx$

$ye^{-x^3} = x + C$

$y = (x+C)e^{x^3}$

15. $\dfrac{dy}{dx} - \dfrac{y}{x} = \dfrac{x}{y}$, Bernoulli

$n = -1, \quad P = -\dfrac{1}{x}, \quad Q = x,$

$e^{\int(-2/x)\,dx} = e^{\ln x^{-2}} = x^{-2}$

$y^2 x^{-2} = \displaystyle\int 2(x)x^{-2}\,dx = \ln x^2 + C$

$y^2 = x^2 \ln x^2 + Cx^2$

16. $\dfrac{dy}{dx} - \dfrac{3y}{x^2} = \dfrac{1}{x^2}$

Integrating factor: $e^{\int -(3/x^2)\,dx} = e^{3/x}$

$ye^{3/x} = \displaystyle\int \dfrac{1}{x^2}e^{3/x}\,dx = -\dfrac{1}{3}e^{3/x} + C$

$y = -\dfrac{1}{3} + Ce^{-3/x}$

17. $(10x + 8y + 2)\,dx + (8x + 5y + 2)\,dy = 0$

Exact: $\dfrac{\partial M}{\partial y} = 8 = \dfrac{\partial N}{\partial x}$

$U(x,\ y) = \displaystyle\int (10x + 8y + 2)\,dx = 5x^2 + 8xy + 2x + f(y)$

$U_y(x,\ y) = 8x + f'(y) = 8x + 5y + 2$

$f'(y) = 5y + 2$

$f(y) = \dfrac{5}{2}y^2 + 2y + C_1$

$U(x,\ y) = 5x^2 + 8xy + 2x + \dfrac{5}{2}y^2 + 2y + C_1$

$5x^2 + 8xy + 2x + \dfrac{5}{2}y^2 + 2y = C$

18. $(y + x^3 + xy^2)\,dx - x\,dy = 0$

$y\,dx - x\,dy + x(x^2 + y^2)\,dx = 0$

$\dfrac{y\,dx - x\,dy}{x^2 + y^2} + x\,dx = 0$

$\arctan \dfrac{x}{y} + \dfrac{1}{2}x^2 = C_1$

$2\arctan \dfrac{x}{y} + x^2 = C$

19. $(2x - 2y^3 + y)\,dx + (x - 6xy^2)\,dy = 0$

Exact: $\dfrac{\partial M}{\partial y} = -6y^2 + 1 = \dfrac{\partial N}{\partial x}$

$$U(x,\ y) = \int (2x - 2y^3 + y)\,dx$$

$$= x^2 - 2xy^3 + xy + f(y)$$

$$U_y(x,\ y) = -6xy^2 + x + f'(y) = x - 6xy^2$$

$$f'(y) = 0$$

$$f(y) = C_1$$

$$U(x,\ y) = x^2 - 2xy^3 + xy + C_1$$

$$x^2 - 2xy^3 + xy = C$$

20. $3x^2y^2\,dx + (2x^3y + x^3y^4)\,dy = 0$

$$3x^2y^2\,dx + x^3(2y + y^4)\,dy = 0$$

$$\int \frac{3}{x}\,dx + \int \left(\frac{2}{y} + y^2\right)dy = 0$$

$$\ln x^3 + \ln y^2 + \frac{1}{3}y^3 = C_1$$

$$3\ln x^3y^2 + y^3 = C$$

21. $\qquad dy = (y\tan x + 2e^x)\,dx$

$$\frac{dy}{dx} - (\tan x)y = 2e^x$$

Integrating factor: $e^{\int -\tan x\,dx} = e^{\ln|\cos x|} = \cos x$

$$y\cos x = \int 2e^x \cos x\,dx = e^x(\cos x + \sin x) + C$$

$$y = e^x(1 + \tan x) + C\sec x$$

22. $y\,dx - \left(x + \sqrt{xy}\right)dy = 0$

Integrating factor: $\dfrac{1}{2x^{1/2}y^{3/2}}$

$$\frac{y\,dx - x\,dy}{2x^{1/2}y^{3/2}} - \frac{1}{2y}\,dy = 0$$

$$\sqrt{x/y} - \ln\sqrt{y} = C$$

23. $(x - y - 5)\,dx - (x + 3y - 2)\,dy = 0$

Exact: $\dfrac{\partial M}{\partial y} = -1 = \dfrac{\partial N}{\partial x}$

$$U(x,\ y) = \int (x - y - 5)\,dx = \frac{1}{2}x^2 - xy - 5x + f(y)$$

$$U_y(x,\ y) = -x + f'(y) = -x - 3y + 2$$

$$f'(y) = -3y + 2$$

$$f(y) = -\frac{3}{2}y^2 + 2y + C_1$$

$$U(x,\ y) = \frac{1}{2}x^2 - xy - 5x - \frac{3}{2}y^2 + 2y + C_1$$

$$x^2 - 2xy - 10x - 3y^2 + 4y = C$$

24. $\qquad y' = 2x\sqrt{1 - y^2}$

$$\int \frac{1}{\sqrt{1 - y^2}}\,dy = \int 2x\,dx$$

$$\arcsin y = x^2 + C$$

$$y = \sin(x^2 + C)$$

25.
$$x + yy' = \sqrt{x^2 + y^2}$$

$$\int \frac{x\,dx + y\,dy}{\sqrt{x^2 + y^2}} = \int dx$$

$$\sqrt{x^2 + y^2} = x + C$$

$$x^2 + y^2 = x^2 + 2Cx + C^2$$

$$y^2 = 2Cx + C^2$$

26.
$$xy' + y = \sin x$$

$$y' + \left(\frac{1}{x}\right)y = \frac{1}{x}\sin x$$

Integrating factor: $e^{\int (1/x)\,dx} = e^{\ln x} = x$

$$yx = \int \sin x\,dx = -\cos x + C$$

$$yx + \cos x = C$$

27.
$$yy' + y^2 = 1 + x^2$$

$$y' + y = \frac{1}{y}(1 + x^2), \quad \text{Bernoulli}$$

$$n = -1, \quad P = 1, \quad Q = 1 + x^2, \quad e^{\int 2\,dx} = e^{2x}$$

$$y^2 e^{2x} = \int 2(1 + x^2)e^{2x}\,dx$$

$$= \left(x^2 - x + \frac{3}{2}\right)e^{2x} + C$$

$$y^2 = x^2 - x + \frac{3}{2} + Ce^{-2x}$$

28.
$$2x\,dx + 2y\,dy = (x^2 + y^2)\,dx$$

$$\int \frac{2x\,dx + 2y\,dy}{x^2 + y^2} = \int dx$$

$$\ln(x^2 + y^2) = x + C$$

$$\ln(x^2 + y^2) - x = C$$

29.
$$(1 + x^2)\,dy = (1 + y^2)\,dx$$

$$\int \frac{1}{1 + y^2}\,dy = \int \frac{1}{1 + x^2}\,dx$$

$$\arctan y - \arctan x = C_1$$

$$\tan(\arctan y - \arctan x) = \tan C_1$$

$$\frac{y - x}{1 + xy} = C$$

30.
$$y' = \frac{x^4 + 3x^2 y^2 + y^4}{x^3 y}$$

$$x^3 y\,dy = (x^4 + 3x^2 y^2 + y^4)\,dx$$

Homogeneous

$$y = vx, \quad dy = v\,dx + x\,dv$$

$$vx^4(v\,dx + x\,dv) = (x^4 + 3v^2 x^4 + v^4 x^4)\,dx$$

$$v^2\,dx + vx\,dv = (1 + 3v^2 + v^4)\,dx$$

$$\int \frac{v}{(1 + v^2)^2}\,dv = \int \frac{1}{x}\,dx$$

$$\frac{-1}{2(1 + v^2)} = \ln|x| + C_1$$

$$\frac{-x^2}{x^2 + y^2} = \ln x^2 + C$$

31.
$$y' - \left(\frac{a}{x}\right)y = bx^3$$

Integrating factor: $e^{-\int (a/x)\,dx} = e^{-a\ln x} = x^{-a}$

$$yx^{-a} = \int bx^3(x^{-a})\,dx = \frac{b}{4 - a}x^{4-a} + C$$

$$y = \frac{bx^4}{4 - a} + Cx^a$$

32.
$$y' = y + 2x(y - e^x)$$

$$y' - (1 + 2x)y = -2xe^x$$

Integrating factor: $e^{\int -(1+2x)\,dx} = e^{-(x+x^2)}$

$$ye^{-(x+x^2)} = \int e^{-(x+x^2)}(-2xe^x)\,dx = e^{-x^2} + C$$

$$y = e^x(1 + Ce^{x^2})$$

33. $y' - 2y = e^x$

Integrating factor: $e^{\int -2\,dx} = e^{-2x}$

$ye^{-2x} = \int e^{-2x}e^x\,dx + C = -e^{-x} + C$

$y = Ce^{2x} - e^x$

Initial condition: $y(0) = 4$

$$4 = C - 1 \Rightarrow C = 5$$

Particular solution: $y = 5e^{2x} - e^x$

34. $y' + \dfrac{2y}{x} = -x^9 y^5$, Bernoulli

$n = 5, \quad P = \dfrac{2}{x}, \quad Q = -x^9,$

$e^{\int -4(2/x)\,dx} = e^{\ln x^{-8}} = x^{-8}$

$y^{-4}x^{-8} = \int -4(-x^9)(x^{-8})\,dx + C$

$$= 2x^2 + C$$

$x^8 y^4 (2x^2 + C) = 1$

Initial condition: $y(1) = 2$

$$16(2 + C) = 1 \Rightarrow C = -\dfrac{31}{16}$$

Particular solution: $x^8 y^4 \left(2x^2 - \dfrac{31}{16}\right) = 1$

35. $\qquad x\,dy = (x + y + 2)\,dx$

$\dfrac{dy}{dx} - \left(\dfrac{1}{x}\right)y = \dfrac{x+2}{x}$

Integrating factor: $e^{\int -(1/x)\,dx} = e^{-\ln|x|} = \dfrac{1}{x}$

$y\left(\dfrac{1}{x}\right) = \int \dfrac{x+2}{x^2}\,dx = \ln|x| - \dfrac{2}{x} + C$

$y = x\ln|x| - 2 + Cx$

Initial condition: $y(1) = 10$

$$10 = -2 + C \Rightarrow C = 12$$

Particular solution: $y = x\ln|x| - 2 + 12x$

36. $ye^{xy}\,dx + xe^{xy}\,dy = 0$

$y\,dx + x\,dy = 0$

$xy = C$

Initial condition: $y(-2) = -5$

$$(-2)(-5) = C \Rightarrow C = 10$$

Particular solution: $xy = 10$

37. $\qquad \ln(1 + y)\,dx + \left(\dfrac{1}{1+y}\right)dy = 0$

$\int dx + \int \dfrac{1}{(1+y)\ln(1+y)}\,dy = C_1$

$x + \ln|\ln(1+y)| = C_1$

$\ln|\ln(1+y)| = C_1 - x$

$\ln|1 + y| = e^{C_1 - x} = Ce^{-x}$

Initial condition: $y(0) = 2$

$$\ln 3 = C$$

Particular solution: $\ln|1 + y| = (\ln 3)e^{-x}$

38. $(2x + y - 3) dx + (x - 3y + 1) dy = 0$

Exact: $\dfrac{\partial M}{\partial y} = 1 = \dfrac{\partial N}{\partial x}$

$$U(x,\ y) = \int (2x + y - 3)\, dx = x^2 + xy - 3x + f(y)$$

$$U_y(x,\ y) = x + f'(y) = x - 3y + 1$$

$$f'(y) = -3y + 1$$

$$f(y) = -\frac{3}{2}y^2 + y + C_1$$

$$U(x,\ y) = x^2 + xy - 3x - \frac{3}{2}y^2 + y + C_1$$

$$2x^2 + 2xy - 6x - 3y^2 + 2y = C$$

Initial condition: $y(2) = 0$

$$8 + 0 - 12 - 0 + 0 = C \Rightarrow C = -4$$

Particular solution: $2x^2 + 2xy - 6x - 3y^2 + 2y = -4$

39. $\qquad y' = x^2 y^2 - 9x^2$

$$\int \frac{1}{y^2 - 9}\, dy = \int x^2\, dx$$

$$\frac{1}{6} \ln \left| \frac{y - 3}{y + 3} \right| = \frac{1}{3}x^3 + C_1$$

$$\ln \left| \frac{y - 3}{y + 3} \right| = 2x^3 + C$$

Initial condition: $y(0) = \dfrac{3(1 + e)}{1 - e}$

$$C = \ln \left| \frac{\dfrac{3(1 + e)}{1 - e} - 3}{\dfrac{3(1 + e)}{1 - e} + 3} \right| = 1$$

Particular solution: $\ln \left| \dfrac{y - 3}{y + 3} \right| = 2x^3 + 1$

$$\frac{y - 3}{y + 3} = e^{2x^3 + 1}$$

$$y = \frac{3(e^{2x^3 + 1} + 1)}{1 - e^{2x^3 + 1}}$$

40. $\qquad 2xy' - y = x^3 - x$

$$y' - \left(\frac{1}{2x} \right) y = \frac{1}{2}(x^2 - 1)$$

Integrating factor: $e^{\int -(1/2x)\, dx} = e^{\ln x^{-1/2}} = x^{-1/2}$

$$yx^{-1/2} = \int \frac{1}{2}x^{-1/2}(x^2 - 1)\, dx$$

$$= \frac{1}{5}x^{5/2} - x^{1/2} + C$$

$$y = \frac{1}{5}x^3 - x + C\sqrt{x}$$

Initial condition: $y(4) = 2$

$$2 = \frac{64}{5} - 4 + 2C$$

$$C = -\frac{17}{5}$$

Particular solution: $y = \dfrac{1}{5}x^3 - x - \dfrac{17}{5}\sqrt{x}$

41.
$$(x - C)^2 + y^2 = C^2$$
$$x^2 - 2Cx + C^2 + y^2 = C^2$$
$$\frac{x^2 + y^2}{x} = 2C$$
$$\frac{x(2x + 2yy') - (x^2 + y^2)}{x^2} = 0$$
$$2x^2 + 2xyy' - x^2 - y^2 = 0$$
$$y' = \frac{y^2 - x^2}{2xy}$$

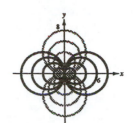

The negative reciprocal of y' is the slope of the orthogonal trajectories.
$$\frac{dy}{dx} = \frac{2xy}{x^2 - y^2}$$
$$2xy\,dx + (y^2 - x^2)\,dy = 0$$
Homogeneous
$$x = vy, \quad dx = v\,dy + y\,dv$$
$$2vy^2(v\,dy + y\,dv) + (y^2 - v^2y^2)\,dy = 0$$
$$\int \frac{2v}{1 + v^2}\,dv + \int \frac{1}{y}\,dy = 0$$
$$\ln(1 + v^2) + \ln|y| = \ln K_1$$
$$y^2 + x^2 = K_1 y$$
Circles: $x^2 + (y - K)^2 = K^2$

42. $y - 2x = C$
$$y = 2x + C$$
$$y' = 2$$

The negative reciprocal of y' is the
slope of the orthogonal trajectories.
$$\frac{dy}{dx} = -\frac{1}{2}$$
$$y = -\frac{1}{2}x + K$$
Lines: $x + 2y = K$

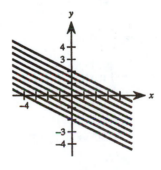

43. a.

$$\frac{ds}{dh} = \frac{k}{h}$$

$$\int ds = \int \frac{k}{h}\, dh$$

$$s = k\ln h + C_1 = k\ln Ch$$

b. Since $s = 25$ when $h = 2$ and $s = 12$ when $h = 6$, it follows that $25 = k\ln 2C$ and $12 = k\ln 6C$, which implies

$$C = \frac{1}{2}e^{-(25/13)\ln 3} \approx 0.0605 \quad \text{and} \quad k = \frac{25}{\ln 2C} = \frac{-13}{\ln 3} \approx -11.8331.$$

Therefore, s is given by the following.

$$s = -\frac{13}{\ln 3}\ln\left[\frac{h}{2}e^{-(25/13)\ln 3}\right]$$

$$= -\frac{13}{\ln 3}\left[\ln\frac{h}{2} - \frac{25}{13}\ln 3\right] = -\frac{1}{\ln 3}\left[13\ln\frac{h}{2} - 25\ln 3\right] = 25 - \frac{13\ln(h/2)}{\ln 3}, \quad 2 \le h \le 15$$

44.

$$\int \frac{1}{x}\frac{dx}{dt}\, dt = \int \frac{1}{y}\frac{dy}{dt}\, dt$$

$$\int \frac{dx}{x} = \int \frac{dy}{y}$$

$$\ln x + \ln C = \ln y$$

$$\ln Cx = \ln y$$

$$y = Cx$$

45.

$$\frac{dN}{dt} = kN(L - N)$$

$$\int \frac{dN}{N(L-N)} = \int k\, dt$$

$$\frac{1}{L}\int\left[\frac{1}{N} + \frac{1}{L-N}\right] dN = \int k\, dt$$

$$\ln|N| - \ln|L - N| = L(kt + C_1)$$

$$\frac{N}{L-N} = e^{Lkt + C_2} = Ce^{Lkt}$$

$$N = \frac{LCe^{Lkt}}{1 + Ce^{Lkt}}$$

When $t = 0$, $N = 100$. Thus, $100 = \frac{LC}{1+C} \Rightarrow C = \frac{100}{L-100}$, and thus, $N = \frac{100Le^{Lkt}}{(L-100) + 100e^{Lkt}}$.

When $t = 4$, $N = 200$. Thus, $200 = \frac{100Le^{4Lk}}{(L-100) + 100e^{4Lk}} \Rightarrow k = \frac{1}{4L}\ln\left[\frac{2(L-100)}{L-200}\right]$.

Therefore, $N = \frac{100Le^{(t/4)\ln[(2(L-100))/(L-200)]}}{(L-100) + 100e^{(t/4)\ln[(2(L-100))/(L-200)]}} = \frac{L}{\left(\frac{L}{100} - 1\right)e^{-(t/4)\ln[(2(L-100))/(L-200)]} + 1}$.

46. $\dfrac{dA}{dt} - rA = -P$

For this linear differential equation, we have $P(t) = -r$ and $Q(t) = -P$. Therefore, the integrating factor is $u(x) = e^{\int -r\,dt} = e^{-rt}$ and the solution is

$$A = e^{rt} \int -Pe^{-rt}\,dt = e^{rt}\left(\dfrac{P}{r}e^{-rt} + C\right) = \dfrac{P}{r} + Ce^{rt}.$$

Since $A = A_0$ when $t = 0$, we have $C = A_0 - (P/r)$ which implies that

$$A = \dfrac{P}{r} + \left(A_0 - \dfrac{P}{r}\right)e^{rt}.$$

47. $A_0 = 500{,}000, \quad r = 0.10$

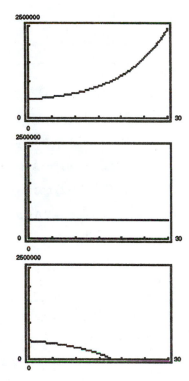

   a. $P = 40{,}000$

$$A = \dfrac{40{,}000}{0.10} + \left(500{,}000 - \dfrac{40{,}000}{0.10}\right)e^{0.10t} = 100{,}000(4 + e^{0.10t})$$

The balance continues to increase.

   b. $P = 50{,}000$

$$A = \dfrac{50{,}000}{0.10} + \left(500{,}000 - \dfrac{50{,}000}{0.10}\right)e^{0.10t} = 500{,}000$$

The balance remains at \$500,000.

   c. $P = 60{,}000$

$$A = \dfrac{60{,}000}{0.10} + \left(500{,}000 - \dfrac{60{,}000}{0.10}\right)e^{0.10t} = 100{,}000(6 - e^{0.10t})$$

The balance decreases and is depleted in $t = (\ln 6)/0.10 \approx 17.9$ years.

48. $\quad A = \dfrac{200{,}000}{0.14} + \left(1{,}000{,}000 - \dfrac{200{,}000}{0.14}\right)e^{0.14t}$

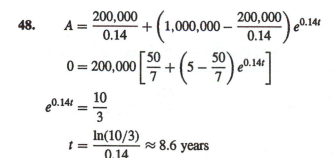

$$0 = 200{,}000\left[\dfrac{50}{7} + \left(5 - \dfrac{50}{7}\right)e^{0.14t}\right]$$

$$e^{0.14t} = \dfrac{10}{3}$$

$$t = \dfrac{\ln(10/3)}{0.14} \approx 8.6 \text{ years}$$

49. $y'' + y = x^3 + x$

$m^2 + 1 = 0$ when $m = -i,\ i.$

$$y_h = C_1 \cos x + C_2 \sin x$$

$$y_p = A_0 + A_1 x + A_2 x^2 + A_3 x^3$$

$$y_p' = A_1 + 2A_2 x + 3A_3 x^2$$

$$y_p'' = 2A_2 + 6A_3 x$$

$$y_p'' + y_p = (A_0 + 2A_2) + (A_1 + 6A_3)x + A_2 x^2 + A_3 x^3$$

$$= x^3 + x$$

$$A_0 = 0,\quad A_1 = -5,\quad A_2 = 0,\quad A_3 = 1$$

$$y = C_1 \cos x + C_2 \sin x - 5x + x^3$$

50. $y'' + 2y = e^{2x} + x$

$m^2 + 2 = 0$ when $m = -\sqrt{2}\,i,\ \sqrt{2}\,i.$

$$y_h = C_1 \cos \sqrt{2}\,x + C_2 \sin \sqrt{2}\,x$$

$$y_p = Ae^{2x} + B_0 + B_1 x$$

$$y_p' = 2Ae^{2x} + B_1$$

$$y_p'' = 4Ae^{2x}$$

$$y_p'' + 2y_p = 6Ae^{2x} + 2B_0 + 2B_1 x = e^{2x} + x$$

$$A = \tfrac{1}{6},\quad B_0 = 0,\quad B_1 = \tfrac{1}{2}$$

$$y = C_1 \cos \sqrt{2}\,x + C_2 \sin \sqrt{2}\,x + \tfrac{1}{6}e^{2x} + \tfrac{1}{2}x$$

51. $y'' + y = 2\cos x$

$m^2 + 1 = 0$ when $m = -i,\ i.$

$$y_h = C_1 \cos x + C_2 \sin x$$

$$y_p = Ax \cos x + Bx \sin x$$

$$y_p' = (Bx + A)\cos x + (B - Ax)\sin x$$

$$y_p'' = (2B - Ax)\cos x + (-Bx - 2A)\sin x$$

$$y_p'' + y_p = 2B\cos x - 2A\sin x = 2\cos x$$

$$A = 0,\quad B = 1$$

$$y = C_1 \cos x + (C_2 + x)\sin x$$

52. $y'' + 5y' + 4y = x^2 + \sin 2x$

$m^2 + 5m + 4 = 0$ when $m = -1,\ -4.$

$$y_h = C_1 e^{-x} + C_2 e^{-4x}$$

$$y_p = A_0 + A_1 x + A_2 x^2 + B_0 \sin 2x + B_1 \cos 2x$$

$$y_p' = A_1 + 2A_2 x + 2B_0 \cos 2x - 2B_1 \sin 2x$$

$$y_p'' = 2A_2 - 4B_0 \sin 2x - 4B_1 \cos 2x$$

$$y_p'' + 5y_p' + 4y_p = (4A_0 + 5A_1 + 2A_2) + (4A_1 + 10A_2)x + 4A_2 x^2 - 10B_1 \sin 2x + 10B_0 \cos 2x = x^2 + \sin 2x$$

$$A_0 = \tfrac{21}{32},\quad A_1 = -\tfrac{5}{8},\quad A_2 = \tfrac{1}{4},\quad B_0 = 0,\quad B_1 = -\tfrac{1}{10}$$

$$y = C_1 e^{-x} + C_2 e^{-4x} + \tfrac{21}{32} - \tfrac{5}{8}x + \tfrac{1}{4}x^2 - \tfrac{1}{10}\cos 2x$$

53. $y'' - 2y' + y = 2xe^x$

$m^2 - 2m + 1 = 0$ when $m = 1, 1.$

$y_h = (C_1 + C_2 x)e^x$

$y_p = (v_1 + v_2 x)e^x$

$\quad v_1'e^x + v_2'xe^x = 0$

$\quad v_1'e^x + v_2'(x+1)e^x = 2xe^x$

$\quad v_1' = -2x^2$

$\quad v_1 = \int -2x^2 \, dx = -\tfrac{2}{3}x^3$

$\quad v_2' = 2x$

$\quad v_2 = \int 2x \, dx = x^2$

$y = \left(C_1 + C_2 x + \tfrac{1}{3}x^3\right)e^x$

54. $y'' + 2y' + y = \dfrac{1}{x^2 e^x}$

$m^2 + 2m + 1 = 0$ when $m = -1, -1.$

$y_h = (C_1 + C_2 x)e^{-x}$

$y_p = (v_1 + v_2 x)e^{-x}$

$\quad v_1'e^{-x} + v_2'(xe^{-x}) = 0$

$\quad v_1'(-e^{-x}) + v_2'(-x+1)e^{-x} = \dfrac{1}{e^x x^2}$

$\quad v_1' = -\dfrac{1}{x}$

$\quad v_1 = \int -\dfrac{1}{x} \, dx = -\ln|x|$

$\quad v_2' = \dfrac{1}{x^2}$

$\quad v_2 = \int \dfrac{1}{x^2} \, dx = -\dfrac{1}{x}$

$y = (C_1 + C_2 x - \ln|x| - 1)e^{-x}$

55. $y'' - y' - 2y = 0$

$m^2 - m - 2 = 0$

$(m-2)(m+1) = 0$

$m_1 = 2, \quad m_2 = -1$

$y = C_1 e^{2x} + C_2 e^{-x}$

Initial conditions:

$\quad y(0) = 0: \ 0 = C_1 + C_2$

$\quad y'(0) = 3: \ 3 = 2C_1 - C_2 \Rightarrow C_1 = 1, \quad C_2 = -1$

Particular solution: $y = e^{2x} - e^{-x}$

56. $y'' + 4y' + 5y = 0$

$m^2 + 4m + 5 = 0$

$m = -2 \pm i$

$y = C_1 e^{-2x} \cos x + C_2 e^{-2x} \sin x$

Initial conditions:

$\quad y(0) = 2: \ 2 = C_1$

$\quad y'(0) = -7: \ -7 = -2C_1 + C_2 \Rightarrow C_2 = -3$

Particular solution: $y = 2e^{-2x} \cos x - 3e^{-2x} \sin x$

57. $y'' + 4y = \cos x$

$m^2 + 4 = 0 \Rightarrow m = \pm 2i$

$\quad y_h = C_1 \cos 2x + C_2 \sin 2x$

$\quad y_p = A \cos x + B \sin x$

$\quad y_p' = -A \sin x + B \cos x$

$\quad y_p'' = -A \cos x - B \sin x$

$y_p'' + 4y_p = (-A \cos x - B \sin x) + 4(A \cos x + B \sin x) = \cos x$

$\quad 3A \cos x + 3B \sin x = \cos x \Rightarrow A = \tfrac{1}{3}$ and $B = 0$

$\quad y_p = \tfrac{1}{3} \cos x$

$\quad y = y_h + y_p = C_1 \cos 2x + C_2 \sin 2x + \tfrac{1}{3} \cos x$

Initial conditions: $y(0) = 6: \ 6 = C_1 + \tfrac{1}{3} \Rightarrow C_1 = \tfrac{17}{3}$

$\qquad\qquad\qquad\quad y'(0) = -6: \ -6 = 2C_2 \Rightarrow C_2 = -3$

Particular solution: $y = \tfrac{17}{3} \cos 2x - 3 \sin 2x + \tfrac{1}{3} \cos x$

58. $y'' + 3y' = 6x$

$m^2 + 3m = 0 \Rightarrow m_1 = 0$ and $m_2 = -3$

$$y_h = C_1 + C_2 e^{-3x}$$

$$y_p = Ax^3 + Bx^2 + Cx + D$$

$$y_p' = 3Ax^2 + 2Bx + C$$

$$y_p'' = 6Ax + 2B$$

$$y_p'' + 3y_p' = (6Ax + 2B) + 3(3Ax^2 + 2Bx + C)$$

$$= 9Ax^2 + (6A + 6B)x + (2B + 3C) = 6x, \quad A = 0, \quad B = 1, \text{ and } C = -\tfrac{2}{3}$$

$$y_p = x^2 - \tfrac{2}{3}x + D$$

$$y = y_h + y_p = C_1 + C_2 e^{-3x} + x^2 - \tfrac{2}{3}x + D = C_3 + C_2 e^{-3x} + x^2 - \tfrac{2}{3}x$$

Initial conditions: $y(0) = 2$: $2 = C_3 + C_2$

$$y'(0) = \tfrac{10}{3}: \tfrac{10}{3} = -3C_2 - \tfrac{2}{3} \Rightarrow C_2 = -\tfrac{4}{3} \text{ and } C_3 = \tfrac{10}{3}$$

Particular solution: $y = \tfrac{10}{3} - \tfrac{4}{3}e^{-3x} + x^2 - \tfrac{2}{3}x = \tfrac{1}{3}(10 - 4e^{-3x} + 3x^2 - 2x)$

59. By Hooke's Law, $F = kx$, $k = F/x = 64/(4/3) = 48$. Also, $F = ma$ and $m = F/a = 64/32 = 2$. Therefore,

$$\frac{d^2y}{dt^2} + \left(\frac{48}{2}\right)y = 0$$

$$y = C_1 \cos(2\sqrt{6}\,t) + C_2 \sin(2\sqrt{6}\,t).$$

Since $y(0) = \tfrac{1}{2}$ we have $C_1 = \tfrac{1}{2}$ and $y'(0) = 0$ yields $C_2 = 0$. Thus, $y = \tfrac{1}{2}\cos(2\sqrt{6}\,t)$.

60. From Exercise 59 we have $k = 48$ and $m = 2$. Also, the damping force is given by $(1/8)(dy/dt)$.

$$2\left(\frac{d^2y}{dt^2}\right) = -\frac{1}{8}\frac{dy}{dt} - 48y$$

$$y'' + \frac{1}{16}y' + 24y = 0$$

$$16y'' + y' + 384y = 0$$

The characteristic equation $16m^2 + m + 384 = 0$ has complex roots

$$m = -\frac{1}{32} \pm \frac{\sqrt{24{,}575}\,i}{32} = -\frac{1}{32} \pm \frac{5\sqrt{983}}{32}i.$$

Thus,

$$y(t) = e^{-t/32}\left[C_1 \cos\left(\frac{5\sqrt{983}}{32}t\right) + C_2 \sin\left(\frac{5\sqrt{983}}{32}t\right)\right].$$

Initial conditions: $y(0) = \dfrac{1}{2} \Rightarrow C_1 = \dfrac{1}{2}$

$$y'(0) = 0 \Rightarrow \frac{5\sqrt{983}}{32}C_2 - \frac{C_1}{32} = 0 \Rightarrow C_2 = \frac{\sqrt{983}}{9830}$$

Particular solution: $y(t) = e^{-t/32}\left[\dfrac{1}{2}\cos\left(\dfrac{5\sqrt{983}}{32}t\right) + \dfrac{\sqrt{983}}{9830}\sin\left(\dfrac{5\sqrt{983}}{32}t\right)\right]$

61. $(x-4)y' + y = 0$. Letting $y = \sum\limits_{n=0}^{\infty} a_n x^n$:

$$xy' - 4y' + y = \sum_{n=0}^{\infty} na_n x^n - 4\sum_{n=1}^{\infty} na_n x^{n-1} + \sum_{n=0}^{\infty} a_n x^n$$

$$= \sum_{n=0}^{\infty}(n+1)a_n x^n - \sum_{n=1}^{\infty} 4na_n x^{n-1} = \sum_{n=0}^{\infty}(n+1)a_n x^n - \sum_{n=-1}^{\infty} 4(n+1)a_{n+1}x^n = 0$$

$$(n+1)a_n = 4(n+1)a_{n+1}$$

$$a_{n+1} = \frac{1}{4}a_n$$

$$a_0 = a_0, \quad a_1 = \frac{1}{4}a_0, \quad a_2 = \frac{1}{4}a_1 = \frac{1}{4^2}a_0, \quad \cdots, \quad a_n = \frac{1}{4^n}a_0$$

$$y = a_0 \sum_{n=0}^{\infty} \frac{x^n}{4^n}$$

62. $y'' + 3xy' - 3y = 0$. Letting $y = \sum\limits_{n=0}^{\infty} a_n x^n$:

$$y'' + 3xy' - 3y = \sum_{n=2}^{\infty} n(n-1)a_n x^{n-2} + 3x\sum_{n=1}^{\infty} na_n x^{n-1} - 3\sum_{n=0}^{\infty} a_n x^n = 0$$

$$\sum_{n=0}^{\infty}(n+2)(n+1)a_{n+2}x^n = \sum_{n=0}^{\infty}(3-3n)a_n x^n$$

$$a_{n+2} = \frac{3(1-n)a_n}{(n+2)(n+1)}$$

$a_0 = a_0$

$a_1 = a_1$

$a_2 = \dfrac{3}{2 \cdot 1}a_0$

$a_3 = 0$

There are no odd-powered terms for $n > 1$.

$a_4 = -\dfrac{3}{4 \cdot 3}\left(\dfrac{3}{2 \cdot 1}a_0\right) = -\dfrac{3(3)a_0}{4!}$

$a_6 = -\dfrac{3(3)}{6 \cdot 5}\left(-\dfrac{3(3)a_0}{4!}\right) = \dfrac{3^3(3)a_0}{6!}$

$a_8 = -\dfrac{3(5)}{8 \cdot 7}\left(\dfrac{3^3(3)a_0}{6!}\right) = -\dfrac{3^4(5 \cdot 3)a_0}{8!}$

$a_{10} = -\dfrac{3(7)}{10 \cdot 9}\left(-\dfrac{3^4(5 \cdot 3)a_0}{8!}\right) = \dfrac{3^5(7 \cdot 5 \cdot 3)a_0}{10!}$

$$y = a_0 + \frac{3}{2}a_0 x^2 + a_0 \sum_{n=2}^{\infty} \frac{(-1)^{n+1}3^n[3 \cdot 5 \cdot 7 \cdots (2n-3)]}{(2n)!}x^{2n}$$